PROPERTY OF
GLEN HACKETT

GROUND-WATER MICROBIOLOGY AND GEOCHEMISTRY

GROUND-WATER MICROBIOLOGY AND GEOCHEMISTRY

FRANCIS H. CHAPELLE
United States Geological Survey
Columbia, South Carolina

JOHN WILEY & SONS, INC.
New York / Chichester / Brisbane / Toronto / Singapore

This text is printed on acid-free paper.

Copyright © 1993 by John Wiley & Sons, Inc.

All rights reserved. Published simultaneously in Canada.

Reproduction or translation of any part of this work beyond that permitted by Section 107 or 108 of the 1976 United States Copyright Act without the permission of the copyright owner is unlawful. Requests for permission or further information should be addressed to the Permissions Department, John Wiley & Sons, Inc., 605 Third Avenue, New York, NY 10158-0012.

This publication is designed to provide accurate and authoritative information in regard to the subject matter covered. It is sold with the understanding that the publisher is not engaged in rendering legal, accounting, or other professional services. If legal advice or other expert assistance is required, the services of a competent professional person should be sought. *From a Declaration of Principles jointly adopted by a committee of the American Bar Association and a Committee of Publishers.*

Library of Congress Cataloging in Publication Data:

Chapelle, Frank.
    Ground-water microbiology and geochemistry / Francis H. Chapelle.
      p.   cm.
    ISBN 0-471-52951-6 (cloth)
     1. Water, Underground—Microbiology. 2. Geomicrobiology.
  3. Geochemistry. I. Title.
QR105.5.C53 1992
576—dc20   92-5589

Printed in the United States of America

10 9 8 7 6 5 4 3 2 1

PREFACE

This book is concerned with how microbial and geologic processes interact to influence ground-water chemistry. It is designed specifically for the geoscientist or the advanced student of the geosciences who has a special interest in ground-water quality issues. The need for this book comes from the gap that presently exists between applied ground-water geochemistry and applied microbiology. Although current geochemistry texts widely acknowledge that microbial processes affect ground-water chemistry, they give little detail as to how the growth, metabolism, and ecology of microorganisms bring about these effects. Conversely, although microbiology texts cover microbial processes in great detail, they give virtually no hint as to how a geoscientist might apply this information to practical ground-water chemistry problems. This book is designed specifically to bridge this gap.

A professor of mine once described ground-water geochemistry as "the study of everything that is interesting." That description, while perhaps a bit of an exaggeration, does illustrate the interdisciplinary nature of this field. There is a dizzying array of academic disciplines—geology, hydrology, biology, mathematics, chemistry, and physics, to name a few—on which one draws in the study of ground-water geochemistry. At the same time, this diversity is precisely what makes it so interesting. From this point of view, incorporating useful concepts and techniques from microbiology is in keeping with an established tradition. This can only serve to enrich the intellectual appeal, deepen the technical scope, and enhance the practical applications of ground-water geochemistry.

This book is divided into three parts. Part I is an overview of basic microbiology. Most geoscientists have not had formal training in microbiology and often are not familiar with microbiologic techniques and nomenclature. Although it is not intended to be an exhaustive treatise or a substitute for formal training, it does lay the groundwork for later discussions dealing with the growth and metabolism of subsurface bacteria.

Part II focuses specifically on microbial processes in pristine ground-water systems—that is, systems that are not impacted by human-related chemical stresses—and on how these processes influence water chemistry. In some ways, this is the most important subject in the book, since the vast majority of ground-water reservoirs have not been chemically contaminated by human activities.

Part III deals with microbial processes in aquifers that have been chemically contaminated by human activities. This subject is presently one of great interest, because bioremediation (using microorganisms to detoxify pollutants) is a potentially useful and cost-effective technology. Like any technology, however, successful application depends on how well critical factors can be identified, controlled, and used to advantage. The nature of these factors, both biologic and hydrologic, are the focus of Part III.

Subsurface microbiology is necessarily a composite between classical microbiology and hydrogeology. One cannot meaningfully consider the organic or inorganic chemical composition of ground water without addressing the microbial processes that mediate many important reactions. On the other hand, one cannot consider microbial processes or the characteristics of microorganisms present in the subsurface without also considering the chemical, physical, and hydrologic environment to which these microorganisms are adapted. The common ground between hydrogeologists and microbiologists studying microbial processes in subsurface environments is the chemical quality of ground water. This, therefore, is the focus of this book.

ACKNOWLEDGMENTS

Preparation of this book was funded by the Regional Aquifer Systems Analysis (RASA) program of the U.S. Geological Survey. Special thanks are extended to Ren Jen Sun, who very early on recognized the importance of microbial processes in regional water chemistry studies and accordingly provided support for such studies. Help from several individuals in the U.S. Geological Survey was instrumental in preparing this book. The historical connection between microbiology and the geosciences was largely provided by discussions with William Back. In addition, this book could not have been written without the friendship and administrative support provided by Rodney N. Cherry and Glenn G. Patterson. The author would like to extend special thanks to James Ray Douglas for preparing the illustrations for this book.

CONTENTS

PREFACE
ACKNOWLEDGMENTS

I   OVERVIEW OF MICROBIOLOGY   1

1   HISTORY, GEOLOGY, AND MICROBIOLOGY   2

   1.1 Geology—An Observational Science   /   3
   1.2 Microbiology—An Experimental Science   /   13
   1.3 Ground-Water Chemistry and Subsurface Microbiology   /   18
       1.3.1 Subsurface Microbiology and the Geosciences   /   19
       1.3.2 Subsurface Microbiology and Microbial Ecology   /   22
   1.4 Integrating Microbiology and Ground-Water Chemistry in Studies of Aquifer Systems   /   24

2   MICROORGANISMS PRESENT IN THE GROUND-WATER ENVIRONMENT   25

   2.1 The Procaryotes   /   26
   2.2 The Eucaryotes   /   29
       2.2.1 Eucaryotes in Ground-Water Systems   /   30
   2.3 The Archaebacteria   /   31
   2.4 The Viruses   /   33
   2.5 Bacteria in Ground-Water Systems   /   35
       2.5.1 Classifying Bacteria   /   36
             Criteria Used to Classify Bacteria   /   *36*

        2.5.2   Gram-Negative Bacteria Found in Ground-Water Systems / 40
                Aerobic Gram-Negative Rods / 40
                Aerobic Gram-Negative Cocci / 42
                Facultatively Anaerobic Gram-Negative Rods / 42
                Anaerobic Gram-Negative Rods / 43
        2.5.3   Gram-Positive Bacteria in Ground-Water Systems / 44
                Gram-Positive Cocci / 45
                Coryneform Bacteria / 45
                Spore-Forming Rods / 45
    2.6   Perspectives / 46

3   BACTERIAL GROWTH   47

    3.1   Bacterial Reproduction / 47
    3.2   Population Growth Kinetics / 49
    3.3   Environmental Conditions and Bacterial Growth / 51
          3.3.1   Temperature / 52
          3.3.2   Water / 54
          3.3.3   Molecular Oxygen / 55
          3.3.4   pH / 55
          3.3.5   Osmotic Pressure / 56
    3.4   Techniques for Culturing Bacteria / 57
          3.4.1   Design of Growth Media / 58
                  Carbon Sources / 58
                  Nitrogen Sources / 60
                  Phosphorous and Inorganic Nutrients / 60
                  Electron Acceptors / 61
                  Selective Growth Using Culture Media / 61
          3.4.2   Isolating Bacterial Strains from Environmental Samples / 63
                  Rich versus Dilute Media / 66
    3.5   Enumerating Bacteria / 66
          3.5.1   Viable Counting Procedures / 67
          3.5.2   Direct Counting Procedures / 68
    3.6   Summary / 69

4   BACTERIAL METABOLISM   71

    4.1   Thermodynamics and Bacterial Metabolism / 71
    4.2   ATP Synthesis—Storing Energy / 73
    4.3   Electron Transport Systems—Releasing Energy / 74
    4.4   Chemiosmosis—Harnessing Energy from Electron Transport / 77
    4.5   The Role of Enzymes / 78

4.6 Energy-Releasing Pathways of Geochemical Importance / 81
    4.6.1 Lactate and Acetate Fermentations / 81
    4.6.2 Ferredoxins and the Production of Hydrogen and Acetate in Fermentation / 82
    4.6.3 Methanogenic Pathways / 85
    4.6.4 Sulfate Reduction / 87
    4.6.5 Fe(III) Reduction / 88
    4.6.6 Nitrate Reduction / 89
    4.6.7 Oxygen Reduction-Aerobic Metabolism / 91
4.7 Biosynthesis / 92
    4.7.1 Amino Acids / 93
    4.7.2 Carbohydrates / 94
    4.7.3 Lipids / 96
4.8 Chemolithotrophy / 96
    4.8.1 Hydrogen Oxidizers / 97
    4.8.2 Sulfide Oxidizers / 98
    4.8.3 Iron Oxidizers / 99
    4.8.4 Ammonia-Oxidizing (Nitrifying) Bacteria / 100
    4.8.5 Autotrophic CO_2 Fixation / 100
4.9 Metabolic Control of Geochemical Processes / 102
4.10 Summary / 103

5 BACTERIAL GENETICS — 105

5.1 DNA—Its Structure and Organization / 106
5.2 RNA—Its Structure and Organization / 108
    5.2.1 Transcription / 108
    5.2.2 Translation—Making Proteins / 109
5.3 Gene Expression and Regulation / 112
    5.3.1 Induction / 113
    5.3.2 Repression / 113
5.4 Mutations / 114
    5.4.1 Mutagenic Agents / 115
    5.4.2 Transposable Genetic Material / 116
5.5 Natural Genetic Exchanges / 116
    5.5.1 Recombination / 117
        Transformation / 117
        Conjugation / 117
        Transduction / 118
5.6 Genetic Engineering / 118
    5.6.1 Plasmids / 119
    5.6.2 Vectors / 120
        Plasmid Vectors / 121

Phage Vectors / *121*

5.7 Applications of Recombinant DNA Technology / 121
   5.7.1 Insulin Production / 122
   5.7.2 Pollution Control Technology / 122
      Aliphatic Petroleum Hydrocarbons / *122*
      Aromatic Hydrocarbons / *123*
5.8 DNA Technology in Subsurface Microbiology / 124
   5.8.1 DNA Homology and Diversity / 124
   5.8.2 Plasmids and DNA Probes / 125
   5.8.3 Release of Engineered Microorganisms / 127
5.9 Summary / 128

6 MICROBIAL ECOLOGY OF GROUND-WATER SYSTEMS   130

6.1 Scope of Subsurface Microbial Ecology / 132
6.2 Methods in Subsurface Microbial Ecology / 135
   6.2.1 Culture Methods / 136
   6.2.2 Direct Observation / 141
   6.2.3 Biochemical Marker Techniques / 142
   6.2.4 Activity Measurements in Microcosms / 146
   6.2.5 Geochemical Methods / 149
   6.2.6 Ecological Modeling / 151
6.3 Microbial Diversity and Niches in Aquifer Systems / 156
   6.3.1 Measurement of Diversity / 156
   6.3.2 Niches and Sources of Microbial Diversity / 158
   6.3.3 Stress and Microbial Diversity / 160
6.4 Population Interactions / 161
   6.4.1 Neutralism / 162
   6.4.2 Commensalism / 162
   6.4.3 Synergism and Symbiosis / 163
   6.4.4 Competition / 164
   6.4.5 Antagonism, Parasitism, and Predation / 167
6.5 r and K Strategies in Microbial Ecosystems / 168
   6.5.1 r and K Strategies in the Aquifer Environment / 169
6.6 Summary / 170

II MICROBIAL PROCESSES IN PRISTINE GROUND-WATER SYSTEMS   173

7 ABUNDANCE AND DISTRIBUTION OF BACTERIA IN THE SUBSURFACE   174

7.1 Classification of Subsurface Environments / 174

7.2 The Unsaturated Zone / 179
    7.2.1 The Unsaturated Zone as a Microbial Habitat / 180
        Moisture and Gas Content / 181
    7.2.2 Biomass Measurements in Soil Microbiology / 182
        Direct Microscopy / 182
        Chemical Techniques / 183
        Activity Measurements / 184
        Other Measures of Biomass and Activity / 185
    7.2.3 Distribution of Bacteria in the Unsaturated Zone / 186
        The Soil Zone / 186
        The Intermediate Unsaturated Zone / 186
        The Deep Unsaturated Zone / 187
7.3 Local Flow Systems / 189
    7.3.1 Local Flow Systems as a Microbial Habitat / 190
    7.3.2 Distribution of Bacteria in Local Flow Systems / 191
7.4 Intermediate Flow Systems / 195
    7.4.1 Intermediate Flow Systems as a Microbial Habitat / 197
    7.4.2 Distribution of Bacteria in Intermediate Flow Systems / 198
    7.4.3 Microbial Processes in Confining Beds / 202
7.5 Regional Flow Systems / 203
    7.5.1 Early Observations from Petroleum Reservoirs / 204
    7.5.2 Distribution of Bacteria in Regional Flow Systems / 205
        Culture Techniques / 205
        Geochemical Methods / 206
7.6 Summary / 207

8 MICROBIOLOGIC SAMPLING OF SUBSURFACE ENVIRONMENTS   208

8.1 Sampling the Unsaturated Zone / 209
    8.1.1 Hand Augering / 209
    8.1.2 Air Drilling and Coring / 210
8.2 Sampling Local Flow Systems / 211
    8.2.1 Split-spoon Sampling / 211
    8.2.2 Push-tube (Shelby Tube) Sampling Methods / 212
    8.2.3 Aseptic Technique with Split-spoon and Push-tube Sampling / 214
        Tool Contamination / 214
        Down-hole Contamination / 215
8.3 Sampling Intermediate and Regional Flow Systems / 215

8.3.1 Mud Rotary Drilling / 215
8.3.2 Drilling Fluids / 216
   Density / 217
   Viscosity / 218
   Yield Point / 219
   Gel Strength / 219
   Fluid Loss Control / 220
   Lubricity / 220
8.3.3 Mud Rotary Coring / 220
   Equipment Used for Mud Rotary Coring / 221
   The Role of the Driller / 222
   Drilling Fluid Technology and Coring / 224
8.4 Drilling Fluid Contamination of Cored Sediments / 224
   8.4.1 Down-hole Saturation Contamination / 224
   8.4.2 Core Seepage Contamination / 225
   8.4.3 Core Fracture Contamination / 225
   8.4.4 Evaluating Drilling Fluid Contamination / 226
      Drilling Fluid as a Tracer / 227
      Chemical Additive Tracers / 228
      Particulate Tracers / 229
      Biologic Tracers / 230
8.5 Sampling Ground Water For Microorganisms / 230
8.6 Summary / 231

9 BIOGEOCHEMICAL CYCLING IN GROUND-WATER SYSTEMS 233

9.1 The Oxygen Cycle / 234
   9.1.1 Oxygen Cycling in Ground-Water Systems / 235
9.2 The Carbon Cycle / 238
   9.2.1 The Integrated Carbon, Oxygen, and Hydrogen Cycles / 240
   9.2.2 Carbon Cycling in Ground-Water Systems / 241
      Local Flow Systems / 242
      Intermediate and Regional Flow Systems / 244
9.3 The Nitrogen Cycle / 244
   9.3.1 Nitrogen Cycling in Ground-Water Systems / 247
      Animal Excrement and Manure / 247
      Sewage Effluents / 248
      Nitrogen Fertilizers / 249
      Municipal Wastes / 250
      Distinguishing Sources of Nitrogen Contamination / 251

Nitrogen Accumulation Due to Dry-Land Farming Practices / 251

9.4 The Iron Cycle / 253
   9.4.1 Iron Cycling in Aquatic Sediments / 254
   9.4.2 Iron Cycling in Ground-Water Systems / 255
      Wells and Iron-Oxidizing Bacteria / 256
      High Iron Concentrations in Ground Water / 257

9.5 The Sulfur Cycle / 258
   9.5.1 Sulfur Cycling in Ground-Water Systems / 259
      Sulfide Oxidation in an Oxygenated Aquifer / 260
      Sulfate Reduction in Sulfate Mineral-Free Aquifers / 260
      Sulfate Reduction in Sulfate Mineral-Bearing Aquifers / 262

9.6 Summary / 263

10 GEOCHEMICAL MODELING AS A TOOL FOR STUDYING MICROBIAL PROCESSES IN GROUND-WATER SYSTEMS   264

10.1 Considerations in Geochemical Modeling / 265
   10.1.1 Hydrologic Considerations / 266
   10.1.2 Mineralogic Considerations / 267
   10.1.3 Thermodynamic Considerations / 267
   10.1.4 Kinetic Factors / 268

10.2 Methodology of Geochemical Modeling / 271
   10.2.1 Speciation Calculations / 271
      The Aqueous Model / 272
      Speciation Calculations with the Aqueous Model / 273
      Oxidation-Reduction Reactions / 274
      Computer Programs for Making Speciation Calculations / 275
   10.2.2 Mineral Equilibrium Calculations / 276
   10.2.3 Material Balance Calculations / 277
      Mass Balance / 277
      Electron Balance / 278
      Isotope Balance / 279
      Computer Programs for Making Material Balance Calculations / 280
   10.2.4 Logic of Geochemical Modeling / 281
      Formulating Hypotheses of Microbial Processes / 282

- 10.3 Geochemical Modeling and Microbial Processes in Aquifer Systems / 283
  - 10.3.1 CO$_2$ Production in the Floridian Aquifer / 283
  - 10.3.2 Sulfate Reduction and Methanogenesis in the Fox Hills–Basal Hell Creek Aquifer / 286
  - 10.3.3 Sulfate Reduction in the Madison Aquifer / 289
    Application to Microbiologic Studies / 293
- 10.4 Summary / 294

III MICROBIAL PROCESSES IN CONTAMINATED GROUND-WATER SYSTEMS — 295

11 MICROBIAL ACCLIMATION TO GROUND-WATER CONTAMINATION — 296

- 11.1 Microbial Response to Environmental Changes / 297
- 11.2 Mechanisms of Acclimation / 298
  - 11.2.1 Induction / 299
    Induction of Hydrocarbon-Degrading Enzymes / 299
  - 11.2.2 Catabolite Repression / 300
    Catabolite Repression of Xenobiotic Oxidation / 301
  - 11.2.3 Genetic Mutations / 301
  - 11.2.4 Acclimation to Available Electron Acceptors / 302
- 11.3 Factors Affecting Microbial Acclimation / 304
  - 11.3.1 Rates of Acclimation / 305
  - 11.3.2 Concentration Effects / 306
  - 11.3.3 Cross-Acclimation of Xenobiotic Compounds / 308
  - 11.3.4 Chemical Structure of Xenobiotics / 308
- 11.4 Acclimation to Xenobiotics in Ground-Water Systems / 311
  - 11.4.1 Acclimation Response in a Contaminated Aquifer / 311
  - 11.4.2 Acclimation Response in Pristine Aquifer Sediments / 313
  - 11.4.3 Acclimation of Eucaryotic Microorganisms / 315
  - 11.4.4 Acclimation in Bioremediation Technology / 316
- 11.5 Acclimation to Metal Toxicity / 317
  - 11.5.1 Metal Detoxification Mechanisms / 317
    Metal Binding / 317
    Biotransformations / 318
    Metal Deposition / 318
  - 11.5.2 Plasmid-Encoded Metal Resistance Mechanisms / 318

11.5.3   Acclimation to Mercury Toxicity   /   320
11.6   Summary   /   320

12   BIODEGRADATION OF PETROLEUM HYDROCARBONS IN SUBSURFACE ENVIRONMENTS   322

12.1   Composition of Crude Oil   /   323
12.2   Petroleum Refining and Fuel Blending   /   323
12.3   Movement and Separation of Petroleum Hydrocarbons in Ground-Water Systems   /   325
    12.3.1   Density-Driven Migration of Hydrocarbons   /   326
    12.3.2   Solubility and Hydrocarbon Separation in Ground-Water Systems   /   327
12.4   Microbial Degradation of Aliphatic Hydrocarbons   /   328
    12.4.1   Methane Oxidation   /   329
    12.4.2   Oxidation of n-alkanes   /   330
        Beta-oxidation   /   *331*
        Methyl Group Oxidation   /   *331*
    12.4.3   Alkene-Oxidation   /   332
    12.4.4   Branched Aliphatics   /   332
12.5   Microbial Degradation of Alicyclic Hydrocarbons   /   334
    12.5.1   Pathways for Cyclohexanol Degradation   /   334
        Application to Environmental Studies   /   *335*
12.6   Microbial Degradation of Aromatic Hydrocarbons   /   335
    12.6.1   Benzene Degradation   /   336
        Ortho and Meta Cleavage of Catechol   /   *337*
    12.6.2   Degradation of Alkyl Benzenes   /   338
        Aerobic Degradation of Aromatic Hydrocarbons by Subsurface Bacteria   /   *339*
    12.6.3   Degradation of Polycyclic Aromatic Compounds   /   339
    12.6.4   Anaerobic Degradation of Aromatic Hydrocarbons   /   340
        Degradation of Benzoate   /   *341*
        Degradation of Toluene and Benzene Under Methanogenic Conditions   /   *341*
        Degradation of Toluene Under Fe(III)-Reducing Conditions   /   *342*
        Degradation of Alkyl Benzenes Under Denitrifying Conditions   /   *344*
        Degradation of Phenols and Cresols   /   *345*
12.7   Microbial Degradation of Petroleum Hydrocarbons in Ground-Water Systems   /   347
    12.7.1   Aerobic Degradation of BTX Compounds   /   348

12.7.2 Anaerobic Degradation of Aromatic Hydrocarbons / 350
12.8 Microbial Processes and the Remediation of Petroleum Hydrocarbon Contamination in Ground-Water Systems / 351
    12.8.1 Field Implementation of Hydrocarbon Bioremediation / 352
        Remediation of Jet Fuel Contamination / 353
12.9 Summary / 356

13 MICROBIAL DEGRADATION OF HALOGENATED ORGANIC COMPOUNDS IN GROUND-WATER SYSTEMS   358

13.1 Chemistry and Uses of Halogenated Organic Compounds / 358
    13.1.1 Aliphatic Compounds / 359
    13.1.2 Monocyclic Aromatic Compounds / 361
    13.1.3 Polychlorinated Biphenyls / 364
    13.1.4 Organochlorine Insecticides / 364
    13.1.5 Chlorinated Herbicides / 365
    13.1.6 Chlorinated Phenols / 366
13.2 Microbial Degradation of Halogenated Organic Compounds / 367
    13.2.1 Aliphatic Compounds / 367
        Aerobic Degradation / 368
        Anaerobic Degradation / 369
    13.2.2 Monocyclic Aromatic Compounds / 371
        Aerobic Degradation / 371
        Anaerobic Degradation / 372
    13.2.3 Polychlorinated Biphenyls / 373
    13.2.4 Organochlorine Insecticides / 375
    13.2.5 Chlorinated Herbicides / 375
    13.2.6 Chlorinated Phenols / 376
        Aerobic Degradation / 376
        Anaerobic Degradation / 376
13.3 Degradation of Halogenated Organic Compounds in Ground-Water Systems / 377
    13.3.1 Aliphatic Compounds / 377
        TCE Contamination at a Metal-Plating Plant / 378
        Degradation Products of TCA in Landfill Leachate / 379
    13.3.2 In Situ Bioremediation of Chlorinated Aliphatic Compounds / 380
        Savannah River Site Demonstration Project / 383
    13.3.3 Degradation Patterns of Alkyl Halide Insecticides / 384

     EDB Contamination in Hawaii / 384
     Pesticide Contamination, Long Island / 384
  13.3.4 Degradation Patterns of Chlorobenzenes / 385
  13.3.5 Degradation of Chlorinated Herbicides / 385
  13.3.6 Degradation of Chlorophenolic Compounds / 386
13.4 Summary / 388

REFERENCES / 391

INDEX / 405

PART I

OVERVIEW OF MICROBIOLOGY

CHAPTER 1

HISTORY, GEOLOGY, AND MICROBIOLOGY

History is filled with many odd coincidences, but perhaps one of the oddest is the early association of microbiology with the study of ground water. The scientific investigation of microorganisms can be traced to about 1677 and the studies of Anton van Leeuwenhoek. Leeuwenhoek was a Dutch merchant and amateur scientist who had an amazing talent for grinding fine optical lenses. The microscopes that he contrived with his lenses (Fig. 1.1) could magnify objects as much as 300 times. This gave him the ability, for the first time in human history, to observe directly bacteria-sized microorganisms.

Leeuwenhoek was an extremely curious individual, and he immediately began to examine all sorts of materials including tooth scrapings, blood, and semen. To his amazement, just about everything he examined was teeming with various microscopic creatures. He was especially interested in examining water samples from various sources. In one of many letters to the Royal Society of London, he described the presence of bacteria-like microorganisms in water drawn from a well. Thus, the study of microorganisms associated with ground water can be traced back to the very beginnings of microbiology.

This is an odd coincidence mainly because the development of microbiology and geology have followed such different paths. Table 1.1 shows a parallel chronology of some milestones in the development of these two disciplines. Note that the majority of milestones in the development of geology relate directly to unraveling the history of the earth and reflect an observational approach. The milestones in microbiology, on the other hand, relate to solving practical medical, industrial, and agricultural problems, and reflect an experimental approach.

It is important that these differences in approach, which have left such a strong imprint on each discipline, be explicitly understood. Geoscientists are often reluctant to rely solely on experimental methods to elucidate geologic phenomena. This reluctance is firmly grounded in the realization that complex geologic systems resist simulation under laboratory conditions. Microbiologists, on the other hand, are equally reluctant to accept observational evidence as being conclusive. This

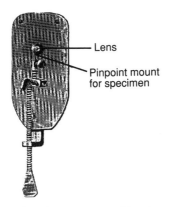

FIGURE 1.1. An early microscope used by Anton van Leeuwenhoek.

reflects accumulated years of experience that appearances are often deceptive. Effective cooperation between geoscientists and microbiologists, which is critical for studying microbial processes in ground-water systems, requires that each understand the other's point of view. This, in turn, is most readily appreciated by considering the history of each discipline.

1.1 GEOLOGY—AN OBSERVATIONAL SCIENCE

From its beginnings, the primary emphasis of geology has been to unravel the history of the earth and to understand the nature of such puzzling phenomena as earthquakes, volcanos, and fossils. Early peoples developed elaborate mythologies dealing with geologic phenomena. To the Achaean Greeks of the Late Bronze Age (c. 1200 B.C.), for example, earthquakes were associated with the whims of the god Poseidon. When angry, he would crush cities by shaking the earth. He could also summon earthquakes to help the causes of those he favored.

It is no accident that, after ten unsuccessful years of the Trojan War, the Achaeans built the fabled Trojan Horse in a last-ditch effort to defeat the city of Troy. The horse was the symbol of Poseidon and was meant as an offering to the god. Whether or not the Achaeans actually hid inside it in order to sneak into the city is immaterial. The fact is, as has been documented by archaeological excavation, the destruction of Troy at the end of the Bronze Age (c. 1270 B.C.) coincided with a major earthquake. This earthquake, which enabled the Achaeans to sack and burn the city, was clearly interpreted by them as the intervention of Poseidon.

Interestingly, the biblical city of Jericho was also destroyed by earthquake at about the same time. The destruction of Jericho, an event which removed a formidable obstacle to the Israelite invasion of Canaan, was also attributed to divine intervention, although from a different deity.

The presence of fossilized clams, snails, and other obviously marine creatures in rocks hundreds of meters above sea level had been considered a mystery from very early times. Early Christian observers generally attributed them to the effects

TABLE 1.1 Milestones in the Development of Geology and Microbiology

Date	Geology	Date	Microbiology
Pre-1A.D.	Geologic phenomena are attributed to actions of gods. Poseidon is thought to cause earthquakes when angered. Volcanos are interpreted as the workshops of Hephaestus, the god of fire and metal working	Pre-1A.D.	Fermentations are used at least as early as 2400 B.C. in making beer and bread.
B.C. 500–1500	Fossils observed in rocks are attributed to Noah's flood.		
1500	Leonardo daVinci observes fossiliferous rocks interbedded with fossil-free rocks and concludes that they could not have been formed by a single flood.		
		1546	Girolamo Fracastoro hypothesizes that disease is caused by germs being passed from one person to another.
1670	Nicholas Steno formulates the principles of superposition and original horizontality. Robert Hook suggests fossils could be used to establish chronology.	1670	Anton van Leeuwenhoek build usable microscope and makes first observation of bacteria-sized microorganisms
1749	Georges deBuffon questions literal significance of the Six Days of creation and suggests the earth is as old as 3 million years.	1770	Lazzarro Spallanzani disputes theory of spontaneous generation and demonstrates that putrefaction is caused by microorganisms.
1787	A. G. Werner founds Neptunist school that regarded granites and basalts as chemical precipitates from seawater. The neptunists are opposed by vulcanists who propose an igneous origin for granites and basalts.		
1795	James Hutton publishes ''Theory of the Earth'' in which he states that earth's history cycles between periods of deposition and periods of uplift and erosion.	1810	Chemist Jons Jakob Berzelius proclaims fermentation to be a strictly inorganic phenomenon that occurs without microorganisms.

1811	Georges Cuvier and William Smith, working independently, show a relationship between fossil assemblages and successions of sedimentary rocks. Cuvier espouses a "catastrophic" origin, such as Noah's flood, for the observed sequences.
1830	Charles Lyell publishes *Principles of Geology*, a compendium of observations showing that geologic process observed today can be assumed to have been operating in the past. This view is termed "uniformitarianism" and directly opposes the "catastrophic" view of Cuvier.
1830	Charles Cagniard de Latour performs experiment showing fermentation to be caused by microorganisms. This view was criticized by chemists on the grounds that heat sterilization changes some essential component in air.
1845	M. J. Berkley shows that Irish potato blight is caused by a mold. This is the first clear demonstration that microorganisms can cause disease.
1850	Henry Cliffton Sorby uses a petrographic microscope to observe sedimentary rocks.
1859	Charles Darwin publishes *The Origin of Species*, proposing that species are not immutable. This provides a framework for understanding why fossil assemblages change with time. Edwin Drake's oil well drilled in Pennsylvania.
1859	Charles Darwin publishes *The Origin of Species* and establishes evolution as a basic concept in biology.
1860	James Hall proposes that accumulation of sediments depresses the earth's crust, forming basins of great depth.
1861	Louis Pasteur shows that fermentations are caused by microorganisms and ends the controversy concerning spontaneous generation.
1867	"Pasteurization", or heating, to kill microorganisms in the wine, beer, and milk is used for the first time.
1873	James D. Dana coins the term "geosyncline" for thick sequences of sediments. He rejects Hall's sediment-loading hypothesis and proposes that geosyncline are a result, not a cause, of basin down warping.

TABLE 1.1 (continued)

Date	Geology	Date	Microbiology
		1876	Robert Koch enumerates "Koch's postulates" and firmly establishes the germ theory of disease.
		1881	Joseph Lister and Robert Koch develop methods for isolation of bacteria in pure cultures.
		1885	Louis Pasteur develops a rabies vaccine.
		1890	Sergei Winogradsky isolates autotrophic nitrifying bacteria and shows that they transform ammonia to nitrate in soils and manure. Martinus Beijerinck describes symbiotic and nonsymbiotic nitrogen fixation by bacteria, and isolates sulfate-reducing bacteria.
1905	Physicist Ernest Rutherford suggests using radioactive minerals to obtain absolute age of rocks.		
		1910	Paul Ehrlich establishes chemotherapy as a tool for treating bacterial infections, such as syphilis.
1912	Alfred Wegener proposes that continents are slowly drifting relative to each other and that they split off from a single "supercontinent."		
1913	Frederick Soddy clarifies transformation of radioisotopes. This leads to reliable radiometric dating methods that demonstrate the earth to be billions of years old.		
1929	Arther Holmes invokes the mechanism of thermal convection in the mantle as the driving force for "continental drift." He envisions "spreading centers" where new crust is formed and "regions of compression" where crust is returned to the mantle.	1929	Alexander Fleming reports antibacterial action of *Penicillium*, a species of mold.

1935	Wendell Stanley crystallizes tobacco mosaic virus and shows it is largely composed of protein.
1940	The electron microscope is invented, antibiotics are first used in medicine.
1953	Structure of DNA is discovered by J. Watson and F. Crick. Hans Krebs receives the Nobel Prize for elucidation of the tricarboxylic acid cycle.
1960	Francis Jacob and Jacques Monad propose the operon theory to explain how genetic information controls protein synthesis.
1961	Peter Mitchell proposes the chemiosmotic model for ATP synthesis.
1940	The discovery is made of bands of positive and negative magnetic anomalies on either side the Mid Atlantic ridge.
1953	Evidence accumulates that the earth's magnetic field has reversed polarity periodically through geologic time.
1963	F. J. Vine and D. H. Mathews in England, and L. Morley and A. Lorochelle in Canada independently propose that the Mid-Atlantic magnetic anomalies correspond to bands of rock that were magnetized during episodes of reversed polarity. This evidence supports the concept of sea-floor spreading and leads to wide acceptance of Plate Tectonics.
1970	Harold Boyer and Stanley Cohen use plasmids and recombinant DNA methodology to clone genes.
1970	Plate tectonics is widely used to explain distribution of mountain ranges and deep-sea trenches.

of Noah's flood. This interpretation was first seriously questioned by Leonardo da Vinci around 1500 A.D. Leonardo was fascinated by water and spent a lifetime carefully observing the action of flowing water in his native Tuscany. These observations are beautifully recorded in the waterfalls, rivers, and springs incorporated into most of his paintings. In addition to springs and rivers, the hills of Tuscany are characterized by the abundant occurrence of marine fossils in limestone. By comparing his own observations of how floods carried debris to what he observed in alternating beds of sedimentary rocks, he became convinced that there were numerous flooding episodes recorded in the rocks, not just the single event mentioned in the Bible.

Leonardo da Vinci was a secretive man and not one to share his observations by publishing them. More than 150 years passed before Nicholas Steno repeated Leonardo's observations and came to very similar conclusions. Steno recognized that the shell-bearing strata beneath the site of ancient Rome—rocks that were used in the city's construction—had to be older than the city itself ($\sim$ 3,000 years). However, Steno couldn't shake the biblical notion that the earth was only as old as written human history, or about 4,000 years. This obviously didn't leave much time for the rocks to form. If one accepted this time scale, one was forced to the conclusion that the rocks must have formed catastrophically. This view was roughly consistent with the biblical story of creation and thus *catastrophism*, came to guide much geologic thought over the next couple of centuries.

Unlike Leonardo, Steno published his observations in 1669. The principle of superposition (younger sediments overlie older sediments) is first attributed to Steno. Similarly, the principle of original horizontality (deformed sedimentary rocks must have been originally deposited in flat-lying beds) and original lateral continuity (faulted sedimentary rocks were originally deposited in continuity) are first attributed to Steno. These principles are the basis of modern stratigraphy.

A hundred years after Steno, the Frenchman Georges de Buffon realized that the thick sections of sedimentary rocks observed in the Paris Basin could not have been deposited in a few thousand years. Daring greatly, in 1749 he publicly suggested that the earth was as old as 75,000 years (his private writings indicate that he thought that the real age of the earth was more like 3 million years). This suggestion took some courage, because he was, in effect, challenging the 4,000-year-old age given in the Bible. The church was not broaching any opposition, however, and he was forced to retract his statements publicly. Buffon represented a major turning point in geologic thought, however, because he planted the idea that the Bible was not necessarily a reliable source of insight into geologic phenomena.

About the same time that Buffon was thinking about the age of the earth, two major schools of geologic thought arose in Europe. The "Vulcanists," principally Italians, proposed that the oldest rocks on earth (which were called the "primitive") were ultimately of volcanic origin. The "Neptunists," on the other hand, who were principally Germans, viewed the primitive rocks as being chemical precipitates from a primeval ocean. For a while, the Neptunists held sway, largely because of the persuasive teaching of their leader, A. Gottlieb Werner. Werner, who taught at the Freiburg Mining Academy in Saxony, was not a field geologist. Instead of collecting samples himself, he preferred to send students to do this work and confined himself to looking at rocks in his laboratory. This reluctance to do fieldwork seriously biased many of his observations. Werner himself never gave up his Neptunist views. However, as several Italian geologists actually witnessed the formation of igneous rocks from active volcanos, the Vulcanist view gradually replaced the Neptunist view.

The Neptunist position also suffered from the writings of the Scottish naturalist James Hutton, who published his *Theory of the Earth* in 1795. Hutton, unlike Werner, believed in going into the field, carefully observing the rocks, and forming conclusions based solely on those observations. The Neptunists' view of the formation of the earth's original crust was largely one of quiescence. They envisioned a vast ocean with granites and basalts gradually precipitating from seawater. In the hills around Edinburgh, however, Hutton could see clear evidence that "primitive" rocks had been deformed and thrust above the level of the sea. Rather than a static earth, where rocks precipitated out of seawater and remained largely unaltered thereafter, Hutton envisioned cycles of sediment accumulation, deformation, uplift, and subsequent erosion. Hutton's greatest contribution was the idea that the earth is internally dynamic and ever changing. This view clearly implied that the earth had to be very old, certainly older than a few thousand years.

Hutton also questioned the view that the earth had been largely shaped by single "catastrophic" events such as the biblical flood. Hutton reasoned that the processes of erosion and deposition that could be presently observed must have occurred in the past. Thus, the best guide to understanding the geologic record was to understand the nature of geologic process that could be presently observed. This view came to be call *uniformitarianism*—a view that was most eloquently developed by a British lawyer turned geologist named Charles Lyell.

Charles Lyell had little patience with the attitude that earth history reflected a series of mysterious catastrophes, such as Noah's flood or the disappearance of mysterious continents (i.e., Atlantis). He preferred a knowable, predictable earth, in which processes proceeded at a slow but steady pace. Given enough time, he reasoned, even gently flowing streams could erode deeply into the landscape. Lyell's extreme uniformitarianism was largely a reaction to the biblical catastrophists and greatly appealed to geologists of the nineteenth century. However, in the twentieth century, it has become apparent that certain catastrophic events (asteroid impacts, for instance) *have* occurred periodically throughout geologic time and that they have greatly influenced the history of the earth.

Meanwhile, in the early nineteenth century, it became clear that different sedimentary strata were characterized by different fossil assemblages. In England, this was the period when great networks of canals were constructed to facilitate moving coal, iron ore, and other commodities. One of the men involved in this huge effort, an engineer named William Smith, developed a deep interest in the fossils exposed by canal construction. Smith realized that rocks could be correlated from place to place based solely on fossil content. He wrote of the "wonderful order and regularity with which nature has disposed of these singular productions (fossils) and assigned to each its class and its peculiar Stratum." Eventually, Smith got to the point where he could recognize most of the strata in the English countryside over long distances. In 1815, Smith completed publication of the first comprehensive geologic map of England. This map is still recognized as a masterpiece of cartography.

At the same time that Smith was working in England, two Frenchmen, Georges Cuvier and Alexandre Brongniart, made similar observations in the vicinity of

Paris. Like Smith, they compiled and published a geologic map based upon the fossils in the rocks. The ability to distinguish strata on the basis of fossils, coupled with the realization that older rocks had to be overlaid by younger rocks, allowed geologists, for the first time, to establish relative chronologies. Two English geologists, Adam Sedgwick and Roderick Murchison, proceeded to construct such a chronology for the sedimentary rocks exposed in Wales. By 1835, Sedgwick had named the oldest fossil-bearing rocks Cambrian (for the ancient roman name for Wales) and Murchison had named the rocks immediately above Silurian (after an ancient Welch tribe).

Unfortunately, Murchison and Sedgwick could not agree on the boundary between the two units. Murchison suspected Sedgwick of trying to enlarge the Cambrian at the expense of the Silurian, and Sedgwick had similar suspicions about Murchison. The resulting row between the two men was finally resolved by naming the disputed interval the Ordovician. Successively younger rocks were named Devonian (after Devonshire) and Carboniferous (due to the presence of coal beds). Later, Murchison recognized rocks in Russia that were younger than the Carboniferous and named them Permian.

By the late nineteenth century, the *relative* ages of major sedimentary rock units in Europe were well known. No one, however, had much of an idea of their *absolute* ages. In 1854, the German thermodynamicist Herman von Helmholtz inferred that the earth could be no more than 20 to 40 million years old based on the luminosity of the sun. The English physicist William Thomson, better known as Lord Kelvin, embellished von Helmholtz's arguments with estimates of cooling rates of the primitive earth and came up with ages ranging from 20 to 75 million years.

The discovery of radioactivity in 1895 changed the debate considerably. Since radioactivity could heat the earth well beyond the length of time estimated by Lord Kelvin, it was quickly realized that the earth must be much older than formerly thought. In 1905, Ernest Rutherford, who first proposed a structure of the atom, suggested that rates of radioactive decay could be used to date rocks. By the 1920's, when the full series of uranium decay products had been worked out by the chemist Frederick Soddy, it became possible to assign absolute ages to rock units in addition to relative ages. Isotopic dating made it clear that the earth was not just millions of years old, but *several billion* years old.

Meanwhile, geologists had other knotty problems to worry about. By the mid-nineteenth century, it was clear that many mountain chains, the Alps being an obvious example, consisted of sedimentary rocks that were thousands of meters thick. How could such thick sediment sequences, which were thicker than the deepest part of the oceans, form? Furthermore, what forces could result in the uplift and deformation of such huge masses of rock? In 1860, the American geologist James Hall proposed that, as sediments accumulate in sedimentary basins, the increased weight causes downwarping of the crust, allowing even more sediments to accumulate. (This hypothesis, termed isostasy, appealed to many geologists because it neatly explained away the sediment thickness problem.) Unfortunately, it couldn't explain the uplift and deformation of the sediments. In 1873, the American geologist James Dana coined the term geosyncline to describe the massive accumulations of sediment associated with basins. He rejected Hall's isostasy hypothesis and argued instead that the basins were the *result* and not the *cause* of crustal downwarping. However, Dana offered no mechanism to account for the inferred downwarping.

The origin of mountain building events was clearly related in some fashion to the accumulation of sediments in geosynclines. This was evident from ongoing efforts to map the geology of Europe and the United States. Furthermore, these mapping efforts showed that there had been, not one, but several episodes of mountain building throughout geologic time. What could explain these findings?

As early as 1620, Francis Bacon had remarked on the parallel configuration of the continental margins on each side of the Atlantic Ocean. This seemed to imply that the continents had at one time been contiguous and then had drifted apart. In 1858, a geologist named Antonio Snider actually published maps depicting the continents as drifting apart. In 1912, a German meteorologist named Alfred Wegener cited the remarkable similarity between rocks, geological structures, and fossils on opposite sides of the Atlantic Ocean as evidence for continental drift. For the next several years, Wegener doggedly made the case for the continental drift hypothesis in the face of severe criticism from most geologists. Wegener's argument suffered first from the fact that he himself had no credentials in geology. Second, Wegener could not put forward an adequate explanation as to the forces driving the continents apart.

A major contribution was made in 1928 by the British geologist Arthur Holmes. Holmes suggested that convection currents in the semi-solid mantle, which were driven by heat from the decay of radioactive elements, could move continents and create deep-sea trenches (Fig. 1.2). This model is remarkably similar to our modern concept of plate tectonics. Holmes was a widely respected geologist, and his ideas lent considerable credibility to the idea of continental drift. Holmes himself, however, considered his paper to be merely an interesting idea and admonished

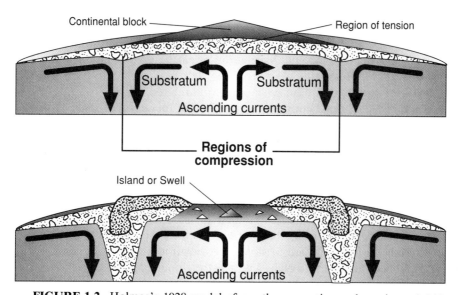

FIGURE 1.2. Holmes's 1928 model of mantle convection and continental drift.

his readers to remember that "purely speculative ideas of this kind, specially invented to match the requirements, can have no scientific value until they acquire support from independent evidence."

The independent evidence needed to establish this concept was a long time in coming. During World War II, the British and U.S. navies had developed extremely sensitive airborne magnetometers to detect German submarines. After the war, oceanographers adapted these magnetometers to be towed behind their research ships. By repeatedly steaming back and forth over the ocean, it gradually became clear that there were broad bands of positive and negative magnetic anomalies. Curiously, these bands were symmetrical around the newly discovered mid-oceanic ridge. What could possibly explain these findings?

In the 1950's, it became evident that the polarization of the earth's magnetic field had changed repeatedly over geologic time. In 1963, two Englishmen, F. J. Vine and D. H. Mathews, and two Canadians, L. Morley and A. Larochelle, independently made a connection between the magnetic anomalies on the seafloor and magnetic polarization reversals. Specifically, they proposed that the positive and negative magnetic zones corresponded to bands of oceanic basalt magnetized during ancient episodes of reversed polarity in the earth's magnetic field. This, in turn, implied that the ocean floor was progressively spreading at the mid-oceanic ridge, a process remarkably similar to the suggestions of Arthur Holmes. These findings were later confirmed by dating the rocks on the seafloor. The youngest rocks were at the mid-oceanic ridge and became progressively older away from the ridge. The demonstration of seafloor spreading firmly established the reality of continental drift.

With the general acceptance of *plate tectonics*, the model that accounted for the phenomenon of continental drift, numerous puzzling geologic questions were suddenly answered. Deep oceanic trenches, a phenomenon first observed in the nineteenth century, were areas where oceanic crust was being thrust underneath lighter continental crust. Sediments filling such trenches could be thousands of meters thick, explaining the presence of geosynclines. The collision of plate boundaries was a ready mechanism for deforming these sediments and thrusting mountains high above sea level. Finally, the motions of the earth at plate boundaries could explain the occurrence and distribution of earthquakes and volcanos. As a unifying model, plate tectonics has been remarkably successful in tying together numerous observations that otherwise seemed unrelated.

The point of this brief history is to emphasize the observational nature of geology. The early geoscientists who made significant contributions-Steno, Buffon, Cuvier, Hutton, Smith and others-all had one thing in common. They strongly believed that the only way to understand the earth was by careful observation, followed by careful interpretation. The establishment of the most important unifying concept in geology-plate tectonics-was based entirely on observational evidence. Experimental evidence, indicating that rocks were insufficiently deformable for continents to move at all, largely muddied the issues involved. Other instances in which the progress of geology has faltered-such as the neptunist teachings of Werner-can largely be attributed to sloppy observation. As a direct result of this history, the belief that accurate observation is the principal means for understanding geologic phenomena is deeply rooted in the earth sciences.

1.2 Microbiology—An Experimental Science

Human beings have an interesting ability to make use of phenomena that they do not understand. The classic example of this is the use that ancient peoples made of microorganisms in order to manufacture yogurt, bread, beer, and wine. The ancients had no idea that these useful processes were microbially mediated, and yet they developed quite sophisticated techniques for using them. Early Egyptian brewers, for example, originally used yeasts on wheat husks to carry out beer fermentation. However, by about 2400 B.C., they were using virtually pure cultures of yeast in order to produce beer of predictable quality. This was all done by trial and error, and the brewers were certainly unaware that they were culturing microorganisms. It does seem, however, that they knew that their brewing made use of some "living" force. Similarly, people used salting and drying techniques to preserve food long before the role of microorganisms in food spoilage was understood.

With our modern understanding of medicine, it's difficult to imagine just how terrifying infectious diseases must have been to early peoples. Not only were the consequences of disease to be feared, but also the fact that no one knew their causes meant that people were largely helpless to protect themselves. Nevertheless, again by trial and error, effective means of dealing with infectious diseases were developed. The practice of "quarantine" for people with certain diseases dates to biblical times. For some diseases, such as cholera or plague, quarantine could be effective. Ironically, however, the disease to which this technique was most commonly applied in the ancient world—leprosy—is one to which quarantine was only marginally effective in confining.

The advent of the black plague, in the mid-fourteenth century, had an enormous impact on medical thinking in Europe. From its outbreak in early 1348 in Marseilles, France, the disease spread northward, reaching Paris on June 30. Later that year, it appeared in Bristol, England, and by 1349 had engulfed all of Europe. The systematic nature of the plague's spread was not lost on the horrified Europeans. It followed trade routes on both land and sea. Clearly, the plague was caused by some communicable agent. This observation was put to immediate use by the inhabitants of fourteenth-century Sicily who required the men in arriving ships to stay at anchor for 40 days before coming ashore. Unfortunately, this quarantine was ineffective against rats that apparently swam ashore at their leisure, bringing disease-harboring fleas with them.

Two centuries later, in 1546, an Italian physician named Girolamo Fracastoro published a treatise on contagious diseases entitled *De Contagione*. In his book, Fracastoro hypothesized, largely on philosophical grounds, that disease could be caused by "germs" being passed from one person to another. Furthermore, Fracastoro recognized that contagious diseases were similar to the putrifaction of wine. This was the first hint that the two processes might have a common cause. Fracastoro also seemed to understand that certain germs caused only certain diseases and that different diseases could be spread by different means.

The actual study of microorganisms began in 1665 with the publication of Robert Hook's *Micrographia or Some Physiological Descriptions of Minute Bodies Made by Magnifying Glasses with Observations and Inquiries Thereupon*. In this work,

Hook provided drawings of various species of fungi. However, Hook's crude magnifying glass could only resolve fairly large microorganisms.

Anton van Leeuwenhoek was able to grind much better lenses than were available to Hook and was able to construct a microscope capable of resolving bacteria. It's interesting to note that van Leeuwenhoek was a close friend of the great Dutch physicist Christian Huygens. Huygens was deeply interested in the behavior of light and developed a wave theory to account for its behavior. It is probable that the quality of Leeuwenhoek's lenses benefited substantially from the theoretical insights of Huygens.

The discoveries of Hook and van Leeuwenhoek opened the way for investigation of the nature of microorganisms. One controversy that emerged and that lasted throughout the eighteenth and into the nineteenth century concerned the origin of microorganisms. Aristotle (384-322 B.C.) had taught that animals could originate spontaneously from soil, plants, or even other unlike animals. To many, it seemed perfectly logical that microorganisms could be generated spontaneously.

The modern debate on spontaneous generation of microorganisms began when in 1749 an Englishman named John Needham experimented with meat exposed to hot ashes. Needham observed that bacteria that apparently were not present at the beginning of the experiment later appeared. He concluded that the bacteria must have been generated by the meat. A contemporary of Needham named Lazzarro Spallanzani observed something quite different. When Spallanzani boiled beef broth for an hour and then sealed the flask, no microbes appeared. When informed of Spallanzani's results, Needham was unmoved. Needham insisted that air was essential to the spontaneous generation of microbes and by sealing the flasks, Spallanzani was simply excluding a necessary ingredient to spontaneous generation.

An argument closely related to the spontaneous generation debate was whether or not microorganisms were responsible for "putrefaction" (fermentation) of organic compounds, such as sugars or proteins. Several otherwise quite capable chemists, including Justus von Liebig, Jons Jakob Berzelius, and Friedrich Wohler, were convinced that fermentations were chemical phenomena that had nothing to do with microorganisms.

The idea of air being necessary for spontaneous generation, as opposed to being a medium carrying microorganisms, was a particularly difficult problem to address. In the nineteenth century, this problem was taken up by Franz Schulze and Theodor Schwann. Schulze passed air through strong acid solutions and into boiled broth, and microbes failed to appear. Schwann passed air through red-hot tubing into broth and again microbes failed to appear. Advocates of spontaneous generation were unmoved. The acid and the heat was changing something in the air needed for growth, they argued. About 1850, two German scientists named Schroeder and von Dusch performed an even more convincing experiment by passing air through cotton filters into boiled broth. Amazingly, however, the proponents of spontaneous generation and chemical fermentation remained unconvinced.

In 1859, a Frenchman named Pouchet published an extensive study "proving" that spontaneous generation did occur. In some ways, Pouchet's paper proved to be pivotal, because the data and logic irritated a fellow Frenchman, Louis Pasteur. Pasteur had been trained as a chemist and was adept at experimental design. He prepared a flask with a long, narrow gooseneck opening that allowed air into the flask but excluded airborne particles. When Pasteur reported his results in 1864, he exhibited the flourish for which he became famous, saying:

> For I have kept from them, and am still keeping from them, that one thing which is above the power of man to make; I have kept from them the germs that float in the air, I have kept from them life.

Pasteur went on to chide the supporters of spontaneous generation:

> They who allege it (spontaneous generation) have been the sport of illusions, of ill-made experiments, vitiated by errors which they have not been able to perceive and have not known how to avoid.

It's worth noting that the substance of Pasteur's jab at his opponents was that they were poor experimentalists. Pasteur recognized that the essence of proper experimentation was first to have a testable hypothesis. Given a hypothesis, one could design experiments to test its validity. Pasteur followed this same procedure throughout his long and productive career.

Pasteur's work on spontaneous generation occurred some years after he became actively involved in microbiological studies. In 1854, while he was a professor at the University of Lille, a local industrialist approached him concerning a problem in local wine production. It seemed that, when grape juice was being fermented in vats, some of the vats produced good wine, whereas others produced sour wine. Pasteur observed wine samples from both kinds of vats under the microscope and saw that the good ones were characterized by a single kind of budding yeast cell. In the bad vats, on the other hand, there were rod-shaped bacteria. Furthermore, these bacteria produced lactic acid from grape sugar, accounting for the sour taste of the wine.

From these studies, Pasteur reasoned that the quality of the wine could be controlled by seeding the fermentation vats only with the proper kind of microorganism. To do this, however, it was first necessary to kill or inactivate the microorganisms naturally present in the juice. Boiling, the accepted method of sterilization at the time, was not possible as this would change the character of the grape juice. Pasteur found that heating the juice at 145°F for half an hour removed the undesirable bacteria without hurting the flavor of the grape juice. This heating process, which we still refer to as *Pasteurization,* greatly increased the productivity and quality of the local wine industry.

In the 1870's, Pasteur turned his attention to studying the cause of anthrax, a deadly disease of sheep, cattle, and sometimes people. Coincidentally, a German country physician named Robert Koch became interested in this disease at about the same time. Koch was a quiet, meticulous man who became fascinated with manipulating bacterial cultures in the laboratory. It was Koch who first discovered the rather large bacilli present in the blood of cattle that had died of anthrax. Koch carefully transferred these cultures, examining them periodically under the microscope, until he was sure that there was only one kind of microorganism present. He then injected cultures of the bacilli into healthy animals and observed that they too became sick and died. Lastly, Koch was able to again isolate the bacilli from the blood of the dead animal. This constituted the first rigorous demonstration that a bacterium could cause disease in animals. Koch published these results in 1876.

The steps that Koch followed in demonstrating the causative agent of anthrax have come to be known as "Koch's postulates". They are:

1. The organism should be present in all animals suffering from the disease and should be absent from all healthy animals.
2. The organism must be grown in pure culture outside the diseased animal.
3. When the pure culture is used to inoculate a healthy animal, the animal must develop the symptoms of the disease.
4. The organism must again be isolated from the experimentally infected animal and shown to be identical to the original isolate.

Koch's work was the final confirmation of the *germ theory of disease*. This theory, at least in rudimentary form, can be traced back to the writings of Fracastoro in the sixteenth century. The theory was formally proposed in 1762 by Anton von Plenciz of Vienna. Plenciz stated that living agents were the cause of disease and went further to state that particular germs caused particular diseases. The establishment of the germ theory of disease by Koch was a milestone in the development of microbiology.

Robert Koch understood at a very early stage in his research that being able to grow and to manipulate bacteria in the laboratory was crucial. Being a born tinkerer who was very clever with his hands, Koch was able to refine techniques for studying bacteria. For example, he found that by smearing bacteria on a glass slide and adding certain dyes, individual bacterial cells were much easier to observe. It was Koch (or possibly his female laboratory assistant, Hesse) who learned how to add gelatin and agar as solidifying materials to culture media in order to grow isolated colonies of bacteria. These colonies could then be sequentially transferred to fresh media until a pure culture was obtained. The development of solid media made it much easier to study bacteria in the laboratory. Using these techniques, Koch isolated the bacteria that caused pulmonary tuberculosis in 1884.

Louis Pasteur, while begrudging Koch (and Germans in general) a share of the limelight, immediately understood the importance of Koch's techniques and began applying them in his own research. In 1880 Pasteur isolated the organism that caused cholera in chickens. In order to prove to a skeptical public that this organism really was the cause of the disease, Pasteur arranged for a public demonstration of an experiment that he had performed many times. He inoculated healthy chickens with cultures of the bacteria, but to his chagrin, the chickens failed to get sick.

Going back over his notes, he discovered that, instead of being injected with freshly grown bacteria, the chickens had been inoculated with a culture that was several weeks old. Evidently, only freshly grown bacteria had the ability to cause the disease. The real surprise came when, a few weeks later, Pasteur injected freshly grown inoculum into a new group of chickens, as well as chickens that had had the old inoculum. Those chickens that had previously been exposed to the old inoculum remained healthy whereas unexposed chickens promptly developed cholera.

This was a sensational discovery for two reasons. First, it explained the phenomenon of immunization that had been discovered in 1798 by Edward Jenner. Jenner had found that inoculating people with material from cowpox boils could provide immunity against smallpox. Pasteur realized that the chicken cholera experiment explained how immunization worked. Pasteur acknowledged this in naming the new procedure "vaccination," which was based on *vacca,* the latin word for *cow.*

Second, this discovery suggested a means for providing immunity against other diseases. Working with old or "attenuated" cultures, Pasteur soon developed an immunization for anthrax. The crowning achievement of Pasteur's career, however, was the development of a vaccine for the viral disease, rabies. Pasteur knew that rabies was not caused by a bacterium, as it could not be observed under the microscope. However, he found that injecting an extract of the dried brains of infected animals provided immunity to dogs. This remained untested in humans until a young boy named Joseph Meister was bitten by a rabid wolf in 1885. The boy's parents brought him to Pasteur who injected Meister with successively more virulent pieces of ground rabbit spinal cord. To Pasteur's delight, and possibly to his surprise, Meister did not develop rabies.

The triumph of the rabies vaccine led to widespread financial support for Pasteur's research and the establishment of the Pasteur Institute. The establishment of this institute led to the widespread production and utilization of vaccines in medicine and agriculture. Just as important, however, it provided the means for training students in Pasteur's techniques.

One of the students who spent time at the Pasteur Institute was a young Russian named Sergei Winogradsky. Winogradsky took the pure culture techniques developed for investigating infectious diseases and applied them to the investigation of other bacterially mediated processes. As early as 1877, it had been proposed that the process of nitrification-forming nitrate ions from ammonia was bacterially mediated. Winogradsky took up this problem and succeeded in isolating pure cultures of nitrifying bacteria. This, in turn, showed for the first time that microorganisms could obtain energy for growing by oxidizing inorganic (in this case, ammonia) as well as organic compounds. Winogradsky later described the microbial oxidation of sulfur, hydrogen sulfide, and ferrous iron.

The early twentieth century was a period of explosive growth in the application of microbiology to medical, industrial, and agricultural problems. During this period, literally thousands of microbes were described and cataloged. This created, as one might suspect, a nightmare of microbial taxonomy (classification). In 1894, the Gram staining procedure was incorporated as a key feature of bacterial classification systems. Beginning in 1915, Robert Buchanan proposed a classification system that was fundamentally different from zoological or botanical systems. Buchanan's system incorporated morphological, biochemical, and pathogenic characteristics in order to classify bacteria. Buchanan's efforts led to the establishment of a Committee on Characterization and Classification of Bacterial Types, of which the first chairman was David Bergey. The first edition of *Bergey's Manual of Determinative Bacteriology* was published in 1923. Subsequent editions of *Bergey's Manual* have been the basis of bacterial taxonomy ever since.

It was during the mid-twentieth century that the basics of bacterial metabolism and physiology were worked out. Albert Kluyver, who was director of the Delft Polytechnical School (a legacy of Anton van Leeuwenhoek), examined microbial metabolism, recognizing the importance of intermediate products in transferring energy. He established that hydrogen transfer was the basis of all metabolic processes. In addition, Kluyver recognized the importance of central metabolic pathways in microbial metabolism. The most important of these pathways, the Tricarboxylic Acid (TCA) cycle and the Calvin cycle, were worked out in the 1930's and 1940's. Interestingly, the Nobel Prizes for discovery of the TCA and Calvin cycles were both awarded in chemistry. This is indicative of the critical role played by biochemistry in understanding microbial processes.

The discovery of the double helix structure of DNA by James Watson and Francis Crick in 1953 set the stage for new studies into bacterial genetics. In 1960, the *operon theory* for explaining how genetic information controls protein synthesis was proposed by Fancois Jacob and Jacques Monad. The genetic code, the cell's "dictionary" for how DNA base sequences specify amino acid sequences in proteins, was broken by Robert Holley, H. G. Khorana, and Marshal Nirenberg in the early 1960's. In the 1970's, advances in microbial genetics led to "genetic engineering" technology, in which genes for specific proteins could be inserted into bacteria. An early use of this technology was the production of human insulin from bacteria.

This brief overview of the history of microbiology does scant justice to the important fields of virology, phycology (study of algae), mycology (study of fungi), protozoology (study of protozoa), or immunology, all of which are important subdisciplines of microbiology. In fact, modern microbiology is best thought of as a collection of specialized subdisciplines unified mainly by the microscopic nature of the organisms being studied.

The point of this history is to emphasize the experimental nature of microbiology. Observations of microorganisms had been made for 150 years before the rigorous experimental methods of Pasteur and Koch were introduced. With experimental methods replacing simple observations, the plodding pace of advancement that had dominated for so long was replaced by an explosion of information. Microbiology went from a virtual non science in 1850 to a mature, highly productive science in 1900. As a direct result of this history, the belief that experimentation is the principal means for understanding microbial phenomena is deeply rooted in the practice of microbiology.

1.3 GROUND-WATER CHEMISTRY AND SUBSURFACE MICROBIOLOGY

Given the very different histories and different traditions of geology and microbiology, it's not particularly surprising that they have stayed largely separate during much of the twentieth century. As the scopes of each discipline enlarged over time, however, it was inevitable that they would eventually reach a point of intersection. The primary issue that provided this point of intersection was the chemical quality of ground water.

Characteristically, however, each discipline approached the ground-water quality issue from different viewpoints. Geoscientists were initially concerned with understanding the natural water chemistry of pristine aquifer systems. It was observations associated with oil field brine geochemistry that first suggested that microbial processes in deep subsurface environments could be of major importance. This, of course, is perfectly in keeping with the observational tradition of the geosciences.

Microbiologists were also drawn into the study of subsurface environments by water quality issues. In this case, however, concern about chemical contamination due to waste disposal practices was the primary motivating factor. Furthermore, development of aseptic drilling techniques and experimental evidence for the presence and activity of subsurface microorganisms was required before the presence of subsurface microorganisms were widely accepted in the microbiological community. This, in turn is in keeping with the experimental nature of the microbiological sciences.

1.3.1 Subsurface Microbiology and the Geosciences

In 1919, a geologist named Sherburne Rogers, working for the U.S. Geological Survey, published a comprehensive study of the Sunset-Midway oil field in California. As part of this study, analyses of the brines associated with the hydrocarbons were made. These analyses pointed out that the oil field brines had a significantly different composition than shallower ground water in the area. Specifically, shallow ground waters were characterized by relatively high sulfate and relatively low bicarbonate concentrations. The brines associated with oil, on the other hand showed high concentrations of bicarbonate but an almost total absence of sulfate.

Rogers interpreted this as evidence that, where ground water and hydrocarbons coexisted, the oxidation of hydrocarbons resulted in the reduction of sulfate with the production of bicarbonate. Furthermore, Rogers suggested that sulfate-reducing bacteria were involved in these reduction processes. At the time, however, there was no direct evidence that such microorganisms were actually present in hydrocarbon reservoirs.

This evidence was provided by a geologist named Edson S. Bastin from the University of Chicago who took up the problem of sulfate reduction associated with oil field brines. The first thing Bastin did was to test the hypothesis that oil could abiotically reduce sulfate at low temperatures. Previous investigators had attempted to reduce sulfate at 300°C, using glucose and starch as a reducing agent, without success. However, Bastin methodically went about duplicating these results. He set up several experiments in which different salts of sulfate were combined with crude oil, coal, and oil shales under aseptic conditions. After incubations that were allowed to proceed for up to 170 days, no production of hydrogen sulfide or bicarbonate was detected.

The absence of evidence for the abiotic reduction of sulfate by crude oil obviously implied that microorganisms were involved in the sulfate reduction observed in oil field brines. To test this possibility, Bastin enlisted the aid of a bacteriologist, Frank Greer, who was also at the University of Chicago. Greer prepared a magnesium sulfate—lactate medium and adjusted the salt content to that of oil field brines found in the Illinois Basin. These brines were in many ways similar to those of the Sunset-Midway field, being characterized by low sulfate content. Next, water from 19 oil wells was collected and used to inoculate the media. In ten days, 17 of the 19 samples showed the active growth of sulfate-reducing bacteria.

In 1926, Bastin reported his findings in the journal *Science*, observing:

> From the work here reported it is evident that anaerobic bacteria of the sulphate-reducing type are present in abundance in some of the waters associated with oil in productive fields and it is very probable that they are responsible for the low sulphate content of these waters.

Over the next 50 years or so in the United States, the basic idea that bacteria are present in deep subsurface environments and that they have important effects on ground-water chemistry was gradually broadened. Cedarstrom (1946), following Bastin's lead, suggested that bacterial sulfate reduction was a principal cause of high-bicarbonate ground water in the Atlantic Coastal Plain. Unlike Bastin, however, Cedarstrom did not produce any direct evidence that bacteria were actually present in the systems they were studying. While conducting studies on the origins of petroleum and associated pore fluids, Claude Zobell described microbial processes in deeply buried marine sediments (Zobell, 1947).

Around this time, it became common for geoscientists, Cedarstrom, for example, to attribute ground-water chemistry changes to microbial processes—a practice that brought a rebuke from Zobell, who insisted that actual evidence of microorganisms was also necessary. The unfortunate practice of attributing ground-water chemistry changes to the activity of microorganisms, without any evidence that such microorganisms were even present, eventually provoked a backlash. Margaret Foster, a chemist working for the U.S. Geological Survey, carefully considered Cedarstrom's hypothesis. She noted that the bicarbonate concentration increases documented by Cedarstrom were not accompanied by decreases in sulfate concentrations and were therefore unlikely to have been caused by sulfate reduction.

Foster performed a series of experiments, showing that such water chemistry patterns could be reproduced without sulfate-reducing bacteria (Foster, 1950). Rather, she suggested that sedimentary organic material subjected to "dynamo-chemical processes" (i.e., increased temperature and pressure) would progressively eliminate carboxyl groups and produce carbon dioxide abiotically. This carbon dioxide could then react with carbonate minerals to form the observed high-bicarbonate ground water. Foster's paper was highly cited in the geochemical literature and was enormously influential. Consequently, the previous tradition that microbial processes were important in ground-water chemistry began to be replaced by a conviction that inorganic processes were predominant.

Curiously, at about the time when consideration of microbial processes in ground-water chemistry was falling out of favor in the United States, Russian scientists were coming to very different conclusions. In 1962, a Russian scientist M. I. Gurevich wrote:

> A study of the microflora of the ground waters of the artesian basins in western Siberia lowland has shown that sulfur oxidizing and denitrifying bacteria that oxidize sulfur are widespread in these deep artesian aquifers, even at depths exceeding 1.8 km.

Gurevich, as did Bastin, made the connection between the presence of viable bacteria and the chemistry of ground water:

1.3 GROUND-WATER CHEMISTRY AND SUBSURFACE MICROBIOLOGY

There is now no doubt that the chemical composition of ground water may change not only by inorganic processes, but also through the activity of microorganisms.

One reason that consideration of microbial processes fell out of favor in the United States during the 1960's and 1970's was that ground-water chemists focused on other equally important lines of inquiry. The concept of hydrogeochemical facies, as outlined by William Back (1966), showed that water chemistry evolution followed regular, predictable patterns that depended on ground-water flow paths. Studies by Garrels and MacKenzie (1967) showed that basic concepts of inorganic chemistry and phase equilibria could explain many of the patterns observed.

The equilibrium approach was especially important because it provided a theoretical basis for understanding ground-chemistry patterns. This approach was also emphasized by John Hem (1959) in his studies of iron and manganese geochemistry. However, in the 1960's, the cumbersome algebra and arithmetic associated with equilibrium calculations limited the use of this approach by geochemists performing field studies. With the introduction of computers, however, these calculations could be quickly and efficiently performed (Truesdell and Jones, 1974). As a consequence, the equilibrium approach became the basis of most subsequent ground-water chemistry studies. At about the same time, advances in mass spectrometry technology began to enable ground-water geochemists to use radioisotopes (Pearson and White, 1967) and stable isotopes (Pearson and Hanshaw, 1970) in hydrochemical studies.

The use of these new quantitative methods in field studies were enormously successful, but they also pointed up some puzzling phenomena. For example, quantitative geochemical modeling techniques developed by L. N. Plummer and his associates clearly documented that some deep confined aquifers were partially open systems with respect to carbon dioxide (Plummer, 1977). What could be the source of this carbon dioxide? Many investigators invoked the "dynamochemical," inorganic explanation given by Foster (1950) as the source of carbon dioxide. However, studies of deep injection of liquid wastes (Leenheer Malcolm, and White, 1976; Godsy and Ehrlich, 1978), studies of shallow waste disposal practices (Ehrlich et al., 1982), and studies of ground water produced from deep wells (Godsy, 1980; Dockins et al., 1980) began to show that microorganisms were indeed present in subsurface environments.

The debate as to whether the observed source of carbon dioxide in aquifers was biotic or abiotic in many ways resembles the nineteenth-century debate as to the origin of fermentation. The idea that microbial metabolism was the principal source of carbon dioxide in deep aquifer systems was revived by Chapelle et al. (1987). However, as was the case in the nineteenth-century biotic-abiotic fermentation debate, it was apparent that observational evidence was insufficient to settle the issue. It took several experimental studies, conducted by different laboratories and using different methods (Chapelle et al., 1988; Phelps et al., 1989a), before the biologic origin of the carbon dioxide in deep aquifer systems was accepted by the geochemical and microbiologic communities.

From a historical perspective, the resolution of the deep aquifer-carbon dioxide debate was pivotal to the geochemical community. First of all, here was evidence that microorganisms were involved in a geochemical process that impacted many aquifer systems. Second, it gave an example of how experimental microbiologic techniques could be applied in a practical way to the study of hydrologic systems. This realization, together with evidence from studies of shallow aquifer systems being performed by microbiologists, catalyzed the acceptance of these concepts in the geochemical community.

1.3.2 Subsurface Microbiology and Microbial Ecology

Soil microbiology has been an important subdiscipline of microbiology since the pioneering studies of Winogradsky and Beijerinck in the early twentieth century. However, soil microbiology traditionally has been preoccupied with soil fertility as it relates to agricultural practice. Consequently, there was virtually no microbiologic investigations that extended consideration below the root zone of plants.

In the 1970's, however, it became increasingly evident that certain waste disposal practices were contaminating subsurface environments below the root zone. There was considerable concern about the effects of these waste disposal practices on ground-water quality, and this concern began to draw microbiologists toward studying subsurface environments. In 1973 two Environmental Protection Agency microbiologists, William Dunlap and James McNabb, driven by concern about contaminated aquifers, prepared a literature search that turned up many of the early references from the geologic and petroleum literature. At this point, the historical differences between microbiology and geology began to make an impact. The problem was that, by microbiologic standards, many of these studies seemed to be of dubious reliability. Normal microbiologic standards of aseptic technique did not appear to have been followed in most of these studies. Furthermore, the possibility that microorganisms could have been introduced to the deep formations through the drilling process or through the finished wells made the microbiologists deeply suspicious of the early work from the petroleum industry.

Microbiologists, who were trained to rely on rigorous experimental procedures, began working on ways to evaluate the possibility that drilling procedures or wells introduced microorganisms into previously sterile subsurface environments. One particularly clever study was described by Olson et al. (1981). These investigators were studying ground water from the deeply buried Madison aquifer of Montana, which produced fairly hot (47-54°C) ground water. It was found that sulfate-reducing bacteria were present that would grow at 50°C but not at 22°C. Accordingly, this suggested that the microorganisms were adapted to the temperature of the aquifer and therefore were certainly indigenous to it. Water from wells was used by a number of microbiologists to isolate microorganisms from pristine (Godsy, 1980) and contaminated aquifers (Godsy and Ehrlich, 1978; Ehrlich et al., 1982).

The development of aseptic procedures for sampling deeply buried sediments was of particular importance in studying subsurface microbiology. Dunlap et al. (1977) described a method by which a bore-hole was drilled with augers, sediment samples were collected with a split-spoon sampler, and the outer (i.e. contaminated) portion of the core removed with a paring device. With some modifications, these sampling procedures were used to collect aseptic samples for a variety of experiments and procedures by Wilson et al. (1983), Ghiorse and Balkwill (1983), and White et al. (1983). These aseptic coring procedures have been widely accepted and are presently standard techniques for sampling shallow water table aquifers.

Auger drilling is not suitable for coring sediments more than about 30 meters deep. For such deeper sediments, mud-rotary, or air-drilling methods are necessary. This greatly complicates the process of obtaining uncompromised sediment samples. Chapelle et al. (1987, 1988) used a combination of mud-rotary drilling and split-spoon coring to obtain sediments as deep as 200 meters. Phelps et al. (1989b) described sampling procedures using mud-rotary coring. Development of these sampling procedures made it possible to investigate microbial processes in deeply buried sediments without being restricted to water samples from wells.

Many of the microbiologists who were involved in early investigations of subsurface environments were microbial ecologists by training. It's not surprising, therefore, that many of these studies focused on ecological issues, such as microbial diversity (White et al., 1983; Hirsch and Rades-Rohkohl, 1983) and nutritional status (White et al., 1983; Balkwill and Ghiorse, 1985). The most important focus, however, was on the ability of subsurface microorganisms to degrade potential pollutants (Wilson et al., 1983; Ehrlich et al., 1982).

Consideration of microbial ecology issues, however, pointed out some limitations of relying solely on aseptically cored sediments for investigating subsurface microbiology. Specifically, coring deep holes using mud rotary methods was extremely expensive. Obtaining enough core holes to cover adequately the large areas of most deep aquifer systems was an economic and practical impossibility. This led microbiologists to consider the use of ground-water chemistry in ecological studies of the deep subsurface.

Again, however, microbiologists brought a different perspective to the problem. It was well known from studies of surface aquatic sediments that certain unstable intermediate products of microbial metabolism, such as hydrogen gas and organic acids, provided information as to the distribution of terminal electron-accepting processes. This raised the possibility that measuring certain metabolites in ground water could provide information as to the distribution of microbial processes in the subsurface. The use of hydrogen concentrations as a tool for delineating iron-reducing, sulfate-reducing, and methanogenic zones of aquifers was first proposed by Lovley and Goodwin (1988). This technique, which has been used in subsequent hydrologic studies (Chapelle and McMahon, 1991; Chapelle and Lovley, 1992) is an effective combination of both microbiologic and hydrologic methods.

In addition to helping delineate the distribution of metabolic processes in subsurface environments, ground-water chemistry has helped in studying the microbial ecology of subsurface environments. Because of the nutrient-limited nature of most aquifer systems, it was expected that rates of microbial metabolism would be very low in the deep subsurface. This was confirmed by experiments with cores of deep sediments that showed the turnover of radiolabeled carbon compounds (Phelps et al., 1989a) was relatively low. However, because the sediments had been disturbed by coring and because artificial nutrients had been used in the assays, it was difficult to extrapolate these experimental results to in situ rates of metabolism. Chapelle and Lovley (1990) used a combination of radiotracer experiments and geochemical modeling to delineate actual in situ rates of microbial metabolism in a deep aquifer system. This approach again showed that a combination of microbiologic and geochemical techniques could be effective in addressing issues of microbial ecology.

1.4 INTEGRATING MICROBIOLOGY AND GROUND-WATER CHEMISTRY IN STUDIES OF AQUIFER SYSTEMS

It is evident from the preceding sections that the history of subsurface microbiology has been one of convergence between the different disciplines of geology and microbiology. Observations of ground-water chemistry had suggested, from at least the 1920's, the possible importance of microbial processes. However, an observational approach was not sufficient to demonstrate that this was, in fact, the case. Rather, rigorous sampling methods and experimental techniques were required to demonstrate adequately the role of microbial processes.

On the other hand, experimental methods themselves had significant disadvantages, such as extreme cost, disturbance of sediments, and an inherent small scale that was inappropriate for large hydrologic systems. In the end, an integration of microbiology and the geosciences has proved to be the most effective means of studying subsurface microbiology. This integrated approach, which requires that microbiologists and geologists understand each other's unique and important perspectives, holds much promise for unraveling important environmental problems in terrestrial subsurface environments.

CHAPTER 2

MICROORGANISMS PRESENT IN THE GROUND-WATER ENVIRONMENT

The kinds of microorganisms encountered in the biosphere are as varied and diverse as the kinds of environments present on earth. If one considers the diversity of potential habitats in the biosphere (surface sediments in fresh water or salt water bodies, subsurface sediments in either aerobic or anaerobic environments, the water column of deep or shallow water bodies, hot hydrothermal waters, frozen sediments in the Arctic and Antarctic plains, extreme pressure in deep ocean waters, and within the bodies of higher plants and animals) it's little wonder that microorganisms display such astonishing diversity. In spite of this bewildering array of microorganisms, they all belong to one of four basic groups: the procaryotes, the eucaryotes, the archaebacteria, and the viruses.

The procaryotes are distinguished from eucaryotes on the basis of cellular architecture. The procaryotic cell is characterized by the lack of a true nucleus (*pro* meaning "early" or "primitive" and *karyo* meaning "nucleus") and includes the bacteria and the cyanobacteria. Cyanobacteria were formerly called "blue-green algae", but now are recognized as procaryotes. The eucaryotic cell has a true nucleus (*eu* meaning "true") and includes algae, fungi, and protozoa. The term "procaryote" also reflects the likelihood that procaryotes evolved earlier in earth's history than did eucaryotes.

The archaebacteria, which include the methane-producing bacteria, were once considered to be procaryotic organisms. Studies in the last 20 years, however, have shown that archaebacteria are not procaryotes and that they represent a previously unrecognized kingdom. Archaebacteria are restricted to anaerobic environments, such as organic-rich sediments and the intestines of higher animals. Geologically, the archaebacteria are extremely important since they inhabit virtually all subsurface environments, are an important source of commercial methane, and have important impacts on ground-water chemistry.

Viruses are distinct from other types of microorganisms in that they are obligate parasites; that is, they do not have the capability to live and reproduce without having a host cell to provide energy. Viruses are important to subsurface microbiology primarily because ground water may transport viruses and thus spread infectious diseases. It is almost certainly true that viruses use subsurface bacteria as hosts and therefore are probably present wherever bacteria are present. That topic, however, has yet to be explored systematically.

2.1 THE PROCARYOTES

Procaryotic microorganisms are characterized by their distinctive and relatively simple cellular structure. Figure 2.1 shows the kinds of structures that are typically observed in procaryotes.

The bacterial chromosome consists of a single molecule of DNA that, in spite of many loops and twists, is arranged into a closed circle. This closed circle arrangement is found only in procaryotes and the archaebacteria. The DNA is otherwise identical to that of other organisms and is characterized by its double helix structure. The DNA carries genetic information needed for the cell to carry out metabolism and growth, as well as carrying the information needed for replication. Each bacterial chromosome consists of anywhere between 2,000 and 10,000 units of heredity called *genes*. Genes are segments of the DNA strand that code for a particular protein or polypeptide. Most of the time, procaryotic cells have just one copy of their chromosomal DNA and are therefore referred to as being *haploid*. Just before cell division, however, two or more copies of the bacterial chromosome may be present. These masses of DNA are sometimes visible under the microscope and are referred to as *nucleoids*.

Procaryotes may also contain smaller circles of DNA, distinct from chromosomal DNA, that are termed *plasmids*. Plasmids are not involved in cell replication but are nevertheless very important. Plasmids code for enzymes or other proteins that have specific functions in helping the organism deal with its environment. For example, plasmids often code for proteins that detoxify or otherwise neutralize antibiotics. Plasmids may also code for proteins that aid in the decomposition of particular organic compounds, enabling the bacteria to use those compounds as an energy source. Much research has gone into identifying plasmids that code for the decomposition of toxic chemicals, because such capability could increase the

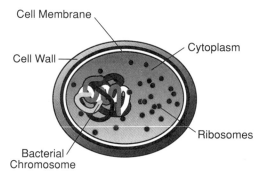

FIGURE 2.1. Cellular structure of procaryotic microorganisms.

effectiveness of bioremediation strategies. This topic is discussed in greater detail in Chapter 5.

Ribosomes are small, dark structures that are embedded in the cytoplasm of the cell. Ribosomes are protein assembly structures and provide a surface upon which amino acids can be brought together and covalently bonded in the proper sequence. Ribosomes in procaryotes are made from two subunits. The smaller subunit is called the 30S subunit and the larger one is called the 50S subunit. The S stands for "Svedberg" units, a measure of the rate of sedimentation in an ultracentrifuge and hence an indirect measure of molecular size. Because the 50S subunit settles faster than the smaller 30S subunit, it is proportionally more massive. Each subunit consists of an RNA molecule, which on its own may be either 23S, 16S, or 5S, and associated proteins that aid in bonding amino acids.

The cytoplasm of procaryotes always contains individual ribosomes. Often, however, ribosomes are arranged into complexes that are called *polysomes*. These polysomes work simultaneously to assemble different polypeptides which may constitute parts of one protein. Sometimes polysomes are embedded into the cell membrane.

The cell membrane, also termed the *cytoplasmic membrane,* acts as the boundary between the interior of the cell and the outside environment. As such, the cell membrane has numerous functions that regulate the chemical environment inside and outside the cell. Figure 2.2 shows a schematic diagram of the cell membrane. It consists of approximately 60% protein and 40% phospholipids. The phospholipids are arranged into a bilayer, in which the hydrophobic (non water-soluble) portions point outward. This arrangement helps the cell to regulate its water balance. Embedded in the phospholipids are proteins that exhibit a variety of arrangements. The purpose of cell membrane proteins is to regulate the transport of chemicals into and out of the cell. Some proteins, for example, act as "ports" for bringing simple sugars into the cell, where they can be utilized for energy. Other proteins act as "switches," transferring electrons in the cell's electron transport system.

The cell membrane is semipermeable and allows some substances, such as water, to cross in response to concentration gradients. This type of transport occurs spontaneously and does not require the cell to expend energy. Other substances are actively carried across the cell membrane by specialized proteins.

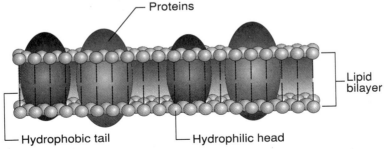

FIGURE 2.2. Schematic diagram of the cell membrane, showing the arrangement of proteins and phospholipids.

This type of transport is termed *facilitated transport* and requires the cell to expend energy.

The cell membrane is enclosed within the cell wall (Fig. 2.1) which gives the cell rigidity and helps to protect it against osmotic stress. A substance called *peptidoglycan,* which is unique to procaryotic microorganisms, provides much of the cell wall's structural strength. Peptidoglycan is a three-dimensional polymer of sugars and amino acids that are cross-linked with short peptide bridges.

There are two major types of cell walls (Fig. 2.3). The *gram-positive* cell wall consists of an inner membrane with a relatively thick layer of peptidoglycan covering it. There are also varying amounts of teichoic acids, polymers of sugar, alcohols and phosphates, present in gram-positive cell walls. This thick peptidoglycan layer has the characteristic that it retains the crystal violet pigment in Gram's stain, even when washed with ethyl alcohol, hence the term "gram positive".

The *gram-negative* cell wall (Fig. 2.3) has a layer of phospholipids and lipoproteins outside a thinner peptidoglycan layer. The gram-negative cell wall does not retain Gram's stain when washed with ethyl alcohol. In the gram-negative cell wall, there is a space between the cell membrane and the peptidoglycan layer, termed the *periplasmic space*. The periplasmic space is absent from gram-positive organisms and reflects a basic difference in how substrate-degrading enzymes are utilized in the two types of microorganisms.

Procaryotic microorganisms often have external coatings on their cell walls. These coatings are called the *glycocalyx* and consist of either polysaccharides or proteins. If the coating is hard and dense, it is often referred to as a *capsule*. If the coating is soft and pliable, it is often termed a *slime layer*. Cells growing on culture media often produce a thick glycocalyx, and this is what gives individual bacterial colonies a smooth appearance. Cells that do not produce abundant glycocalyx often produce colonies that appear rough.

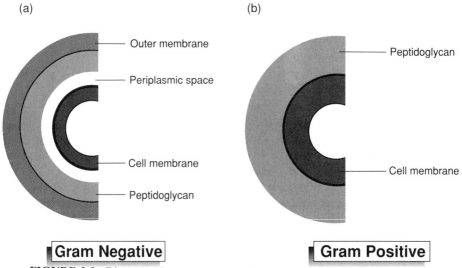

FIGURE 2.3. Diagram of a (a) gram-negative and a (b) gram-positive cell wall.

One important function of the glycocalyx is that it facilitates the attachment of bacteria to surfaces in the environment. For example, *Streptococcus mutans*, a bacterium that lives in the mouths of human beings, produces a thick glycocalyx slime designed to allow it to stick to its host's teeth (thus avoiding being swallowed and killed). *S. mutans* produces this polysaccharide glycocalyx from sucrose (refined sugar), so that eating a candy bar is a boon for slime production. The "cottony" feeling in your mouth that becomes noticeable 10 or 15 minutes after eating candy comes from the slime produced from sucrose that coats your teeth. This slime, also called plaque, contributes greatly to tooth decay.

In the same way that *S. mutans* uses a glycocalyx to cling to teeth, bacteria in sediments use these coatings to cling to mineral surfaces. This ability helps bacteria gain access to nutrients associated with sediment particles and provides a stable environment for subsequent reproduction.

2.2 THE EUCARYOTES

Algae, fungi, and protozoa have cells that are much different from procaryotic cells and are classified as eucaryotes. In contrast to the relatively simple structures present in procaryotic cells, the eucaryotic cell is extremely complex. The eucaryote is characterized, as mentioned earlier, by the presence of a distinct nucleus (Fig. 2.4). Many, but not all, eucaryotes have a cell wall, and all eucaryotes have a cell membrane. Also present are large folded structures, called the *endoplasmic reticulum*, oval membrane disks called *golgi bodies*, and organelles called *mitochondria, lysosomes,* and *chloroplasts* (Fig. 2.4).

The cell wall in eucaryotes performs many of the same functions as that in procaryotes, such as protecting against osmotic shock and giving tensile strength to the cell. The cell walls of eucaryotes are formed from polysaccharides, but peptidoglycan is absent. The cell membrane in eucaryotes is similar to that of procaryotes, the main differences being in the composition of the lipids. The

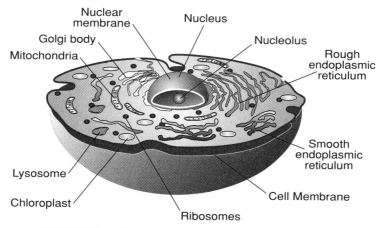

FIGURE 2.4. Structures present in eucaryotic cells.

eucaryotic cell membrane carries out facilitated and active transport, as in procaryotic cells. Many of the metabolic functions carried out in procaryotic cell membranes, however, have been taken over by specialized organelles in the eucaryote.

In the eucaryotic cell, energy production occurs in the mitochondria, photosynthetic activity is carried out by chloroplasts, and digestive enzymes are stored in lysosomes. All of these functions are carried out in the cell membranes of procaryotes. Protein assembly in eucaryotes is carried out by ribosomes, as in procaryotes. However, eucaryotic subunits are larger (40S and 60S) than in procaryotes. Furthermore, eucaryotic ribosomes are often arranged along the surface of a folded membrane, the endoplasmic reticulum. This structure aids in the synthesis of very complex proteins that require numerous steps during assembly. Further processing of enzymes is performed by golgi bodies that separate from the endoplasmic reticulum.

Algae are photosynthetic eucaryotes and have the distinction of producing most of the world's oxygen. As such, algae contain photosynthetic chloroplasts. Algae have cell walls composed of cellulose, pectin, or silica, and may live singly or in colonies. There are six groups of algae recognized: green algae, euglenids, diatoms, dinoflagellates, brown algae, and red algae.

Unlike the photosynthetic algae, the fungi are heterotrophic; that is, they are decomposers and obtain energy and nutrients from preexisting organic carbon sources. The fungi have developed mechanisms for degrading almost any kind of organic carbon compound found in nature. Most fungi are obligate aerobes, meaning that they must have oxygen in order to respire. However some fungi, notably the yeasts, are capable of fermentation.

The special function of fungi in the environment is to recycle the remains of plant and animal debris. In this way, fungi often compete directly with heterotrophic bacteria for available resources. The greater complexity of eucaryotic structure, which allows the synthesis of many degrading enzymes lacking in the procaryotes, often gives fungi a competitive advantage over bacteria. Fungi, for example, are particularly adept at degrading lignins in decaying plants, a class of compounds that bacteria attack, but less efficiently. Also, the metabolic flexibility of fungi allow them to live in particularly stressful environments, such as hypersaline lakes. This resistance to osmotic stress is the main reason that molds, which are a type of fungi, are able to grow on preserved foods, such as jams, that are resistant to attack by bacteria.

The protozoa are single-celled heterotrophic eucaryotes that are characterized by extremely complex cell structure. The most familiar protozoa to geoscientists are foraminifera, radiolaria, and dinoflagellates, all of which are prominently represented in the fossil record. Some protozoans are predatory and are able to attack and ingest other microorganisms. Many predatory protozoans are specifically adapted to feeding on bacteria, and this grazing is often a limiting factor in bacterial populations in both surface and ground-water systems. Most protozoa, however, belong to floating communities, called zooplankton.

2.2.1 Eucaryotes in Ground-Water Systems

The presence and distribution of eucaryotic microorganisms in ground-water systems has been studied seriously only recently. The first microbiologic investigations of shallow water table aquifers indicated that procaryotes were the dominant microorganisms present and that eucaryotes might be absent altogether (White et al., 1983; Balkwill and Ghiorse, 1985). Subsequent studies, using more sensitive techniques, were able to show the presence of limited populations of eucaryotes in ground-water systems.

Sinclair and Ghiorse (1987) used a most-probable-number counting procedure to document the presence of protozoa in the shallow pristine water-table aquifer at Lula, Oklahoma. Many of the horizons sampled, to a depth of 8 meters, had fewer than 0.2 protozoa/gram of dry sediment. A gravelly bed at a depth of 7.5 meters, however, exhibited between 2 and 5 protozoa/gram of dry sediment. Comparable numbers were found in other sandy horizons. Surface soils from this site, in contrast, were characterized by counts of protozoa in the 102-106 cells per gram dry sediment. So, while protozoa were demonstrably present, they were present in very low numbers. A later study of buried-valley aquifer sediments in Kansas (Sinclair et al., 1990) showed a similar pattern of protozoa abundance.

While the presence of eucaryotic cells is low in pristine aquifer sediments, there is evidence that, in aquifers contaminated by organic chemicals, the abundance of eucaryotes may be much higher. For example, Madsen, Sinclair, and Ghiorse (1991), in a study of an aquifer that was contaminated by polyaromatic hydrocarbons, showed that numbers of protozoa were much higher in the plume of contamination than outside the plume. This was interpreted as reflecting much greater growth rates of bacteria in the contaminated zone which in turn supported a significantly higher population of protozoan grazing on the bacteria. In contrast, little relationship was found between the presence of fungi and contamination. On the basis of this data, Madsen, Sinclair, and Ghiorse (1991) suggested that the population size of protozoans could be used as evidence for chemical contamination of aquifers.

2.3 THE ARCHAEBACTERIA

The distinction between procaryotic and eucaryotic microorganisms was first defined in terms of the internal structures that were visible with a microscope (Figs. 2.1 and 2.4). On this basis, as well as on the basis of differences between the biochemistry of their enzyme systems, biologists constructed a tree in which all cells could be classified under either the procaryotic or eucaryotic "stems."

This dichotomous classification was questioned in the 1970's by Carl R. Woese and his associates who were working at the University of Illinois. Woese's group was interested in being able to quantify genetic links between microorganisms. In order to do this, they decided to look at the nuceotide sequences of ribosomal RNA. The choice of ribosomal RNA was based on the fact that it was relatively easy to isolate from bacterial cells and that sequencing discrete portions of it was technically feasible. There are three kinds of ribosomal RNA in bacteria. The "large" ribosomal unit has a Svedberg unit value of 23S and is approximately 2,900 nucleotide units long. There is also a "small" one that is 5S and is only 120 nucleotides long. Intermediate between these two is the 16S ribosomal RNA that is about 1,540 nucleotides long. The 23S was too long to be conveniently sequenced in the early 1970's whereas the 5S did not contain enough nucleotides to give much information. Thus, Woese's group settled on the 16S as the RNA unit of choice and began sequencing the 16S RNA of many different types of procaryotes and eucaryotes.

As the sequencing proceeded, a surprise emerged. As expected, eucaryotes were most closely related to eucaryotes. Among the procaryotes, however, it appeared as if the methane-producing bacteria were distinct from other types of bacteria. In fact, the methanogens were no more closely related to the other bacteria than they were to eucaryotes. The implications of this finding were straightforward. There were not just two but three stems in the tree (Fig. 2.5). Woese named these three stems the eubacteria (true bacteria), the eucaryotes (true nucleus), and the archaebacteria (ancient bacteria). This nomenclature reflected Woese's belief that archaebacteria evolved very early in earth's history. More recently, Woese has suggested that the use of the word *bacteria* in archaebacteria is misleading and has suggested the term "archaeotida" instead. This question of nomenclature is unresolved at the time of this writing.

In addition to their distinctive 16S RNA sequences, archaebacteria exhibit other differences from the procaryotes. The lipids of both eucaryotes and procaryotes consist mainly of two straight-chain fatty acids bound at one end to a glycerol molecule through an ester linkage (-CO-O-). In archaebacteria, on the other hand, the glycerol and the chains have an ether (-O-) link. There are several other molecular distinctions between procaryotes and the archaebacteria.

The archaebacteria include three very different kinds of microorganisms: methanogens (microbes that produce methane from carbon dioxide and hydrogen), extreme halophiles (microbes that live in concentrated salt brines), and thermoacidophiles (microbes that live under extreme conditions of heat and low pH). Of these types, however, the methanogens are the most common and have the most impact on geologic processes in the subsurface. Early in the earth's history, methanogens could have existed almost anywhere. Today they live only where oxygen has been excluded and where hydrogen and carbon dioxide are available. This generally means that they live in close association with fermentative bacteria, such as clostridia, that metabolize decaying organic matter and give off carbon dioxide and hydrogen as waste products. Methanogens are found in anaerobic sediments of surface water bodies, in sewage treatment plants, and in the intestines

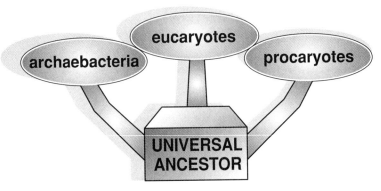

FIGURE 2.5. The three stems in the Tree of Life.

of most animals. Methanogens are also widely distributed in deep subsurface environments, where they can effectively metabolize organic matter after mineral electron acceptors, such as ferric oxyhydroxides and sulfate, have been exhausted.

2.4 THE VIRUSES

Viruses are very small and very simple infectious agents. They are too small to be seen with a light microscope, and their presence was discovered indirectly. In 1886, an American bacteriologist, A. E. Mayer, described a mottling disease of the tobacco plant, called "tobacco mosaic disease." He noted that the disease could be transmitted by injecting the sap of infected plants into healthy plants. A few years later, O. Iwanowski discovered that the causative agent of tobacco mosaic disease was filterable; that is, the ground-up extract of diseased plants could infect healthy plants even when the extract was passed through a very fine filter. Iwanowski's experiments were reproduced by Martinus Beijerinck in 1898. Clearly, the infectious agent was not a bacterium. These experiments had the effect of extending microbiology beyond the study of procaryotes and eucaryotes into the study of infectious agents, viruses, which were too small to be seen.

Exactly what kind of infectious agent viruses were was largely a mystery in the early part of the twentieth century. A major contribution to virology was made in 1935 when the American Wendell Stanley showed that the tobacco mosaic virus could be crystallized. This discovery led to studies of viral size and shape using X-ray defraction techniques. It was not until the advent of the electron microscope, however, that viral particles could be directly observed.

What is a virus? The best way to answer this question is to say what it is not. Viruses are not cells, for example, because they cannot reproduce themselves independently and because they have no independent metabolism. One of the classic (and unresolved) debates in microbiology is whether or not viruses are living organisms. When it was first shown that "inanimate" crystals of the tobacco mosaic virus could cause the disease, many people argued that viruses were merely toxic chemicals. On the other hand, these particular toxic chemicals appeared to have the ability to reproduce themselves; was this not a clear characteristic of life? Not necessarily, came the rejoinder. Many crystalline substances "grow" from supersaturated solutions. Was this not "reproduction"? This debate has never been settled to everyone's satisfaction, and most people have come to the conclusion that there is not much to be gained by arguing about it. It does point out, however, some of the unique features of viruses.

At the simplest level, a virus is simply a genome, genetic information stored on DNA or RNA, wrapped in an exterior coat of protein. In the tobacco mosaic virus (Fig. 2.6a), a single strand of RNA is surrounded by protein subunits (called capsomeres) that link together to form the protein coat (called a *capsid*). Viral particles are called *virions* and are characterized by a regular geometrical arrangement of the capsomeres. Common shapes of virions are spheres, cylinders, wedges, or prisms. Some virions are covered by an envelope and are termed *enveloped;* others have no envelope and are termed *naked*.

Viruses that are parasitic to bacteria are called *bacteriophages* and have a somewhat more complicated stucture (Fig. 2.6b). The *head* of the virus contains

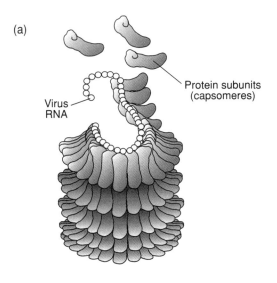

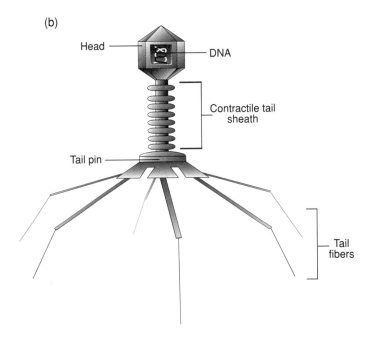

FIGURE 2.6. Structures present in (a) tobacco mosaic virus and (b) parasitic bacteriophages.

the nucleic acid genome (in this case, DNA) that is set upon the contractile tail sheath. At the base of the tail sheath is the *tail pin* and the *tail fibers*. These stuctures enable bacteriophages, such as the T4 bacteriophage, to attack a bacterium. First, the tail fibers find and attach to a target protein on the cell wall. Second, the cell wall is pierced by the tail pin. Finally, the tail sheath contracts and injects DNA into the bacterium. The DNA promptly takes over the cell's metabolism and redirects it to manufacture more virus particles.

Viruses are often classified according to the nucleic acid that makes up its genome. The DNA viruses include the herpesvirus, which causes herpes simplex types I and II, and adenoviruses, which cause symptoms of the common cold. The RNA viruses include enteroviruses, which cause polio and gastrointestinal illness; rhinoviruses, which also cause symptoms of the common cold; and the HIV virus, which causes the deadly disease AIDS in humans.

The study of viruses in water science has focused largely on the mobility of pathogenic viruses from sewage effluents. Infectious hepatitis is the most closely studied disease that has been documented as being transmitted by a waterborne virus. Many epidemics of hepatitis have been traced to fecal contamination of drinking water or fecal contamination of shellfish that are subsequently consumed by humans. Although the disease hepatitis has been closely studied, not much is known about the virus itself. This is largely because of the extreme difficulty in studying the hepatitis virus in the laboratory.

There is considerable debate as to the possibility of pathogenic viruses being transported in ground water. Fecal contamination of shallow water table aquifers could result in the transmission of infectious hepatitis. However, solid evidence that such transmission has occurred is generally lacking. Studies have shown, however, that viruses can persist in ground-water environments. For example, Keswick, Wang, and Gerba (1982) showed that polioviruses, coxsackieviruses, and rotaviruses survive much longer in subsurface environments—on the order of weeks or months—than had generally been assumed. A field study by Wellings et al. (1975) gave direct evidence that viruses could survive up to 28 days in ground-water systems. These survival times are sufficient for viruses to move through shallow aquifers to wells from, for example, septic system effluent. Because the kinds of cells needed for the replication of human pathogenic viruses are present in low concentrations in ground-water systems, it has often been assumed that long term survival and transport of such viruses is unlikely. Additional studies on the persistence of viruses in subsurface environments are needed in order for this assumption to form the basis of public health policies.

2.5 BACTERIA IN GROUND-WATER SYSTEMS

This book is concerned largely with bacteria, that is, procaryotic microorganisms, and how they influence the chemistry of ground water. In this context, the archaebacteria, even though they are not procaryotic microorganisms, will be considered together with bacteria. While eucaryotic microbes are present in ground-water systems, their abundance is typically three or four orders of magnitude less than that of bacteria. Also, because most eucaryotes are restricted to oxygenated environments, they are excluded from many deep subsurface environments. Similarly, viruses are present in some subsurface environments. Because they have no independent metabolism, however, their impact on ground-water chemistry is probably limited to the effects that they have on indigenous bacteria. Thus, the emphasis on bacteria and bacterial processes in this book reflects the relative importance of the different types of microorganisms on the chemistry of ground water.

2.5.1 Classifying Bacteria

The classification of living organisms is called *taxonomy*. The purpose of bacterial taxonomy is really twofold. First, its purpose is to be able to distinguish between different bacteria. Second, and equally important, its purpose is to identify similarities between different bacteria so that phylogenic relationships can be determined. Because of these two aims, classifications of bacteria are *hierarchical;* that is, groups of closely related *species* are arranged into genera, similar genera are arranged into *families*, families into orders, and so on. The names given to the microorganisms reflect this hierarchical classification scheme.

Bacteria are named according to a binomial nomenclature, with each distinct species given a name consisting of two words. Take, for example, the bacterium *Escherichia coli*. The first part of the name, *Escherichia* is the genus to which the microorganism belongs and is always capitalized. The second part of the name, *coli,* is the name of the species and is not capitalized. It is proper usage to italicize or underline the proper names of organisms. The root of the genus name is either a Latin or a Greek word that may in some way be descriptive of the organism or be in honor of a particular person. It is common practice to abbreviate the names of microbes to the first letter of the genus, followed by the species name. *Escherichia coli,* therefore, is commonly refered to as *E. coli*. *E. coli* belongs to a family of related genera that inhabit the intestines of higher animals and are therefore called the Enterobacteriaceae, or "intestine bacteria." The Enterobacteriaceae belong to the order Eubacteria, or "true bacteria." This order includes the greatest number of bacterial species that inhabit ground-water systems.

Criteria Used to Classify Bacteria Bacteria are classified on the basis of cell morphology, type of cell wall, cell growth, biochemical transformations carried out by the cell, nutrition, and nucleic acid sequences (Table 2.1). All of these characteristics, of course, are determined by the genetic information stored in the cell's DNA. Because of this, DNA *homology,* or the establishing of the similarity of base sequences in DNA between different strains of bacteria, is being increasingly used to characterize bacteria.

The *gram stain* is used to determine cell shape and type of cell wall. In this process, the bacteria are obtained in pure culture, smeared onto a glass slide, and fixed (heated) so that the smear will adhere to the slide. The fixed smear is subjected to the following solutions in this order: crystal violet, iodine solution, ethyl alcohol (a decolorizing agent), and safranin. Bacteria that have peptidoglycan as their outer cell wall retain the crystal violet, are stained a deep purple color by this procedure, and are termed *gram positive*. Bacteria that have an outer cell wall

TABLE 2.1 Some Characteristics Used to Classify Bacteria

Microscopic characteristics
  Morphology
    cell shape
    cell size
    arrangement of cells
    arrangement of flagella
    capsule
    endospores
  Staining reactions
    gram stain
    acid fast stain

Growth characteristics
  Appearance in liquid culture
  Colonial morphology
  Pigmentation

Biochemical characteristics
  Cell wall constituents
  Pigment biochemicals
  Storage inclusions
  Antigens
  RNA molecules

Physiological characteristics
  Temperature range and optimum
  Oxygen relationships
  pH tolerance range
  Osmotic tolerance
  Salt requirement and tolerance
  Antibiotic sensitivity

Nutritional characteristics
  Energy sources
  Carbon sources
  Nitrogen sources
  Fermentation products
  Modes of metabolism (autotrophic, heterotrophic, fermentative, respiratory)

Genetic characteristics
  DNA %G + C
  DNA hybridization

Reprinted from Atlas (1984), with permission courtesy of the Macmillian Publishing Co.

consisting of lipopolysaccharides and proteins do not retain the crystal violet stain, are colored red by the counter stain safranin, and are termed *gram negative*.

The shape and arrangement of the cells (Fig. 2.7) are also observed using the gram stain technique. A *coccus* (plural, *cocci*) is a spherical cell (Fig. 2.7a), a *bacillus* (plural, *bacilli*) is a rod-shaped cell (Fig. 2.7b), a *spirillum* (plural, *spirilla*) is a helical rod (Fig. 2.7c), and a *vibrio* is a V-shaped cell (Fig. 2.7d). It is common to refine these basic shapes by terming different cells long rods, short rods, or whatever term the individual observer deems important. The arrangement of cells is also important. Some cells, notably those of the genus *Streptococcus,* are characterized by cells arranged in chains. Others, such as *Micrococcus,* are typically found in grape-like clusters. Some bacteria tend to occur in pairs or in clusters of four, called *tetrads*. All of these characteristic cell arrangements are useful in taxonomy.

Motility in bacteria is conferred by the presence of extremely thin, hair-like appendages, called *flagella,* that protrude from the cell wall (Fig. 2.8). Because flagella confer motility and because not all species of bacteria are flagellated, it follows that there are motile and nonmotile species. As such, the presence or absence of motility is a convenient characteristic for the classification of bacteria.

Flagella are hooked structures that are imbedded in granular bodies just beneath the cell wall in the cytoplasm. As the flagellum is rotated, much like a motorboat's propeller, the bacterium is propelled forward, allowing it to "swim." There are many different arrangements of flagella found in nature. If a single flagellum is found on one or both ends of a rod-shaped bacterium, it is termed a polar flagellum. All bacteria with a polar flagellum are classified into the order Pseudomonadales. Other types of rod-shaped bacteria have many flagella arranged over much of the cell. These are termed *peritrichous flagella,* and all bacteria showing this arrangement are classified in the order Eubacteriales.

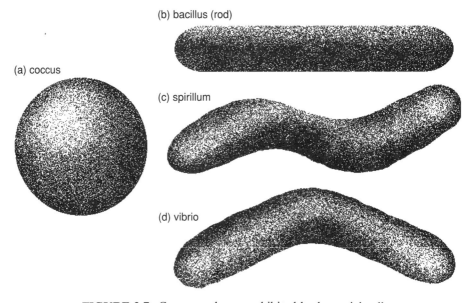

FIGURE 2.7. Common shapes exhibited by bacterial cells.

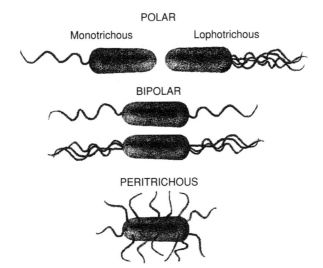

FIGURE 2.8. Different arrangements of flagella in bacterial cells.

Nutrition, or the types of substances used for carbon and energy sources, is another logical basis upon which to classify bacteria. *Heterotrophs* are organisms that use organic carbon as energy and carbon sources. *Lithotrophs* use inorganic carbon, such as carbon dioxide or bicarbonate, as carbon sources and an external source of energy. *Chemolithotrophs* are a type of lithotroph that obtain energy by oxidizing reduced inorganic chemicals, such as ferrous iron or sulfides. *Photolithotrophs* are lithotrophs that are able to obtain energy from light.

Metabolism, the ways in which bacteria utilize their nutrients, is also used in classification. In order to obtain energy from a substrate, microorganisms remove electrons and transfer them to other chemicals that serve as *electron acceptors*. The use of inorganic chemicals, such as oxygen, ferric iron, or sulfate as electron acceptors is called *respiration*. The use of organic chemicals as an electron acceptor is called *fermentation*. Bacteria that use oxygen as an electron acceptor are called *aerobes*. Bacteria that can use only oxygen as an electron acceptor are called *obligate aerobes*. Bacteria that use oxygen when it is available but are also capable of fermentation are termed *facultative anaerobes*. Microorganisms that grow only in the absence of oxygen are called *obligate anaerobes*.

Perhaps the ultimate basis for classification is the composition and structure of DNA that codes for all of the characteristics—size, shape, nutrition, and metabolism—exhibited by microorganisms. One common method of comparing the DNA of different microorganisms is to compare the relative abundance of the nucleotides guanine plus cytosine (G + C) to total nucleotide abundance. For example, based just on G + C, the enteric bacteria *Escherichia, Shigella,* and *Salmonella* (48-52% G + C) can be easily distinguished from most species of *Pseudomonas* (66-68% G + C).

Measuring G + C has the advantage that it is relatively easy to to do. However, because most of the information in DNA is coded by the *sequence* of nucleotides, not just by their relative abundance, this method is not ideal. Sequencing the order in which nucleotides occur is a much more powerful tool for classifying bacteria, but it is also much more difficult and expensive to perform.

The RNA present in bacteria also have been extensively used for classification purposes. The most common approach has been to separate and sequence the 16S ribosomal RNA from bacterial cells (Woese, 1981). Because there are many fewer nucleotides in the 16S RNA than in the DNA itself, this is a much more manageable task and has been extensively used as a classification tool. This tool is especially useful for determining the "relatedness" of different bacterial species.

2.5.2 Gram-Negative Bacteria Found in Ground-Water Systems

Gram-negative bacteria are found extensively in ground-water systems. For example, in a shallow water table aquifer in Oklahoma (Balkwill and Ghiorse, 1985), about two thirds of the bacteria isolated were gram negative. In a carbonate aquifer in the Atlantic Coastal Plain (Chapelle et al., 1988) about 70% of the isolates recovered were gram negative. Finally, in clastic aquifers of the Atlantic Coastal Plain (Balkwill, 1989), anywhere from 60% to 90% of the isolates recovered were gram negative. This group of microorganisms, therefore, are important components of the terrestrial subsurface flora.

Aerobic Gram-Negative Rods Aerobic gram-negative rods are widely distributed in soils and shallow ground-water systems. Most of these bacteria have an enzyme called *cytochrome oxidase* which enables the organism to transfer electrons onto molecular oxygen, thus using oxygen as a terminal electron acceptor. Such microorganisms are termed *oxidase positive*. In the shallow aquifer near Lula, Oklahoma (Balkwill and Ghiorse, 1985), all of the bacteria isolated and characterized were oxidase positive, meaning that they carried this particular enzyme. In deeper anaerobic aquifers, as one might expect, the presence of gram-negative, aerobic rods becomes less common.

Members of the genera *Pseudomonas, Azotobacter, Rhizobium, Alcaligenes, Flavobacterium,* and *Bordetella* are representative of aerobic gram-negative rods. *Pseudomonas* appears to be particularly common in ground-water systems. Of 29 strains of bacteria identified to the genus level in the carbonate Floridan aquifer system of South Carolina, 11 could be assigned to the genus *Pseudomonas* (Chapelle et al., 1988). At the Lula, Oklahoma site (Balkwill and Ghiorse, 1985), 3 of the 6 isolates characterized appeared to be *Pseudomonas*.

One reason pseudomonads are common in shallow water table aquifers is that they are extraordinarily versatile in the kinds of organic substrates on which they can grow. In addition, many species do not require specific vitamins, growth factors, or amino acids and readily live on a number of different carbon sources. In one study, for example, a species of *Pseudomonas* was shown to grow on 127 different organic carbon compounds. In fact, the only reason that study stopped at 127 was that the investigators ran out of ideas as to what to feed it! This adaptability makes it an ideal candidate for inhabiting shallow aerobic aquifers, where organic carbon substrates are largely limited to dissolved organic carbon leaching out of the soil zone above.

Pseudomonads are always respirative and do not have the ability to ferment if conditions become anaerobic. Some species of *Pseudomonas,* however, do have the ability to use alternative pathways of respiration. For example, some strains are able to use nitrate as an alternative electron acceptor if oxygen concentrations become limiting for growth. In addition, it is possible that some strains of *Pseudomonas* are capable of dissimilatory Fe(III)-reduction. The flexibility of these microorganisms, both in terms of the organic substrates utilized and the electron acceptors used, make them well adapted for the variable conditions encountered in the terrestrial subsurface.

Azotobacter and *Rhizobium* are notable in that some members of these genera are capable of *nitrogen fixation;* that is, these bacteria take inorganic nitrogen gas, N_2, and transform it into organic nitrogen compounds. Because all of the higher plants and animals require organic nitrogen for protein synthesis, nitrogen fixation is extremely important in agricultural microbiology. Bacteria are the only organisms on earth capable of this critical function.

Members of the genus *Rhizobium* form symbiotic associations with the roots of legume plants, such as soybeans. Because of this symbiotic strategy, members of this genus are probably not widespread in subsurface environments below the root zone. *Azotobacter,* unlike *Rhizobium,* are free-living and are known to inhabit soils. As nitrogen fixation activity has been observed in samples from shallow water table aquifers (D. Yoch, written communication, 1988) and from deeper aerobic aquifers (Fredrickson et al., 1989), it seems probable that representatives of the genus *Azotobacter* are present in such systems. The only identified member of *Azotobacter* that has been identified in ground-water systems, however, was associated with drilling fluid rather than with uncontaminated sediments (Chapelle et al., 1988). It has yet to be shown, therefore, that members of the genus *Azotobacter* inhabit deep subsurface environments.

Flavobacterium is widely distributed in soils and water. They have not been reported as being isolated from cored aquifer sediments, but they have been reported in ground water produced from aerobic aquifers. Of 500 isolates recovered from ground water produced from an alluvial aquifer in Arizona, between 3 and 10% were identified to be *Flavobacterium* (Stetzenbach, Delley, and Sinclair, 1986). It is not possible to determine unequivocally if these bacteria occur naturally in the aquifer or if they colonized the well subsequent to drilling. However, because these bacteria are so common in soils, it's likely that they are present in some aerobic ground-water systems. An interesting attribute of *Flavobacterium* is that they are nutritionally diverse and are able to utilize many kinds of organic compounds as substrates. Some isolates of *Flavobacterium* are even reported to hydrolyze lignin, a metabolic strategy most often associated with the fungi.

Other aerobic gram-negative rods include the genera *Agrobacterium, Mycrocyclus, Gallionella,* and *Caulobacter. Agrobacterium* is notable mostly in that it contains a tumor-inducing characteristic that produces abnormal growths on some plants. *Agrobacterium* has not been reported from cored sediments but has been isolated from ground-water samples (Hirsch and Rades-Rohkohl, (1983). *Mycrocyclus* has been recovered from cored sediments of the Atlantic Coastal Plain (Chapelle, unpublished data) and from samples of ground water (Hirsch and Rades-Rohkohl, 1983).

Gallionella is a chemolithotrophic bacterium that obtains energy by oxidizing dissolved ferrous iron to ferric oxyhydroxides. Since it is an obligate aerobe and because ferrous iron is largely absent from aerobic water, *Gallionella* generally lives where aerobic and anaerobic waters mix. This commonly occurs at the bottom of streambeds receiving discharge from anaerobic aquifers. Unfortunately, wells that tap anaerobic aquifers also provide such an aerobic-anaerobic interface. The growth of *Gallionella* in wellbores is a significant problem because the ferric oxyhydroxides that it produces will eventually clog the well.

Caulobacter is interesting morphologically because it is characterized by a stalk that is a continuation of the long axis of the cell. *Caulobacter* is an *oligotroph*, or a bacterium adapted to conditions of extreme nutrient deprivation. These adaptations include the ability to enter a vegetative state, which reduces energy requirements, and the ability to store energy in the form of poly-beta-hydroxybutyric acids. Given the oligotrophic (nutrient-poor) conditions of most aquifer systems, *Caulobacter* is probably widely distributed in shallow aquifer systems. Members of this genus have been isolated from ground-water samples by Hirsch and Rades-Rohkohl (1983).

Brucella, Bordetella, and *Francisella* are other genera of aerobic gram-negative bacteria. These genera contain species that are important human pathogens but are almost certainly of minor importance in ground-water systems.

Aerobic Gram-Negative Cocci There are relatively few genera of aerobic gram-negative cocci among the bacteria. Most of these genera, which include *Neisseria, Maxorella,* and *Acinetobacter,* are notable in that they are important human pathogens. *Neisseria gonorrhoeae,* for example, causes the human venereal disease gonorrhea, and *Neisseria meningitidis* causes a particularly severe form of meningitis. Neisseria-like microorganisms have been recovered from sediments of aerobic aquifers (Jimenez, 1990), and several *Maxorella* species have been recovered from groundwater produced from aerobic aquifers (Stetzenbach, Delley, and Sinclair 1986).However, there is no evidence that the species observed are pathogenic.

Members of the genus *Acinetobacter* are commonly found in soils and water, and so it is reasonable to expect that they also inhabit aerobic ground-water systems. In an analysis of 65 bacterial isolates from aerobic aquifers of the Atlantic Coastal Plain in South Carolina, 12% of the isolates exhibited chromosomal DNA that contained from 38 to 49 mole % guanine + cytosine, which is in the range characteristic of *Acinetobacter* species (Jimenez, 1990). This evidence, plus the fact that many of these isolates also exhibited biochemical characteristics of *Acinetobacter,* shows that this genus is part of the biota in that hydrologic system. Species of the genus *Acinetobacter* have also been reported in well water produced from deep aerobic aquifers. Of 500 different isolates recovered from well water produced from a deep (450-500 feet) aerobic aquifer in Arizona (Stetzenbach, Delley, and Sinclair, 1986), 54% were identified as *Acinetobacter.*

Facultatively Anaerobic Gram-Negative Rods The enteric bacteria, or Enterobacteriaceae, are the most important members of this group of bacteria. They have been carefully studied because of their importance in human and animal health. The single most studied microorganism in microbiology, *Escherichia coli,* belongs to the family Enterobacteriaceae. Other members of this family include *Shigella, Salmonella, Klebsiella,* and *Enterobacter,* all of which include members that are human pathogens. These bacteria all have the capability of respiratory metabolism if molecular oxygen is available. If molecular oxygen is not available, these bacteria are able to switch to fermentation as an alternative metabolic strategy. This enables them to grow under either aerobic or anaerobic conditions.

The presence of enteric bacteria in ground water implies the presence of fecal contamination, either from humans or animals, that has been transported from land surface. As such, *E. coli* is widely used as an indicator of sewage contamination of ground water (Keswick, Wang, and Gerba, 1982). Some members of the Enterobacteriaceae, however, are indigenous to soils. An example of this type of soil microorganism is *Aerobacter aerogenes*.

Because the Enterobacteriaceae are so closely related and because they are of such medical and commercial importance, a considerable amount of work has gone into distinguishing between species. One key taxonomic characteristic is the kind of fermentation products produced by the microorganisms. Two broad patterns are recognized: the *mixed-acid* fermentation and the *butylene glycol* fermentation. In mixed-acid fermentation, three acids—lactic, acetic, and succinic—are formed, as well as ethanol, CO_2, and H_2. In butylene glycol fermentation, smaller amounts of acid are formed, with butylene glycol, ethanol, CO_2, and H_2 being the main end products. Mixed-acid fermentation yields equal amounts of CO_2 and H_2, whereas that of butylene glycol yields more CO_2 than H_2. This is because mixed-acid fermenters produce CO_2 and H_2 only from decomposition of formate

$$HCOOH \rightarrow H_2 + CO_2$$

by means of the formic hydrogenlyase enzyme system. The butylene glycol fermenters also produce CO_2 and H_2 from formic acid, but they produce additional CO_2 during the reactions that lead to butylene glycol. *E. coli* is an example of a mixed-acid fermenter, and *Aerobacter aerogenes* is an example of a butylene glycol fermenter.

The presence of facultatively anaerobic gram-negative rods have not been extensively described in ground-water systems. *Chromabacterium,* a member of this group, has been recovered from a shallow water table aquifer in Lula, Oklahoma (Balkwill and Ghiorse, 1985) and from a shallow water table aquifer in the Atlantic Coastal Plain (Chapelle et al., 1988). However, because these kinds of bacteria are relatively common in soils, it is probable that they are also widely distributed in shallow ground-water systems. Data on the presence of this type of bacteria are uniformly lacking from deeper, anaerobic aquifer systems.

Anaerobic Gram-Negative Rods Truly anaerobic bacteria are able to grow without molecular oxygen as an electron acceptor. In fact, many anaerobic bacteria do not tolerate the presence of oxygen. When molecular oxygen is dissolved in water, hydrogen peroxide is formed by the reaction

$$H_2O + 1/2\ O_2 \rightarrow H_2O_2.$$

Hydrogen peroxide formed by this reaction is a strong oxidant and is highly toxic to all bacteria. Aerobic bacteria, which must live in oxygenated environments, produce an enzyme called *catalase,* which breaks down the peroxide to oxygen and water and keeps the peroxide from damaging the cell. Anaerobic bacteria generally lack catalase and therefore have no defense against the toxic effects of peroxide.

Several anaerobic, gram-negative, rod-shaped bacteria belong to the family Bacteroidaceae and include the genera *Bacteroides* and *Fusobacterium*. Species of *Bacteroides* inhabit the oral cavities and gastrointestinal tracts of mammals.

Many species of sulfate-reducing bacteria, notably *Desulfovibrio,* are gram-negative anaerobic rods. These microorganisms can use H_2 or simple organic compounds such as acetate as energy sources and sulfate as a terminal electron acceptor. *Desulfovibrio* is not autotrophic; that is, it cannot obtain all of its carbon from CO_2 but, rather, requires organic compounds.

Sulfate-reducing bacteria are widely distributed in ground-water systems. They have been reported from petroleum reservoirs (Bastin, 1926; Davis, 1967), and anaerobic ground-water systems (Dockins et al., 1980; Chapelle et al., 1987). Interestingly, sulfate-reducing bacteria, which are poorly tolerant of molecular oxygen, have also been reported from aerobic ground-water systems (Jones, Beeman, and Suflita, 1989). It is probable that anaerobic microenvironments in predominantly aerobic aquifers support the growth of these organisms.

Because of their widespread occurrence, sulfate-reducing bacteria are one of the most important types of microorganisms in terms of their impact on ground-water geochemistry. An important limitation of sulfate-reducing bacteria is that they are not capable of completely metabolizing carbohydrates to CO_2 and water with the reduction of sulfate. Rather, they are dependent on associated fermentative bacteria to oxidize partially the carbohydrates to simple organic acids and hydrogen (H_2). These organic acids and H_2 are then coupled with sulfate reduction in order to provide energy for growth. Because they consume the products of fermentation, sulfate-reducing bacteria prevent the buildup of fermentation products, which in turn aids the metabolism of fermentative microorganisms. Sulfate-reducing bacteria and fermenters thus have a symbiotic relationship with each other, each contributing to the energy metabolism of the other.

An important physiologic type of anaerobic gram-negative rods are the Fe(III)-reducing bacteria. This particular type of respiration, using Fe(III) oxyhydroxides as a sole terminal electron acceptor, had been long recognized by microbiologists. Curiously, however, it wasn't until 1987 that the first organism, informally called strain GS-15, that uses this strategy was isolated in pure culture from river sediments and characterized (Lovley and Phillips, 1988). A few years later, the same techniques used to isolate GS-15 were applied to sediments cored from a number of anaerobic aquifers in the coastal plain of South Carolina (Lovley, Chapelle, and Phillips, 1990). Not surprisingly, an abundance of gram-negative microorganisms capable of Fe(III) reduction were recovered. It is not presently known, however, whether these microorganisms are closely related to GS-15.

2.5.3 Gram-Positive Bacteria in Ground-Water Systems

There are not as many kinds of gram-positive bacteria as there are gram-negative bacteria. The gram-positive bacteria include, however, many important human pathogens and have therefore been closely studied. Gram-positive bacteria also have been widely reported from many ground-water systems and may play important roles in modifying ground-water chemistry.

Gram-Positive Cocci The most common gram-positive cocci include the genera *Micrococcus*, *Staphylococcus*, and *Streptococcus*. Several species of *Staphylococcus* and *Streptococcus* are important human pathogens but are not known to live naturally in ground-water systems. Streptococci are fermentative, do not respire, and do not produce the H_2O_2-neutralizing enzyme catalase. One particular streptococcus, *Streptococcus faecalis*, is a normal inhabitant of the large intestines of humans and animals. As such, fecal strep is widely used as an indicator organism for fecal contamination of water supplies.

Staphylococcus and *Micrococcus* are indistinguishable based just on morphology, because both tend to form grape-like clusters when gram-stained. *Micrococcus*, however, is a common soil bacterium and has been reported to occur in ground-water systems (Chapelle et al., 1988). *Staphylococcus*, on the other hand, is a normal inhabitant of human skin and is not thought to occur naturally in ground-water systems. It is important to be able to distinguish between these organisms in studies of ground-water systems, because the presence of *Staphylococcus* may indicate contamination due to sample handling. *Micrococcus*, on the other hand, may indeed be part of the ambient microflora. There are two primary differences between *Staphylococcus* and *Micrococcus*. *Staphylococcus* is capable of fermentation, whereas *Micrococcus* is strictly respirative. Also, because *Staphylococcus* must live in a relatively high-salt environment (human skin), it is able to grow in brines that are up to 8 % salt. *Micrococcus*, however, cannot tolerate salt in growth media. So, even though these types of bacteria are morphologically similar, it is relatively easy to distinguish between them.

Coryneform Bacteria The coryneform bacteria have variable shapes that grade from cocci to club-shaped irregular rods. They are aerobic, nonmotile, and often show pronounced intracellular granules. Such bacteria have been isolated from the shallow water-table aquifer at Lula, Oklahoma (Balkwill and Ghiorse, 1985) and from a shallow carbonate aquifer in South Carolina (Chapelle et al., 1988).

The genus *Arthrobacter* contains many species of bacteria that are widely distributed in soils and water. The coryneform bacteria described from the Lula site exhibited many of the characteristics of *Arthrobacter*, including pronounced intracellular granules and development of a septum during cell division. These characteristics led Balkwill and Ghiorse (1985) to assign these microorganisms to the genus *Arthrobacter*. It is probable that members of this genus are widely distributed in aerobic ground-water systems.

Spore-Forming Rods The ability to form *endospores*, or specialized structures that develop when environmental conditions become adverse, is characteristic of some gram-positive bacteria. *Sporulation* (spore formation) is a complex process that is initiated when the cell senses the presence of adverse conditions such as a lack of substrates or excessive heat or dryness. First, the bacterium makes a copy of its chromosomal DNA and forms a membrane called the *spore septum* around the DNA. Thick coats of peptidoglycan are deposited around the septum, together with a substance called dipicolinic acid, to form the *cortex*. These materials shield the genome of the bacteria from drying and heat stress. In this form, the genome of the bacterium can maintain itself virtually indefinitely.

Two important genera of spore formers are *Bacillus* and *Clostridium*. *Bacillus* is widely distributed in soils and is capable of both fermentative and respirative (aerobic) metabolism. *Clostridium* is equally widely distributed in soils but is always fermentative. *Clostridium* is very important in the decomposition of organic material under anaerobic conditions.

Members of the genus *Bacillus* have been isolated from a shallow aerobic ground-water system by Chapelle et al., 1988. Curiously, however, species of *Bacillus* have not been widely reported from other studies of aerobic aquifers. Members of the genus *Clostridium* are probably widely distributed in anaerobic ground-water systems. They have been reportedly isolated from ground-water samples by Hirsch and Rades-Rohkohl (1983). In addition, there is abundant evidence that fermentative bacteria produce organic acids and reduced gases in anaerobic aquifer systems (McMahon and Chapelle, 1991). The organisms that perform these fermentations almost certainly include members of the genus *Clostridium*.

Recently, a spore-forming, thermophilic, sulfate-reducing bacterium has been isolated from oilfield waters produced from the North Sea (Rosnes, Torsvik, and Lein, 1991). These bacteria grew autotrophically on H_2-CO_2 and heterotrophically on a number of fatty acids and alcohols. The spores were extremely heat-resistant and could survive at temperatures of up to 131°C for limited periods of time. This suggests that spore-forming bacteria may be quite common in deep subsurface environments, particularly those in which heat tolerance may be important.

2.6 PERSPECTIVES

This brief overview of the kinds of microorganisms found in ground-water systems serves primarily to emphasize the diversity of microbial life. Many ground-water systems are characterized by the presence of protozoa, algae, fungi, and viruses in addition to bacteria and archaebacteria. However, as relatively few aquifer systems have had their microbial flora extensively characterized, it's difficult to generalize about the kinds of microorganisms that will be present in any given system. More extensive studies to investigate these interesting questions are certainly warranted.

CHAPTER 3

BACTERIAL GROWTH

The chief characteristic that separates living from nonliving entities is their ability to grow and reproduce. In this chapter, the growth and reproduction of bacteria will be discussed. The basis of this discussion is to understand the factors and conditions that encourage, or alternatively discourage, the growth of individual cells and populations of cells.

It is useful to begin by differentiating between the growth of individual cells and the growth of populations of cells. The growth of individual cells is defined as an increase in cellular mass, an increase in the number of cellular components, and an increase in size. The growth of populations of cells, however, results in an increase in the total number of cells. Strictly speaking, the term *cell growth* refers to the growth of an individual cell. However, because the growth of individual cells invariably leads to reproduction and population growth, there is an obvious functional relationship between the two.

This functional relationship between cell growth and population growth is important for practical reasons. Observing and quantifying the growth of individual cells is fairly difficult because of their small size. However, it is relatively easy to follow the growth of populations. Thus, the most common measure of cell growth is in fact to measure the increase in cell populations.

3.1 BACTERIAL REPRODUCTION

Most bacterial cells reproduce by means of *binary fission,* in which individual cells split to form two daughter cells. The way in which this occurs is shown schematically in Figure 3.1. First, a parent cell undergoes elongation, a process that is more readily apparent in rod-shaped cells than in cocci, and chromosomal DNA is reproduced. Second, there is internal reorganization in which each half of the dividing cell receives its complement of DNA, and a septum forms in

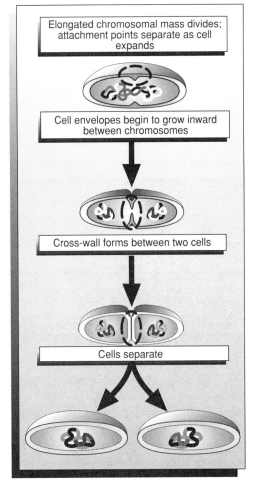

FIGURE 3.1. Sequence of steps leading to binary fission in bacterial cells.

between. Third, the septum develops into a transverse cell wall, and the daughter cells separate.

The time that it takes an individual cell to produce two daughter cells, which is often called its *doubling time,* is highly species dependent. Some enteric bacteria, notably *E. coli,* are capable of very rapid reproduction and have doubling times on the order of 20 minutes. Other bacteria, such as *Micrococcus,* characteristically grow more slowly and have doubling times on the order of hours. While the fastest doubling time that a bacterium can achieve is largely genetically determined, actual doubling times are dependent on a variety of environmental conditions. Under conditions of nutrient deprivation or in the presence of toxic substances, the division of bacteria can come to a virtual halt.

There has been relatively little work done to elucidate doubling times of bacteria that live in ground-water systems. Investigations of soil bacteria show that doubling times are largely dependent on soil temperature, soil moisture, and substrate availability. In ground-water systems, temperature and moisture content are, for all practical purposes, constant. Thus, it might be expected that doubling times would largely reflect the availability of substrates for cell growth.

There is some evidence that this is indeed the case. In studies of a shallow sand aquifer that had been contaminated with sewage effluent, Harvey and George (1987) presented evidence that, based on the incorporation of tritium-labeed thymidine into cellular DNA, doubling times in the contamination plume were on the order of days. Outside the contamination plume where concentrations of organic carbon were much lower, doubling times were much lower, in some cases on the order of months or years. Thorn and Ventullo (1988) calculated growth rates for bacteria in shallow pristine aquifers of Wisconsin, Michigan, and Oklahoma. Based on this data, doubling times on the order of 1 to 180 days may be calculated with averages on the order of 10 to 20 days. These rates are roughly consistant with the results of Harvey and George (1987). Because of the wider range of aquifer conditions studied, however, Thorn and Ventullo's data give a hint of the variability of growth rates encountered in shallow aquifer systems. This variability is not at all surprising and, as is the case with soils, is certainly a characteristic of these types of systems. While showing the same sort of variability in growth rates as exhibited by soils, rates in shallow aquifers are ubiquitously slower than those in soils.

Studies on doubling times of bacteria that live in deep subsurface environments that are largely cut off from the surface have yet to be performed. It is quite likely, however, that these doubling times will prove to be among the slowest of any terrestrial microbial habitat.

3.2 POPULATION GROWTH KINETICS

A consequence of binary fission being the predominant reproductive strategy of bacteria is that population growth is *exponential;* that is, if one were to plot the increase of a bacterial population with time, the curve approximates a function of the form

$$A(t) = A_o \exp(kt)$$

where $A(t)$ is the number of cells at any time t, A_o is the initial number of cells, and k is the first-order growth constant.

Take, for example, the case shown in Figure 3.2, in which a single cell divides with a doubling time of half an hour. After half an hour there are two cells; after 1 hour there are four cells; after 1.5 hours, eight cells, and so on. Initially, the population growth is fairly slow. However, because the rate of growth is dependent upon the number of cells present, after a few hours population growth rapidly accelerates. If growth were to continue unchecked, in few weeks there would be enough bacteria present literally to cover the earth.

Obviously, because the world is not covered with bacteria, the growth of bacterial populations in nature never proceeds unchecked. Environmental factors, such as the lack of substrates, the production of toxic waste products, infection by viruses, or predation by protozoa always serve to limit population growth.

In the laboratory, the growth of bacterial populations typically follows a charac-

50 BACTERIAL GROWTH

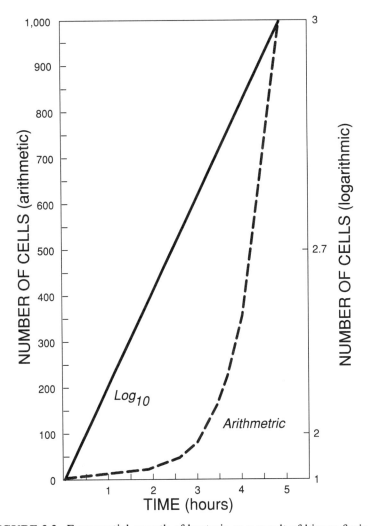

FIGURE 3.2. Exponential growth of bacteria as a result of binary fission.

teristic pattern (Fig. 3.3). Initially, if fresh media is inoculated with a bacterial culture, there is a period in which there is little population growth. This period of slow growth is termed the *lag phase* and reflects the time needed by the bacteria to adjust to the new conditions. After the lag phase, the population begins a period of rapid growth that is termed the *exponential phase,* reflecting the characteristic exponential nature of population increase. Eventually, the nutrients present in the media become depleted and/or there is a buildup of toxic-waste products; the rate of reproduction equals the death rate, and population growth ceases. The bacterial cells shift utilization of scarce resources away from cellular growth, and instead, concentrate on maintaining themselves in a viable state. This is termed the *stationary phase*. As time passes and nutrients are completely used up, the population will begin to die off. This is termed the *death phase*.

In soils where there is a constant flux of nutrients being added from the surface (fertilization) and removed (plant growth), the bacterial population is very dynamic, shifting rapidly between the exponential, stationary, and death phases. It is much less clear how populations change in ground-water systems. In shallow systems that are closely connected with the surface, rapid shifting between exponential growth and death phases may possibly occur. In deeper systems in which changes in nutrient availability occur very slowly, it is likely that population dynamics are more sluggish.

It is clear that bacteria in most deep subsurface environments are not in a state of exponential growth. However, it is possible that an unusual substrate-providing event, such as the intrusive migration of hydrocarbons associated with petrogenesis, could result in a "bloom" of bacteria in exponential growth phase.

Most deep subsurface environments are strongly nutrient deprived, and bacterial populations are probably in either stationary or death phase. In either case, the time frame associated with these phases is vastly expanded from what is observed in the laboratory (Fig. 3.3). In the laboratory it is often observed that the stationary phase predominates for a few days, and that the death phase might predominate for a few weeks or even months. In the deep subsurface, however, with sediments and rocks that are typically millions of years old, it is clear that bacterial populations are able to maintain themselves in states approximating stationary phase for long periods of time. It is debatable whether the term "stationary" or "death" phase more properly describes the state of deep subsurface bacterial populations. It may well be that at the time scales considered in these environments, identifying one or the other states is simply not feasible.

3.3 ENVIRONMENTAL CONDITIONS AND BACTERIAL GROWTH

The ability of microorganisms to grow in nature is largely a function of the environmental conditions in which they find themselves. An understanding of how such conditions affect microbial growth is essential to understanding their observed

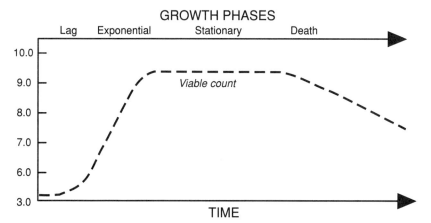

FIGURE 3.3. The different phases of growth typically observed in bacteria growing in culture media.

distribution and abundance in nature. It is characteristic of microbial populations that not all species respond positively to the same set of conditions. In fact, the conditions that are necessary for the growth of one particular strain may be toxic to other strains. More commonly, microbes have a range of tolerance to environmental conditions, and as long as conditions remain inside that envelope, they can effectively grow. In some cases, microorganisms exhibit remarkable ability to grow under adverse conditions. Bacteria in the deep subsurface are particularly notable for their tolerance of difficult growth conditions.

3.3.1 Temperature

The temperature of an environment is an important factor in the growth of microorganisms. Life in microbes, as in all living systems, is supported by a series of enzymatically catalyzed chemical reactions. As temperature increases, the rates of chemical reactions tend to increase. So, to a point, an increase in temperature is favorable for biochemical processes. However, the enzymes and nucleic acids that organize and facilitate these processes are heat-sensitive. If temperatures increase too much, the structures of some proteins may become distorted to the point that they no longer catalyze life-sustaining reactions. Similarly, if the structure of nucleic acids is altered by excessive heat, the information stored in them may become distorted. This, in turn, prevents the production of enzymes needed for growth.

It is universally observed that particular strains of microorganisms exhibit a characteristic envelope of temperature tolerance (Fig. 3.4). First, there is a *temperature minimum,* below which the microorganism cannot grow. As the temperature increases above this minimum, the growth rate increases until a *temperature optimum* is reached. This is the temperature at which the organism can grow at its maximum rate. As the temperature increases above the optimum, there is typically a sharp decrease in the growth rate. The *temperature maximum* is the highest temperature at which a microorganism can grow effectively. The temperature minimum, optimum, and maximum are sometimes referred to as the *cardinal temperatures* and are characteristic of particular strains. Cardinal temperatures are not, however, completely fixed and may be affected by other environmental factors, such as pH and salinity.

Microorganisms that have relatively low cardinal temperatures, those between 0° and 20° centigrade, are termed *psychrophiles*. As one would expect, psycrophilic microorganisms live in environments where cold temperatures typically are found. For example, the bottom waters of the deep ocean basins exhibit temperatures in the 1–3° C range. Bacteria that inhabit these environments are quite capable of growth at these low temperatures, although they may exhibit temperature optima in the 10–20° C range.

Freezing generally has the effect of stopping microbial growth, but it does not always lead to death. Freezing will be fatal to microorganisms if the formation of ice crystals disrupts essential cellular structures, such as the cell membrane. However, if such cellular damage is avoided, the cell often simply becomes inactive and will remain in that state virtually indefinitely. Because ice crystals are formed less efficiently when freezing occurs rapidly, fast-freezing techniques are commonly used to preserve bacteria in the laboratory.

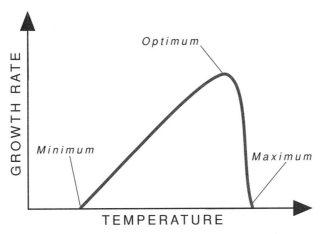

FIGURE 3.4. The envelope of temperature tolerance exhibited by microorganisms.

Microorganisms with temperature ranges in the 20–40° C range are called *mesophiles*. This is the temperature range of many natural environments, and it is safe to say that the majority of microorganisms that have been characterized are mesophiles. Also, this is the temperature range maintained by warm-blooded animals, and so, many pathogenic bacteria have their optima in this range. It is interesting to note that the temperature maximum is always closer to the temperature optima than the temperature minima (Fig. 3.4). Some researchers think that the mammalian characteristic of producing higher temperatures (i.e., fever) in response to illness is an adaptation designed to take advantage of this characteristic. Increasing body temperature above the temperature optima of the infecting microorganisms is an effective method of fighting some infections.

Most of the bacteria recovered from subsurface environments and characterized are mesophiles and generally grow most efficiently from 20 to 30° C, as this is the most common temperature range of ground-water systems used for water supply. However, many microbiology laboratories habitually maintain the temperatures of their incubators at 37° C, which is body temperature for humans. These laboratories have sometimes had the unfortunate experience of having interesting bacterial strains killed by these "elevated" temperatures.

Microorganisms that grow at temperatures above 45° C are said to be *thermophilic*. Because of the relatively high temperatures associated with very deep aquifer systems and petroleum reservoirs, thermophilic bacteria are of particular interest in the study of ground-water chemistry.

There is no general agreement as to the highest documented temperature associated with microbial growth. At this writing, investigators culturing bacteria isolated from hot vents on the ocean floor claim to see bacterial growth as hot as 125° C. The optima for most thermophiles that have been characterized, however, is generally in the 55–60° C range.

There is a large body of liturature on thermophilic bacteria isolated from the deep subsurface. For example, Zobell (1958) decribed sulfate-reducing bacteria from formation waters that ranged from 60° to 105° C. A number of Russian scientists have made similar observations from oil field brines. Zinger (1966) described sulfate-reducing bacteria in formation water that was about 70° C; Rozanova and Khudyakova (1974) recovered a sulfate reducing bacteria from water reported to be 84° C; Rozanova and Nazina (1979) reported an organism designated *Desulfovibrio thermophilus* in formation water above 80° C. Those same authors reported a strain of *Desulfotomaculum nigrificans* in water of 60–70° C. Because hot ground water typically cools as it is brought to the surface, it is often difficult to be precise as to the *in situ* temperature.

One of the most interesting reports of thermophilic bacteria in the deep subsurface is found in the work of Olson et al., 1981. Olson and her coworkers isolated a number of strains of sulfate-reducing and methanogenic bacteria from water produced from the Madison aquifer of Montana. The bacteria that Olson isolated grew well at 50° C, the temperature of the water at the wellhead (which is lower than the in situ temperature), but could not grow at 22° C. Because the formation water was recovered within three weeks of drilling, Olson realized that this was almost certainly insufficient time for introduced organisms to acclimate to higher temperatures. This, in turn, was good evidence that the bacteria recovered were indeed part of the indigenous flora of the Madison aquifer and had not been introduced during drilling procedures.

Petroleum geologists often consider that 100° C is the maximum temperature at which bacteria can be active and thus be involved with petrogenesis or methanogenesis. This assumption appears to be based more on convention than on actual data. The destruction of bacteria at 100° C (i.e., by boiling) is more related to the coexistence of two phases, liquid and vapor, that occur at "normal" (i.e., one atmosphere) pressures. Because water in the deep subsurface is under pressure and hence does not boil as easily, there is no a priori reason why bacteria cannot be as active at 150° C as at 50° C. The only constraint is that the organisms would have to be equipped with enzymes and nucleic acids capable of efficient operation at these temperatures. Thus, while hard evidence is lacking to show that bacteria in the deep subsurface are capable of activity at temperatures above 100° C, it is highly likely that such microorganisms exist. In fact, such thermophilic bacteria may be instrumental in such important processes as petrogenesis, methanogenesis, and reservoir porosity enhancement.

3.3.2 Water

All microorganisms require water in order to carry out life processes, and all but a few require liquid water in order to grow. Many microorganisms, however, are especially designed to resist drying. The thick cell wall of *Staphylococcus,* which is adapted to living on human skin, is very efficient at preserving moisture in the cell and allows it to thrive in that relatively dry environment. Similarly, many soil microorganisms are highly resistant to drying, an adaptation crucial in an environment that is subject to successive wet and dry periods.

At first glance, it might seem that, in ground-water systems drying would not be a limiting factor to microbial growth. However, there are several ways in which moisture availability may be a problem for bacteria in aquifer systems. The most obvious one is the unsaturated zone-the zone that exists between the soil zone and the saturated water table-where water between pore spaces coexists with gases. In these gas-filled interstices, cell drying is at least a possibility, and one that microbes living in them must address. Another way that drying may be a problem in ground-water systems is in the very fine-grained sediments that comprise confining beds. Because of the high-surface area of clay minerals in these sediments, their generally compact arrangement, and the affinity for water to clay mineral surfaces, water actually available to bacteria may be limited. It is commonly observed that clayey confining beds exhibit much less microbial activity than adjacent sandy aquifers (Phelps et al., 1989a; Chapelle and Lovley, 1990). One of the possible causes of this phenomenon is the nonavailability of water due to the high clay content.

3.3.3 Molecular Oxygen

The presence of molecular oxygen, either in the gaseous or dissolved form, is an absolute requirement for some microorganisms that are termed *obligate aerobes*. These microorganisms produce energy by removing electrons from organic carbon or inorganic electron donors, and by transfering these electrons to molecular oxygen in order to produce energy. Other microorganisms, called *facultative anaerobes* prefer to use molecular oxygen if it is available, but if it is not available, they are able to use alternative electron acceptors for metabolism. Some kinds of obligate aerobes, while requiring oxygen in order to grow, cannot tolerate oxygen in high concentrations. These are termed microaerophilic bacteria.

While oxygen is the preferred electron acceptor for many kinds of bacteria, oxygen (or rather its byproduct hydrogen peroxide) is toxic to many other bacteria. For these *obligate anaerobes,* the presence of molecular oxygen either inhibits growth or actually kills the microorganism. Obligate anaerobes vary widely in their tolerance to oxygen. Some kinds of anaerobes, such as methanogens, tolerate oxygen very poorly and must be cultured under strict anaerobic conditions. Other kinds of anaerobes, such as some sulfate reducers, can tolerate exposure to oxygen but generally cannot grow in its presence.

In many ground-water systems, oxygen is present in and near recharge areas where oxygen is delivered with infiltrating rainwater. In these systems, aerobic microorganisms are the predominant species present. However, because of the heterogeneity of subsurface sediments, particularly the discontinuous distribution of organic carbon, small pockets termed *microenvironments,* may be present that lack molecular oxygen. These anaerobic microenvironments may then become the focus of anaerobic metabolism. Thus, the presence of dissolved oxygen in ground-water systems does not generally exclude the presence of anaerobic metabolism.

3.3.4 pH

Bacteria in nature are capable of living in a surprisingly wide range of pH. Some bacteria, notably sulfide-oxidizing bacteria, are perfectly at home in an acidic pH range of 3.0 to 4.0. This is not terribly surprising, in that sulfide-oxidizing bacteria produce sulfuric acid (H_2SO_4) as a by-product of their metabolism. Clearly, the ability to tolerate acid is a requirement of that sort of life-style. Other bacteria are quite capable of growth at alkaline pH of 10.0 or above. Many of these bacteria live in unusual environments, such as alkaline lakes, where such pH's occur naturally.

Regardless of the pH that they are adapted for, bacteria usually maintain an intercellular pH in the 7.5 range. Thus, in order for bacteria to live in environments where pH conditions are significantly different from 7.5, they must employ specialized proton-transporting mechanisms in order to remain viable. The nature of these mechanisms is of considerable interest to microbiologists, because, as we will see later, bacteria utilize proton gradients across their cell membranes in order to harness energy.

Actively growing bacteria often produce a variety of waste products that can significantly affect the pH of the surrounding environment. When laboratory media for growing bacteria are prepared, this fact is taken into account by providing pH buffers, such as carbonates or phosphates, that tend to stabilize the pH of the media.

Extremes of pH are not common in pristine ground-water systems. In systems in which there is active oxidation of sulfides, natural pH's may be as low as 4.0 or 5.0. But pH's that are lower than this are extremely rare. Similarly, very high pH's are not common in ground-water systems. In carbonate-buffered systems, pH's may be found as high as 9.0 but rarely higher. The relatively restricted range of pH's found in ground-water systems generally reflects the buffering effects of the framework carbonate or silicate minerals that make up the aquifers. Because of the relatively restricted pH range typical of ground-water systems, it seems unlikely that pH presents a significant problem to bacterial growth in pristine systems.

In ground-water systems contaminated by toxic wastes, more extreme pH conditions may be found. In aquifers contaminated by municipal waste leachates, which often contain elevated concentrations of organic acids, pH's as low as 3.0 have been reported (Baedecker and Back, 1979). Similarly, ground-water systems contaminated by some industrial wastes, such as sludges from cement manufacturing, may exhibit pH's as high as 11.0. In these cases, pH stress may indeed present a significant environmental problem for indigenous bacteria.

3.3.5 Osmotic Pressure

Cells can gain or lose water by osmosis depending on the relative concentrations of dissolved constituents inside or outside the cell. If the outside of the cell has higher concentrations, the environment is said to be *hypertonic*. If the outside of the cell has lower concentrations, the environment is said to be *hypotonic*. If the concentrations are the same inside and outside, the conditions are said to be *isotonic*.

A cell in a hypertonic environment will tend to lose water into its surroundings, which could cause it to shrink and dehydrate. Conversely, a cell in a hypotonic environment will tend to gain water, which could cause it to lyse. Clearly, osmotic pressure is a critical condition that the cell must deal with in order to live and grow.

As discussed earlier, a primary function of the cell wall is to regulate the osmotic flow of water out of or into the cell. Not surprisingly, microorganisms that live in exceedingly saline environments-the halophiles, for example-have cell walls that have much different properties than those found in more dilute aqueous environments.

Because subsurface environments include a wide range of salinities (from practically rainwater to 30 % brine solutions) subsurface bacteria exhibit a wide range of tolerance to osmotic conditions. It is very important for investigators to take such conditions into account when they are designing the growth conditions of enrichment media. One particularly good example of why it is important to consider osmotic sensitivity was given by Wunsch (1988). Wunsch was studying water quality problems associated with coal seam aquifers in Kentucky. One common problem was abnormally high concentrations of dissolved barium in the ground water. In addition, there was often a characteristic "rotten egg smell" associated with the ground water. Wunsch knew that concentrations of barium in water are largely controlled by concentrations of sulfate. Because sulfate tends to precipitate barium as the mineral barite, any process that actively removed sulfate from the system would tend to increase concentrations of barium in the ground water. These two bits of evidence suggested to Wunsch that sulfate-reducing bacteria were present and active in the coal seam aquifers, and that their activity was the root cause of the high barium concentrations in ground water. However, in order to show that this was the case, Wunsch needed proof that such bacteria were, in fact, present.

Wunsch approached a colleague who had worked extensively on sulfate-reducing bacteria in oil fields. The colleague assured him that the presence of such bacteria was easily shown by simply inoculating an appropriate growth media with samples of the water in question. Furthermore, the colleague volunteered to prepare the media. Some weeks later, armed with 20 or 30 vials of the media, Wunsch sampled the wells in question, innoculated the sulfate-reducing media, and allowed them to incubate. After several weeks of incubation, not one of the vials showed any evidence of bacterial growth.

Puzzled, Wunsch returned to his colleague to inquire if he had perhaps inoculated the media improperly. During the course of the conversation, the colleague mentioned that the media contained about 100,000 mg/L total dissolved solids (TDS), which approximated the composition of the oil field brines that he was accustomed to. The water that Wunsch was studying, however, had chloride concentrations of less than 1,000 mg/L TDS. Could it be that the bacteria in Wunsch's system simply couldn't cope with the osmotic stress when transferred to the high TDS media? In order to find out, the colleague made up a new batch of media and Wunsch dutifully sampled his wells again. This time, much to Wunsch's satisfaction, the sulfate-reducing bacteria *Desulfovibrio* grew from all but two of the wells that he sampled. Furthermore, those two wells recently had been chlorinated in order to minimize the hydrogen sulfide smell problem. The reason that *Desulfovibrio* had not grown before was that the microorganisms were incapable of dealing with the osmotic shock dealt them when they were placed in the high-TDS media. This incident clearly demonstrates the importance of matching the osmotic conditions of growth media to the osmotic conditions prevailing in the ground-water system under study.

3.4 TECHNIQUES FOR CULTURING BACTERIA

Suppose a fisherman went to his favorite lake for a nice relaxing day of fishing. When he gets there, he decides that he'd like to bring home some nice fat bass for his family to have for supper. Accordingly, he opens his tackle box and selects a lure that looks, acts, and moves like a crawfish—a favorite food of bass. Using this lure, he promptly gets a strike and in due course reels in a "lunker" bass. This is repeated until he has as many bass as his family can use.

Having accomplished his goal of catching bass, he decides that now what he needs is a stringerfull of bluegills in order to make a succulent fish chowder. So, he removes the crawfish lure and replaces it with a lure that simulates a swimming insect—a favorite food of bluegills. Sure enough, as soon as he changes lures, he stops catching bass and begins catching bluegills.

The point here is that both types of fish were present in the lake, but in order to catch one or the other, the fisherman needed to use the proper bait for the particular fish he wanted to catch. This is very similar to the problem that microbiologists face when they are attempting to isolate a particular bacterial species from an environmental sample. One difference is that, instead of trying selectively to catch one species out of four or five like our fisherman, the microbiologist must typically differentiate between hundreds of microbial species. Culture media is the microbiologist's "lure" for "catching" (i.e., isolating) the particular species of interest.

3.4.1 Design of Growth Media

The first step in isolating a particular type of bacteria from an environmental sample is to select the appropriate culture medium. Culture media consist of at least five separate functional components. First, there must be a carbon source provided in a form that the bacteria are capable of assimilating. Second, the medium must contain a source of nitrogen so that the growing cells can synthesize proteins. Third, inorganic ions—phosphate, potassium, sulfur, trace metals, etc.—needed by the organism must be present. Fourth, the appropriate vitamins must be supplied. Fifth, the medium or the surroundings of the medium must contain an appropriate electron acceptor to allow the microorganism to respire (if it is a respiring microorganism).

Carbon Sources Microorganisms are capable of using a wide variety of carbon-bearing compounds as carbon sources for cell growth. Simple sugars, polysaccharides, organic acids, and proteins are the most common carbon sources that are utilized in preparing growth media. Glucose, the most common hexose (six-carbon sugar) in both the cell and in nature, is very commonly used as a carbon source in growth media. Glucose is a *monosaccharide* (Fig. 3.5a) and can be used by the cell as a biosynthetic precursor for many other sugars. Thus, by providing glucose in the media, the cell is able to synthesize many of the other compounds necessary for growth. Because glucose and other saccharides can be partially oxidized by fermentative bacteria without the reduction of external electron acceptors, they are sometimes referred to as *fermentable* substrates.

It is also common to provide *disaccharides,* sugars composed of linked monosaccharides as fermentable substrates in growth media. For example, the disaccharide *lactose* (milk sugar) is composed of glucose and galactose connected by glycosidic linkages (Fig. 3.5b). In order to utilize dissacharides, the cell must first cleave the linkage connecting the monosaccharides by using specialized enzymes. For example, bacteria that are able to utilize lactose produce the enzyme beta-galactosidase. A simple method for discriminating strains of bacteria that produce

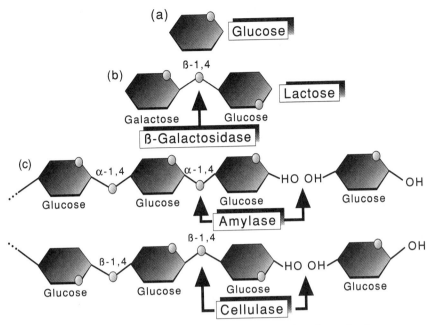

FIGURE 3.5. Structure of some common sugars, and the site of attack of some common degradative enzymes.

beta-galactosidase from strains that do not is to grow them alternatively on media containing glucose as a carbon source and then transfer them to media containing lactose as a carbon source. *Polysaccharides,* or compounds formed by linking many monosaccharides (Fig. 3.5c), are a common substrate for microorganisms in natue and may be used in growth media. Bacteria capable of producing the enzyme *cellulase,* an enzyme that is specific for beta linkages between sugars, can easily be distinquished from those that produce *amylase,* an enzyme specific for alpha linkages, by using different polysaccharides in media. Because of their insoluble nature, polysaccharides are somewhat more difficult to work with and are not commonly used in preparing growth media.

Organic acids, such as acetate or formate, may also be used as carbon sources in growth media. Because fermentative bacteria cannot partially oxidize these compounds, they are sometimes refered to as *non-fermentable substrates*. Amino acids and proteins are other forms of carbon commonly used in growth media. Many pathogenic strains of *Streptococcus* have the capability of hydrolizing proteins, whereas nonpathogenic strains of the same genus lack this capability. Growth on media containing proteins as a carbon source is sometimes used as a criteria for differentiating between these pathogenic and nonpathogenic strains.

For many environmental applications, it is of particular interest to determine if indigenous bacteria are capable of utilizing *xenobiotics,* or carbon compounds not commonly found in nature, as carbon sources. For example, carbon sources such as petroleum hydrocarbons, phenolic compounds, or chlorinated hydrocarbons may be used in preparing growth media. This method has been used to demonstrate the presence or absence of petroleum hydrocarbon-oxidizing bacteria in field studies (Ehrlich, Schroder, and Martin, 1984). A more common method of demonstrating xenobiotic utilization, however, is through the use of microcosm experiments, which will be discussed later.

Nitrogen Sources Nitrogen is needed by microorganisms for protein and nucleic-acid synthesis. Nitrogen can be obtained from either organic or inorganic sources. The most common inorganic sources are ammonia and nitrate. When nitrate is used as a nitrogen source, it must first be reduced to ammonia. This reduction is catalyzed by the enzyme *nitrate reductase,* a molybdenum-containing flavoprotein. Ammonia is assimilated into the cell's metabolism with the mediation of glutamic dehydrogenase, which forms glutamic acid. Nitrogen for cell growth can also be provided in growth media from organic sources, such as amino acids or proteins.

Phosphorus and Inorganic Nutrients Phosphate is utilized by microorganisms for synthesizing nucleic acids. In addition, the energy currency of the cell is largely controlled by adenosine triphosphate (ATP). Thus, phosphate is an important component of growth media and is generally provided as a potassium salt (K_2HPO_4 or KH_2PO_4).

Potassium is universally required by microorganisms. Many enzymes involved with protein synthesis are specifically activated by potassium. As just mentioned, potassium is often provided in growth media as salts of phosphate.

Magnesium is required in order to activate certain enzymes. Magnesium also is involved in the production and activity of ribosomes, nucleic acids, and cell membranes. Gram-positive bacteria require about ten times more magnesium then typical gram-negative bacteria, an observation that probably reflects differences in cell membrane function. Although calcium and magnesium are closely related in the periodic table, calcium cannot replace magnesium for many functions. Calcium ions are important in the function of cell walls and in the production of bacterial endospores.

Sodium is not required by all bacteria, particularly bacteria adapted to freshwater environments. Many species of marine bacteria seem to have a requirement for sodium, however, and it is typically provided in culture media for marine environmental samples.

Iron is an essential trace nutrient for all microorganisms. Iron can exist in either metallic form with an oxidation state of zero (Fe^o); in the ferrous state with an oxidation state of plus two ($Fe(II)$); or in the ferric state with an oxidation state of plus three ($Fe(III)$). Because iron is readily oxidized from $Fe(II)$ to $Fe(III)$ and reduced from $Fe(III)$ to $Fe(II)$, it is widely used as an electron carrier in cytochromes. It is therefore a critical component of the electron transport system of many microorganisms.

Other trace elements necessary for bacterial growth and therefore typically provided in culture media include copper, zinc, cobalt, manganese, and molybdenum. Molybdenum is a component of some important metal protein complexes, such as nitrate reductase. Cobalt is needed for the synthesis of vitamin B_{13}. Zinc is a structural component of certain enzymes in that it helps hold protein subunits together. Manganese is an activator of some enzymes that act on phosphate-containing compounds and is a functional component in other enzymes.

Many microorganisms are not capable of synthesizing organic compounds necessary for their growth. These compounds, known as *growth factors* or *vitamins*, must be supplied from external sources. These growth factors are highly species-specific. For example, the fermentative bacteria *Clostridium butyricum* cannot synthesize biotin, which is necessary for carboxylation reactions. Similarly, the lacctobacilli and many other bacteria require riboflavin for their electron transport systems and vitamin B_{12} to assist with molecular rearrangement reactions. Metabolizing organic acids as an energy and carbon source requires pantothenic acid, which must be supplied externally for many bacteria. Microorganisms that require particular growth factors to be successfully cultured are termed *fastidious*. In culturing fastidious microorganisms in the laboratory, growth factors must be supplied in the culture media.

Electron Acceptors Microorganisms obtain energy by transferring electrons from *electron donors,* such as organic carbon, or reduced inorganic compounds, such as sulfides, to compounds that accept electrons. In fermentations, organic carbon acts as both the electron donor and the electron acceptor, producing a number of incompletely oxidized by-products. In respiration, electrons are transferred directly to inorganic compounds that act as *terminal electron acceptors*. Electron acceptors are compounds that are relatively oxidized and include molecular oxygen, nitrate, Mn(IV), Fe(III), sulfate, or carbon dioxide. Thus, in order for respirative microorganisms to grow in culture, one or more of these compounds must be provided in the culture media.

In order to grow aerobic or facultatively anaerobic microorganisms, oxygen is provided simply by culturing in a petri dish or a test tube that is open to the air. However, the presence of molecular oxygen often inhibits the utilization of other potential electron acceptors. To promote the growth using electron acceptors other than oxygen, media must be prepared under *anaerobic* conditions, or under conditions from which oxygen has been excluded. Nitrate and sulfate are generally provided to culture media as electron acceptors from salts, such as sodium nitrate or sodium sulfate. However, Mn(IV) and Fe(III) compounds are relatively insoluble and are often provided in a suitable solid form such as colloidal hydroxides. Some soluble Fe(III) compounds, such as ferric citrate, may also be used.

Selective Growth Using Culture Media A culture medium that is designed to grow one particular type of microorganism to the exclusion of others is called a *selective medium*. Selective media are extensively used in medical microbiology in order to obtain the causative agents of disease. For example, sodium tetrathionate is toxic to the normal flora of the human intestinal tract, but it is not toxic to bacteria of the pathogenic genera *Salmonella* or *Shigella*. Thus, if these pathogens are suspected of causing intestinal disease, sodium tetrathionate is added to media used to culture stool specimens—a procedure that selectively excludes growth of the more numerous normal flora.

Another technique for separating bacterial species is by the use of *differential media*. Differential media contain dyes, pH indicators, or other components that make a colony of a particular bacterial strain look different from other strains. This colony can then be selectively transferred to fresh media, effectively separating it from other strains that might be present. A common method of differentiating pathogenic from nonpathogenic strains of *Streptococcus,* for example, is to grow the suspect strains on media that contains red blood cells. Because pathogenic *Streptococcus* stains typically destroy red blood cells, a colony of pathogenic *Streptococcus* will form a clear zone around itself that can be readily distinquished from the red background of the blood agar.

Numerous other techniques can be used in order to grow selectively one kind of microorganism to the exclusion of others. Salts, antibiotics, specific enzyme inhibitors, the presence or absence of electron acceptors, and the use of different carbon sources are all tools that can be used in designing selective media.

Table 3.1 lists the formulas for two types of selective media commonly used in investigations of subsurface bacteria. Both media use acetate as a sole carbon and

TABLE 3.1 Formulas for a Defined (Medium F) and Non-defined (Medium S) Media.

	Medium F (Selective for Fe(III)-Reducers)	Medium S (Selective for Sulfate Reducers)
Electron donor (g/L)	Na · Acetate, 6.8	Na · Acetate, 6.8
Electron Acceptor (g/L)	$Fe(OH)_3$, 100	$NaSO_4$, 1.0
		$MgSO_4$, 2.0
Minerals (g/L)	NH_4Cl, 1.5	NH_4Cl, 1.0
	NaH_2PO_4, 0.6	KH_2PO_4, 0.5
	KCl, 0.1	$CaCl_2$, 1.0
	$NaHCO_3$, 2.5	
	$CaCl_2$, 0.1	
	$MgSO_4$, 3.0	
	$MnSO_4$, 0.5	
	NaCl 1.0	
	$FeSO_4$ 0.1	
	$CaCl_2$ 0.1	
	$CoCl_2$ 0.1	
	ZnCl 0.13	
	$CuSO_4$ 0.001	
	$AlK(So_4)_2$ 0.01	
	H_3BO_2 0.01	
	Na_2MoO_4 0.025	
	$NiCl_2$ 0.024	
	$NaWO_4$ 0.025	
Vitamins (mg/L)	biotin, 2.0	yeast extract, 3.5 g/L
	folic acid, 2.0	
	pyridoxin. HCl, 10.0	
	riboflavia, 5.0	
	thimaine, 5.0	
	nicotinic acid, 5.0	
	pantothenic acid, 5.0	
	B_{12}, 0.1	
	p-aminobenzoic acid 5.0	
	Thioctic acid 5.0	
Redox buffers (g/L)	none, purge with $N_2:CO_2$ to remove oxygen	Ascorbic acid, 1.0
		Na · thioglycolate, 0.1
pH	Adjust to 7.0	Adjust to 7.0

energy source. Because acetate is a nonfermentable substrate, these media will tend to exclude growth of strictly fermentative bacteria. However, medium S contains sulfate as a sole electron acceptor, whereas medium F contains poorly crystalline ferric oxyhydroxide as a sole electron acceptor. Thus, medium S will systematically exclude growth of microorganisms that require molecular oxygen, nitrate, or Fe(III) as terminal electron acceptors. Similarly, Medium F will systematically exclude growth of microorganisms that require oxygen, nitrate, or sulfate. [for growth.] A common use of such selective media is to evaluate the metabolic potential of bacteria in subsurface sediments (Lovley, Chapelle, and Phillips, 1990).

Another difference between these two types of media is the manner in which trace minerals and vitamins are provided. In medium F, which is known as a *defined medium,* each trace mineral or vitamin is added in known amounts. In medium S, vitamins and trace minerals are provided by adding *yeast extract,* a compound known to contain essential nutrients but in undefined amounts. In general, defined media are used in order to delineate the exact nutritional requirements of a microorganism. When microbiologists are dealing with microorganisms that have unknown nutritional requirements, they often use yeast extract as a catch-all ingredient.

Finally, the two media differ in the manner in which they establish redox conditions. Medium F is flushed with a mixture of $N_2:CO_2$ from which all traces of oxygen have been removed. This effectively removes all oxygen from the medium thereby establishing strict anaerobic conditions. In medium S, redox buffers (ascorbic acid, thioglycolate) are added to lower the redox potential of the medium.

3.4.2 Isolating Bacterial Strains from Environmental Samples

Environmental samples, such as water from lakes, aquatic sediments, ground water, or sediments from aquifer systems, almost never contain just a single type of microorganism. Such environments contain mixed populations, of which each species fills a particular niche. The first problem in isolating bacteria from samples of these environments is to physically separate the strains present.

In practice, this separation is generally accomplished through a multistep procedure. First, a selective medium is used that allows or encourages growth of the particular kind of microorganism sought. Growth on the selective medium implies that such microorganisms are indeed present. However, since multiple strains may be present, a dilution procedure is used to separate the strains. It is important to remember that all culture media are selective to some degree, and that the lack of growth of particular microorganisms *does not* imply that those microorganisms are absent from the sample. It may simply mean that the culture medium is not appropriate for that organism.

For aerobic and facultatively anaerobic bacteria, by far the most common dilution procedure is the *streak-plate* method. In this method, the growth medium is solidified by addition of about 2% *agar* to the media. Agar is a polysaccharide gelling agent that is derived from certain kinds of algae. The agar medium mix is heated to dissolve the agar and then *autoclaved,* that is, sterilized by heating at 250 F under 15 atmospheres of pressure for 15-30 minutes, and then poured into sterile petri dishes. When the medium is cooled it solidifies, providing a surface of nutrients on which aerobic bacteria may be grown.

Mixtures of bacterial strains can be diluted by taking a small amount of the inoculum, or sample to be diluted, and streaking it onto the agar medium using an *inoculating loop* (Fig. 3.6). The process of streaking the sample spreads individual cells over the medium. As these individual cells reproduce, they form *colonies,* in which each cell is identical to the first cell. This streak dilution procedure may be repeated several times in order to make sure that the isolate is indeed pure. The purity of the isolate is generally checked by making a gram stain and examining it under the microscope.

Mixtures of bacterial strains can also be separated by *serial dilution* in liquid media (Fig. 3.7). In this procedure, water or sediment is added to a tube of liquid medium, mixed, and then 1 milliliter (ml) of the mixture transferred to 9 mls of fresh medium making a 1 : 10 dilution. This dilution procedure is repeated until the original sample has been diluted as much as $1 : 10^{10}$. The goal is to dilute the sample until just a single cell is transferred to the next tube, resulting in a pure culture. In practice, this procedure of successive dilution must be repeated numerous times in order to obtain a pure culture. It is also common practice to combine the serial dilution with the streak dilution procedure in order to obtain pure cultures.

An example of how these techniques have been used to study bacteria from

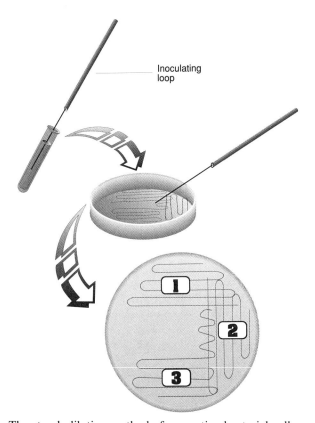

FIGURE 3.6. The streak dilution method of separating bacterial cells on solid media.

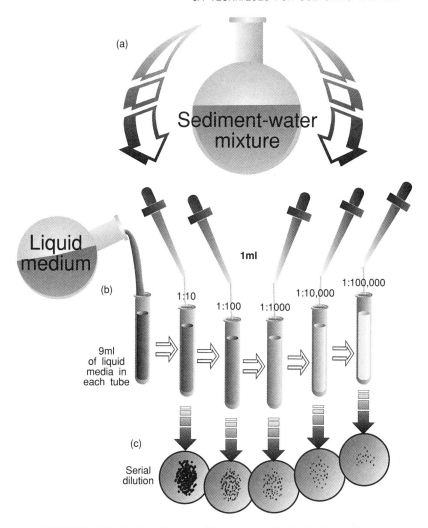

FIGURE 3.7. Serial dilution of bacterial cells in liquid media.

subsurface environments was given by Balkwill and Ghiorse (1985) in a study of a shallow water-table aquifer in Lula, Oklahoma. Because the nutritional requirements of indigenous bacteria were largely unknown, three types of culture media were prepared. The first medium consisted of 10 grams (g) peptone, 5 g trypticase, 10 g yeast extract, 10 g glucose, 5 g $MgSO_4$, 0.6 g $CaCl_2$, and 15 g agar per liter of distilled water. This medium is known as PTYG (peptone, trypticase, yeast extract, glucose) agar and is fairly rich in available nutrients. It contains several different carbon and energy sources, as well as trace nutrients. However, because the subsurface environment was thought to be relatively nutrient deficient, a 1 : 100 dilution of the PTYG agar was also prepared. Finally, because it was possible that the microorganisms were adapted specifically for the carbon sources available in the soil at that site, a growth medium was prepared in which the extract of the local surface soil (the liquid supernate left over after autoclaving a water soil suspension) was used as a carbon source.

These media were inoculated by making serial dilutions of the subsurface sediments and spreading them onto each type of medium. This procedure was performed in triplicate, and the inoculated plates incubated aerobically at about 25° C (the temperature of the local ground water). After two weeks, the plates were examined for colony growth, and selected colonies were transferred to fresh media. After several transfers, the bacteria were gram-stained in order to verify culture purity, and the strains subsequently were characterized.

This general procedure, with minor variations imposed by local conditions, has been widely used in order to characterize the microflora of subsurface environments (Chapelle et al., 1988; Sinclair and Ghiorse, 1987; Balkwill, 1989).

Rich versus Dilute Media   The components of culture media are just one of several important attributes that must be considered when microorganisms are cultured from environmental samples. Because subsurface environments are typically *oligotrophic* (nutrient deficient) compared with many surface environments, many investigators have suggested that dilute media are more effective for growing indigenous microorganisms than the rich media often used in the laboratory.

There is experimental evidence that this is indeed the case. In the Lula, Oklahoma, study just mentioned, sediments from the deeper saturated zone ($\sim$ 15 feet depth) contained bacteria that grew more readily on dilute PTYG agar and on soil extract agar than on the nutritionally rich full-strength PTYG agar. And, as before, this general pattern has been observed in other studies of subsurface microorganisms. In conditions of extreme nutrient deprivation, having cell walls and membranes that are very permeable to potential substrates would be a considerable advantage. Selection for that trait would therefore be expected. However, if these cells were transferred to very rich culture media, the permeable cell walls might be unable to protect effectively against osmotic shock, resulting in cell lysis and death.

The mechanisms leading to more efficient growth of some subsurface bacteria on dilute media have yet to be systematically explored. However, this trait must be taken into consideration by microbiologists when designing culture media for studies of subsurface bacteria.

3.5   ENUMERATING BACTERIA

Techniques for enumerating bacteria are widely used in laboratory studies of bacterial population growth. These techniques enable microbiologists to quantify rates of population growth under different environmental conditions. For example, the optimum pH range of a particular isolate may be delineated by setting up multiple tubes of liquid media that are identical in all respects except for pH. The tubes are inoculated at the same time, and concentrations of cells are enumerated over time. The optimum pH range for that particular organism is then clearly indicated by the observed rates of cell growth. In field studies of subsurface microorganisms, considerable resources have been applied to counting the number of bacteria present. These studies have generally borrowed techniques from soil microbiology for making cell counts.

There are two main types of cell-counting procedures. There are *viable counts*, which use culture techniques to estimate the number of cells capable of growth and reproduction. There are also *direct counts,* which attempt to enumerate all of the cells that are present, whether they are viable or not. As we shall see, each of these procedures gives different types of information, and neither one can give a "total" count of the microorganisms actually present. It is critical, when cell count data are evaluated, that the types of techniques used be carefully considered.

3.5.1 Viable Counting Procedures

By far, the most widely applied procedure for counting viable cells is the *plate count*. This procedure is applicable to both sediment samples and water samples, although the details differ somewhat. For sediment samples, the most common protocol is to place sediment in a test tube with sterile distilled water and to shake the tube vigorously in order to dislodge bacteria attached to sediment surfaces. Sometimes chemicals, such as sodium pyrophosphate, are added to the sediment slurries in order to encourage bacteria to desorb from sediment surfaces, although the effectiveness of such chemicals is not well documented. Once the slurry has been prepared, it is serially diluted, and aliquots of each dilution are spread onto agar-based culture media and incubated. If the sediment contains large amounts of bacteria, then plates inoculated with a concentrated slurry will grow hundreds of colonies. Such high numbers of colonies cannot be reliably counted, and plates inoculated with more dilute slurry are enumerated instead. Generally, an ideal number of colonies to count on each plate is between 30 and 300. Because of sample nonhomogeneity, it is important to inoculate more than one plate for each sediment dilution. Typically, plates are inoculated in triplicate. Once the colony counts are made, the original sediment slurry is dried, weighed, and results reported in cells per gram of dry sediment plus or minus one standard deviation.

Enumerating cells in water samples using the plate count technique is identical to that just described, except that, of course, the bacteria do not need to be desorbed from sediment surfaces. A variation of the plate count technique that is often applied to water samples is called the *membrane filtration method*. In this method, the water sample to be counted is passed through a membrane filter under vacuum. The filter is then placed on an appropriate medium, and the trapped microorganisms are allowed to grow into colonies. These colonies are then counted. This technique is particularly well suited for water samples that contain low concentrations of bacteria because relatively large volumes of water can be filtered.

Another viable counting technique that is widely applied to sediment and water samples alike is the *most-probable-number* (MPN) technique. In this method, one milliliter of water sample is diluted into 99 milliters of liquid growth media, giving a 1 : 100 dilution. The 1 : 100 dilution is itself then diluted 1 : 100 giving a 1 : 10,000 dilution. This procedure is repeated as many times as is appropriate, and the tubes are incubated. The media in tubes that received viable cells soon turn cloudy due to cell growth. However, the media in tubes that were diluted to the point that no cells were received remain clear. Thus, if growth is observed in the 1 : 10,000 tubes but not in the 1 : 1,000,000 tubes, there is a high probability that the concentration of cells in the original sample was greater than 10,000 cells per ml but less than 1,000,000 cells per ml. When this procedure is performed with an appropriate number of replicate tubes for each dilution (usually three), statistical procedures can be used to give a probable cell concentration in the original sample.

Several drawbacks to these viable procedures are immediately obvious. First, not all bacteria may be dislodged from sediment surfaces when samples are slurried. In fact, there is good evidence that many bacteria are not dislodged. Second, the only bacteria counted with these procedures are those that are capable of growth on the media and under the incubation conditions provided. Fastidious or slow-growing bacteria will be undercounted relative to nutritionally diverse and fast-growing microorganism. Also, if the incubations are carried out under aerobic conditions, obligate anaerobes will not be counted with this procedure. Given these problems, which may differ from sample to sample, it is clear that plate counts give a *minimum* estimate as to the number of bacteria present.

3.5.2 Direct Counting Procedures

A widely used technique for counting bacteria in environmental samples is the *direct count*. This involves placing a known volume of a microorganism-containing liquid on a microscope slide and individually counting the number of cells present. This technique was developed for enumerating cells in water. However, it has been widely applied to sediment samples as well by attempting to desorb cells from sediment particle surfaces.

The most commonly applied direct-count method in subsurface microbiology studies is the acridine orange direct count, or *AODC*, method. When it is applied to sediment samples, cells are desorbed from sediment particles, as discussed earlier, and the cell-containing supernate is stained with acridine orange. Acridine orange is a *fluorescent* stain, so-called because it fluoresces when subjected to ultraviolet light of the appropriate wavelength. The cells are then passed through a membrane filter and are enumerated using an *epifluorescent microscope,* which illuminates the sample with ultraviolet light so that the fluorescing cells can be observed visually.

In addition to allowing enumeration of cells, the AODC method gives some information as to the growth status of individual cells. Acridine orange binds to nucleic acids, both RNA and DNA, in the cell. When the cell binds to RNA, it emits a characteristic orange light, hence the name "acridine orange". When the cell binds to DNA, the emitted light tends to be somewhat greenish in color. Rapidly growing cells contain more RNA relative to DNA than inactive cells and therefore appear orange in color. Cells that are not growing rapidly contain more DNA than RNA and appear greenish in color. Thus, the AODC method can help the investigator judge if cells in environmental samples are rapidly growing or not.

A modification of the AODC method combines some of the features of viable counts with direct-counting procedures. In this method, cells in the sample to be counted are treated with nalidixic acid, an antibiotic that allows cellular growth but inhibits cellular division, nutrients to encourage cell growth, and incubated. Because viable cells can grow but not divide, they tend to enlarge enormously. For example, a micron-long cell might grow to be 10 microns in length under these conditions. Viable cells—those cells capable of active growth—therefore enlarge, whereas nonviable cells do not. This technique, which is called the *direct viable count* method (DVC), can be used to estimate what portion of a bacterial population is viable and what portion is nonviable. As with the viable cell-counting methods discussed earlier, however, this method can count only those "viable" cells that are capable of growth on the nutrients offered and under the given incubation conditions.

There are several practical limitations to the effectiveness of AODC and DVC procedures. The most serious has to do with separating the cells from the sediment. Because acridine orange is a *cationic dye*—that is, it has a net positive charge—it binds readily to negatively charged clay particles. If sediment samples contain a large amount of clay, staining of the suspended clays can render a sample uncountable. Just as troublesome, it is often difficult to distinguish a fluorescing clay particle from a bacterial cell, and so inaccuracies in the counts can be a significant problem. Finally, this procedure works best on sediment samples that contain anywhere from 10^6 to 10^8 cells per gram. Many sediments from the deep subsurface exhibit cell abundances that are less than 10^6 cells per gram. Thus, these direct-counting procedures do not work well on many sediments. Most investigators who are working with subsurface sediments consider direct cell counts to be highly qualitative and not an unequivocal measure of microbial abundance.

Other direct-counting procedures that are widely used in microbiology laboratories but that have seen only limited use in studies of subsurface bacteria include *electronic cell counters* and *turbidity measurements*. Electronic cell counters were devised for routine counting of red blood cells in hospital laboratories. They are very fast and convenient, although somewhat expensive. Also, they tend to clog at high cell densities. Electronic counters measure the electrical resistance of fluid passing through a small ($\sim$ 5 micron) hole. As a cell passes through the hole, the resistance increases sharply, and this increase is recorded on a counting device. Given the volume of liquid passing through the instrument, and given the electrical "pulses" recorded, the number of cells can be quickly computed. As with most direct-counting methods, the counter cannot distinguish between mineral particles and cells. Thus, this method is not easily applied to counting cells in sediment suspensions.

Turbidity measurements are widely used in cell growth studies, because they are simple, direct, and nondestructive. A cell suspension looks turbid because light is scattered by the cells in the liquid. The amount of light scattered is directly proportional to the number of cells in suspension. Thus, light passing through a cell suspension can be measured with a photometer and calibrated to cell numbers. However, as turbidity measurements are specific to the kind of medium being used and are specific to bacteria in pure culture, this technique is not generally applicable to environmental samples. It is used, however, to characterize the growth behavior of microorganisms isolated in pure culture from environmental samples.

3.6 SUMMARY

This chapter has described how individual bacterial cells and populations of cells grow. Most bacterial cells reproduce by binary fission, thus producing an exponential increase in cell numbers in the absence of limiting conditions. In nature, however, there are always limits on bacterial growth. These limits may be the lack of carbon, nitrogen, phosphorus, or inorganic nutrients in the environment.

Similarly, environmental conditions, such as temperature, pH, or osmotic conditions, may be more or less favorable for growth of a given bacterial strain.

Techniques for culturing bacteria from environmental samples are based on reproducing optimum growth conditions for the strains of interest. Bacteria present in shallow ground-water systems are most efficiently grown on relatively dilute culture media, reflecting the oligotrophic conditions of most aquifers. A variety of techniques are available for enumerating bacteria in subsurface environments. Culture media techniques are most commonly used to estimate numbers of viable cells present, whereas direct microscopic techniques are generally used for obtaining total counts of cells.

CHAPTER 4

BACTERIAL METABOLISM

A reasonably precise definition of a *living entity* is "an unstable, organized complex that maintains a steady-state condition through the constant utilization of energy." The ability to utilize energy in an organized way, therefore, is the hallmark of life processes.

The term *metabolism* denotes the complex series of energy-utilizing chemical reactions carried out by the cell. There are two broad types of metabolic processes. First, *catabolism* extracts energy from organic compounds by breaking them down into component parts, thereby releasing energy. *Anabolism* does just the reverse, using energy to build organic compounds by fitting the component parts together in the proper sequence. The metabolic functions of the cell include extracting energy from certain compounds, storing that energy temporarily within the cell, using the energy for construction of cell parts-synthesis of enzymes, nucleic acids, and polysaccharides-repairing damage, and generally maintaining the status quo. An enormous amount of energy is necessary in order support such extensive functions. Some bacteria, for example, are capable of metabolizing an amount of nutrients equivalent to their own weight in just a matter of seconds. In this chapter, the basic principles of energy-utilization in bacterial cells are discussed.

4.1 THERMODYNAMICS AND BACTERIAL METABOLISM

All energy transformations, in either living or nonliving systems, must conform to the laws of thermodynamics. The *first law* of thermodynamics is simply one of energy balance: In any chemical reaction, the amount of energy present in the reactants must equal the amount of energy in the products. For example, if one mole of glucose is completely oxidized, 686 kilocalories (kcal) of energy are produced along with carbon dioxide and water:

$$C_6H_{12}O_6 + 8O_2 \rightarrow 6CO_2 + 12H_2O + 686 \text{ kcal} \tag{1}$$

Thus, by the first law, a cell using one mole of glucose as an energy source can obtain not more than 686 kcal of energy.

This does not mean, however, that all of that 686 kcal of energy is available to the cell. The *second law* of thermodynamics states that some portion of the energy, called *free energy*, will be available to do work, whereas another portion, called *entropy*, will be unavailable to do work. Studies of energy utilization in living cells show that, of the 686 kcal released by oxidizing one mole of glucose, a maximum of only about 280 kcal is actually available as free energy and can be utilized by the cells. In other words, at least 60% of the energy released by glucose oxidation goes toward increasing entropy and is lost as heat.

The glucose oxidation reaction just discussed has a net release of free energy and is known as an *exergonic* reaction. Reactions that must absorb free energy in order to go to completion, such as ATP synthesis, are known as *endergonic* reactions. The amount of useful energy liberated or taken up during a reaction is called the *Gibbs free energy change* (ΔG) of the reaction. At *standard conditions* (i.e., all reactants present at concentrations of 1.0 M, pH equal to 7.0, one atmosphere of pressure, and temperature of 25° C), the Gibbs free energy of the reaction is denoted as $\Delta G°$. Energy-releasing exergonic reactions have a negative ΔG, whereas energy consuming endergonic reactions have a positive ΔG.

Living cells combine, or *couple*, exergonic and endergonic reactions in order to drive useful life functions. This is accomplished by employing intermediate reactants that temporarily store energy obtained from an exergonic reaction. These intermediate reactants then transfer the energy to the site where the endergonic reaction takes place. For example, consider these two reactions:

$$A \rightarrow B \qquad \Delta G° = -100 \text{ kcal} \qquad (2)$$

$$C \rightarrow D \qquad \Delta G° = 50 \text{ kcal} \qquad (3)$$

Reaction (2) is exergonic and can proceed spontaneously. Reaction (3), however, requires energy input in order to proceed. By adding two intermediate reactants (X and Y), the energy released from reaction (2) can be used to drive reaction (3).

$$A + X \rightarrow B + Y \qquad \Delta G° = -20 \text{ kcal} \qquad (4)$$

$$C + Y \rightarrow D + X \qquad \Delta G° = -30 \text{ kcal} \qquad (5)$$

In reaction (4), 80 of the 100 kcal of free energy available from reaction (2) is used to convert X to Y. This results in an exergonic reaction that can freely proceed but also effectively stores 80 kcal in compound Y. When compound Y is added to reaction (3), the overall free energy of reaction becomes ($-80 + 50$), or -30 kcal, again creating another exergonic reaction, but one that synthesizes compound D. Thus, by coupling exergonic and endergonic reactions through the use of energy-storing intermediate reactants, it is possible to synthesize compounds that otherwise could not be synthesized.

This reaction-coupling strategy is the basis for metabolism in living cells. Energy is derived from catabolic (exergonic) reactions, stored in high-energy intermediate compounds, and used in anabolic (endergonic) reactions that allow the cell to grow. Clearly, the energy-storing intermediate reactants are fundamentally important to the cell, and these will be considered next.

4.2 ATP SYNTHESIS—STORING ENERGY

Living cells use a number of compounds for temporarily storing energy and for using this energy to couple exergonic and endergonic reactions. These energy-storing compounds include adenosine triphosphate (ATP), guanosine triphosphate (GTP), acetyl-coenzyme A, and others. By far, the most important of these is ATP. ATP performs very much the same function in the cell that money plays in an economy. It is the "currency" by which all or most metabolic "transactions" are valued. The formation of ATP stores energy, and the hydrolysis of ATP releases energy.

The basic reaction in forming ATP is shown schematically in Figure 4.1. In this reaction, adenosine diphosphate (ADP) acquires an additional phosphate group to form ATP. This type of reaction is called *phosphorylation* and requires energy input. This energy is stored in the phosphate bond, which is sometimes referred to as a *high-energy bond*. The reverse of this reaction, which involves breaking the high-energy bond, liberates energy.

In addition to using intermediate compounds for storing energy, living cells also use intermediate compounds for storing electrons that are generated in energy-releasing oxidation reactions. Enzymes that remove electrons from energy-rich compounds (called *dehydrogenases*) often have electron-storing intermediate compounds as their *coenzymes*. One of these coenzymes is called *nicotine adenine dinucleotide (NAD)*. NAD can exist in a reduced (NADH + H$^+$) or oxidized (NAD+) form, with the electron-transferring reaction written as

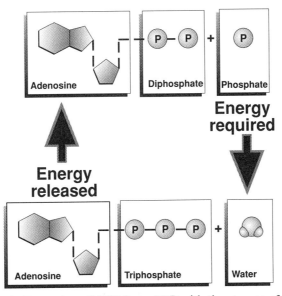

FIGURE 4.1. Formation of ATP from ADP with the storage of energy.

$$NAD^+ + 2H^+ + 2e^- \rightarrow NADH + H^+ \tag{6}$$

Because ATP is an intermediate compound in an overall energy-releasing oxidation reaction, actual phosphorylation of ADP occurs only in the larger context of electron-transferring, energy-releasing oxidation reactions. Consider, for example, the oxidation of glucose to ethanol and carbon dioxide

$$C_6H_{12}O_6 \rightarrow 2CH_3CH_2OH + 2\ CO_2 \tag{7}$$

This overall reaction is exergonic and releases 57 kcal of energy per mole of glucose oxidized. How does the cell store and utilize this energy? The cell performs this process in stepwise fashion using ATP and reduced NAD as intermediates.

The steps involved in one *pathway,* or sequence of biochemical processes, used by cells to oxidize glucose are shown in Figure 4.2. First, glucose is phosphorylated by ATP to form glucose-6-phosphate. This initial phosphorylation *activates* the glucose so that the subsequent reactions become possible. Glucose-6-phosphate undergoes an *isomerization,* or rearranging, reaction to form fructose-6-phosphate. This molecule is again phosphorylated to form fructose-1,6-diphosphate. This molecule is then split, through the mediation of the enzyme aldolase, forming two glyceraldehyde-3-phosphate molecules. Note that, up to this point, there has been no oxidation and no energy produced. In fact, two molecules of ATP have been spent. Now, the first oxidations take place, with glyceraldehyde-3-phosphate being converted to 1,3 diphosphoglyceric acid and with NAD accepting two electrons. In this step of the reaction sequence, a new high-energy phosphate bond is produced. The energy in this bond is then transferred and stored in ATP by forming 3-phosphoglyceric acid. Another ATP molecule is synthesized later in the reaction sequence when phosphoenol pyruvic acid is converted to pyruvic acid.

This sequence of reactions, ending with the synthesis of pyruvic acid (Fig. 4.2) is called *glycolysis* or the *Embden-Meyerhof-Parnas* (EMP) pathway, and it is basic to the metabolism of many microorganisms. The overall reaction of the glycolytic pathway is

$$\text{glucose} + 2\text{phosphate} + 2\text{ADP} + 2\ \text{NAD} \rightarrow 2\ \text{pyruvate} + 2\ \text{NADH} + 2\text{CO}_2 + 2\text{ATP} \tag{8}$$

Note that even though two ATP's were needed to initiate the reaction sequence, the cell receives a net gain of 2 ATP's. This mode of generating ATP is *fermentative*—that is, electrons are not transferred to a mineral electron acceptor (such as oxygen)—and is referred to as *substrate-level phosphorylation*. This distinguishes it from respirative pathways that employ electron transport systems in ATP synthesis. Electron transport systems will be discussed in the next section.

4.3 ELECTRON TRANSPORT SYSTEMS—RELEASING ENERGY

In the glycolysis pathway discussed earlier, 1 molecule of glucose yields 2 molecules of ATP. This is only a fraction of the free energy actually available in glucose. If all of the free energy available from glucose oxidation is used, 38 molecules of ATP could be synthesized. However, in order for complete oxidation to occur,

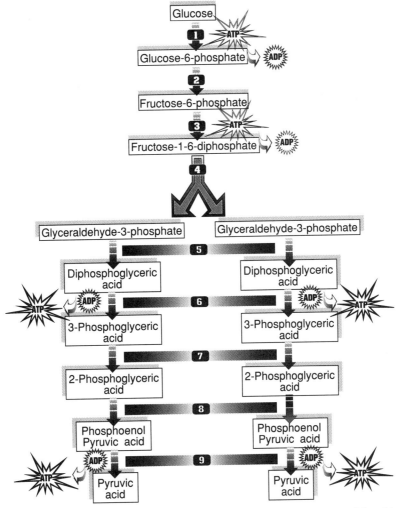

FIGURE 4.2. The Embden-Meyerhof-Parnas (EMP) pathway for the partial oxidation of glucose.

there must be materials available to accept the electrons generated. Inorganic compounds that can accept electrons and can allow complete oxidation of organic substrates are called *electron acceptors*. The most common electron acceptors used by microorganisms in natural environments are oxygen, nitrate, ferric iron, sulfate, and carbon dioxide. Other oxidized compounds that are abundant in unusual natural environments, such as manganate, selenate, or uranyl anions, also can act as electron acceptors for microorganisms. The process of coupling oxidation of organic chemicals with the reduction of external electron acceptors is called *respiration*.

In order to transfer electrons from organic molecules to electron acceptors, cells employ *electron transport systems*. These systems, as the name implies, physically carry electrons. More important, however, electron transport systems are able to conserve some of the energy released in oxidation-reduction reactions by synthesizing ATP.

Electron transport systems in microorganisms consist of *hydrogen carriers*, such as flavoprotein and coenzyme Q, and *electron carriers*, such as cytochromes and iron-sulfur proteins. There are several different cytochromes, designated cytochrome a, b, c, and d, respectively. Each of these cytochromes have slightly different electronegativities (affinity for electrons) so that electrons can be passed from one to the other sequentially. Cytochromes and flavoproteins are associated with each other within the cell at sites called *electron transport particles*. At these sites, electrons are efficiently passed from molecule to molecule in an organized fashion.

An example of an *electron transport chain*, that utilized by *E. coli*, is shown in Figure 4.3. In the cytoplasm, an NADH molecule is oxidized to NAD with the concurrent reduction of a flavoprotein in the cell membrane. The flavoprotein then transfers two H^+ ions out of the cell and reduces an FeS protein. The FeS protein transfers the electrons back to a flavoprotein, which then picks up two more H^+ ions from the cytoplasm. The electrons and $H+$ are then carried out of the cell membrane by coenzyme Q with the reduction of cytochrome b_{556}. Cytochrome b_{556} transfers its electrons to cytocrome o, which carries the electrons back to the cytoplasm. The last step in this chain of processes occurs in the cytyoplasm, with the transfer of electrons from cytochrome o to molecular oxygen, forming water. The energy released by this series of oxidation-reduction reactions performs work by transporting $H+$ ions out of the cell. The potential energy stored by these $H+$ transport processes is then converted into ATP by the cell using *chemiosmosis*, which will be discussed in the next section.

When NADH is the electron donor and oxygen is the electron acceptor, about three ATP's are generated for each oxygen atom and each NADH molecule. Thus, processing NADH through an electron transport system enables the cell to capture

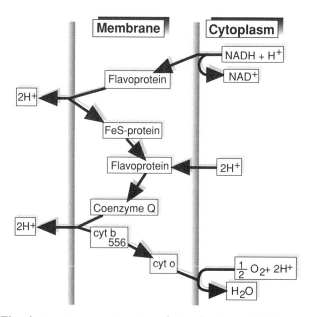

FIGURE 4.3. The electron transport system of *E. coli*. (From Haddock and Jones, 1986).

and to use far more of the energy present in glucose than would be the case with substrate level phosphorylation.

4.4 CHEMIOSMOSIS—HARNESSING ENERGY FROM ELECTRON TRANSPORT

The electron transport mechanisms involved in energy production (Fig. 4.3) show schematically how energy is released by moving electrons in sequences of oxidation reduction reactions. However, the electron transport diagram does not explain the mechanism by which energy thus released is utilized by the cell. This question—perhaps the most important in microbial metabolism—took more than 40 years of determined investigation to answer. By the mid- 1980's, however, it became clear that cells employ a mechanism called *chemiosmosis* to convert electrical energy (i.e., electrons moving through the electron transport system) to chemical energy (ATP).

Figure 4.4 shows the essential characteristics of the chemiosmosis model. As electrons move through the electron transport chain, their energy is used to transport protons from inside the cell membrane to outside the cell membrane. This active transport of protons sets up a *proton gradient* between the outside and

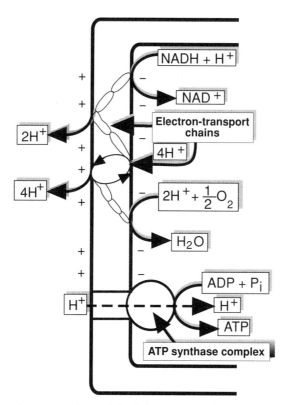

FIGURE 4.4. The chemiosmosis model of ATP synthesis. (Modified from Gottschalk, 1986, with permission courtesy of Springer-Verlag).

inside of the cell membrane. Like any osmotic gradient, this proton gradient has potential energy associated with it that is termed the *proton motive force*. The cell captures this potential energy through the use of complex membrane-bound enzymes that are collectively referred to as the *ATP synthase complex* or simply ATP-ase. The ATP-ase complex acts much like a water gate, through which running water is used to drive a waterwheel. As protons diffuse through the ATP synthase complex, the potential energy is captured and stored chemically as ATP. In the case of aerobic respiration, the electrons are transferred to molecular oxygen, forming water.

4.5 THE ROLE OF ENZYMES

We have seen that microorganisms must conform to the laws of thermodynamics when they are carrying out metabolic processes. Only reactions with a negative free energy occur spontaneously and can be used by the cell. Reactions with a positive free energy can be made to occur spontaneously, however, by coupling them with energy-releasing reactions through energy-storing intermediate compounds such as ATP.

Thermodynamics determines which reactions are energetically favorable and which are not. However, the rate at which the reactions are actually carried out do not depend on free energy change but, rather, on reaction mechanism. For example, if one were to mix glucose with oxygenated water in the laboratory, these chemicals would not immediately react to form CO_2 and water, even though that reaction is energetically favorable. At the low temperatures in the laboratory (25°C), collisions of oxygen and glucose molecules occur too infrequently and without enough force for the reaction to proceed very quickly. At a higher temperature, say 1000°C, this reaction proceeds very quickly. But, at the temperatures favored by living cells, glucose oxidation simply does not occur quickly enough by itself to be of much use in energy production.

Living systems, which depend on hundreds of chemical reactions per second in order to sustain life, solve this problem through the use of *enzymes*. Enzymes are proteins that catalyze chemical reactions in living cells. They work by combining chemically with a substrate or a combination of substrates and by bringing them into a configuration so that a particular reaction may occur. For example, we've seen that, if glucose and oxygen are chemically mixed in the laboratory at room temperature, oxidation proceeds at an immeasureable rate. However if glucose oxidase, an enzyme that brings glucose and oxygen together so that they can react, is added to the solution, glucose is quickly oxidized to gluconic acid with the production of hydrogen peroxide. The mechanism of this reaction is similar to many enzymatically catalyzed reactions. First, glucose, oxygen, and water combine on the surface of glucose oxidase to form a complex:

$$\text{glucose} + O_2 + H_2O + \text{glucose oxidase} \rightarrow \text{glucose-}H_2O\text{-}O_2\text{-glucose oxidase complex}$$

By binding with the glucose oxidase, the three reactants are brought together in a configuration that allows them to combine chemically. Once they have combined, they no longer "fit" on the surface of the enzyme, and the products are released:

glucose-H$_2$O-O$_2$-glucose oxidase complex → glucose oxidase + gluconic acid + H$_2$O$_2$

This general behavior is common to many enzymatically catalyzed reactions and is often summarized by the equations

$$\underset{\text{enzyme}}{E} + \underset{\text{substrate}}{S} \rightarrow \underset{\text{enzyme substrate complex}}{ES}$$

and

$$ES \rightarrow \underset{\text{enzyme}}{E} + \underset{\text{products}}{P}$$

Inspection of these equations suggest that the rate of reaction will be sensitive to a number factors. First of all, if the concentration of substrate increases, one would expect that the rate of reaction would increase. However, because the reaction rate is also dependent on the amount of enzyme present, it can only increase until all of the available enzyme is fully employed. This behavior is called *saturation kinetics*, referring to the fact that as substrate concentrations become large, all of the enzyme is fully utilized (saturated), and reaction rates can no longer increase (Fig. 4.5).

This behavior is generally expressed mathematically by means of the *Michaelis-Menton equation*, which may be written in the form

$$v = V_{max}S/(K_m + S)$$

where K_m is the substrate concentration at which reaction rate (v) is half of the maximum rate (V_{max}). This equation states that the reaction rate will increase in proportion to increasing S up to a particular point, at which saturation is achieved and the reaction rate becomes constant (Fig. 4.5).

Enzymes are classified according to the kinds of reactions that they catalyze. *Oxidases* and *reductases* catalyze oxidation reduction reactions, *lyases* remove functional groups from organic compounds, and *transferases* move functional groups from molecule to molecule. *Hydrolases* hydrolize polymers that are linked together by covalent bonds. *Ligases* perform the reverse of this function by combining monomers together into polymers. Ligases may also be called *polymerases*.

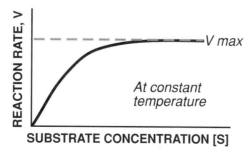

FIGURE 4.5. The dependence of enzymatic reaction rates on substrate concentration.

Enzymes are also sometimes classified on the basis of where in the cell they are used. Enzymes that are used in the cytoplasm are termed *cytoplasmic enzymes.* Similarly, enzymes that occur in the cell membrane may be referred to as *membrane-bound enzymes.* Many cells, particularly gram-positive cells, produce *exoenzymes,* which are passed out of the cell to help break up potential substrates into components that can then be transported into the cell.

Many enzymes require a nonprotein component called a *coenzyme* in order to become active. These coenzymes often contain vitamins or nucleotides, and their special function is often to transfer components from the enzyme to the substrate. NAD, which we've discussed earlier, is a coenzyme that transfers electrons and protons. ATP (in addition to storing chemical energy) is a coenzyme that transfers phosphate groups, and coenzyme A transfers acetyl groups. Some enzymes also require the presence of cofactors, which are inorganic ions, such as iron or magnesium. Such cofactors often form the reactive center of the enzyme and assist in bringing the substrates together in the proper configuration so that reaction occurs.

The regulation of enzymatic activity is crucial to the cell, and this is accomplished by a number of mechanisms. If the cell encounters a potential substrate, enyzmes specific for that substrate can be *induced,* or newly manufactured, to utilize that substrate. The classic example of enzyme induction is that of the lactose operon of *E. coli.* An *operon* is that portion of the cell's DNA that codes for specific proteins. In the absence of lactose in the cell, a protein molecule, called a *repressor protein,* covers the lactose operon and prevents the expression of three enzymes needed to break down lactose. However, in the presence of lactose, lactose binds with the repressor protein and removes it from the lactose operon. This allows expression of the genes and results in the production of the necessary enzymes. However, once lactose has been used up and the cell no longer needs the enzymes, the repressor protein recombines with the lactose operon, and enzyme production stops.

Inductive-repressive regulation of enzymatic activity is very efficient from the standpoint of optimizing the cell's resources. In the case of the lactose operon, it is inefficient for the cell to spend energy continually in order to produce lactose-utilizing enzymes, unless lactose is actually present. However, induction takes time. The characteristic lag time observed when cells are placed in a medium to which they are not adapted is due in part to time required for induction of enzymes to utilize the new substrate. In order to respond quickly to environmental changes, other enzyme-regulating strategies are also employed.

In addition to induction and repression, the activity of enzymes can be altered in the cell by a number of other strategies. For example, if the enzyme encounters a molecule that resembles its substrate, the enzyme might bind the "imposter" to its active site rather than to the normal substrate. Because the imposter molecule competes with the normal substrate, the activity of the enzyme decreases. This is termed *competitive inhibition.* Alternatively, an organic molecule might bind a nonactive site of an enzyme in such a way as to change its shape. If this change of shape is sufficient, the enzyme may be inactivated. This is called *noncompetitive inhibition.*

Rates of biochemical processes in cells can also be regulated by altering the shape, and therefore the activity, of the enzyme. Binding sites on enzymes that are involved in regulating enzyme activity are called *allosteric effector sites.* In some cases, the end product of an enzymatic reaction sequence binds with an enzyme and acts as an *allosteric inhibitor*. This strategy allows the cell to shut off production of a particular product quickly without having to repress enzyme production at the genetic level. This type of regulation is called *feedback inhibition* and is widely used to regulate biochemical pathways. Alternatively, the allosteric binding site may serve to activate a particular enzyme. This is referred to as *feedback activation*. Both feedback inhibition and feedback activation are widely used as enzyme-regulating strategies in cells.

Regulating the activity of enzymatic reactions is crucial to the cell. The goal is to match the supply of enzymes to the supply of substrates in order to maximize metabolic efficiency. Enzyme activity is closely regulated by balancing enzyme production with enzyme need and by employing feedback inhibition and feedback activation strategies. Because of the dynamic nature of enzyme activity regulation, the cell is able to adjust to rapidly changing conditions in its environment.

4.6 ENERGY-RELEASING PATHWAYS OF GEOCHEMICAL IMPORTANCE

We have seen that microorganisms are able to obtain energy for cell growth by coupling endergonic and exergonic reactions into metabolic pathways. In the absence of an external electron acceptor, fermentative pathways, such as glycolysis may be used. When electron acceptors are available, microorganisms employ electron transport systems and chemiosmosis to oxidize organic compounds completely to CO_2 and to store the energy obtained as ATP.

There are hundreds of known metabolic pathways, painstakingly worked out over the years, by which microorganisms produce and transform energy. In this section, some of these pathways of particular importance to microbial processes in ground-water systems are considered.

4.6.1 Lactate and Acetate Fermentations

As illustrated in Figure 4.2, the principal end product of glycolysis is pyruvate. This pyruvate may then be partially oxidized to a number of organic compounds with the production of additional ATP. The variety of these fermentations is virtually endless, and some of the possible endproducts are shown in Figure 4.6. Of these pathways, the production of formate, acetate, and lactate are important in ground-water geochemistry as these products are widely used by respirative bacteria.

One pathway by which both acetate and lactate are produced is known as the *Bifidum pathway* and is shown schematically in Figure 4.7. This pathway is named after the microorganism *Bifidobacterium bifidum* that carries out this particular fermentation. First, glucose is phosphorylated to fructose-6-phosphate, as in glycolysis. One molecule of fructose-6-phosphate is then split into acetyl-phosphate and erythose-4-phosphate. The acetyl-phosphate is then converted to acetate with the production of ATP. The remaining molecules of fructose-6-phosphate and erythose-4-phosphate are then combined and led through a number of intermediate steps to produce more acetyl-phosphate and glyceraldehyde-3-phosphate. The acetyl-phosphate produces more acetate, whereas the glyceraldehyde is oxidized

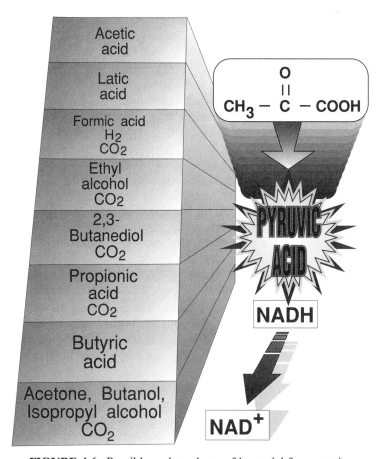

FIGURE 4.6. Possible end products of bacterial fermentations.

to lactate, both with the production of ATP. In this fermentation, therefore, three moles of acetate and two moles of lactate are produced for each two moles of glucose.

4.6.2 Ferredoxins and the Production of Hydrogen and Acetate in Fermentation

Hydrogen is an important substrate for many terminal electron-accepting bacteria, particularly Fe(III) reducers, sulfate reducers, and methanogens. Hydrogen is produced in a number of fermentations, notably by members of the genus *Clostridium*. In order to carry out these hydrogen-producing fermentations, the clostridia employ *ferredoxins*. Ferredoxins are iron-sulfur proteins that serve as electron carriers. They have the unusual property that the redox potential of the reduced-oxidized ferredoxin couple is virtually identical to that of the H_2/H^+ couple. Thus, even in an environment that is saturated with hydrogen gas, reduced ferredoxin can transfer electrons to hydrogenase and hydrogen can be evolved. This enables the clostridia to continue fermentations in extremely reducing environments.

The first step in these hydrogen fermentations is the production of pyruvate through glycolysis. Pyruvate then goes through a multistep process by which

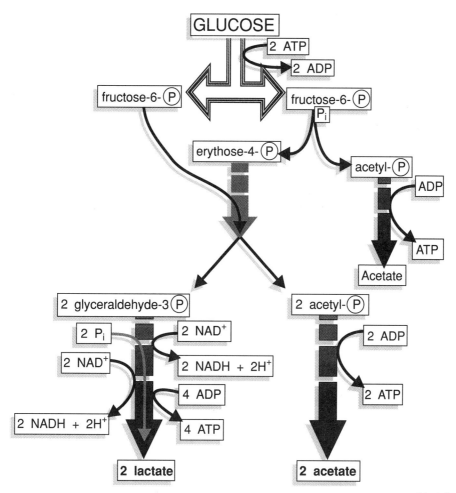

FIGURE 4.7. Production of acetate and lactate via the Bifidum pathway. (Modified from Gottschalk, 1986, with permission courtesy of Springer-Verlag.)

acetyl-phosphate and hydrogen are formed. This multistep process, shown in Figure 4.8, is known as the *phosphoroclastic reaction*. First, pyruvate is decarboxylated by pyruvate-ferredoxin oxidoreductase and combined with thiamine pyrophosphate-containing oxidoreductase (TPP-E) to form a hydroxyethyl-TPP-E complex (HETPP-E). HETPP-E then reduces ferredoxin with the production of acetyl-coenzyme A. Hydrogen gas is then evolved, and acetyl-coenzyme A is phosphorylated to acetyl-phosphate.

These types of hydrogen-producing fermentations are very important in the carbon flow of anaerobic aquifer systems. It is common for anaerobic sediments collected from deep aquifer systems to exhibit hydrogen production when incubated in the laboratory (McMahon and Chapelle, 1991).

Some organisms produce acetate as their predominant fermentation product. These *acetogenic* bacteria are possibly important to anaerobic metabolism in deep

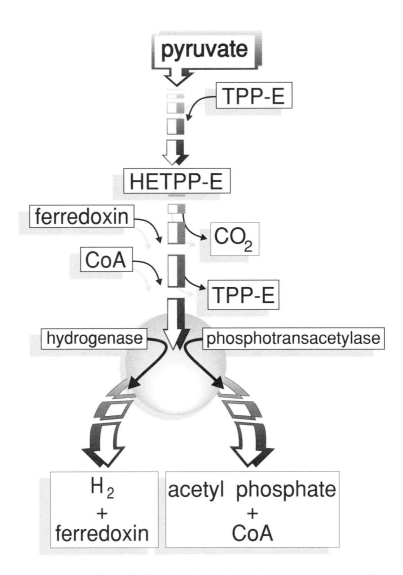

FIGURE 4.8. Production of molecular hydrogen in the phosphoroclastic reaction.

aquifer systems because much of the carbon flow goes through acetate. As with the hydrogen-producing bacteria, ferredoxins serve as important electron carriers.

An acetate fermentation pathway is shown in Figure 4.9. Again, pyruvate from glycolysis is the starting material. First, pyruvate is oxidized to aceytl-coenzyme A, which is subsequently used to produce ATP, with acetate as the end product. Carbon dioxide liberated by this reaction may be combined with hydrogen to form formate as an intermediate product. The enzyme that catalyzes this reaction, formate dehydrogenase, is structured to produce formate and not just oxidize it, like the membrane-bound formate dehydrogenase found in many other organisms.

4.6 ENERGY-RELEASING PATHWAYS OF GEOCHEMICAL IMPORTANCE

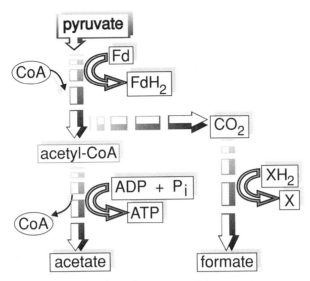

FIGURE 4.9. Production of acetate and formate from pyruvate.

Formate may be further cycled through a number of steps with the formation of methyltetrahydrofolate. This molecule combines with a vitamin B_{12}-based enzyme to produce a substrate enzyme complex termed methyl-B_{12}-E. In a separate reaction, CO_2 may be reduced by hydrogen with the formation of carbon monoxide (CO) as the intermediate product. This CO carboxylates the methyl-B_{12}-E complex with the production of acetyl-coenzyme A and acetate. The enzyme that catalyzes CO formation, CO dehydrogenase, is unique to the acetogenic bacteria and, like the ferredoxins, is characterized by a very low redox potential.

The hydrogen and acetogenic pathways described earlier are fundamental to microbial metabolism in deep aquifer systems. When sediments from deep aquifer systems are incubated in the laboratory, the most abundant fermentation products that are produced are acetate, formate, and hydrogen. All of these are involved in the acetogenic pathway shown in Figure 4.9. Thus, the presence of acetogenic and hydrogen fermentations neatly accounts for observed fermentation products. However, studies on the production of other fermentation products, such as the higher molecular weight organic acids such as lactate, propionate and butyrate, have not been performed on sediment from the deep subsurface. It is possible that these fermentations are also important in some subsurface environments.

4.6.3 Methanogenic Pathways

Fermentative and respirative microorganisms in anaerobic aquatic sediments and subsurface environments live together in *symbiotic* association. Specifically, fermentative bacteria degrade complex sedimentary organic matter using acetogenic or hydrogen-producing pathways similar to those discussed in the previous sections. The products of these fermentations, hydrogen, carbon monoxide, acetate, and formate, still contain usable energy. This energy is then tapped by respirative microoganisms by coupling the oxidation of fermentation products to the reduction of electron acceptors. Methanogenic respiration, in which the carbon present either in CO_2 or in acetate serves as the terminal electron acceptor, is one of the most important respirative pathways found in anaerobic subsurface environments.

There are two distinct pathways of methanogenesis that occur in natural environments. The first of these is called the *CO_2 reduction pathway*. As is true with many microbial processes, the overall stoichiometry of the reaction is deceptively simple and may be written as

$$4 H_2 + CO_2 \rightarrow CH_4 + 2 H_2O.$$
$$\Delta G° = -32.4 \text{ kcal}$$

In fact, methanogenic microorganisms accomplish this overall reaction in at least four steps (Fig. 4.10a) in which hydrogens are sequentially reacted with the carbon. In each hydrogen transfer, a specialized protein factor, F420, acts as the hydrogen carrier. In the final step, a methyl-coenzyme M methylreductase (CoM-CH3) complex is formed and the carbon is reduced to methane. The fourth step in this sequence is coupled to the generation of a proton motive force and subsequent synthesis of ATP.

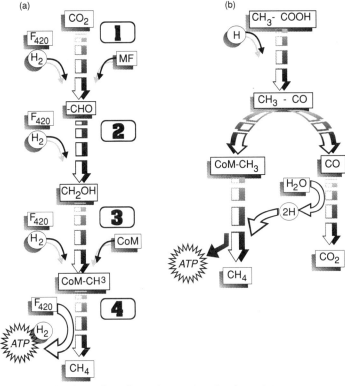

FIGURE 4.10. Production of methane from (a) reduction of CO_2 and from (b) acetate.

A second methanogenic pathway is reduction of acetate according to the stoichiometry:

$$CH_3COOH \rightarrow CH_4 + CO_2.$$
$$\Delta G° = -31 \text{ kcal}$$

This process is sometimes referred to in the literature as "acetate fermentation" but should not be confused with the production of acetate by fermentative acetogenic bacteria, as described in section 4.7.2.

Many of the details of the acetate fermentation methanogenic pathway have yet to be worked out. One likely schematic of these processes, however, is shown in Figure 4.10b. First hydrogen combines with acetate to form a carbon monoxide (CO)-bearing complex (CH_3-CO). This complex then supplies the reducing equivalents needed to transform the methyl-coenzyme M methylreductase (CoM-CH_3) complex to methane, with the generation of ATP. An interesting feature of this reaction is that CO, and not CO_2, appears to be the primary cleavage product of the pathway.

It is clear from the free energy of these reactions (-32.4 and -31 kcal, respectively), that these are not very thermodynamically favorable reactions. In fact, methanogenic microoganisms only predominate in subsurface environments that lack other inorganic electron acceptors. However, many subsurface environments lack such electron acceptors and methanogenesis, by either CO_2 or acetate reduction, is the predominant respirative process.

4.6.4 Sulfate Reduction

Sulfate reduction, like methanogenesis, is a fundamentally important process in the microbiology of deep subsurface environments. Sulfate reducers, in fact, were the first kinds of bacteria that were actually isolated from deep environments such as oil reservoirs and aquifers.

Sulfate-reducing bacteria are dependent upon fermentative bacteria to supply them with the simple organic compounds that their metabolism is designed to exploit. Formate, acetate, lactate, and hydrogen are the principal substrates upon which sulfate reducers commonly grow. Of these, the lactate-utilizing pathway is best known. The essential features of lactate metabolism by the sulfate-reducing bacteria *Desulfovibrio sp.* is shown schematically in Figure 4.11. A particularly interesting feature of this metabolic scheme is that the hydrogenase and its electron acceptor cytochrome c_3 is located in the periplasmic space. Other electron carriers and lactate dehydrogenase occur in the membrane, and all enzymes for pyruvate oxidation, sulfate reduction, and a second hydrogenase are located in the cytoplasm.

The first step in this process is the oxidation of lactate to acetate with the liberation of hydrogen. This hydrogen is then transported through the membrane to the periplasmic space. The electrons are then transferred to cytochrome c_3 and transported back into the cytoplasm. The hydrogen ions remaining in the periplasmic space form a proton motive force which is subsequently used by the cell for ATP synthesis. Once back in the cytoplasm, the electrons are transferred

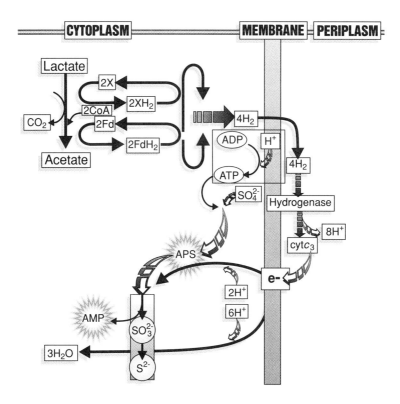

FIGURE 4.11. Biochemistry of sulfate reduction with lactate as a carbon source. (Modified from Gottschalk, 1986, with permission courtesy of Springer-Verlag.)

to sulfate. Note that, for the sulfate to act as an electron acceptor, sulfate must first be activated by ATP to form adenosine-5-phosphosulfate (APS).

4.6.5 Fe(III) Reduction

From a geochemical standpoint, Fe(III) reduction is one of the most important microbially mediated redox processes that occur in ground-water systems. Fe(III) oxyhydroxides are present in a broad range of hydrologic systems, and the reduction of these compounds, coupled with the oxidation of organic matter is a principal cause of undesirably high concentrations of dissolved iron in ground water. In fact, producing and maintaining water treatment facilities that are designed specifically to treat high-iron ground water is a multimillion dollar business in the United States alone. Additionally, Fe(III) reduction is an important process in the oxidation of anthropogenic chemicals artificially introduced into the environment (Lovley et al., 1989) and is important in biogenic localization of some kinds of chemical spills.

Of all the anaerobic respiration pathways, Fe(III) reduction is the one about which the least is known. The principal reason for this is that until 1987 there were no Fe(III)-reducing microorganisms isolated in pure culture, and therefore this metabolic processes could not be systematically explored. With the isolation of strain GS-15 by Lovley and Phillips (1988), investigations into the metabolism of Fe(III) reduction became feasible.

At the present time, the only details known about Fe(III) reduction as carried out by GS-15 are that (1) there is a membrane-bound iron reductase that transfers electrons to solid Fe(III) oxyhydroxides outside the cell; (2) the electron transport mechanism involves NAD as an electron carrier, as well as at least one cytochrome of the b type; and (3) NAD^+ is reduced to NADH in the tricarboxylic acid (TCA) cycle. Details of the TCA cycle are discussed in section 4.6.7.

A possible pathway for the oxidation of acetate with the reduction of Fe(III) is shown in Figure 4.12. First, acetate is oxidized in the TCA cycle with the reduction of NAD+ to NADH. These electrons are then transferred from NADH to cytochrome b, transported through the cell membrane, and combined with Fe(III) oxyhydroxides in contact with the cell membrane through the mediation of iron reductase.

The details of this scheme are very preliminary and will almost certainly be revised by future work. Also, it is beginning to become evident that there are many kinds of Fe(III)-reducing microorganisms in subsurface environments, and their metabolic pathways may differ substantially from that of GS-15.

4.6.6 Nitrate Reduction

A wide variety of microorganisms are able to couple oxidation of organic substrates with the reduction of nitrate in order to obtain energy for growth. This process is of great economic importance in the practice of agriculture since it removes an important plant nutrient, nitrate, from the soil. Nitrate reduction is also an important process in the degradation of organic matter in marine sediments. In ground-water systems, nitrate reduction tends to be an important microbial process when human influences are present (i.e. landfill leachates, agricultural chemicals, septic effluents). Nitrate reduction is rarely an important respirative process in pristine ground-water systems.

When molecular nitrogen (N_2) is the major product of nitrate reduction, as in

$$\text{glucose} + 4.8\ NO_3^- + 4.8\ H^+ \rightarrow 6\ CO_2 + 2.4\ N_2 + 8.4\ H_2O,$$

the process is called *denitrification*. Many different genera of bacteria, including particular species of *Alcaligenes, Bacillus, Pseudomonas,* and *Thiobacillus,* are capable of denitrification. Many bacteria reduce nitrate only as far as nitrite

$$\text{glucose} + 12NO_3^- \rightarrow 6CO_2 + 6H_2O + 12NO_2^+$$

or reduce nitrate to ammonia

$$\text{glucose} + 3NO_3^- + 3H^+ \rightarrow 6CO_2 + 3NH_3 + H_2O$$

A schematic diagram showing a possible reaction sequence in denitrification is shown in Figure 4.13. According to this sequence, nitrate is reduced to nitrite by

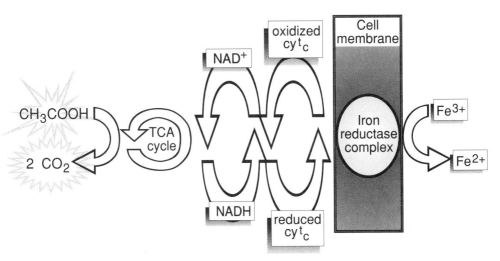

FIGURE 4.12. Model of electron transport in the dissimilatory Fe(III)-reducer GS-15.

means of a specialized enzyme, *nitrate reductase*. This nitrite is then bound to an enzyme-ferrous complex ($X - Fe^{2+} - N = O^+$). An additional nitrite molecule is added to this complex, which is then reduced to oxyhyponitrate through the mediation of *nitrite reductase,* and then reduced again to nitrous oxide. Finally, nitrous oxide is reduced to N_2 by *nitrous-oxide reductase*.

Nitrate reductase is a particularly interesting enzyme. It contains a *molybdenum cofactor* that is noncovalently bound to the protein portion of the molecule. Nitrate reductase is membrane-bound and contains cytochromes of the c and b type. The

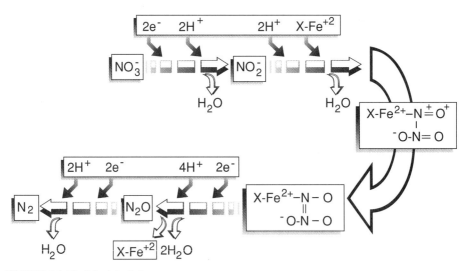

FIGURE 4.13. Model of electron transport in nitrate reduction. (Modified from Gottschalk, 1986, with permission courtesy of Springer-Verlag.)

activity of nitrate reductase is inhibited by the presence of molecular oxygen. For this reason, nitrate-reducing microorganisms are restricted to anaerobic or microaerophilic (oxygen <2%) environments. This inhibition of nitrate reduction by oxygen is the principal reason that nitrate accumulates in aerobic ground water that receives recharge from fertilized farmland, sometimes to the point of becoming dangerous for human consumption. In anaerobic ground-water systems, denitrification rarely allows the accumulation of nitrate in solution.

4.6.7 Oxygen Reduction-Aerobic Metabolism

The most energetically favorable mechanism by which microorganisms oxidize organic carbon material is *aerobic metabolism,* in which molecular oxygen is used as a terminal electron acceptor. There are dozens of known pathways that produce ATP with the reduction of molecular oxygen, but the most important and most closely studied is called by a variety of names, including the *tricarboxylic acid* (TCA) cycle, the *citric acid cycle,* or the *Krebs* cycle (named after one of the investigators involved in working out the cycle). A schematic diagram of the essential features of the TCA cycle is shown in Figure 4.14. While the TCA cycle is most common in aerobic bacteria, it also occurs in some anaerobic bacteria, such as the Fe(III) reducer GS-15 (Fig. 4.12).

The starting material for the TCA cycle, as in the other respirative pathways that we've considered, is pyruvate from glycolysis. Pyruvate is decarboxylated to form acetyl-coenzyme A, which temporarily stores energy, much like ATP, and which carries a two-carbon acetyl group. This acetyl group is combined with oxaloacetic acid to form a six-carbon molecule, citric acid, after which this cycle takes one of its names. Citric acid is referred to as a *tricarboxylic acid* because of the three carboxyl groups that it contains, and this is the source of the term "tricarboxylic acid" (TCA) cycle.

The six-carbon compound citric acid is then oxidized with the release of CO_2 and with the reduction of NAD^+ to NADH. The five-carbon entity remaining is then oxidized, again producing CO2 and NADH. The energy released in this oxidation is captured as GTP (Guanosine Triphosphate) which is used by the cell as a high-energy compound much like ATP. Continuing on, the four-carbon entity, succinic acid is oxidized to fumaric acid, this time with the electrons being captured by FAD, a lower-energy electron carrier than NAD. The final product of the cycle is oxaloacetic acid, which initiates the entire process over again when combined with acetyl-coenzyme A. Figure 4.14 shows only an outline of the biochemistry envolved in the citric acid cycle. The net products, however, include NADH, FADH, ATP, and GTP. NADH and FADH then carry electrons to the electron transport system of the cell, where additional ATP is formed via chemiosmosis.

A comparison of the energy available from substrate level phosphorylation (2 ATP's) and oxidative phosphorylation through the citric acid cycle (38 ATP's) shows why it is so much in a cell's interest to have an electron transport apparatus. It vastly increases the amount of energy that is available for use. Some microorganisms, such as *E. coli,* are capable of both aerobic respiration and anaerobic fermentation. Not surprisingly, these bacteria much prefer to respire and are strictly fermentative only if oxygen becomes depleted in their environment.

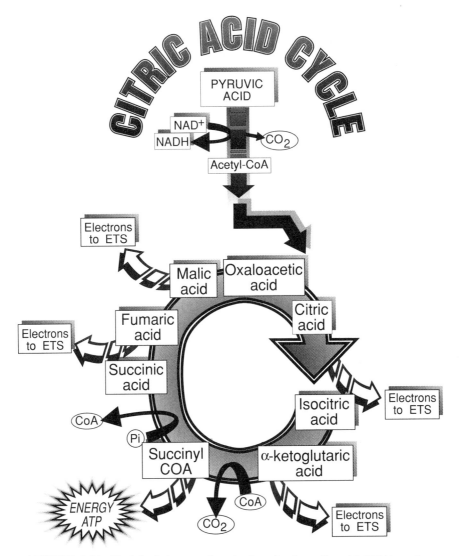

FIGURE 4.14. Model of electron flow in the tricarboxylic acid (TCA) cycle.

4.7 BIOSYNTHESIS

The ultimate reason for metabolism in microorganisms is to produce and to store energy into a form (ATP) that can then be used for cellular growth. Because microbial cells are largely composed of macromolecules, such as proteins, polysaccharides, lipids, and nucleic acids, cellular growth requires the synthesis of these compounds.

Biosynthesis pathways have been closely studied in *E. coli,* and this organism serves as a convenient model for the kinds of biosynthetic processes that occur in other cells. The basic strategy employed by the cell is to use the intermediate products of glucose metabolism as the starting materials for larger macromolecules. For example, pyruvate, which is produced in glycolysis (section 4.2), can be transformed to sugar phosphates, which in turn can be linked together to form polysaccharides. Similarly, pyruvate can be transformed into fatty acids, which, when linked together, produce lipids.

All of these synthetic reactions are by themselves endergonic. Thus, in order to effect cell growth, biosynthetic reactions must be coupled with energy-producing reactions, using ATP as the intermediate energy-storing compound. As might be expected, some macromolecules require more energy for biosynthesis than others. Table 4.1 shows the ATP requirement for the synthesis of 1 gram of cell material broken down by cellular component. This illustrates that much of the total energy required to produce 1 gram of cell material (34.8 mmols ATP) is used just to make proteins. This makes intuitive sense if one considers the variety of enzymes used by the cell and the architectural complexity of each of these. The synthesis of nucleic acids and proteins is discussed in Chapter 5.

4.7.1 Amino Acids

Amino acids form the building blocks of protein molecules used by microbial cells. Many amino acids are taken up directly by the cell as substrates and used directly to build proteins. However, not all necessary amino acids are available in the environment, particularly the nutrient-poor environment of ground-water systems, and must be synthesized by the cell.

There are numerous pathways for the synthesis of amino acids that have been worked out by biochemists over the years. An example of this kind of pathway is shown in Figure 4.15 which shows the synthesis of several amino acids from oxaloacetate and pyruvate. Recall that pyruvate is synthesized in gycolysis (Fig. 4.2) and that oxaloacetate is synthesized in the tricarboxylic acid cycle (Fig. 4.14). Thus, those metabolic pathways not only provide energy production for the cell but also may be used to provide materials needed for amino acid production and, ultimately, protein synthesis.

TABLE 4.1 ATP Requirement for Synthesis of Macromolecules from Glucose

Macromolecule	ATP Required to synthesize macromolecule (mmol ATP/g of cells)
polysaccharide	2.1
protein	19.1
lipid	0.1
RNA	3.5
DNA	0.9

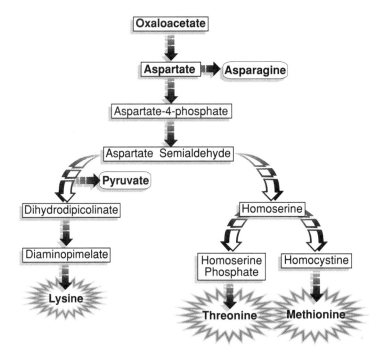

FIGURE 4.15. An example of amino acid synthesis, with pyruvate and oxaloacetate as starting materials. (Modified from Gottschalk, 1986, with permission courtesy of Springer-Verlag.)

4.7.2 Carbohydrates

Microbial cells use carbohydrates for a number of important life functions, most notably in the synthesis of the lipopolysaccharides and peptidoglycan that make up the cell wall. Polysaccharides built from carbohydrates are also used by bacterial cells to attach and adhere to surfaces in their environment, an important characteristic of bacteria that inhabit subsurface environments. Pathways of carbohydrate biosynthesis are therefore anabolic processes that are fundamentally important to the cell.

We have seen how cells are able to extract energy from carbohydrates (glucose) via glycolysis (Fig. 4.2), with a net production of pyruvate. The cell is also able to perform the reverse of this net process, that is, begin with pyruvate and synthesize glucose. The production of glucose from pyruvate is termed *gluconeogenesis* and involves many of the same intermediate products found in glycolysis. Gluconeogenesis does not, however, simply involve reversing each step of glycolysis. Rather, several specialized enzymes are involved.

Pyruvate, obtained from amino acid or fatty acid metabolism, forms the starting material for gluconeogenesis (Fig. 4.16). This pyruvate is then converted to phosphoenolpyruvate (PEP) via the mediation of phosphoenolpyruvate carboxykinase. Recall that, in glycolysis, the reverse of this process is catalyzed by a different

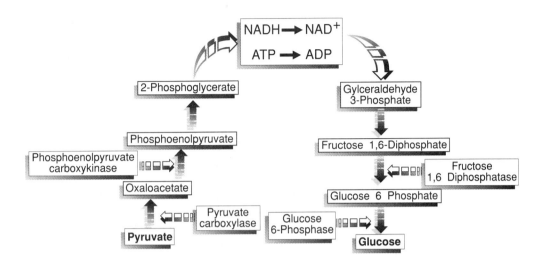

FIGURE 4.16. Synthesis of glucose from pyruvate via gluconeogenesis.

enzyme, pyruvate kinase. The steps leading from PEP to fructose 1,6-diphosphate are identical to the steps in the glycolitic pathway. However, the conversion of fructose 1,6-diphosphate to fructose-6-phosphate and the conversion of glucose-6-phosphate involve different enzymes than in glycolysis.

The different enzymes involved in the glycolytic and gluconeogenic pathways allow the cell to regulate the flow of carbon efficiently. For example, if there is a net deficit of ATP in the cell, the production of enzymes necessary for glycolysis is favored, resulting in the net production of ATP. If, on the other hand, the cell has ample ATP, the production of enzymes involved in gluconeogenesis are stimulated, and synthesis of glucose becomes possible. Thus, the cell will only try and synthesize glucose for cell growth if there is sufficient energy available. This type of feedback mechanism for the production or inhibition of particular metabolic pathways is basic to energy regulation in the cell.

The synthesis of glucose is important, because glucose forms the starting material for many important structural components of the cell. For example, glucose is involved in the production of the hexose amines UDP-N-acetylglucosamine and UDP-N-acetylmuramic acid, which in turn form the building blocks of the cell wall polymer peptidoglycan. The starting material for this biosynthetic pathway is fructose-6-phosphate, which as we have seen is itself synthesized in the glycolitic pathway (Fig. 4.2) or the gluconeogenic pathway (Fig. 4.16). Fructose-6-phosphate then has amine, acetyl, and phosphate groups attached sequentially to form UDP-N-acetylglucosamine. Addition of phosphoenolpyruvate at the third carbon atom of UDP-N-acetylglucosamine, with subsequent reduction of NADPH (a coenzyme closely related to NADH) forms UDP-N-acetymuramate. These two molecules are then transported out of the cell membrane and linked together by peptide bridges in an alternating, cross-linked pattern to form peptidoglycan.

One interesting feature of this topic is that several antibiotics, notably penicillins and cephalosporins, derive their antibiotic properties from inhibiting peptidoglycan synthesis. Penicillin interferes with the peptide cross-linking of UDP-N-acetylglucosamine and UDP-N-acetymuramate. This weakens the cell wall and, if the cell is actively growing, will result in cell lysis. Thus, penicillin will only kill actively growing cells.

4.7.3 Lipids

Lipids are used by microorganisms for a variety of functions in the cell. Lipids are widely used for energy storage in bacteria, and phospholipids are important components of the cell wall. The basic architecture of all lipids involves fatty acids and glycerol molecules bound together by *ester linkages*. The synthesis of fatty acids, therefore, is basic to lipid synthesis.

The overall scheme of fatty acid biosynthesis involves the sequential addition of carbon units derived from acetyl CoA. First, acetyl CoA reacts with CO_2 with the consumption of ATP to produce malonyl CoA. Malonyl CoA then binds with a protein known as acyl carrier protein (ACP). The malonyl-ACP complex then successively contributes two carbon units for the synthesis of fatty acids. As a general rule, fatty acid synthesis involves the consumption of ATP and the use of reducing power in the form of NADPH. Fatty acids, therefore, store considerable chemical energy.

Phospholipids, which form an important component of the outer cell membrane of many bacteria, are synthesized by the addition of fatty acids to glycerol phosphate. Acetyl CoA reacts sequentially with glycerol-3-phosphate to form phosphatidate. Molecules of phosphatidate are then linked together via the mediation of ACP to form elongated phospholipid chains.

4.8 CHEMOLITHOTROPHY

Microorganisms employ three basic strategies for obtaining energy to support life processes. *Heterotrophy,* or utilizing organic compounds as electron donors, is the most widespread strategy for microorganisms living in subsurface environments. *Photosynthesis,* which is extensively used by microorganisms in aquatic and terrestrial surface environments, is obviously not an option for microbes living in the subsurface. The third basic energy-obtaining strategy is *chemolithotrophy*. Chemolithotrophy, which literally means "chemical rock eater," refers to the use of reduced inorganic chemicals such as H_2, H_2S, Fe^{2+}, or NH_3, as electron donors in metabolism.

The ability to obtain electrons from inorganic sources is a great ecological advantage to microbes in environments that are isolated from sunlight. This advantage, of course, is that chemolithotrophs need not be dependent on a photosynthetic food chain to produce the organic carbon compounds needed for energy and growth. This also presents chemolithotrophs with a problem, however. This problem is that they must be able to synthesize organic compounds from CO_2 in order to produce cellular components. Many chemolithotrophs are able to assimilate organic compounds in addition to deriving electrons from inorganic sources. Other chemolithotrophs, however, have biochemical mechanisms for fixing CO_2 (i.e., reducing it to organic carbon compounds) and are able to grow with CO_2 as their sole carbon source.

4.8.1 Hydrogen Oxidizers

Hydrogen, as discussed previously (section 4.3), is produced in a variety of microbial fermentations under anaerobic conditions. The interfaces of anaerobic and aerobic environments may therefore be characterized by the mixing of H_2- and O_2-bearing waters. The reaction

$$2H_2 + O_2 \rightarrow 2H_2O$$

yields 112 kcal of energy and thus has the potential to supply energy for microbial growth. Another important environment in which H_2 oxidizers often thrive is in tectonically active areas. In these areas, H_2 associated with deep hydrothermal activity is sometimes brought to the surface along fault and fracture traces where O_2 is available.

Most hydrogen-oxidizing microbes contain a membrane-bound *hydrogenase* that feeds electrons into the electron transport system, with subsequent ATP synthesis via chemiosmosis. An interesting feature of organisms that use the membrane-bound hydrogenase is that they carry out *reverse electron transfer,* forming NADH from NAD (Fig. 4.17). This NADH is then used by the cell to fix organic carbon compounds from CO_2. This process of CO_2 fixation will be discussed later in more detail. Other hydrogen-oxidizing bacteria, however, are unable to fix CO_2 and also require sources of organic carbon.

Examples of hydrogen-oxidizing bacteria include members of the genus *Hydrogenomonas*. These bacteria are gram-negative rods characterized by the presence of a polar flagella. In addition to growing on H_2, these bacteria are able to grow on some organic acids and alcohols. Hydrogen oxidizers are widely distributed in soils and surface waters. In addition, hydrogen oxidizers have been reported from deep subsurface sediments by Fredrickson et al. (1989).

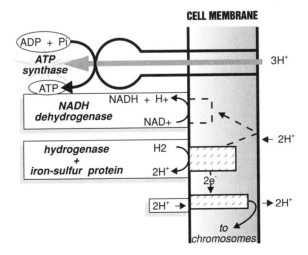

FIGURE 4.17. Reverse electron transfer by hydrogen-oxidizing bacteria. Modified from Gottschalk, 1986, with permission courtesy of Springer-Verlag.

4.8.2 Sulfide Oxidizers

Reduced sulfur compounds, H_2S and elemental sulfur, can be used by a variety of chemolithotrophic bacteria as a source of energy. The reaction

$$2H_2S + O_2 \rightarrow 2H_2O + 2S$$

yields 83 kcal of energy and the reaction

$$2S + O_2 + 2H_2O \rightarrow 2H_2SO_4$$

yields 236 kcal of energy. Thiosulfate ($S_2O_3^=$) may also act as an electron donor.

Representative examples of sulfide-oxidizing bacteria include members of the genus *Thiobacillus*. These bacteria are obligately aerobic, gram-negative, polarly flagellated rods. Many species of *Thiobacillus* are obligate autotrophs, meaning that they are unable to grow with organic carbon as an electron and carbon source. A few species, however, do grow on simple organic compounds. The biochemistry of sulfide oxidation as performed by *Thiobacillus* is summarized in Figure 4.18. Reduced sulfur compounds are complexed with the sulfhydryl group of the tripeptide glutathione. The reduced sulfur is then oxidized to sulfite through the mediation of the enzyme sulfide oxidase. The sulfite is then further oxidized to sulfate with the generation of ATP.

Many members of the genus *Thiobacillus* grow best at approximately neutral pH, a characteristic that may become a disadvantage in some environments, since a byproduct of sulfide oxidation is sulfuric acid. One species in particular, however, *T. thiooxidans*, is adapted to growth at low pH's (2-5) and thrives in the acidic conditions often produced by extensive sulfide oxidation.

Chemolithotrophic sulfur oxidation occurs in a number of important environments. For example, the metabolism of sulfide-oxidizing bacteria in coal spoil piles, with the net production of sulfuric acid, is the principal cause of *acid mine drainage* characteristic of strip-mined areas. The recent spectacular discovery of "black smokers" on the seafloor of tectonically active regions is another important

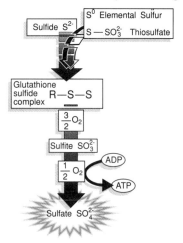

FIGURE 4.18. The biochemistry of sulfide oxidation in *Thiobaccillus*.

habitat of sulfide-oxidizing bacteria. The term "black smoker" refers to the characteristic black clouds that emanate from seafloor vents. These black clouds result from the rapid oxidation of sulfides, which are present in the hydrothermal waters, by means of oxygen present in the seawater. This provides ideal conditions for the growth of sulfide-oxidizing bacteria. The growth of these sulfide oxidizers provides the base of a food chain that includes grazing protozoa, filter feeders, crabs, and fish.

Sulfide-oxidizing bacteria appear to be especially common in subsurface environments. For example, out of 45 samples of sediments cored from deep subsurface sediments (30-260 M) of the South Carolina Coastal Plain, 27 showed active growth of sulfide oxidizers (Fredrickson et al., 1989). The numbers of sulfide oxidizers present varied between about 10^1 and 10^3 cells per gram (dry weight) of sediment. Interestingly, there was evidence that sulfide oxidizers coexisted with viable sulfate-reducing bacteria in these sediments (Jones Beeman, and Suflita, 1989). It's difficult to judge from this evidence if both of these physiologic types of bacteria are active in situ, but it raises the interesting possibility of an active sulfur cycle in subsurface sediments.

4.8.3 Iron Oxidizers

The iron-oxidizing bacteria are well-known among hydrologists and are generally referred to as "iron bacteria". Iron-oxidizing bacteria are responsible for the slimy coatings of ferric iron that sometimes encrust wells, and for this reason, they are the most notorious microbiological pests in the water well industry.

Iron-oxidizing bacteria obtain energy for growth from the exergonic reaction

$$4 Fe^{2+} + 4 H^+ + O_2 \rightarrow Fe^{3+} + 2 H_2O$$

which releases 160 kcal of energy. Iron oxidation obtains relatively little energy per mole of substrate oxidized (40 kcal/mole Fe^{2+}) compared to sulfide oxidation (150 kcal/mole H_2S) or even hydrogen oxidation (56 kcal/mole H_2). Thus, iron oxidizers are obliged to turn large amounts of Fe^{2+} to Fe^{3+} in order to obtain sufficient energy to sustain growth. Because Fe^{3+} forms highly insoluble ferric oxyhydroxides, iron-oxidizing bacteria can quickly clog a water well. This problem is particularly severe if the iron-oxidizing bacteria *Gallionella* is the principal organism present. *Gallionella* produces a sheath made of a slimy material that not only helps clog wells but also is quite unpleasent to look at. Although *Gallionella* is perfectly harmless to humans, few people would be willing to drink water from a well in which this organism was seen to be growing.

Many members of the genus *Thiobacillus,* which are principally known for their ability to oxidize sulfides, also oxidize Fe^{2+}. For example, the species *Thiobacillus* ferrooxidans is distinquished from its obligately sulfide-oxidizing cousin *Thiobacillus* thiooxidans by its ability to grow on Fe^{2+} as well as on H_2S.

Iron-oxidizing bacteria are widely distributed in aquatic sediments. They are particularly well suited for such environments as streambeds which receive base flow from anaerobic, Fe^{2+}-containing ground water. The mixing of this high-iron ground water with oxygen-containing surface water provides an ideal habitat for these microorganisms. Evidence for the activity of such iron-oxidizing bacteria in streambeds is often noticable as a reddish-brown floc that coats sediment grains.

In ground-water systems, the presence of iron oxidizing bacteria in aquifer sediments is indicated by their common presence in water wells. This evidence is not unequivocal since they may be introduced with the drilling process. However, the widely observed presence of sulfide oxidizers and the fact that many sulfide oxidizers also are capable of oxidizing iron suggest that they are a common component of the subsurface microflora.

4.8.4 Ammonia Oxidizing (Nitrifying) Bacteria

The decomposition of plant and animal remains under anaerobic conditions often results in the production of reduced nitrogen compounds such as ammonia (NH_3) and the ammonium cation (NH_4^+), by means of deamination reactions. These compounds can be oxidized to nitrate in the presence of oxygen and nitrifying bacteria. The processes leading to nitrification are of great economic importance to agriculture. It was known for centuries, for example, that composting manure under aerobic conditions increased the manure's fertilizing properties. Winogradsky was the first to show that this resulted from the action of nitrifying bacteria that converted ammonia present in the uncomposted manure to nitrate.

Nitrification involves two distinct types of bacteria that operate in sequence. First, ammonia is oxidized to nitrite (NO_2^-) by such bacteria as *Nitrosomanas*. After this, nitrite is then further oxidized to nitrate by such bacteria as *Nitrobacter*.

Nitrifying bacteria are common in soils, and the activity of these microorganisms has been intensely studied by soil microbiologists. Fredrickson et al. (1989) showed that nitrifying bacteria are present in deep subsurface sediments of the Atlantic Coastal Plain. Out of 45 samples assayed from three different core holes in South Carolina, 13 contained nitrifying bacteria in concentrations of about 10^1 cells per gram of sediment. In all cases, however, numbers of nitrifiers in subsurface sediments were two to three orders of magnitude less than in the overlying surface soils. In these deep aerobic sediments, ammonia concentrations were fairly low (< 0.5 mg/L), and this lack of available substrate is probabaly the chief limiting factor for nitrifying activity. It seems likely that nitrifying bacteria are widely distributed in aerobic ground-water systems.

4.8.5 Autotrophic CO_2 Fixation

Because many autotrophic bacteria do not or can not obtain energy for growth from organic carbon substrates, they must have an alternative source of organic carbon to support cellular growth. In order to produce the needed organic carbon, autotrophs fix CO_2 by means of a special pathway, called the *Calvin cycle*. The Calvin cycle is utilized by both photoautotrophic bacteria and chemolithotrophic bacteria to convert CO_2 into carbohydrates.

The Calvin cycle (Fig. 4.19) begins with the reaction of CO_2 with ribulose 1,5-diphosphate (RuDP) to form a six-carbon compound. This exergonic reaction is catalyzed by the key enzyme ribulose 1,5-diphosphate carboxylase. The six-carbon compound formed by this reaction then splits into two molecules of 3-phosphoglycerate (PGA). Both PGA molecules are then reduced by dihydroxy-

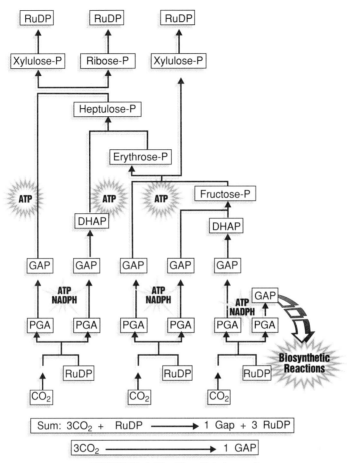

FIGURE 4.19. The Calvin cycle.

acetone phosphate (NADPH), a step that also requires energy in the form of ATP. This step in the process reduces the carbon to the valence state (zero) of carbohydrates. In the Calvin cycle, the coenzyme NADPH performs electron carrier functions as does NADH in metabolic pathways, such as glycolysis. Next, a molecule of RuDP is regenerated, and the cycle repeats. Three turns of the cycle are necessary in order to produce one molecule of glyceraldehyde-3-phosphate. In all, 12 NADPH and 18 ATP are required to synthesize 1 hexose molecule. It's evident from this that the Calvin cycle requires a considerable investment of energy by the autotrophic organism.

The coenzyme NADPH acts as an allosteric activator of RuDP in the Calvin cycle. Thus, if the cell has abundant NADPH, provided either from photosynthesis or from oxidation of inorganic compounds, the cycle will be activated and carbohydrates synthesized. On the other hand, if NADPH is not available from energy-producing reactions, the Calvin cycle is effectively shut off. This allosteric mechanism enables the cells to grow rapidly if a source of oxidizable substrate is present and allows them to conserve energy if such substrates are absent.

4.9 METABOLIC CONTROL OF GEOCHEMICAL PROCESSES

We have seen that living cells extract energy from reduced compounds, such as organic matter, by oxidation; transfer electrons from these oxidation processes to various electron acceptors; and store the energy produced in a form that can be tapped as needed. In order to carry out these functions, energetically favorable reactions ($\Delta G < 0$) are coupled with energetically unfavorable reactions ($\Delta G > 0$). The oxidation of glucose in glycolysis, for example, requires that the cell to first spend 2 molecules of ATP to initiate the reaction sequence and thereby produce a net gain of 2 ATP molecules. These metabolic pathways are regulated by enzymes that catalyze the reactions in a particular sequence and at an appropriate rate.

In studying geochemically important processes that are carried out by microorganisms, it is necessary for researchers to consider limitations imposed by these metabolic processes. Living cells must conform, not only to thermodynamic constraints, but also to biochemical constraints. The mere fact that a particular reaction is energetically favorable is often not sufficient for that reaction to occur. In addition, the cell must have the enzymatic mechanisms to carry the reaction to completion. Because many important geochemical processes are mediated by microorganisms in ground-water systems, it follows that there are biochemical constraints on many geochemical processes as well.

There are many examples of this. Most geochemically important redox reactions-oxygen reduction, nitrate reduction, Fe(III) reduction, sulfate reduction, and methanogenesis-are mediated by microbial biochemical pathways and electron transport systems. In order to understand the occurrence and distribution of these processes in subsurface environments, therefore, it is necessary to understand the biochemical constraints imposed by these mechanisms. Perhaps the best example of enzymatic control of a geochemically important reaction is that of Fe(III) reduction.

High concentrations of dissolved iron (Fe(II)) are very common in anaerobic ground-water systems. Traditionally, the occurrence of Fe(II) in ground water was explained in terms of equilibrium thermodynamics and Eh-pH diagrams (Hem, 1985). Because Fe(III) oxyhydroxides are thermodynamically unstable in the presence of organic compounds under anaerobic conditions, Fe(III) reduction was considered to occur spontaneously and reversibly according to the reaction

$$Fe(OH)_3 + CH_3COOH \rightarrow Fe^{2+} + 2CO_2 + H_2O + 6H^+ \quad (9)$$

It had long been known that particular microorganisms were able to take advantage of this exergonic process in order to obtain some energy for growth. However, with the isolation of strain GS-15, it was apparent that this was a major metabolic strategy of some microorganisms (Lovley and Phillips, 1988). It quickly became apparent that not only could microorganisms obtain energy by iron reduction, but that this reaction required microbial enzymatic catalyzation to proceed at all. In a series of experiments, Lovley, Phillips and Lonergan (1991) showed that none of the most common organic compounds found in natural systems could reduce Fe(III) oxyhydroxides at circumneutral pH (5-9) and low temperatures (25°C), even if the "eH" (redox potential) of the experiment was low. In addition, this reaction did not occur in the presence of a microoganism such as *E. coli* which lacked the essential enzyme system, iron reductase, that mediated the reaction. However, in the presence of GS-15, which did contain iron reductase, these reactions proceeded quickly.

Thus far, this story has proceeded fairly predictably. An important geochemical process, Fe(III) reduction, turns out to require the electron transport system of specialized microorganisms. In this manner, Fe(III) reduction is the same as, say, sulfate reduction. There is, however, a twist. Studies with GS-15 showed that the iron reductase enzyme was *membrane-bound;* that is, it was located in the membrane of the cell. Because Fe(III) oxyhydroxides are solids, Fe(III) cannot diffuse into the cell as an oxyhydroxide. It follows that, for the cell to be able to reduce Fe(III), the cell membrane has to be in direct contact with a solid Fe(III) oxyhydroxide particle. This is, in fact, what is observed for GS-15. If the microorganisms were separated from the solid Fe(III) oxyhydroxides by a semipermeable membrane, no Fe(III) reduction would occur. Thus, iron reduction is subject to several constraints, such as the requirement for physical contact between the Fe(III) and the cell membrane, in addition to the more obvious thermodynamic ones.

It is tempting, sometimes, to think of microoganisms as magic boxes that use fancy biochemistry to perform novel and unlikely processes. Indeed, the ability to couple thermodynamically favorable and unfavorable processes into metabolic pathways enables microbes to do some very unlikely things. This flexibility, however, is obtained at a price. This price is that microbially-mediated processes must conform not only to thermodynamic constraints, but also to constraints imposed by the enzymatic mechanism.

The Fe(III)-reduction example illustrates this point. For Fe(III) reduction to occur in natural systems, microorganisms with the appropriate enzymatic capabilities obviously must be present. However, these microorganisms must also be able to come into physical contact with Fe(III). It follows that Fe(III) present in the crystal structure of silicate minerals, such as biotite, is more difficult for microorganisms to reduce than Fe(III) oxyhydroxides. This last constraint, which limits Fe(III) reduction in many subsurface environments, could never have been deduced in the absence of detailed microbiologic investigations. Similar constraints on other geochemically important microbial processes remain to be worked out. This general topic will be a productive field of research for years to come.

4.10 SUMMARY

The ability to utilize energy in an organized manner is the hallmark of living entities. Metabolism in microorganisms and in the higher plants and animals consists of catabolism, the breaking down of organic molecules to extract energy, and anabolism, using energy to assemble needed proteins and other complex molecules. Energy-producing (exergonic) and energy-consuming (endergonic) reactions are coupled by means of energy-storing intermediate reactants in order to synthesize complex molecules. The most important of these energy-storing intermediate reactants is ATP, but microorganisms use other intermediates, such as GTP, as well.

Many of the energetically favorable reactions used by cells, such as aerobic glucose oxidation, proceed much too slowly at low temperatures to sustain life. Cells solve this problem through the use of enzymes, proteins that catalyze chemical reactions. Enzymes work by combining chemically with a substrate or a combination of substrates, and by bringing them together in a configuration so that reaction can occur. Enzymes are classified according to the kinds of reactions that they catalyze. Oxidases and reductases catalyze oxidation-reduction reactions, lyases remove functional groups from organic compounds, and ligases combine compounds together.

The regulation of enzymatic activity is crucial to the cell, and this is accomplished by means of a number of mechanisms. The production of enzymes can be induced or repressed at the genetic level by the presence or absence of particular substrates. Enzyme systems are also regulated by feedback inhibition mechanisms, in which the product of an enzymatic reaction acts to inhibit the function of that particular enzyme.

The two basic metabolic strategies available to microorganisms are fermentation and respiration. The basis for both strategies is using electron transfers to perform work. In fermentation, however, electrons are retained in organic carbon compounds or molecular hydrogen. In respiration, electrons are transferred to mineral terminal electron acceptors, such as oxygen, nitrate, ferric iron, sulfate, or carbon dioxide. Respiration is much more efficient than fermentation, producing a much greater yield of ATP per mole of substrate oxidized than fermentation.

Microorganisms use energy obtained from catabolic processes to build the proteins, polysaccharides, lipids, and nucleic acids needed to repair damaged cell parts and to grow new cellular material. The basic strategy of the cell in building these compounds is to use intermediate products of metabolism as building blocks. Pyruvate, for example, which may be produced by both fermentative and respirative processes, may be used to build new polysaccharides, proteins, and lipids.

Numerous geochemically important processes in subsurface environments, oxygen reduction, Fe(III) reduction, sulfate reduction, and methanogenesis are carried out exclusively by enzymatically controlled microbial processes. Because of this, these geochemical processes are subject to biochemical as well as thermodynamic constraints. For example, the enzymatic reduction of Fe(III) oxyhydroxides requires physical contact between Fe(III) and a cell membrane for the reaction to go to completion. Similar biochemical constraints on other geochemically important processes are less well understood but are probably equally important.

CHAPTER 5

BACTERIAL GENETICS

In Chapter 4 we saw how the ability to use energy in an organized fashion is the hallmark of living organisms. By coupling energy-yielding and energy-consuming chemical reactions, the cell is able to synthesize the macromolecules of which it is composed and maintain its inherently unstable structure. However, it takes more than just available energy for these complex biochemical reactions to occur. There must be a "master plan" that organizes the sequence of reactions and supervises the construction of cellular components. This plan must contain all of the information needed to build each protein, carbohydrate, lipid, and nucleic acid present in the cell. Furthermore, this plan must be self-regulating, in the sense that only those cellular components needed at a particular time will be actually made and used. Finally, this plan must allow the organism to reproduce itself. In all living organisms, the information for this "master plan" is coded into nucleic acids that are present in each cell.

The term "gene", which refers to a unit of heredity, has been used for about a century. In 1900, Hugo De Vries, K.E. Correns, and Erich Tschermak-Seysenegg discovered the rules of heredity, only to learn that they had been worked out independently 40 years earlier by Gregor Mendel. These investigators discovered, as had Mendel, that separate characteristics are inherited independently. Early geneticists rationalized these observations by postulating the presence of particles, or genes, that carried each characteristic from generation to generation. It is one of the curiosities of history, however, that the term "gene" was used for at least 50 years before anyone had any idea what it was made of.

In the first half of the twentieth century, there were two schools of thought as to what constituted genes. One school favored proteins as carrying genetic information, and the other school favored nucleic acids. It wasn't until the structure of deoxyribonucleic acid (DNA) was deduced in 1953 by Francis Crick and James Watson that it became clear that genes were, in fact, composed of nucleic acids.

5.1 DNA—ITS STRUCTURE AND ORGANIZATION

The function of DNA, deoxyribonucleic acid, in most microorganisms is to store the genetic information needed to synthesize cellular components. One exception to this rule is the case of some viruses (the HIV virus that causes AIDS is an example) that use RNA to store genetic information. The chief property of DNA that suits it to this information-storing role is it's highly ordered structure. Figure 5.1 shows the double-helix structure of DNA. In this structure, sugar molecules (deoxyribose) are linked together by phosphate groups to form two long strands. These two strands are linked together by hydrogen bonds between nitrogenous bases, and the entire structure is twisted into a helix. The structure is analogous to a spiral staircase, with the sugars and phosphates forming the two rails and the bases forming the steps.

The order of the structure is inherent in the positioning of the nitrogenous bases. The *sequence* of bases along each strand of DNA is capable of storing information in a manner exactly analogous to a code. In the Morse code, for example, letters are coded as sequences of dots and dashes. The code for *S* is dot-dot-dot and the

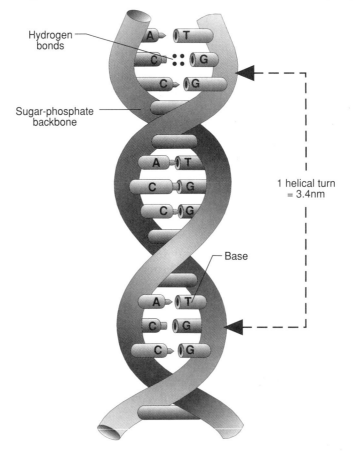

DNA double helix

FIGURE 5.1. Modified from Norton, 1986. Courtesy of the Addison-Wesley Publishing Co.

code for *O* is dash-dash-dash. Thus, the coded expression dot-dot-dot dash-dash-dash dot-dot-dot is immediately recognizable as *SOS*. In the *genetic code,* sequences of bases are used instead of dots and dashes. Specifically, sequences of bases on DNA code for the formation of particular protein molecules needed for cell maintenance and growth.

In living cells, however, simply the ability to store information is not sufficient. Living cells must also have the ability to reproduce, which means that copies of DNA must be made. The ability of DNA to make copies of itself, or *replicate,* is inherent in its double helix structure and on the specificity of base pairs. Adenine on one strand of DNA binds only with thymine on the other strand. Similarly, guanine on one strand can bind only with cytosine on the other strand. Thus, if the DNA molecule is "unwound," each single strand is capable of exactly reproducing its complementary strand.

This process of replication is shown schematically in Figure 5.2. First, a double strand of DNA is unwound through the action of specialized enzymes. Each strand is then replicated by linking the complementary bases A-T and G-C with the mediation of *DNA polymerase.* As the new DNA is synthesized, sections of the new molecule are linked together by enzymes known as *DNA ligases,* enzymes that attach DNA at particular places. DNA ligases also have the important task of "editing" the new DNA, that is, checking to see that the new copy exactly matches the *template* from which it was made. If errors were made, DNA ligases are also capable of making last-minute repairs. The net result of these complex replication processes is two exact copies of the original DNA molecule.

The nucleotide sequences present on a DNA molecule do not all code for specific proteins. Rather, it is now clear that DNA consists of functional areas, termed *operons,* that are separated by "start" and "stop" sequences (Fig. 5.3). The operons are often called *structural regions,* and the start, promoter, and stop sequences are called *regulatory regions.* Furthermore, only one strand of DNA, called the *informational strand,* is actively involved in protein synthesis. The other strand, called the *complementary strand,* is used only for DNA replication.

In bacteria, DNA that is involved in cell replication is called *chromosomal DNA.* Chromosomal DNA in the bacteria is different from that in eucaryotic organisms in that it is organized into closed loops. Chromosomal DNA is present in the nucleoid region of the bacterial cell and is seldom visible through a light microscope. Occasionally, the nucleoid region becomes visible after the cell has doubled its complement of DNA in preparation for cell division.

Bacterial cells often contain DNA that is not part of the chromosomes. This DNA, called *plasmid DNA,* is present as small loops embedded in the cell's cytoplasm. Plasmid DNA is copied much like chromosomal DNA but is not involved directly in cell replication. Rather, plasmid DNA codes for proteins that confer specific attributes to a cell. For example, many bacterial species contain plasmids that confer antibiotic resistance. Other plasmids may give the cell the ability to use specific organic compounds as substrates. Thus, plasmids are directly involved with the ability of an organism to deal with its environment. Plasmids have the important property of being able to transfer between cells. Thus, it is possible for a cell to obtain specific functions by plasmid transfer without modifying its chromosomal genome. This property, which is the basis of much genetic engineering of microorganisms, allows bacteria to respond to environmental stresses efficiently.

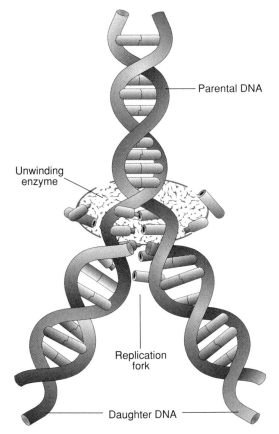

FIGURE 5.2. Replication of DNA. Modified from Norton (1986). Courtesy of the Addison-Wesley Publishing Co.

5.2 RNA—ITS STRUCTURE AND ORGANIZATION

DNA carries the blueprints for all of the components needed by a living cell. However, the translation of this information into specific proteins involves another class of nucleic acids, called ribonucleic acids, or RNA.

There are three major classes of RNA. *Messenger RNA* (mRNA) is *transcribed* directly from DNA and carries the genetic information from each gene to a place where the protein molecule can be assembled. *Transfer RNA* (tRNA) carries amino acids to the site of protein synthesis and delivers them to the appropriate locations on the growing protein molecule. *Ribosomal RNA* (rRNA) forms part of the ribosomes that are the surfaces upon which proteins are assembled.

5.2.1 Transcription

The process of *transcription,* or making mRNA, is shown schematically in Figure 5.3. First, an enzyme called *RNA polymerase* binds to a region on a DNA molecule that starts the coding of a particular genetic sequence. RNA polymerase unwinds a portion of the DNA molecule, and RNA nucleotides base-pair to the exposed

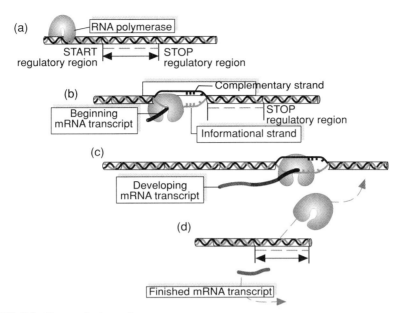

FIGURE 5.3. Transcription of mRNA from the DNA template. Modified from Norton (1986). Courtesy of the Addison-Wesley Publishing Co.

DNA. Unlike DNA replication, however, RNA contains uracil in the place of adenine. Thus, RNA contains no thymine, which binds with adenine in DNA. The process of transcription continues until the RNA polymerase reaches a region on the DNA molecule that codes for the process to stop. The completed mRNA then moves to the ribosome to begin protein assembly.

5.2.2 Translation—Making Proteins

Molecules of tRNA and rRNA are transcribed from particular portions of the DNA molecule in an analogous manner. The function of tRNA is to carry amino acids and to deliver them to the appropriate location on mRNA. To this end, tRNA molecules have sites designed to bind only to specific sequences composed of three bases, called *codons,* on mRNA (Fig. 5.4). The portion of the tRNA molecule that binds to the mRNA is thus called an *anticodon*. The other end of the tRNA molecule has a binding site specific to a particular amino acid. Thus, when a tRNA molecule binds to a particular codon on mRNA, it always carries the amino acid specific for that codon. In this manner, amino acids can be delivered in the exact sequence in which the mRNA is coded.

Like any code, however, the genetic code requires a "key" in order to interpret messages. After the discovery of the structure of DNA in 1953, several groups of researchers worked on this problem. It was Francis Crick and his coworkers who first correctly proposed that three nitrogenous bases coded for a single amino acid, which was an important step in breaking the genetic code. The next step was to determine exactly which amino acids were coded for by each codon. Several research groups contributed to this work, and the 1968 Nobel Prize was jointly

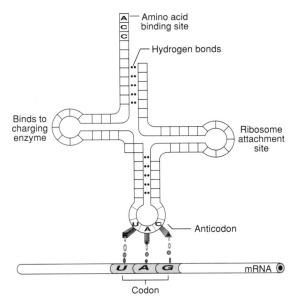

FIGURE 5.4. The structure of tRNA. Modified from Norton (1986). Courtesy of the Addison-Wesley Publishing Co.

awarded to Robert Holley, H. G. Khorana, and Marshall Nirenberg for their work on breaking the genetic code.

Table 5.1 shows the key for the genetic code. Suppose, for example, a base sequence on mRNA is

CUU-AGA-AAA-UUU-AGU-GGG-ACU-UCU-UAA

The translation of this code into a protein chain, according to the key shown in Table 5.1, would be

Leu-Arg-Lys-Phe-Gly-Thr-Ser

Protein synthesis, in addition to being able to translate the genetic code, requires an assembly area where amino acids can be physically linked together. These protein assembly areas are the ribosomes. Ribosomes are assembled from rRNA and associated proteins. The large subunit of a ribosome contains 2 strands of RNA and 32 proteins. The small subunit contains a single RNA strand and 21 proteins. Once the necessary RNA molecules have been synthesized, assembly of the protein can commence.

The process of assembling a protein molecule from the information coded in mRNA is called *translation*. The overall idea is to translate the sequence of nucleotides in mRNA to a sequence of amino acids in the protein. The process of translation is shown in Figure 5.5. First, the mRNA binds to the small subunit of the ribosome. Next, a tRNA molecule carrying the amino acid formylmethionine binds to the large ribosomal subunit. There are only two tRNA binding sites on a ribosome, and a tRNA molecule always initially binds to the first site where a peptide is attached to the amino acid. Once the peptide is attached, the tRNA

TABLE 5.1 Key for Translating the Genetic Code

Codon	Amino Acid	Codon	Amino Acid
UUU	Phe	UCU	Ser
UUC	Phe	UCC	Ser
UUA	Leu	UCA	Ser
UUG	Leu	UCG	Ser
CUU	Leu	AGU	Ser
CUC	Leu	AGC	Ser
CUA	Leu	CCU	Pro
CUG	Leu	CCC	Pro
AUU	Ile	CCA	Pro
AUC	Ile	CCG	Pro
AUA	Ile	ACU	Thr
AUG	Met	ACC	Thr
GUU	Val	ACA	Thr
GUC	Val	ACG	Thr
GUA	Val	GCU	Ala
GUG	Val	GCC	Ala
GCG	Ala	UAC	Tyr
GCA	Ala	CAU	His
UGU	Cys	CAC	His
UGC	Cys	CAA	Gln
UGG	Try	CAG	Gln
CGU	Arg	AAU	Asn
CGC	Arg	AAC	Asn
CGA	Arg	AAA	Lys
CGG	Arg	AAG	Lys
AGA	Arg	GAU	Asp
AGG	Arg	GAC	Asp
GGU	Gly	GAA	Glu
GGC	Gly	GAG	Glu
GGA	Gly	UGA	Nonsense*
GGG	Gly	UAA	Nonsense*
UAU	Tyr	UAG	Nonsense*

molecule moves to the next site, providing space for another tRNA molecule to attach, bringing the next amino acid in the sequence. The two amino acids are then linked by the peptide bond. The first tRNA molecule is released, the second tRNA moves over, and another tRNA molecule bringing the next amino acid arrives.

The sequence in which the amino acids are linked is controlled by the interaction of tRNA with the template on the mRNA. Each codon on the mRNA codes for one, and only one, amino acid (Table 5.1). However, the converse of this is not true, since several different codons code for the same amino acid. Thus, the codons UUU (uracil-uracil-uracil) and UUC (uracil-uracil-cytocine) both code for the amino acid L-phenylalanine. Similarly, the codons GCU and GCC code for L-alanine. The mRNA template is bound to the small subunit of the ribosome; as the tRNA molecules arrive, only those carrying the next amino acid required in the sequence are able to bind to the mRNA and hence the large subunit. Once the

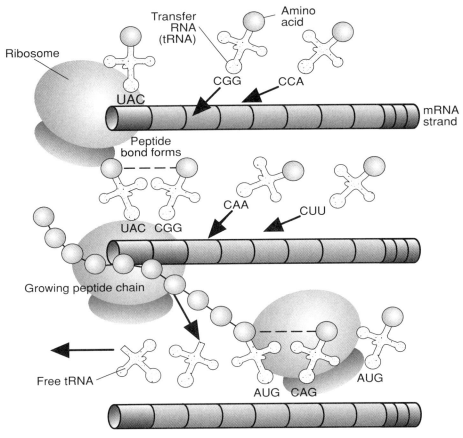

FIGURE 5.5. Translation of the genetic code from mRNA to the production of particular proteins. Modified from Norton (1986). Courtesy of the Addison-Wesley Publishing Co.

appropriate amino acid has been delivered and the peptide linkages made, the mRNA slides to the next codon, allowing the next amino acid to be delivered by tRNA. This process continues until a "stop" codon is encountered on the mRNA (UAA, UAG, or UGA), and the protein is complete. When the "stop" codon is reached, the protein is released from the ribosome. It may then be modified by forming or breaking disulfide bridges or be combined with lipids or polysaccharides before it takes on it's biological role in the cell.

5.3 GENE EXPRESSION AND REGULATION

In addition to having the genetic information to build specific proteins and having mechanisms for transcribing this information and translating to the amino sequences required, it is necessary for the cell to regulate protein production. At one extreme, it is obvious that the cell need not make all the proteins that it is capable of making all the time. At the other extreme, the genetic information does little good if it cannot be "turned on" when required. Thus, the regulation of gene expression is a critical function of the cell.

5.3.1 Induction

There are numerous regulatory mechanisms used by bacterial cells. Perhaps the best studied is that of the lac-operon in *E. coli*. *E. coli* is capable of producing enzymes that allow it to metabolize the sugar lactose. However, because lactose is not always available as a substrate, it would be inefficient for *E. coli* to make these enzymes continually; rather, it is more efficient for the cell to be able to sense the presence of lactose when it becomes available, and subsequently to synthesize the lactose-degrading enzymes. After the lactose has been consumed and the enzymes are no longer needed, it would be advantageous for the cell to cease making the enzymes.

E. coli is, in fact, able to regulate production of these lactose-metabolizing enzymes by means of the lac-operon. The lac-operon, which is present in the chromosomal DNA of the cell, consists of three distinct regions. These regions are called the *regulator,* the *promoter,* and the *operator* regions. The regulator region codes for formation of a protein, called a *repressor protein,* that binds to the promoter region of the operon. This prevents RNA polymerase from binding to that portion of the DNA and prevents transcription of the operator that codes for the lactose-metabolizing enzymes. Thus, as long as the repressor protein is in place, *E. coli* will not be able to metabolize lactose.

The repressor protein is designed, however, to be removed from the DNA strand if an *inducer* binds to it. In the case of the lac-operon, the inducer is allolactose, a derivative of lactose. When allolactose binds to the repressor protein, the protein is removed. RNA polymerase can then bind to the promoter region of the operon and transcribe the operator region. Following transcription, the enzymes for metabolizing lactose are assembled, and lactose is then used as a substrate.

Once the available lactose has been consumed, allolactose is no longer available to bind the repressor protein. The repressor protein binds to the promoter region of the lac-operon, and production of lactose-metabolizing enzymes ceases. The enzymes that degrade lactose are referred to as *inducible enzymes,* since they can be induced and manufactured as required by the cell.

5.3.2 Repression

In addition to being able to induce production of enzymes in order to metabolize specific substrates, it is also frequently advantageous for the cell to be able to repress certain metabolic functions. For example, *E. coli* is able to obtain more energy for growth by using glucose as a substrate than it can obtain from lactose. Thus, if both glucose and lactose are available, *E. coli* would obviously be better off using the glucose first. The cell accomplishes this by means of *catabolite repression.*

Catabolite repression acts by means of the promoter region of the DNA. The cell normally produces a number of specialized proteins, called *catabolite activators.*

Catabolite activators bind to promoter regions of DNA, greatly increase the affinity for RNA polymerase, and promote the synthesis of particular enzymes. In *E. coli,* these promoter activators must be bound to cyclic AMP (adenosine monophosphate) and normally encourage production of enzymes for metabolizing sugars, such as lactose, arabinose, and galactose. However, when concentrations of glucose in the cell are high, the cell produces ATP via the glycolytic pathway, and concentrations of cyclic AMP are correspondingly low. Cyclic AMP is thus unable to bind to the catabolite activators and the transcription of the lactose-arabinose-galactose-metabolizing enzymes is depressed. This, in turn, allows the cell to concentrate on utilizing glucose via glycolysis. Once glucose becomes depleted, cyclic AMP levels increase, the activator proteins are again able to bind to promoter regions, and production of the lactose-arabinose-galactose enzymes may resume.

Bacterial cells exhibit a very complex interplay between induction and repression of particular metabolic patterns. The efficient interplay of these processes is greatly encouraged by competition between bacteria for scarce resources. Obviously, microorganisms with the most efficient combination of inducible enzyme systems, together with appropriate repression mechanisms, will have a substantial competitive advantage in nature. In ground-water systems, where there is fierce competition for limited resources, such mechanisms have undoubtedly selected for extremely efficient microorganisms. The nature of many of these specialized induction-repression mechanisms, however, is only understood at a rudimentary level. This is a topic of considerable interest for future research.

5.4 MUTATIONS

During the process of DNA replication, the bacterial cell employs several biochemical mechanisms to ensure that mistakes are not made. For example, DNA polymerase enzymes in procaryotes, the enzymes that attach complementary nitrogenous bases to the replicating DNA molecule, often exhibit exonuclease activity; that is, they have the ability to depolymerize a nucleic-acid chain. This allows the polymerase to correct errors in base sequences as needed. Similarly, procaryotic organisms have ligase enzymes whose sole function is to repair "nicks" in the newly synthesized DNA molecule.

One environmental factor that can damage DNA is radiation. Exposure to certain kinds of radiation may cause the formation of covalent bonds between adjacent pairs of thymine on the same strand of DNA. These *thymine dimers* are harmful, because they prevent the proper functioning of DNA polymerases. Many bacteria, particularly those commonly exposed to sunlight, have evolved mechanisms for repairing this kind of damage. In one mechanism, endonuclease enzymes remove the dimer, polymerase enzymes fill in the gap, and ligase enzymes attach the two ends of the DNA molecule.

These repair mechanisms, together with the inherent reliability of the base-pairing design of DNA, lead to an extremely low rate of errors during DNA synthesis. It is generally observed that copying errors occur at the rate of less than 1 per 100 million base pairs copied.

In spite of this accuracy, however, mistakes do occur. When a single base in the DNA strand is changed, it is referred to as a *point mutation*. There are several kinds of point mutations. A *substitution mutation* is when a particular base, say guanine (G), is incorrectly replaced by a different base such as thymine (T). This type of error has the potential for changing one codon. For example, if the sequence AAG on a DNA molecule is mistakenly changed to AAT, the resulting mRNA will no longer code for the amino acid phenylalanine but instead will code for leucine. This difference may be sufficient to inactivate the resulting protein. In addition, because several codons code for the same amino acid, all point mutations do not necessarily result in changes to proteins.

In addition to substitutions, it is possible for a particular base to be added or deleted from the sequence, called *addition* and *deletion* mutations, respectively. These kinds of mutations are potentially devastating to the microorganism, because they result in a frame shift in how the genetic code is read. Because the code is read in three-base units, the addition or deletion of one unit changes all of the codons in that particular sequence of DNA. This generally results in the formation of a completely inactive protein.

5.4.1 Mutagenic Agents

Mutations often occur as random events, due only to physical errors made by DNA polymerases. However, certain chemicals and radiation have the potential to interfere with the normal functioning of DNA replication and can cause mutations. For example, under ordinary circumstances, cytosine pairs only with guanine in a DNA molecule. However, hydroxylamine chemically alters cytosine to uracil so that, instead of pairing with guanine, it pairs with adenine. This results in the GC pair being replaced with an AT pair after the DNA molecule reproduces. Hydroxylamine, therefore, is referred to as a *chemical mutagen*.

In addition to chemically altering nitrogenous bases, several chemicals have configurations similar to nucleotides and can replace them in the DNA structure. For example, 5-bromo-uracil can replace thymine in the DNA structure and pair with adenine. This kind of chemical reaction results in substitution mutations and greatly increases rates of mutation.

The ability of particular chemicals to produce mutations is often evaluated by means of the *Ames test*. In this procedure, a strain of the bacterium *Salmonella typhimurium* is exposed to the chemical being tested in a range of concentrations and is grown on a medium that lacks histidine. Under normal conditions, *S. typhimurium* cannot synthesize histidine and will not grow. However, at particular concentrations of mutagenic agents, a mutation occurs that allows the bacterium to synthesize histidine and grow on the medium and to produce visible colonies. The Ames test is widely used to test chemicals for mutagenic activity. Because mutations in the human genome may result in cancer, the Ames test is also widely used to determine if a chemical is a potential carcinogen.

Chemicals are not the only source of mutations in bacteria. Radiation, particularly high-energy radiation such as X-rays, can physically break DNA and cause substitution mutations. In addition, ultraviolet light can create covalent linkages between bases on the same strand of DNA producing thymine dimers. If not repaired, this damage can lead to catastrophic mutation and death of the microorganism. In fact, high doses of ultraviolet light are sometimes used as a sterilization procedure.

5.4.2 Transposable Genetic Material

The locations of particular nucleotide sequences on a chromosome were once thought to be fixed. However, it is now known that particular segments of the bacterial chromosome are capable of moving from place to place. One kind of transposable genetic agent is known as an *insertion sequence*. These genetic agents have what are termed inverted repeated terminal sequences; that is, they have an identical nucleotide sequence on both ends, such as

$$G G G T C T G \ldots \ldots C A G A C C C$$
$$C C C A G A C \ldots \ldots G T C T G G G$$

This configuration, for reasons not well-understood at the molecular level, allows the sequence to move around the chromosome inserting itself at different locations. There appear to be particular enzymes involved in translocating insertion sequences, so that base pairing is not the only factor involved in translocation.

Insertion sequences have several possible effects on the expression of an organism's genome. For example, if an insertion sequence is placed in the middle of an operative gene, the protein coded for by that gene will be altered. Similarly, if the insertion sequence is placed at the promoter region of a gene, expression of that gene will be altered.

Insertion sequences are fairly small genetic elements and do not appear to code for particular proteins. Their function in the biology of cells is not always clear, but they may be involved in specifying sites at which DNA recombination may occur. Larger transposable genetic elements that do code for specific proteins are called *transposons*. Many of the transposons that have been studied code for antibiotic resistance. Furthermore, their ability to detach and to insert themselves in another portion of the DNA molecule has led to their use in genetic engineering applications.

The evolutionary significance of transposable genetic agents in cell biology remains controversial. It is not clear whether these kinds of genetic agents confer a biochemical advantage to individual cells or whether they are simply a mechanism for introducing variability into the expression of genes. It is interesting to note, however, that a new copy of the transposable agent is generated during transposition. This, in some cases, leads to multiple copies of the insertion sequence throughout the genome.

5.5 NATURAL GENETIC EXCHANGES

Most procaryotic cells reproduce by binary fission, in which the two daughter cells formed are genetically identical to the parent. However, genetic uniformity is often not in an organism's best interest. Natural selection, the mechanism that allows populations to adapt to changing conditions, requires a source of genetic variability to act upon. Sexual reproduction, which is practiced by some eucaryotic cells, is one method for introducing genetic variability. However, this mechanism is not available to procaryotic microorganisms. As we saw earlier, some genetic variability is conferred by mutations. However, more common means for introducing genetic variability to bacterial populations involves natural genetic exchanges.

5.5.1 Recombination

The term *recombination* refers to any process by which DNA from two separate cells is combined into a single cell. When recombination occurs, the resulting microorganism retains some of the characteristic of each donor cell. Thus, recombination serves to introduce genetic variability into microbial populations.

When bacterial cells undergo recombination, genetic material from a donor cell is transferred to a *recipient* cell, and the DNA is physically combined. The process by which the combination of DNA from different organisms is carried out is called *integration*. The actual mechanisms involved with integration are very complex and remain the subject of active research. However, there are basic steps involved. First, the delivery of DNA is made from the donor to the recipient cell. Second, the two sets of DNA line up to match sites where the DNA can be cleaved with endonucleases. For integration to occur, the strands of DNA must be roughly *homologous;* that is, the complementary base sequences in the two strands must be similar, although they need not be identical. Third, endonucleases break the two DNA strands at roughly compatible sites. Finally, the donor DNA is joined with the recipient DNA by DNA ligases. Once the DNA has been integrated, it becomes part of the genome of the recipient microorganism and is replicated during cell division. The daughter cells therefore will possess features of both the donor and the recipient cells.

Transformation There are several ways in which DNA is physically transferred among microorganisms, initiating recombination. *Transformation* is the simplest mechanism for DNA transfer. In transformation, DNA is released from the donor cell when it dies and is then taken up by a living cell and integrated into its genome. Transformation has been observed in a number of different genera of bacteria under laboratory conditions and certainly occurs in the environment. Although transformation appears to be most common between closely related strains, it is possible that interspecies transformation also occurs in nature.

Conjugation Plasmids, extrachromosomal loops of DNA, are capable of conferring the ability for some bacteria to "mate," or exchange DNA, with adjacent cells. The process by which bacteria cells exchange DNA by this means is called *conjugation*.

A closely studied example of this novel behavior is the *F (fertility) plasmid* in some strains of *E. coli*. The F plasmid codes for the formation of proteinaeous pili, often called sex pili, that serve to connect two cells. Strains of *E. coli* that have the F plasmid (F^+) are able to form a sex pilus and attach to cells that lack the F plasmid (F^-). Once the connection is made, DNA from the F^+ cell is transferred to the F^- cell and either incorporated into the chromosomal DNA of the recipient cell or maintained as an independent plasmid. When the F plasmid is incorporated into the chromosomal DNA, the resulting strain retains the capability for forming sex pili and is referred to as an Hfr (high frequency of recombination) strain.

Hfr strains maintain the ability to form sex pili and hence the ability to transfer DNA. In this case, however, a portion of the chromosomal DNA, rather than just the F plasmid portion, is transferred. In fact, the portion of DNA transferred often lacks the F plasmid so that the resulting cell usually remains F^-. However, the recipient cell has obtained a large portion of the donor cell's DNA, and daughter cells exhibit a corresponding share of the donor's characteristics.

Transduction   In addition to genetic transfers between bacteria by means of transformation and conjugation, genetic material may also be transferred between bacteria by viruses. Many kinds of viruses, termed *bacteriophages,* are parasites of bacteria. In the process of parasitizing bacteria, viruses may act to combine genetic material from different cells.

When viruses are involved in genetic transfers between bacteria, the process is referred to as *transduction*. When a virus infects a bacterial cell, the viral DNA often integrates with the chromosomal DNA of the bacterium. When the virus has reproduced itself and leaves the cell, it occasionally carries portions of the host's chromosomal DNA. When the virus infects a second host cell, these fragments of DNA are integrated into the new host.

Transduction by bacteriophage lambda, a virus that infects *E. coli* cells, has been extensively studied. Lambda is capable of entering the cell and incorporating its DNA into the genome of its host. When lambda enters a cell, it binds to the host DNA at one particular place, between the genes coding for galactose and biotin production. Normally, when the virus leaves the host DNA, it detaches from the chromosome, leaving it unchanged. Occasionally, however, the genes for galactose or biotin are detached with lambda and can be carried to a new *E. coli* cell. Once in the new cell, the lambda virus carrying, say, the galactose genes is defective and cannot attach to its normal binding site. However, if a normal lambda phage attaches to the binding site, the galactose end of the defective lambda can bind with the normal phage and the normal end of the defective lambda can bind with the chromosomal DNA. This produces an *E. coli* variant that contains two sets of genes coding for the production of galactose.

5.6   GENETIC ENGINEERING

In the previous section, we saw that microorganisms can transfer genetic information among themselves in the absence of sexual reproduction. These genetic transfers are advantageous to individual species, as they increase genetic variability and provide leeway for species to evolve in the face of changing environmental conditions. Since the 1970's, however, microbiologists have learned how to manipulate these natural DNA-transferring mechanisms in the laboratory. This has enabled them to insert specific genetic capabilities into specific microorganisms. This, in turn, enables microbiolgists to "program" cells to produce particular products or to carry out useful functions. The technology that enables humans to manipulate the genomes of bacterial cells in this manner is called *genetic engineering.*

To many people, the term "genetic engineering" brings to mind images of Frankenstein and of scientists creating other horrible monsters. The truth is more mundane. Genetic engineering simply redirects natural processes of transformation, conjugation, or transduction in order to place certain genes in the genome of certain bacteria. This process serves exactly the same function as selective breeding in plants or animals. The difference is that, instead of focusing on the macroscopic exchange of genetic information (breeding), the focus is on exchanging genetic information at the molecular level. Furthermore, all of the techniques for engineering microorganisms are based on naturally occurring processes of genetic transfer.

5.6.1 Plasmids

The occurrence of plasmid DNA in bacterial cells has been recognized since about 1960. The first studies of plasmids were motivated by the observation of antibiotic resistance in strains of pathogenic bacteria. These "R factors" present in the bacteria could be traced by epidemiological (epidemiology is the study of disease patterns) techniques, implying that some sort of infectious agent conferred the antibiotic resistance. It was quickly realized that these agents were very similar to viruses, since they were capable of replicating themselves within bacteria and were capable of moving between bacterial cells. In fact, plasmids capable of transferring between cells by means of conjugation have been termed "epiviruses." Many microbiologists do not make a logical distinction between plasmids, bacteriophages, transposons, or insertion sequences, preferring, rather, to consider them as one family of unnamed parasitic organisms. In this view, the only difference between viruses and plasmids is that viruses may produce many copies of themselves within bacterial cells, whereas plasmids (as well as transposons and insertion sequences) generally replicate as the bacterial cell replicates.

Putting philosophical questions aside, the fact is that many genetic-engineering techniques are based on manipulating plasmid DNA in bacteria. Plasmid DNA has the advantage of being easy to isolate from cells, easy to manipulate as desired, and easy to reintroduce into other cells. The first step is to lyse the cells and concentrate the plasmid DNA. In the second step, the plasmid DNA is cut at particular points by *restriction endonucleases.* These restriction enzymes bind to specific nucleotide sequences and cut the DNA molecule at that point. An important feature of restriction enzymes is that they break the DNA molecule unevenly, leaving "sticky ends," or ends that have the ability to base-pair. Because restriction enzymes act only on specific sequences, different DNA's can be cut so that they have complementary sticky ends. If different DNA's that have been treated with the same restriction enzyme are mixed, the different fragments will bind together at the sticky ends. These fragments can then be covalently bonded together by treating the mixture with DNA ligase.

This procedure, termed *in vitro recombination,* is shown schematically in Figure 5.6. First, a plasmid and a segment of foreign DNA are treated with the same restriction endonuclease. Ideally, the plasmid is cleaved at one site, whereas the

120 BACTERIAL GENETICS

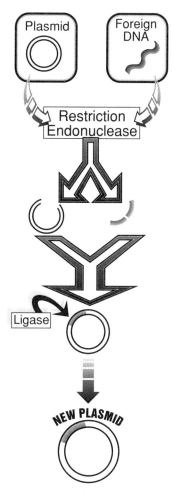

FIGURE 5.6. An example of *in vitro recombination*.

foreign DNA is cleaved at two locations. This allows the foreign DNA to be inserted into the plasmid. The foreign genes are then covalently bonded into the new plasmid by means of DNA polymerases.

5.6.2 Vectors

Once the new molecule of DNA has been synthesized, the next step is to insert it into a cell. As before, these insertion procedures are based on processes carried out naturally by the cell. DNA can sometimes be introduced into cells by transformation. The most common technique for transformation is to treat the recipient bacteria with calcium chloride to make the cell more permeable to DNA. The mechanism of this transformation is neither well understood nor particularly efficient. However, transformation of *E. coli* cells with some small plasmids yields a success rate of only about 1 in 10,000 to 1 in 100,000 DNA molecules. For some applications, this low success rate will suffice. For other applications, however, the new plasmid is carried directly into the target cell by active transport systems called *vectors*. This greatly increases the efficiency of genetic transfer.

Plasmid Vectors Recall that some plasmids, such as the F plasmid in *E. coli,* carry genetic instructions for conjugation and can move from cell to cell by means of pili. These plasmids can be engineered to carry foreign DNA into cells directly. These plasmids, which are very useful in genetic-engineering applications, are called *plasmid vectors*.

The strategy in using plasmids for carrying foreign genetic material into a new cell involves several steps. First, the plasmid is cleaved at an appropriate place and the genes to be transferred are inserted. Next, the plasmid is inserted into a host cell by means of transformation. Once transformed, the plasmid-containing cells are capable of conjugation with adjacent cells and subsequent DNA transfer. Most plasmid vectors are specific for *E. coli*. However, a few plasmids, called *promiscuous plasmids,* are capable of existing in other gram-negative bacteria. Promiscuous plasmids are important, because they extend cloning technology beyond *E. coli* to other genera, such as *Pseudomonas*.

Once inserted, the first step is to identify those cells that contain the new genetic material. A common technique for such identification is to place antibiotic resistance genes on the plasmid vector. After conjugation, the resulting hybrids are plated onto media containing antibiotic. Only cells containing the desired plasmid can grow on the antibiotic-laced media, and they can be readily purified. Once the transformed cells are purified, they are allowed to reproduce, which creates numerous copies of the new plasmid. The process of mass-producing copies of the plasmid is termed *cloning,* as the idea is to obtain identical copies of the DNA. The cloned DNA can then be concentrated and inserted into additional cells as needed.

Phage Vectors There are many different known bacteriophages. However, only lambda has been extensively used as a vector in genetic engineering technologies. A major drawback of using lambda as a vector is that it limits the size of the foreign DNA insert. Two major types of lambda are used in genetic engineering. The first has a single restriction enzyme site within a known gene. This allows insertion of foreign DNA into a particular site. When DNA is cloned into this site, the gene is inactivated, allowing the new phenotype to be identified. The second type of phage vector has two targets for the restriction enzyme. This allows replacement of part of the phage's DNA with foreign DNA. If the foreign DNA has a readily recognizable phenotype or, alternatively, if the deleted DNA sequence is recognizable, the transduced cells can be readily identified.

5.7 APPLICATIONS OF RECOMBINANT DNA TECHNOLOGY

Genetic-engineering technology has created a whole new industry since the mid-1970's. Prior to these technologies, often the only way to mass-produce certain biomolecules was by means of cumbersome chemical techniques, followed by extensive purification procedures. With the advent of recombinant DNA technology, however, it became possible to direct living cells to produce the biochemicals.

5.7.1 Insulin Production

Perhaps the best-known example of how recombinant DNA technology has been used for practical benefit is the production of human insulin. Prior to 1983, all insulin used by human diabetics was manufactured from the pancreases of slaughtered cows or sheep. Unfortunately, cow insulin is slightly different from human insulin, and many diabetics became allergic to it. The genes for making human insulin were recovered from pancreatic cells, recombined with plasmids, and inserted into *E. coli* cells. The plasmids enabled the *E. coli* cells to mass-produce human insulin, which could be recovered by harvesting the cells. Presently, much of the insulin used by diabetics in the United States is produced by this technology.

5.7.2 Pollution Control Technology

There has been much interest in producing genetically engineered cells for environmental applications. One of the first efforts in this direction was the work of Ananda Chakrabarty and his associates in engineering microorganisms to contain hydrocarbon pollution. More recent efforts have been focused on engineering microorganisms to degrade a wide range of xenobiotic halogenated hydrocarbons.

Aliphatic Petroleum Hydrocarbons The steps taken by Chakrabarty in the early 1970's to engineer an aliphatic hydrocarbon-degrading bacterium are a good example of how recombinant DNA techniques have been used in environmental applications. The investigators knew that certain strains of *Pseudomonas* had the capability for degrading particular aliphatic hydrocarbons. Specifically, they isolated a strain of *Pseudomonas aeruginosa,* designated strain AC59, that rapidly degraded C_6, C_8, and C_{10} aliphatic hydrocarbons but that was not capable of degrading C_{12} or C_{14} hydrocarbons. Strain AC59 harbored a plasmid, termed the "OCT plasmid" (for "octane"), that conferred this ability. Next, they isolated another *P. aeruginosa* strain, designated strain AC63, that grew rapidly on C_{12} and C_{14} aliphatic hydrocarbons but that could not grow on C_6, C_8, or C_{10} hydrocarbons. The ability to utilize these shorter molecules was conferred by the DOD plasmid.

Unfortunately, both of these plasmids were non-transmissible. However, it was possible to splice the OCT plasmid with a transmissible CAM plasmid. The CAM-OCT plasmid, when transferred to strain AC63 resulted in a strain capable of metabolizing aliphatic hydrocarbons from C_6 through C_{18}.

Chakrabarty's research is notable for historical and legal reasons, as well as for technical reasons. Chakrabarty planned to use his engineered microorganism to help remediate oil spills. However, once introduced into the environment, the microorganism could conceivably be reisolated by other individuals and used at other spills without Chakrabarty's permission. Therefore, Chakrabarty applied for patent protection for his engineered microorganism. This patent application was challenged, and the issue was placed before the U.S. Supreme Court. In 1980, in *Diamond v. Chakrabarty,* the Supreme Court ruled that the microorganism constituted a "manufactured" product and therefore was subject to patent protection. This ruling gave a tremendous boost to applying recombinant DNA techniques to environmental problems, because it affirmed that those investing in engineering useful microorganisms could expect to have their investments protected.

Aromatic Hydrocarbons Since the early 1970's, an astonishing amount of information has been learned about the manner in which plasmids encode for particular catabolic processes. One particular plasmid that has been extensively studied is the TOL plasmid, which codes for the degradation of toluene, xylene, and related aromatic hydrocarbons. Various configurations of this plasmid occur in species of the genus *Pseudomonas*.

There are at least 11 separate steps in the overall process, each of which are catalyzed by different enzymes. The "upper" pathway enzymes, which are involved in oxidizing the methyl group of toluene to form benzoate, are coded for in one operon. The "lower" pathway enzymes, which are involved in cleaving the ring, are coded in another operon. The pathway is therefore organized into two functional parts: (1) the conversion of the hydrocarbons to aromatic carboxylic acids and (2) the conversion of these carboxylic acids to aliphatic metabolites. The production of the enzymes involved in this pathway are induced by toluene, xylene, and related compounds.

In addition to illustrating the complexity of plasmid-mediated degradation pathways, the TOL plasmid is a good example of how plasmids can be manipulated in order to confer desired traits on bacterial strains. One good example of these kinds of manipulations was worked out in K.N. Timmis' laboratory in the early 1980's. These investigators described how they "recruited" genes from the TOL plasmid for the lower pathway enzymes, and by a series of transfers succeeded in building a new plasmid (Fig. 5.7). They started with a plasmid capable of degrading 3-chlorobenzoate (i.e. 3CB+) and attempted to add genes from the TOL plasmid that would confer the ability to degrade 4-chlorobenzoate (4CB). The problem was to be able to transfer both the degradative genes, D and L, with the promoter gene, S. Without the promoter gene S, the degradative genes would not be expressed and therefore would be of no use. Unfortunately, cleaving the TOL plasmid with the s endonuclease separated the degradative genes from promoter genes.

In order to solve this problem, the S gene was isolated by cleavage with the x endonuclease (fig. 5.7b) and inserted into an intermediate plasmid that already coded for the degradation of 3-chlorobenzoate (Fig. 5.7c). The D L E G genes were cleaved separately and inserted into a different intermediate plasmid (Fig. 5.7d). These genes were then cloned, copied by allowing the *Pseudomonad* strains to reproduce, in preparation for the final step. In the final step, the S gene, together with the genes for degrading 3CB, were removed from one intermediate plasmid with the H endonuclease (Fig. 5.7e) and inserted into the other plasmid at a single H endonuclease site (Fig. 5.7f). The resulting plasmid could catabolize 4-chlorobenzoate in addition to 3-chlorobenzoate. The net result, therefore, was an increase in the degradative capacity for the plasmid.

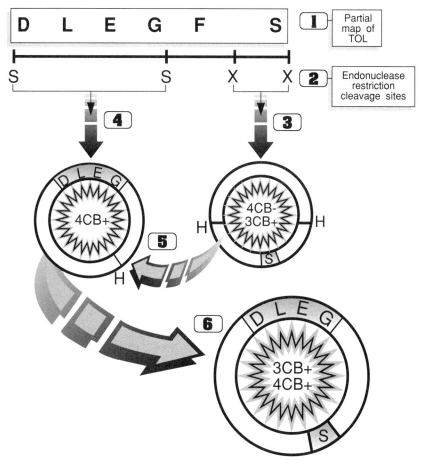

FIGURE 5.7. Construction of a new degradative plasmid using the techniques of genetic engineering.

5.8 DNA TECHNOLOGY IN SUBSURFACE MICROBIOLOGY

The manipulation of bacterial DNA has several applications in subsurface microbiology. For example, we shall see (Chapter 5) how biochemical tests can be used to evaluate diversity in subsurface microbial populations. However, the biochemical capabilities of bacteria (i.e. the ability to metabolize particular compounds) is genetically determined. Thus, a more direct method of establishing diversity among microorganisms is to compare the "relatedness," or homology, of different isolates.

5.8.1 DNA Homology and Diversity

There are a number of methods available for comparing homology between bacterial isolates. One method is to prepare DNA "probes" for particular isolates. In this method, DNA is taken from a particular strain, labeled with [3H]thymidine, and broken into short single-stranded units. Next, DNA is taken from the organism to which homology is to be tested and broken into similar short, single-stranded units. Finally, the radiolabeled "probe" DNA is added to the test DNA. Because single-stranded DNA can only recombine with identical or nearly identical nucleotide sequences, the degree to which the probe DNA combines with the test DNA is proportional to the genetic similarity between the strains.

An application of this technology to bacteria from the deep subsurface has been given by Jimenez (1990). In this case, the question of primary interest was the diversity exhibited by subsurface bacteria in a coastal plain aquifer system. Biochemical testing suggested a high degree of diversity (Balkwill, Fredrickson, and Thomas, 1989). However, as it was not possible to document all of the metabolic capabilities of particular isolates, actual diversity may have been underestimated.

In order to address this possibility, the DNA homology between three strains for which DNA probes were prepared (strains A0481, B0703, and C0397) and forty strains of subsurface isolates were tested. Also tested was one strain of *E. coli* from the American Type Culture Collection. Interestingly, the results of the DNA homology showed considerably more diversity among bacterial strains than was apparent from biochemical testing. For example, strains B0617, B0703, and B0725 were all isolated from the Middendorf aquifer and were all identified as *Pseudomonas aeruginosa* on the basis of biochemical testing. However, two of these strains showed only about 13 % homology with the probe made from B0703, indicating that the strains had been misidentified on the basis of biochemical testing. Furthermore, the homology data indicate considerably more diversity in the isolated bacteria than was indicated from the biochemical testing.

5.8.2 Plasmids and DNA Probes

The ability of plasmids to reproduce is intimately tied to the reproductive success of their host microorganisms. Those plasmids that confer functions that are useful in particular environments will tend to enhance the survival of the microorganism- and thus the plasmid in those environments. Conversely, plasmids that confer functions not needed for a particular environment may actually be a hindrance to microorganisms if energy resources are spent on unproductive functions. Thus, the kind of plasmids inhabiting microorganisms from a particular environment may give some information about the kinds of stresses encountered.

The best example of this general principle comes, not from natural systems, but from hospitals. Hospitals are an environment in which microorganisms are much more likely to encounter antibiotics than, say, pristine aquatic sediments. Not surprisingly, since the advent of antibiotics in 1945 the incidence of plasmid-mediated antibiotic resistance has increased enormously in hospitals. Those plasmids that conferred antibiotic resistance gave their host microorganisms a competitive advantage over strains that lacked the plasmid. Consequently, both the plasmid-bearing microorganism and the plasmid itself were able to thrive where other bacteria were simply killed. This resistance does carry a price, however. In some cases, it has been documented that some hospital microorganisms are so loaded down with resistance-encoding plasmids and so used to not having to compete with other microorganisms that they are unable to survive outside of the hospital environment.

Plasmids are commonly present in soil, aquatic sediment, and aquatic bacteria. For example, about 50% of the marine bacteria living at the air-water interface exhibit the presence of plasmids. In shallow aquatic sediments, generally about 10% of isolates contain plasmid DNA. Since most plasmids present in environmental strains of bacteria have not been characterized, it's very difficult to say what useful functions, if any, these plasmids perform. Some studies of have shown that plasmids are more common in polluted environments than in pristine ones. For example, it has been found that hydrocarbon-polluted waters in the Gulf of Mexico contained a higher incidence of plasmids than pristine waters. However, this is not always observed. Burton, Day, and Bull (1982) found that there was no difference in plasmid abundance in polluted river sediments versus pristine sediments. Clearly, the success of microorganisms in different environments reflects many factors in addition to plasmid occurrence.

There have been relatively few studies of plasmid occurrence and function on subsurface bacteria. Ogunseitan et al. (1987) found that up to 8% of isolates from shallow pristine aquifers contained plasmids. Furthermore, these investigators found some evidence that plasmid occurrence increased (up to 35 %) in ground water that was contaminated with aromatic hydrocarbons.

The first comprehensive study of plasmid occurrence in deep subsurface sediments was conducted by Fredrickson et al. (1988). Interestingly, these investigators found that the frequency of plasmid occurrence tended to increase with depth. In the relatively shallow Upland and Tobacco Road Sand units, the percentage of stains containing plasmids were 14% and 16% respectively. In the deeper sediments of the Black Creek and Middendorf aquifers, the percentage of plasmid-bearing strains increased to 39% and 38% respectively.

Documenting the presence of plasmids in these subsurface bacteria is one thing, but documenting the functions (if any) of the plasmids is quite a different, and extremely difficult, question to address. The first obvious question is whether the plasmids conferred antibiotic resistance. The resistance of the bacteria to various antibiotics was measured by the disc plate method. This involves plating the bacteria onto media and placing discs containing certain concentrations of antibiotics on the plates. Those strains resistant to the antibiotics can grow next to the discs. Conversely, those strains sensitive to the antibiotics cannot grow next to the discs.

A high percentage (70%) of the strains tested by Fredrickson et al. (1988) showed some level of resistance to penicillin whereas a low percentage (5%) showed resistance to tetracycline. Significantly, 100% of the penecillin-resistant strains showed beta-lactamase activity. Beta-lactamase is an enzyme that inactivates penicillin and high levels of beta-lactamase activity are normally associated with plasmids. It is possible, therefore, that some of the antibiotic resistance found in the subsurface isolates was plasmid conferred.

Fredrickson et al. (1988) were particularly interested in whether the observed plasmids conferred unusual catabolic properties on the bacteria. This, as one might imagine, is a particularly difficult question to address. Their approach was to probe the plasmid DNA for homology with a well-characterized TOL plasmid (pWWO). First, the TOL plasmid DNA was concentrated and treated with chemicals to "unwind" the DNA structure. The DNA was then allowed to hybridize with nucleic acids that had been labeled with radioactive phosphate (^{32}P). The radiolabeled, single strands of DNA were then used as "probes"; that is, they were added to single-stranded plasmid DNA and examined for homology, indicating that the plasmid DNA was similar to the TOL plasmid.

A number of the subsurface strains assayed by Fredrickson et al. (1988) exhibited significant homology with the TOL plasmid. While it was unlikely that this indicated that the bacteria had been exposed to toluene or xylene in the recent past, it did suggest a more intriguing hypothesis: The subsurface bacteria were adapted to metabolizing recalcitrant, aromatic hydrocarbons that are present naturally in these sediments. As the enzymes required to mineralize these aromatics could be similar to those required to mineralize toluene, this could indicate similar pathways used by the subsurface bacteria.

The study by Fredrickson et al. (1988) is a good example of how DNA probes are potentially useful for studies of subsurface microbiology. DNA probes can be prepared to assay for the presence or absence of many metabolic processes. This technique, in turn, could prove useful for a variety of applications in subsurface microbiology.

5.8.3 Release of Engineered Microorganisms

There has been considerable debate among microbiologists concerning the possible release of genetically engineered microorganisms (GEMs) into the environment. The principal fear is that GEMs, since they would be new to the immune systems of plants, animals, and humans, could produce catastrophic epidemics. This fear has been greatly compounded by the fact that most of genetic- engineering research and technology involves *E. coli*, a normal component of the human digestive tract. Genetically engineered strains, so goes the argument, might not be recognized and attacked by the human immune system. This could lead, in turn, to massive infection and death.

This kind of scenario plays to some very basic fears that human beings have about infectious diseases. In the 1970's, a popular novel entitled *The Andromeda Strain* told a tale about the ultimate pathogenic microorganism. This strain was totally new to human and animal immune systems, was capable of wildly rapid reproduction, and rendered virtually instant death to whoever it infected. Furthermore, it mutated so rapidly that developing a vaccine for it was out of the question. In the book, however, the microorganism was outwitted by a wily physician who realized that the pH optima for growth was very narrow. Thus, infection could be prevented by slightly altering the pH of blood.

Recent experience with the AIDS virus has demonstrated that concern about extremely virulent microorganisms is warranted. The question is whether engineering microorganisms and releasing them into the environment is a significant threat to human, plant, or animal health.

There are several reasons why such microorganisms are unlikely to produce health threats. First of all, it has been observed since the time of Pasteur that cultures of pathogenic microorganisms tend to lose virulence if they are maintained in a laboratory culture. Strains of microorganisms used for genetic engineering purposes are essentially adapted to the laboratory environment. There is, therefore, very little possibility that such microorganisms would become spontaneously pathogenic. However, this "little possibility" must be carefully evaluated. In the late 1980's, experiments with the survival of genetically engineered bacteria in the soil environment were undertaken at Clemson University. The results of these studies, which are not yet complete, will have much bearing on the possible future release of genetically engineered microorganisms to the environment.

The general observation that laboratory-adapted microorganisms survive poorly when they are released into the environment raises another issue concerning the use of genetic-engineering technology for environmental pollution remediation. The process of engineering microorganisms for specific functions, say PCB degradation, is expensive. Suppose a microorganism is produced that can degrade PCB in the laboratory. What are the chances that that microorganism could survive the rigors of the polluted soil environment to carry out its mission? Considering the cost of developing such microorganisms, this question must also be carefully evaluated before such technologies can be applied.

The general, and entirely proper, attitude taken by the majority of microbiologists is that questions of survival, pathogenic properties, and viability must be carefully evaluated. If the technology proves to offer an acceptable risk/benefit ratio, then the technology should be used. The burden of proof, however, rests with those who would utilize the technology.

5.9 SUMMARY

The information required for the metabolism, growth, and reproduction of bacterial cells is stored in chromosomal and plasmid DNA present in each cell. DNA is capable of reproducing exact copies of itself-a property that allows genetic information to be efficiently passed from generation to generation. Genetic information is coded into DNA by the particular sequence of nitrogenous bases. This information is then translated to produce specific proteins by means of several different kinds of RNA. Messenger RNA is transcribed directly from the DNA template and carries the information to a protein assembly site. Transfer RNA physically carries the specific amino acids to the mRNA template where they are assembled in the required order with the mediation of ribosomal RNA.

The expression of genes is carefully regulated in the cell by a number of mechanisms. The production of certain enzymes can be induced if the substrate for that enzyme becomes available from the environment. Alternatively, if a substrate is not present, production of degradative enzymes can be repressed. It is common for the production and repression of enzymes to be regulated by feedback inhibition loops, in which the substrate for a particular reaction induces enzyme production and where the product of that reaction acts as a repressor. In this manner, cells are able to balance enzyme production with their changing needs.

Genetic variability is introduced to bacterial cells by a number of processes including mutations and transposable genetic material. Genetic exchanges between microorganisms is effected by a number of processes, including transformation (DNA from a dead cell being taken up by a live cell), conjugation (DNA transferred by means of quasi sexual joining of cells), and transduction (DNA transferred by viruses). These natural processes, when manipulated in the laboratory, form the basis for genetic-engineering technologies.

Plasmid DNA is often the focus of genetic-engineering technologies. These relatively short DNA molecules often confer useful biochemical functions, such as the capacity to degrade xenobiotic chemicals. Furthermore, because their DNA can be cut up and spliced into new combinations, it is possible to "engineer" microorganisms with useful characteristics. In one of the first applications of this technology during the 1970's, Chakrabarty engineered microorganisms with the ability to degrade the relatively long hydrocarbons associated with hydrocarbon spills. This research, in turn, led to the extension of patent protection to genetically engineered microorganisms.

Genetic technologies can be used in subsurface microbiology applications in a number of ways. The use of particular gene probes-sequences of DNA coding for particular functions-can be utilized to assay bacteria for particular enzymatic capabilities. The use of DNA homology techniques have been used to evaluate the diversity of microorganisms isolated from subsurface environments. It is possible that the production of genetically engineered microorganisms (GEMs) will be applied to subsurface pollution problems in the near future.

CHAPTER 6

MICROBIAL ECOLOGY OF GROUND-WATER SYSTEMS

Ecology is that branch of biology dealing with relationships between organisms and between organisms and their physical surroundings. An ecological *community* consists of all the organisms-microbes, plants, and animals-that live in an area. A community, together with the physical environment to which it is tied, is called an *ecosystem*. The major focus of ecological research is to determine how communities are structured, how species in the community interact with each other, and how communities interact with their physical surroundings.

In order to simplify the investigation of communities, ecologists usually study some subset of the organisms present. For example, an ecologist might be mainly concerned with a particular taxonomic group, such as birds. Thus, one might read about "bird communities." Similarly, microbial ecologists are mainly concerned with relationships between microorganisms, and so they focus on "microbial communities." Given this considerable simplification, determining how organisms in particular communities interact with each other and their environment might appear to be straightforward. In practice, however, there are often considerable technical and interpretive problems to be overcome.

Consider, for example, a particularly simple ecological question—one concerning blackbirds in marsh environments of the northwestern United States. It is observed that, during the mating season, part of a particular marsh is populated exclusively by Yellow-headed blackbirds and another part of the marsh populated by Red-winged blackbirds. The ecological question is, what factor or combination of factors lead to the observed segregation of the two populations?

In order to answer this question, the ecologist would begin by making observations on how the two species interact. In this case, the ecologist would observe that, in the spring, male Red-winged Blackbirds arrive in the marsh several weeks before its cousin, the Yellow-headed Blackbird. The Red-winged males promptly stake out all of the best nesting sites, which consist of cattail clumps surrounded by deep water. This insures a constant supply of insects to feed upon while the nestlings are growing. However, when the larger male Yellow-headed Blackbirds arrive in a couple of weeks, they forcibly evict the Red-wings from the best sites. This forces the Red-winged Blackbirds to retreat to less desirable sites. By the time the females arrive and mating starts, the Yellowheaded Blackbirds inhabit one part of the marsh-the part with the most food-totally excluding Red-winged Blackbirds. For their part, the Red-headed Blackbirds inhabit an entirely different, less productive, part of the marsh. The two populations do not overlap.

Given these observations, it's perfectly clear how the Yellowheaded Blackbirds exclude Red-winged Blackbirds—they kick them out. Less clear is why the Yellow-headed Blackbirds do not also inhabit the poorer nesting sites. To answer this question, the ecologist would have to observe the rate at which Yellow-headed Blackbirds successfully raise their chicks as a function of the productivity of the nesting site. It turns out that the physical size of the Yellow-headed Blackbirds, in addition to providing the advantage in one-on-one encounters with Red-winged Blackbirds, also carries a price. They are not able to raise their young successfully at nesting sites that do not produce sufficient food. The young Yellow-headed Blackbirds simply require too much food. The Red-headed Blackbirds, however, with their smaller size can successfully support offspring in the poorer nesting sites. The redwings exploit an environment closed to the yellowheads, proving that competition can cut two ways. The yellowheads have a competitive advantage in direct confrontation between the two species. The redheads, on the other hand, have a competitive advantage in exploiting available habitat.

Having made these observations, the next challenge to the ecologist is to determine the evolutionary factors that led to the observed relationship between the two species. The Yellow-headed Blackbirds are bigger, perhaps a result of natural selection favoring the ability to oust Red-winged Blackbirds from high-quality areas. On the other hand, the size of the yellowheads might reflect a completely different factor, such as a relatively low temperature in it's winter range, or perhaps less vulnerability to predators. In this case the Yellow-headed Blackbirds ability to outcompete redwings is unrelated to the actual competition between the species.

For their part, are the Red-winged Blackbirds smaller because they are adapted to inhabiting poorer nesting sites? Alternatively, does the smaller size of the redwings reflect something unrelated to nesting, such as migration habits or food availability in their winter range? These kinds of questions, central to the study of ecology, are very difficult to answer with confidence and require careful study. The relationship between the two blackbird species used here as an example, for instance, is still a matter investigation and debate.

Given the inherent problems associated with unraveling relationships between species such as birds—species that can be observed directly—the formidable difficulties faced by the microbial ecologist studying subsurface environments can be more readily appreciated. The microbial ecologist faces the same difficult questions concerning relationships between microorganisms and their environments, but in this case it is very difficult to directly observe either the microorganisms or the environments that they inhabit. Furthermore, the sheer number of microbial species inhabiting many subsurface environments can be daunting. Microbial ecologists have only recently begun to examine relationships between microorganisms in ground-water systems (Ghiorse and Wilson, 1988). Nevertheless, the special problems that arise in the study of these unique ecosystems place subsurface microbial ecology as an important subdiscipline of both microbiology and ecology. This chapter surveys our present understanding of microbial ecology of ground-water systems.

6.1 SCOPE OF SUBSURFACE MICROBIAL ECOLOGY

The microbial ecology of many natural environments has been carefully investigated over the years. Examples of environments that have received particular attention include soils and aquatic sediments. It is no accident that these environments have been selected for intensive study. The microbial ecology of soils is directly related to the agricultural productivity of farmland. Understanding the microbial ecology of soils and how microbial populations adjust to different agricultural practices has helped boost agricultural production in the twentieth century. Thus, the microbial ecology of soils has been closely studied because it is a problem that has major economic applications.

Similarly, the microbial ecology of aquatic sediments has received considerable attention because of practical reasons. In the nineteenth and early twentieth centuries, rivers and streams in the United States and Europe were routinely used for direct waste disposal. It was observed, however, that if raw sewage was placed in a river, the resulting pollution decreased rapidly downstream. Furthermore it was observed that microbial processes in the fluvial sediments were largely responsible for consuming the dissolved organic carbon and nutrients in the sewage, thus renovating the river. Understanding the nature of these pollution-attenuating processes, as well as understanding their rates and capacities was therefore an important practical problem.

The microbial ecology of subsurface environments has emerged as an important topic because of similar practical considerations. In the 1970's, it became apparent that there was widespread contamination of shallow water table aquifers due to a variety of human activities including waste-disposal, agricultural practices, and fuel storage. Gradually, it became clear that subsurface microorganisms played an important role in the mobility and fate of many contaminants (McNabb and Dunlap, 1975; Freeze and Cherry, 1979; Matthess, 1982). As such, understanding the ecological interactions between microorganisms in the subsurface assumed much more importance than had been placed on it earlier in the twentieth century (Ghiorse and Wilson, 1988).

The scope of microbial ecology as applied to the terrestrial subsurface is closely tied to the occurrence of ground water. While ground water by definition can only be produced from aquifers, confining beds (also called aquicludes) play an important role in shaping ground-water flow patterns and have important effects on ground-water chemistry (Back, 1986). Thus, the microbial ecology of subsurface environments has come to include both water-bearing aquifer sediments and water-retarding aquiclude sediments. In addition, unsaturated sediments overlying water table aquifers generally are included in the scope of subsurface microbial ecology. To some degree, this introduces some overlap with traditional soil microbial ecology. However, subsurface microbial ecologists tend to focus on unsaturated sediments below the root zone, whereas soil microbial ecologists focus on the root zone.

The questions considered in the microbial ecology of the terrestrial subsurface are similar to those considered by aquatic-sediment microbial ecologists. Energy, most of which comes originally from the sun, is trapped by photosynthetic plants in organic carbon compounds and flows through an ecosystem. The subsurface microbial ecologist seeks to eulcidate the pathways through which the flow of carbon and energy passes, and the processes that control the flow. Closely linked to the flow of carbon are the flow of nutrients, such as phosphorus and nitrogen, and electron acceptors that are coupled to carbon oxidation.

One important regulating mechanism of carbon flow in subsurface environments is microbial *food chains*. It is impossible for any single microbe to have the enzymatic capability for degrading all of the organic carbon compounds present in sediments. Thus, it is common for particular microbes to extract energy from organic carbon by performing specialized metabolic processes and transforming the carbon. The transformed carbon compounds may then be used by other microbes as primary substrates.

A simple example of how microbial food chains may regulate carbon and energy flow in a water-table aquifer is shown in Figure 6.1. In this example, the chain begins with the fermentation of lignin under anaerobic conditions, perhaps by *Clostridia*, to produce molecular hydrogen and carbon dioxide. Some of the hydrogen and carbon dioxide produced may diffuse upward to an oxygenated zone near the water table and be utilized by hydrogen-oxidizing bacteria. Under anaerobic conditions in which mineral electron acceptors such as nitrate, Fe(III), or sulfate are

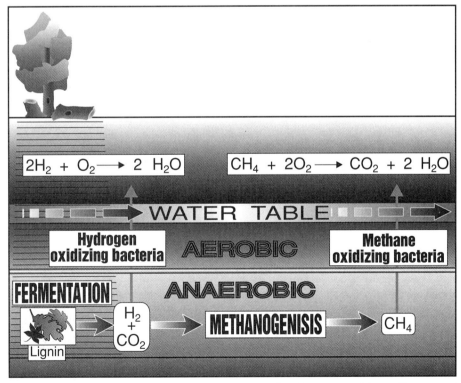

FIGURE 6.1. An example of how carbon and energy flow in a water table aquifer proceeds via food chains.

absent, methanogenic bacteria take up a portion of the hydrogen and carbon dioxide to produce methane. Some of the methane produced in this manner may then diffuse upward to oxygenated water near the water table and be oxidized to carbon dioxide.

Because sediments typically contain many different kinds of organic compounds that are potential substrates for microbes, it is commonly observed that numerous food chains operate parallel to each other. For example, the lignin present in sediments typically consists of many classes of complex organic compounds including aliphatic side chains and carbohydrates linked to aromatic molecules. Metabolizing these different compounds requires specific enzymatic capabilities, not all of which are present in particular fermentative microorganisms. Thus, a number of different species may be present, which not only attack certain classes of compounds but also produce different fermentation products. These different fermentation products, in turn, form different food chains.

The summation of these ongoing, parallel food chains give rise to the community structure of a subsurface environment. This *community structure* may be fairly simple, with only a few species actively metabolizing in a given system. Alternatively, the community structure may be very complex with hundreds of strains of microbes involved in organic matter degradation. Determining the community structure of aquifer systems is a major goal of the subsurface microbial ecologist.

Also of primary importance to the microbial ecologist is determining the *trophic levels* associated with subsurface environments. In classical ecology, the concept of a *trophic pyramid* is widely used (Fig. 6.2a). In this model, the base of the pyramid is made up of photosynthetic primary producers such as plankton or plants. The next level of the pyramid contains the primary consumers, principally plant-eating herbivores. At the same trophic level are the decomposers, microorganisms (fungi, bacteria) that oxidize the organic matter. The next trophic level is filled by secondary consumers (predators) that feed on the primary consumers and decomposers. There may be additional trophic levels filled by progressively higher animals.

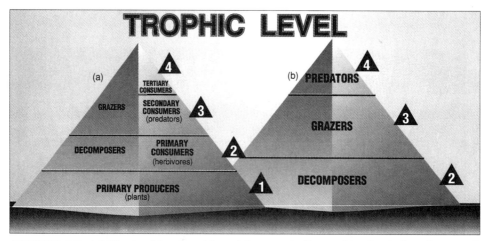

FIGURE 6.2. (a) The trophic pyramid models as typically used in classical ecology, and (b) trophic levels present in ground-water systems.

Subsurface environments are fundamentally different from many ecosystems in that the number of trophic levels present are greatly restricted. For example, there can be no photosynthetic primary production in subsurface environments isolated from sunlight. It is possible for primary production to be based on chemolithotrophy, and this occurs in some subsurface environments. However, that is the exception rather than the rule. In most subsurface systems, there is a greatly attenuated trophic structure (Fig. 6.2b). Because there is no primary production and no primary consumers, decomposers, largely bacteria but also including fungi, form the base of the trophic pyramid. Superimposed on the decomposer level is a secondary consumer level. This trophic level is largely filled by protozoa that may actively graze on the decomposers, although it is possible that some predatory bacteria may be present in some systems. There may be, in some systems, higher trophic levels present. An example of these higher trophic levels in ground-water systems would be the blind fish that inhabit some carbonate aquifers. These fish feed on protozoa that in turn feed upon the decomposers.

Understanding the energy flow, food chains, community structure, and trophic levels of different subsurface environments are important goals in subsurface microbial ecology. The motivation for achieving these goals is varied and depends upon the systems under consideration. For example, in a contaminated shallow water table aquifer, the effects of protozoan predation on decomposer populations-and therefore rates of contaminant biodegradation-might be an important consideration. For a clastic aquifer system consisting of alternating beds of sand and clay, the important ecological question might be how community structure differs between the two sediment types and how these differences affect carbon and energy flow. The focus of an ecological study for any subsurface environment therefore depends largely on the questions being asked. The nature of the relevant questions is then a major factor in selecting appropriate methods of study.

6.2 METHODS IN SUBSURFACE MICROBIAL ECOLOGY

Microorganisms can only be directly observed using a microscope. In order to achieve the resolution required to observe bacteria using either light or electron microscopy, only a very small portion of the microbial habitat can be observed at one time. Aquifer systems that compose the habitat of subsurface microorganisms are often hundreds of square miles in area and may be hundreds of feet thick. Clearly, if the decision is made to observe bacteria in an aquifer directly, it will not be possible to observe the aquifer itself in a meaningful way. This difference in scale between the microorganisms and their habitat is a basic conflict in the study of microbial ecology in subsurface environments. In practice, this conflict can be partially resolved by adjusting the scale of the observation technique used to the scale of the feature of interest. To use an example familiar to geoscientists, one cannot observe cross-bedding by looking at thin sections of sandstone with a petrographic microscope. Such features can be observed only at the scale of a hand specimen or outcrop. Conversely, one cannot observe individual sand grains in hand specimen, one must use a petrographic microscope. In general, therefore, the scale of observation must match the scale of the feature that is of primary interest.

136 MICROBIAL ECOLOGY OF GROUND-WATER SYSTEMS

This scale conflict is especially important in subsurface microbial ecology. In selecting a scale suitable for observing microorganisms, little information concerning the microbial environment (i.e. the aquifer system) can be gleaned. Alternatively, if a scale suitable for observing a large aquifer system is selected, little direct information concerning individual microorganisms can be gleaned. This scale conflict is a fundamental consideration in the choosing of the most appropriate method for studying a given environment.

The most commonly used methods to study the microbial ecology of subsurface environments include the following:

1. Culture techniques (small-scale)
2. Direct observation (micro-scale)
3. Biochemical marker techniques (small-scale)
4. Activity measurements in microcosms (small-scale)
5. Ground-water geochemistry (large-scale)
6. Ecological modeling (scale variable but usually small)

As can be seen, most of these techniques make observations at a small scale (i.e., samples from 1 to 1000 grams of aquifer material) or micro scale (i.e. samples less than 1 gram). The only large-scale observations (volumes of aquifer material greater than 10 square meters) that can be made are those using ground-water geochemistry. As we shall see, each of these approaches and their corresponding scales have particular advantages and disadvantages which must be carefully evaluated by the investigator. As a practical matter, the selection of a method or combination of methods should reflect the questions being asked. Thus, if the presence of microorganisms capable of particular metabolic functions (such as degradation of xenobiotics) is of primary interest, culture techniques or activity measurements (small scale) are appropriate. On the other hand some tasks, such as determining the distribution of predominant electron-accepting processes in regional aquifer systems, require the larger scale of geochemical techniques.

6.2.1 Culture Methods

Procedures for obtaining microorganisms in pure culture and then analyzing pure cultures for their biochemical properties are deeply embedded in the practice of microbiology. Many of the most important advances in medical and agricultural microbiology were made by pure culture methods. These methods can, in some instances, be usefully applied to the microbial ecology of subsurface environments. The basic idea is to isolate as many strains of bacteria as possible from sediments recovered aseptically from the subsurface, and characterize them on the basis of colony morphology, cell morphology, cell arrangement, gram reaction, and physiological or biochemical characteristics. These data can then be used qualitatively and quantitatively to characterize populations of microorganisms and to compare populations from different environments.

An example of using culture techniques for evaluating the microbial ecology of a subsurface environment was described by Kolbel-Boelke, Anders, and Nehrkorn (1988). These investigators isolated 2,700 strains of bacteria from a shallow Pleistocene aquifer in the Lower Rhine region of Germany near the Netherlands. Samples were obtained from aseptically cored sediments, as well as from water samples from wells screened in the aquifer. Sediments were sampled by placing about 25 grams of aquifer material in 100 milliliters of sterile water, shaking them with glass beads for 30 minutes, and spread-plating 0.1 milliliter samples of the resulting suspension onto a peptone-glucose media. Ground water was filtered through sterile cellulose nitrate filters and the filters placed on plates of the media. After 21 days of incubation, individual colonies were picked and streak-plated onto fresh media in order to purify the strains. Each of these strains recovered was then categorized numerically by means of 155 characteristics (Table 6.1) which were scored as positive (1), negative (0), or not tested (2). These characteristics included colony morphology (1-40), cell morphology (41-77), and physiological/biochemical properties (78-155). For example, if the isolate formed a circular colony (character 2), then character 2 would be scored positive and the other colony-shape characters (characters 1,3,4) would be scored negative. Thus, for the character "colony form" (characters 1-4), the isolate would be scored 0100. This process is then repeated for all 155 characters considered in the analysis. These morphological characteristics can be compiled in tabular form so that different populations can be compared.

Kolbel-Boelke, Anders, and Nehrkorn (1988) compared bacteria isolated from ground water pumped from a shallow aquifer (i.e. a ground-water community) to bacteria isolated from sediments cored (i.e. a sediment-bound community) from

TABLE 6.1 Summary of Characteristics used by Kolbel-Boelke, Anders, and Nehrkorn (1988) to Classify Subsurface Bacteria

Characteristic	Criteria	Examples
Colony morphology	Form	Circular, irregular
(characters 1–40)	Edge	Lobate, branching
	Elevation	Concave, convex
	Surface	Smooth, wrinkled
	Optics	Transparent, opaque
	Consistency	Slimy, brittle
	Color	White, yellow
Cell Morphology	Length	—
(characters 41–77)	Length/diameter	—
	Form	Straight rod, curved rod
	Cell end	Rounded, tapered
	Appendages	Stalks
	Movement	Flagella
	Arrangement	Pairs, chains
	Endospore form	Spherical, ellipsoid
	Gram stain	Positive, negative
Physiology and biochemistry	Growth on:	
(characters 78–155)	carbohydrates	Glucose, arabinose
	organic acids	Lactate, acetate
	enzyme activities	Lysine decarboxylase
	hydrolysis	Starch, chitin, casein

the same aquifer (Table 6.2) using these culture methods. The ecological question was, how similar are the two communities? In the ground-water samples, 23.5% of the isolates were gram positive/gram variable and 76.5% were gram negative. In the sediments, the percentage was roughly similar; 31.9% being gram positive/gram variable and 68.1% being gram negative. Based just on gram reaction, therefore, the two communities seem fairly similar. Interestingly, however, 61.2% of the ground water isolates showed flagellar movement whereas only 44.6% of the sediment-bound isolates showed this feature. On the basis of this difference, one might speculate that there is a greater adaptation toward motility in the ground-water community, and less adaptation toward motility in the sediment-bound community. This speculation, which seems intuitively reasonable, could then form the basis for studies designed specifically to test that hypothesis.

Of particular interest is the ability of isolates to metabolize different organic compounds (characters 78–135) under oxidative (O) or fermentative (F) conditions. Thus, if the isolate could oxidize glucose (GLU-O) but could not ferment glucose (GLU-F), could not oxidize arabinose (ARA) or xylose (XYL), could oxidize galactose (GAL), mannose (MNE), fructose (FRU), and lactose (LAC-O), but could not ferment lactose (LAC-F), the strain would be scored 100011110 for characters 78-87. Kolbel-Boelke, et al. (1988) described the potential "physiological activity" of a particular strain as

$$pA = 100 \ [P/(P + N)]$$

where pA is the potential physiological activity, P is the number of substrates for which a strain was "positive" (i.e. could utilize it under the given conditions) and N is the number of substrates for which the strain was "negative." If a strain could oxidize all 58 substrates, it's pA would be 100%, if it oxidized 29 substrates, its pA would be 50%, and so forth.

Similarly, Kolbel-Boelke Anders, and Nehrkorn (1988) described the potential activity of an entire community recovered from a sampled zone using the equation

$$pCA = 100 \ \{[P(a)/[P(a) + N(a)]\}$$

where pCA is the potential community activity, $P(a)$ is the number of strains positive for the ath substrate, and $N(a)$ is the number of strains negative for the ath substrate.

The distribution of potential community activity in both ground-water and sediments showed several notable features. First, there appeared to be considerable diversity in all of the communities studied. In other words, in each community there were strains present that degraded a very narrow range (less than 10 %) of the substrates and strains that degraded a broad range (more than 50 %). Somewhat surprisingly, there were no clear similarities between each of the ground-water communities or between the sediment communities.

When the kinds of substrates degraded by each community are considered, some differences become apparent. For example, a high percentage of isolates in each community exhibited the ability to oxidize glucose (GLU-O). However, only isolates from the sediment communities consistently exhibited the ability to ferment glucose (GLU-F). As ground water is aerobic in this aquifer, it stands to

TABLE 6.2 Comparison of Ground Water and Sediment-Bound Bacterial Communities by Means of Culture Techniques

Character of cell morphology	Average of all isolates	Ground water							Sediment-bound				
			wells							cores			
		Average	1–01	1–03	1–08	1–09		Average	2–02	2–04	2–05	2–06
Gram-positive/variable	28.44	23.5	10.6	57.3	7.7	5.8		31.9	40.1	10.8	34.7	43.1
Gram-negative	71.6	76.5	89.4	42.7	92.3	94.2		68.1	59.9	89.2	65.3	56.9
Gram positive/variable												
Rods	1.4	0.3	0.8	0.3	—	—		2.3	0.8	—	4.7	—
Coccoid rods/cocci	8.6	14.9	—	46.7	—	0.3		4.1	3.4	2.3	2.0	13.1
Coryneform/pleomorph	17.9	8.2	9.4	10.3	7.7	5.5		24.7	34.8	7.2	27.0	29.6
Branched filaments	0.5	0.1	0.4	—	—	—		0.8	1.1	0.3	1.0	0.4
Gram-negative												
Straight rods	52.2	63.1	80.9	34.5	73.6	74.8		44.8	41.9	76.6	31.9	41.2
Curved rods	3.5	6.9	0.8	1.4	9.9	15.5		1.1	1.1	1.2	0.7	2.2
Filaments (unbranched)	3.7	1.7	1.2	4.6	—	—		5.0	12.0	7.3	3.2	—
Coccoid rods/cocci	2.6	2.1	4.5	1.4	2.2	0.9		3.0	0.8	1.5	1.5	10.9
Variable cells	9.5	2.7	2.0	0.8	6.6	3.0		14.3	4.1	2.6	28.0	2.6
Flagellar movement	51.4	61.2	69.1	33.3	57.5	87.1		44.6	27.6	77.4	35.9	42.7
Gliding movement	3.6	1.4	—	4.6	—	—		5.0	11.2	7.3	3.2	—
Pigmentation	29.1	21.0	6.1	19.9	34.2	25.8		34.8	47.0	43.1	32.5	18.3
Stalks	2.2	3.4	—	4.0	3.3	5.5		1.4	0.4	1.2	1.5	—

Source: After Kolbel-Boelke, Anders, and Nehrkorn (1988). Reprinted courtesy of Springer-Verlag New York, Inc.

reason that free-living bacteria would gain little advantage by retaining the ability to ferment. However, because microenvironments in sediments may become anaerobic, sediment-bound microorganisms may gain some advantage by retaining this ability.

Another difference was in the ability of the communities to reduce nitrate (NO3N2). This characteristic was more pronounced in ground-water communities than in sediment communities. At first glance, this might seem inconsistent with the aerobic-anaerobic hypothesis for the distribution of the ability to ferment glucose given earlier. However, because nitrate is a negatively charged ion, it tends to be repelled from negatively charged mineral surfaces and would be concentrated in the pore water. Microbial communities in pore water could thus have more access to nitrate as a substrate and have more use for a nitrate reducing ability.

The basic approach of Kolbel-Boelke, Anders, and Nehrkorn (1988) has been widely used in documenting diversity of subsurface microbial communities. Balkwill, Fredrickson, and Thomas (1989) reported isolating 626 physiologically distinct types of bacteria from deep sediments of the Atlantic Coastal Plain. Like the studies of Kolbel-Boelke and associates, the physiological capabilities of individual strains were investigated. Balkwill's group, however, employed commercially available API Rapid NFT test kits (Analytab Products, Plainview, N.Y.) for this purpose. The use of such rapid-identification test kits greatly reduces the amount of time and effort required to characterize the metabolic potential of isolated microorganisms.

Culture methods for investigating the microbial ecology of subsurface sediments have both advantages and disadvantages. The chief advantage is that culture methods can demonstrate the diversity of microbial communities. The principal conclusion of Kolbel-Boelke, Anders, and Nehrkorn (1988) was that the communities, both waterborne and sediment-bound, were fairly diverse. This result could not have been predicted a priori. This method also has the advantage of being able to compare microbial diversity between different horizons. For example, the community of sediment probe 2-04 was characterized by a very high percentage of gram-negative bacteria, which set it apart from other sediment-bound communities. Thus, the culture method is also effective in documenting heterogeneity within aquifers. Culture methods have been extensively applied in other ground-water systems to document microbial diversity and heterogeneity (Balkwill, 1989).

Culture methods also have significant disadvantages. The most serious of which is the selectivity introduced by using culture media. Many of the bacteria present may not grow with the media and culture conditions offered, or may grow so slowly as to be underrepresented. This was observed to be the case in studies of a shallow water table aquifer in Fort Polk, Louisianna (Ghiorse and Balkwill, 1983). These authors noted that when particles of aquifer material were placed on culture media, small colonies would begin to grow. However, when these cultures were transferred to fresh media, they could not grow. Ghiorse and Balkwill (1983) concluded that a significant number of microorganisms present were not culturable with the methods being used. The net result of such problems in culturing subsurface bacteria is that not only may the information on diversity and community structure be systematically biased, but there is no way of knowing how much this bias has affected the data.

In addition to the selectivity problem, there is the problem of strain duplication. How does one determine whether two strains of isolates are closely related or whether they simply come from two cells of the same strain? If all isolates are treated as different strains, then isolating a large number of cells of the same strain would create the appearance that many strains have similar properties. On the other hand, if one were to take the position that isolates with similar properties belong to the same strain, one takes the risk of obscuring the presence of closely related but different organisms. In practice, this potential problem has been of little importance, since subsurface microbial communities appear to be very diverse. However, this problem must eventually be dealt with.

Another problem with culture methods has to do with the "activity" of bacteria in culture versus under in situ conditions. Just because a particular strain exhibits the ability to reduce nitrate in culture, for example, does not mean that that particular strain is involved in nitrate reduction in situ. In general, it is not feasible to evaluate which processes are important in situ from culture data alone. For these important ecological questions, other techniques must be utilized.

6.2.2 Direct Observation

An important limitation of culture methods in subsurface microbial ecology is that they give little information as to the niches inhabited by microbes or to their nutritional status. In order to deal with these problems, microbial ecologists have employed a number of direct observational techniques to address questions of community biomass, diversity, and nutritional status.

An example of how microbes present in aquifer sediments can be directly observed was given by Ghiorse and Balkwill (1983). These investigators employed a two-step approach. First, they prepared smears of aquifer sediment on microscope slides and stained them with acridine orange. Acridine orange binds with cellular nucleic acids to form a fluorescent dye. When excited by ultraviolet cellular light, the nucleic acid-acridine orange complex emits visible light, which can be observed microscopically. In theory, the only light visible should be stained cells. In practice, however, acridine orange binds with mineral and organic particles present in the sediment, and these AO-mineral complexes may also fluoresce. Distinguishing between bacterial cells and sediment particles is often straightforward, due to size and shape considerations. However, a measure of experience and skill is involved in such determinations.

The second step used by Ghiorse and Balkwill (1983) for directly observing aquifer microorganisms was to separate the microbes from sediment particles, embed the microbes into epoxy, and observe them with transmission electron microscopy (TEM). Microorganisms were released by placing sediment in a waring blender with a solution of 0.1% pyrophosphate and 3.0% glutaraldehyde. The purpose of the pyrophosphate was to help dislodge the microorganisms. The glutaraldehyde was added to fix (kill and preserve) the microorganisms and prevent growth during the mounting procedures.

Using the sediment smear acridine-orange method, it was found that microorganisms were not uniformly distributed in the sediment. Rather, bacteria tended to occur in microcolonies consisting of anywhere from three or four to hundreds of cells. Several other features of interest, such as bacteria present in filaments, were also observed. Both rod-shaped and coccoidal cells were observed, as well as cells that exhibited features of eucaryotes. Additionally, cells that appeared to be in the process of dividing were observed. This indicated that the cells were actively growing in situ immediately before or during the sampling procedures.

These observations are important from an ecological standpoint because they provide direct evidence as to how microbes interact with their environment-evidence that cannot be gleaned from culturing microorganisms. For example, culture studies could not have demonstrated that microcolonies were the predominant mode of occurrence of subsurface microorganisms. Furthermore, while culture studies can show that microorganisms are *capable* of growth and replication, they cannot show that this occurs in situ. The direct observations of Ghiorse and Balkwill (1983), as well as direct observations made in other studies (Wilson et al., 1983; Harvey and George, 1987), show that such growth does occur in situ. Harvey and George (1987) went further and determined by direct observation the percentage of cells that were in the process of dividing. These data were then used to estimate growth rates of subsurface bacteria.

On an even smaller scale, direct observation of cells with TEM can also provide information that is of ecological importance. For example, Ghiorse and Balkwill (1983) showed that many cells contained inclusions of a substance similar to polybeta-hydroxybutyrate (PHB). PHB is used by bacterial cells as an energy-storing compound. This, in turn, indicates that the microorganisms at the Fort Polk site were subjected to, and probably adapted for, starvation stress.

TEM observation also showed that about two-thirds of the bacterial cells had gram-positive cell walls. Initially, this was thought to be a surprising result since culturing microorganisms from subsurface environments often revealed a preponderance of gram-negative cells (Kolbel-Boelke et al., 1983). Interestingly, this observation was repeated at the Lula, Oklahoma site where direct observation of microorganisms showed a preponderance of gram-positive cells (Wilson et al., 1983) but cultures of the same sediments indicated that gram-negative cells predominated. This raises the possibility that the methods used to culture the microorganisms systematically discriminated against gram-positive cells. Exclusive use of culture methods may therefore present a false impression of the relative abundance of gram-positive and gram-negative cells in a system. As pointed out by Ghiorse and Balkwill (1983), however, such determinations can be made only by supplementing culture methods by direct observation of subsurface sediments or by employing biochemical techniques that directly reflect in situ conditions.

6.2.3 Biochemical Marker Techniques

Direct observational techniques for observing microorganisms in deep subsurface sediments suffer primarily from two limitations. First, only a very small volume of aquifer material can be accurately screened due to the high magnification required to resolve microbes. Second, direct observations of microorganisms give only limited information as to the phylogenetic types present. It is feasible, as we have seen, to differentiate rod versus coccoid forms and to discriminate between gram-positive and gram-negative microorganisms. However, in order to evaluate accurately the community structure present, it is desirable to have better resolution of different microorganisms present.

One very powerful tool for providing this better resolution has been introduced by David C. White and his associates (White et al., 1983; Balkwill et al., 1988). The basis of this technique is that different microorganisms exhibit significant biochemical differences in the fatty acids that comprise cell walls and internal storage granules. By directly extracting and characterizing these fatty acids from sediments, information concerning total biomass, cell wall type, presence or absence of eucaryotic organisms, and nutritional status of the microorganism present may be derived.

Phospholipids are part of every bacterial membrane (section 1.2) and maintain a relatively constant proportion of the cell mass. Furthermore, phospholipids are turned over fairly rapidly after cellular death. Because of these two features, the amount of phospholipid present in sediments is a potential measure of the living biomass present. Moreover, the fatty acid composition of the phospholipids can give information as to the kinds of microorganisms present. Eucaryotic microorganisms, for example, are characterized by long-chain polyenoic fatty acids. The presence of such fatty acids in sediment is therefore a direct indicator of the presence of eucaryotic microorganisms. Similarly, anaerobic microorganisms form cis-vaccenic acids in the anaerobic fatty acid desturase pathway. Therefore the presence of these acids indicates the relative importance of anaerobic and aerobic metabolism in a given sediment.

As with direct microscopic observation, biochemical techniques also indicate the relative abundance of gram-negative and gram-positive microorganisms. Gram-negative bacteria average about 14 umol muramic acid/g dry wt and gram-positive bacteria average 38 umol/g muramic acid, a difference that reflects the relative proportion of peptidoglycan in cell walls. Also, gram-positive bacteria often have acidic macromolecules, called *teichoic acids,* associated with the cell wall (*teichoic* comes from the Greek word *teichos,* which means "wall"). These teichoic acids are composed of repeating glycerol or ribitol units connected by phosphate esters. Thus, the abundance of ribitol, in particular, is a rough measure of the relative abundance of gram-positive bacteria. The abundance of gram-negative bacteria can be estimated by the abundance of hydroxy fatty acids in the lipopolysaccharides.

The extraction procedure used to separate lipids from sediment samples was as follows. First, lipids were removed from 45-50 g of sediment using a chloroform-methanol extraction procedure. Part of the recovered lipids were analyzed directly and part were hydrolyzed for analysis of various components such as teichoic acids, lipopolysaccharides, exopolymers, and polymeric hydroxyalkanoic (PHA) acids.

An example of how these techniques can be applied to the microbial ecology of subsurface sediments was given by White et al. (1983). In this example, White and coworkers compared shallow sediments recovered from Fort Polk, Louisiana (discussed in the previous section) with deeper sediments (410 m) recovered from the Bucatunna Clay near Pensacola, Florida. Data from these subsurface environments were also compared to data from modern estuarine and marine aquatic sediments.

Table 6.3 shows estimates of predominant cell wall type and biomass as determined by lipid analysis. Because cells contain about 5.0×10^4 nanomoles (nmol) phospholipid per cell, and because a cell weighs about 10^{-12} grams, biomass (B_m; cells/sediment (dry wt)) may be estimated from the equation

TABLE 6.3 Microbial Biomass Estimates in Sediments of the Bucatunna Clay and Fort Polk Aquifers

	Bucatunna Clay		Fort Polk	
Component	nmol Component/gm Dry Weight	Biomass[d] Cells/gm	nmol Component/gm Dry Weight	Biomass[d] Cells/gm
Phospholipid[a]	0.52	1.0×10^7	0.98	2.0×10^7
Muramic acid[b]	0.62	$1.6 - 4.4 \times 10^7$	2.02	$1.4 \times 10^8 - 5.3 \times 10^7$
Teichoic acid (Ribitol)[c]	2.8	1.3×10^7	N.D.	

[a] Biomass estimate based on 5.0 nanomoles/gm cell.
[b] Biomass estimate based on 1.4×10^5 (gram negative) and 3.8×10^5 (gram positive) nanomoles/gm cell.
[c] 2.1×10^5 nanomoles/gm cell.
[d] Cell mass assumed equal to 10^{-12} gm/cell.

$$B_m = [PL/(5.0 \times 10^4)]/10^{-12}$$

where PL is the mass of phospholipid in nanomoles/gram of sediment. The biomass as calculated by using the data of table 6.3 is 10^7 cells per gram of dry sediment.

Similarly, biomass can be estimated from measured concentrations of muramic acid with the equation

$$B_m = (MA/C_{ma})/10^{-12}$$

where MA is the mass of muramic acid in nanomoles/gram sediment and C_{ma} is a constant equal to the amount of muramic acid per gram of dry cell material. Because gram-negative cells average about 14 micromoles muramic acid/gram (dry wt), and gram positive average about 38 micromoles muramic acid/gram, the data of table 6.3 indicates a biomass of between 6.4×10^7 (if all the cells were gram negative) and 1.6×10^7 (if all the cells were gram positive) cells per gram of dry sediment.

Finally, biomass of gram-positive cells [$B_{m(gp)}$] that contain teichoic acids can be estimated from measured concentrations of ribitol using the equation

$$B_{m(gp)} = [R/2.1 \times 10^5]/10^{-12}$$

where 2.1×10^5 is the amount of ribitol in Staphylococcus aureus (nanomoles per gram of dry cells). Using this equation gives an estimate of 1.3×10^7 cells per gram. It is notable that the estimates of biomass using total phospholipids, muramic acid, and teichoic acids were all in the range of 10^7 cells per gram dry weight for the Bucatunna Clay.

The lipid data generated by White et al. (1983) also provide information on the community structure of the subsurface sediments. Both the Bucatunna Clay and Fort Polk sediments did not contain measurable quantities of polyenoic fatty acids, suggesting that eucaryotes were either absent or present in very low numbers. Differences between the sediments were also apparent. Bucatunna Clay sediments contained higher levels of *cis*-vaccenic acid, which is characteristic of anaerobic metabolism, than did the Fort Polk sediments. This makes sense considering the different hydrology of the two sites. The Fort Polk sediments were relatively shallow, providing a source of dissolved oxygen for aerobic respiration. In contrast, the deep sediments of the Bucatunna Clay were isolated from a source of dissolved oxygen, rendering anaerobic respiration and fermentation the only metabolic strategies available to indigenous bacteria.

Other differences in community structure were apparent in the relative abundance of gram-negative and gram-positive cells. There was more phospholipid present in the Fort Polk sediments than in the Bucatunna Clay sediments. However, teichoic acids were present in Fort Polk sediments and not in Bucatunna Clay sediments. Those data suggest that gram-positive microorganisms were relatively more abundant in the Bucatunna Clay than in the Fort Polk aquifer. Interestingly, this conclusion was also arrived at by Ghiorse and Balkwill (1983) by using direct observation of sediment smears.

Phospholipid analysis is also potentially useful for identifying nutritional stress in a microbial population. When bacteria grow under conditions of unbalanced nutrition, such as carbon resources exceeding supplies of other critical nutrients, polymeric hydroxyalkanoic (PHA) storage polymers tend to accumulate. These products disappear when the critical needs of the microorganisms are met. Therefore, the presence of such storage products are an indicator of nutritional stress in a microbial community. The presence of PHA was used by White et al. (1983) to postulate the presence of such nutritional stress in the Bucatunna Clay and Fort Polk aquifers. Furthermore, the presence of such storage polymers was independently verified by direct observations (Ghiorse and Balkwill, 1983). Clearly, the use of biochemical marker methods in microbial ecology is potentially a very powerful tool and is a promising topic for future research.

6.2.4 Activity Measurements in Microcosms

Culture methods, direct observation, and biochemical marker techniques focus on the nature of organisms present and their physical and biochemical characteristics. Because the kinds of microorganisms present in any one environment are often so large and the relationships between microbial species so complex, these methods are often quite cumbersome. An alternative to these methods, developed initially by soil microbiologists, is to focus on the net *effects* of microbial processes rather than on the kinds of organisms present. The guiding philosophy of this approach is that what the microorganisms do is important ecological information and can supplement the findings of culture studies.

An example of how activity measurements can supplement the use of culture methods in studying the microbial ecology of aquifer systems was given by Chapelle et al. (1988). The system being studied was the carbonate Floridan aquifer and its overlying confining bed, the Hawthorn formation. Ground water produced from the carbonate aquifer at this site contained up to 12 mmol/L of dissolved inorganic carbon-far more than could be accounted for by dissolution of carbonate minerals in a system closed to CO_2. The obvious hypothesis, therefore, was that CO_2 was being actively produced in the sediments of this system. The production of this CO_2 resulted in an open system with respect to CO_2 and vastly more carbonate material dissolution. In order to test this hypothesis, the first and most obvious question was ecological: Were there microorganisms present that were capable of producing CO_2?

To answer this initial question, culture methods similar to those employed by Kolbel-Boelke, Anders, and Nehrkorn (1988) were used to show the presence of a viable bacterial community. Not surprisingly, the results of these culture studies documented the presence of a wide variety of aerobic and facultatively anaerobic bacteria, including strains of *Bacillus, Pseudomonas,* and *Micrococcus.* However, the fact that those microorganisms could be coaxed into active growth on culture media in no way indicated that they actively produced CO_2 in situ, and that was the important question.

In order to address this question, samples of aquifer material were placed in sterile vials under aerobic and anaerobic conditions, and the production of CO_2 was monitored with time (Fig. 6.3a). CO_2 was generated aerobically and somewhat slower anaerobically. However, sediment samples that had been sterilized by gamma radiation showed no CO_2 production.

While these data were consistent with the hypothesis that bacteria produced

CO_2 in the sediments, the possibility that the sterilization technique had in some way interfered with some abiotic CO_2-producing process could not be ruled out. For this reason, a second experiment was performed. In this experiment (Fig. 6.3b), sediment samples were initially incubated anaerobically for seven days. CO_2 production was very low, ruling out the possibility that inorganic degassing of the sediments was producing significant amounts of CO_2 in the experiments. After seven days, oxygen was introduced into the vials, and CO_2 production rapidly increased. However, this observation did not rule out the possibility that inorganic oxidation of organic material was producing CO_2. In order to test this possibility, dicyclohexylcarbodiimide (DCCD), an inhibitor that uncouples chemiosmosis from ATP synthesis in bacteria, was added to the sediment. DCCD immediately shut off CO_2 production and showed unequivocally that bacteria in the sediments were, in fact, responsible for the active CO_2 production. Thus, a combination of culture data and measurements of potential activity (Fig. 6.3) neatly answered an ecological question posed by observing the ground-water chemistry of this system.

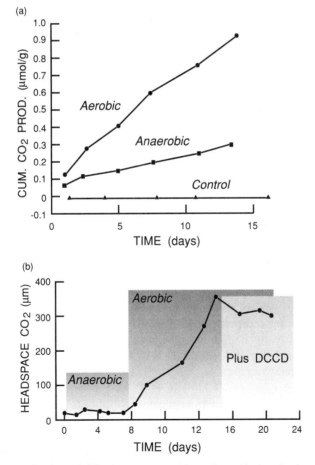

FIGURE 6.3. Production of CO_2 from (a) aerobic and anoxic incubations of aquifer sediments, and (b) inhibition of CO_2 production by DCCD. (Adapted, by permission, from Chapelle et al., 1988).

Aquifer materials placed into laboratory vessels such as test tubes or septated serum vials for measurement of microbial activity are referred to as *microcosms*. The word *microcosm* is defined in the dictionary as "a community or other unity that is representative of a larger unity". The reasoning for microcosms is that, by understanding the activity of a small portion of an aquifer, much can be learned about the aquifer as a whole. Microcosms have been used extensively in studies of subsurface microbial ecology and are especially useful in assaying indigenous microbial communities for specific activities. In the previous example, the activity of interest was potential bacterial CO_2 production in a pristine aquifer. Often, however, the activity of interest deals with the potential of microbial communities to metabolize particular xenobiotic organic compounds.

One of the first uses of microcosm/activity studies to assay the potential of subsurface microbial communities to degrade xenobiotic compounds was described by Wilson et al. (1983). Studies by Wilson and coworkers had previously documented the presence of a viable bacterial population in a shallow aquifer near Lula, Oklahoma. Now the question was whether these bacteria were capable of degrading hydrocarbons such as toluene or trichloroethylene. Because the Lula site had not previously been exposed to such compounds, any degradation of them would reflect the activity of a nonacclimated microbial population. Furthermore, if such nonacclimated populations could degrade these compounds, this information would have important implications as to the potential fate of the xenobiotics in ground-water systems.

Because Wilson's group was particularly interested in volatile organic compounds-compounds that evaporate quickly in open vessels-the design of the microcosms had to allow sampling of pore water without exposure to the air. Aquifer material was aseptically transferred to a 50-ml test tube, and organic compounds were added in concentrations ranging from 85 to 666 $\mu g/L$. The tube was then sealed with a Teflon-lined screw cap and the tube was incubated. After incubation, it was necessary to sample the pore water in the microcosm while minimizing exposure to the air. This was accomplished by attaching a separate tube filled with dilution water to the microcosm by means of a Teflon connector. The dilution water was then mixed with the pore water of the microcosm using a vortex mixer, and the tubes centrifuged until the water cleared. Aliquots of the diluted water were then analyzed.

The results of this study (Wilson et al., 1983) showed that toluene was rapidly degraded with about 90% of the toluene disappearing within a week. Chlorobenzene was degraded by sediments cored from the unsaturated zone but was not degraded significantly by deeper aquifer sediments. Other xenobiotics such as trichloroethylene and tetrochloroethylene were not observed to undergo measureable degradation.

Measuring CO_2 production (Chapelle et al., 1988) or xenobiotic degradation (Wilson et al., 1983) activity in microcosms is suitable for answering specific questions concerning the metabolic potential of microbial populations. They are not suitable for determining what part of a microbial community is involved in the degradation processes. It is common practice, therefore, to establish the presence of a particular metabolic activity using microcosms and, if necessary, follow up with culture methods for isolating particular microorganisms capable of the activity.

6.2.5 Geochemical Methods

A feature shared by culture methods, direct observational methods, biochemical tracer methods, and microcosm methods is that the scale of observation is relatively small. Each of these methods is based on recovery of anywhere from 1 to a 1,000 grams of aquifer material. This scale is fixed largely by practical considerations of core recovery and sample handling in the laboratory. It is very difficult, however, to extrapolate meaningfully from such small-scale samples to the scale of aquifer systems that may cover hundreds of square kilometers in area. One way of enlarging the scale of subsurface microbiological observations is to use the chemistry of ground water as a tracer for particular microbial processes.

The basic premise of using ground-water geochemistry in this fashion is that particular microbial processes impact the chemistry of ground water in certain ways. For example, oxygen respiration in an aquifer that is closed to the atmosphere will result in the consumption of oxygen as ground water flows downgradient. Similarly, sulfate reduction will consume sulfate along aquifer flowpaths. In addition to consuming oxidants or substrates, some microbial processes will mobilize chemical species. Ferric-iron reduction will result in the production of dissolved iron and methanogenesis will produce methane. Thus, the accumulation or depletion of particular chemical species in ground water can be used to indicate the presence or absence of certain microbial processes in aquifer systems.

An example of this approach was given by Lovley and Goodwin (1988). The system being studied was the Middendorf aquifer in South Carolina and the specific question was how predominant terminal electron-accepting processes are distributed. This kind of question is difficult to answer using culture techniques. The fact that sulfate-reducing bacteria may be cultured from aquifer sediments does not prove that such bacteria are actually active in situ. Also, because of the large area covered by this aquifer system, sampling sediments from the subsurface over a wide area was a practical and economic impossibility.

In view of these difficulties, Lovley and Goodwin (1988) took the approach of using ground-water chemistry as an indicator of microbial processes. First, they selected a series of wells oriented along a regional flowpath and sampled them for dissolved oxygen, nitrate, sulfate, and methane. Their data (Fig. 6.4) showed that oxygen and nitrate were present only near the outcrop-recharge areas of the aquifer. The presence of nitrate appeared to be related to agricultural practices near the outcrop area. As dissolved oxygen and nitrate were consumed along the aquifer flowpath, concentrations of dissolved iron increased. Further downgradient, concentrations of dissolved iron decreased and sulfate concentrations also decreased. Even further downgradient, concentrations of methane increased sharply.

These data suggested that the microbial ecology of this aquifer system was characterized by discrete zones, each of which was dominated by particular microbial processes. Thus, in the part of the aquifer containing dissolved oxygen, oxygen reduction was the predominant process with ferric-iron reduction, sulfate reduction, and methanogenesis largely being absent. Similarly, in the part of the aquifer where sulfate concentrations decreased, sulfate reduction was the predominant process, and where methane was being actively generated, methano-

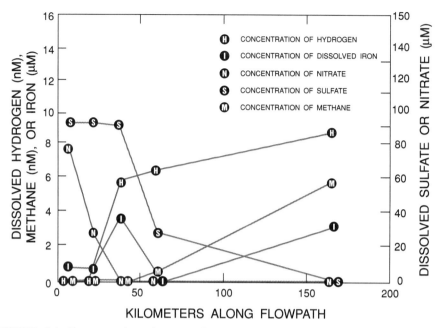

FIGURE 6.4. Concentration changes of oxygen (O), nitrate (N), iron (I), sulfate (S), methane (M), and hydrogen (H) along a flowpath segment of the Middendorf aquifer. (Adapted from Lovley and Goodwin, 1988, with permission courtesy of the American Society for Microbiology).

genesis was the predominant process. The ground-water chemistry data presented the picture of an aquifer segregated into discrete zones dominated by different terminal electron-accepting processes.

The drawback of using ground-water chemistry data in this fashion, however, is that direct evidence of the types and activities of microorganisms present is lacking. One method of providing this data would be to perform culture and activity studies of sediments cored from the different zones. In fact, extensive studies of the Middendorf aquifer have subsequently confirmed that in the aerobic zone, oxygen reduction is by far the predominant process with anaerobic processes such as sulfate reduction or methanogenesis being virtually absent (Hicks and Fredrickson, 1989; Jones, Beeman, and Suflita, 1989). However, in much of the aquifer, culture studies were not feasible due to a lack of cored material.

To deal with this difficulty, Lovley and Goodwin, 1988, used a novel method dealing with concentrations of dissolved hydrogen gas in ground water. Hydrogen (H_2) is an intermediate product of fermentative bacteria and is consumed as an electron donor by ferric iron reducing, sulfate reducing, and methanogenic bacteria. In natural systems, therefore, hydrogen concentrations reach a steady state where production by fermentation is balanced by consumption by respiration. Significantly, the steady-state hydrogen concentration is dependent only upon the physiologic properties of the consuming bacteria and not on the rate of production or consumption. Thus, sediments dominated by ferric iron reduction have different steady-state hydrogen concentrations (0.2 to 1.0 nmolar) than sulfate reducing (1 to 6 nmolar), or methanogenic (7 to 20 nmolar) sediments.

By showing that the parts of the aquifer characterized by sulfate consumption and methane production coincided with H_2 concentrations characteristic of sulfate reduction and methanogenesis respectively (Fig. 6.4), Lovley and Goodwin (1988) provided independent evidence that these water chemistry patterns resulted from microbial processes. This evidence, in turn, suggested that the microbial ecology of the aquifer system was characterized by a succession of predominant terminal electron-accepting processes. Such behavior had been postulated by Champ, Gulens, and Jackson (1979) by analogy with aquatic surface sediments, but had not been well-documented in ground-water systems.

As with other methods for studying the microbial ecology of subsurface aquifer systems, the use of ground water geochemistry has advantages and disadvantages. The chief advantage is that ground water chemistry provides a large-scale view of aquifers that cannot be obtained with other methods. Also, it is relatively inexpensive to sample water chemistry whereas obtaining aseptic samples of deep sediments can be prohibitively expensive. However, ground-water chemistry can give no direct information as to the biomass, diversity, or biochemical potential of the organisms responsible for the processes under consideration. This can be a considerable disadvantage if, for example, the potential degradation of xenobiotic compounds is of primary interest. Furthermore, because ground water chemistry integrates the effects of microbial processes, it gives little useful information as to the heterogeneity of microbial processes.

6.2.6 Ecological Modeling

Thus far, all of the methods that we've considered for evaluating the microbial ecology of subsurface environments have relied on sampling aquifer sediments or associated ground water and performing various procedures designed to answer particular questions. This is the only way in which questions concerning microbial communities in individual aquifer systems can be meaningfully addressed. However, the scope of microbial ecology also includes investigating interactions between populations within a community. One method for studying such interactions is by building idealized mathematical representations of interacting microbial populations. These mathematical representations, or *ecological models*, can provide insights into the population dynamics of microbial communities.

Ecological modeling has been used extensively to evaluate potential effects of competition between two microbial populations. For example, one tool for investigating the growth kinetics of microbial populations is the *chemostat*. A chemostat is a flow-through incubation vessel in which fresh nutrients are continuously added at one end and there is a continuous outflow at the other end. Between the inflow and outflow is a chamber in which bacteria can grow. If the rate of outflow is greater than the rate of bacterial reproduction, then the cells will eventually be washed out of the chemostat. However, if cell reproduction is faster than the rate of cells washing out, the population will increase until it becomes limited by the availability of incoming substrate. The growth of the population thus increases asymtotically, and may be described by the equation

$$\mu = (\mu_{max} S)/(K_s + S) \tag{6.1}$$

where μ is the specific growth rate, μ_{max} is the maximum specific growth rate, S is the concentration of the growth-limiting substrate, and K_s is a substrate uptake constant numerically equal to S at 0.5 μ_{max}. When written in this form, equation 6.1 is generally referred to as the *Monad* equation. Note, however, the similarity with the Michaelis-Menton equation (Chapter 4) that describes enzyme kinetics. However, the Michaelis-Menton equation is derived from a mechanistic analysis of enzyme-catalyzed reactions. Equation 6.1 describing the growth of microbial populations is entirely empirical.

Veldkamp et al. (1984) described how a simple modeling analysis based on equation 6.1 can be used to evaluate competition between microbial species. For example, in a chemostat, it would seem that being able to grow quickly on a substrate would be a considerable advantage. This is true sometimes, but not always. Consider two organisms, A and B, that have different growth characteristics but that grow on the same substrate (Fig. 6.5). In both cases the substrate uptake constant (K_s) for organism B is greater than that for organism A. In the first case (Fig. 6.5a), the maximum growth rate (μ_{max}) is greater for organism A than organism B. In this case, if A and B are introduced into a chemostat, the population of A will always be larger than for that of B regardless of the substrate concentration. Furthermore, B will tend to be washed out of the chemostat as the flow rate is increased. In the second case (Fig. 6.5b), μ_{max} is greater for organism B than for A. In this case, organism A has an advantage at low substrate concentrations. When substrate concentrations are higher and cease to be limiting, however, organism B will predominate. In a chemostat, organism B will be washed out at lower substrate concentrations whereas organism A will be washed out at higher rates of substrate delivery.

This analysis indicates that, at the given conditions, the outcome of competition between microbial species for limited resources is determined by substrate uptake efficiency (K_s) and potential growth rates (μ_{max}). This model assumes that the only point of competition between the two species is for one substrate. This is obviously too great a simplification to be applied directly to microbial populations in nature where competition is for much more than one substrate. However, it does point

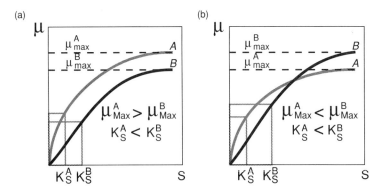

FIGURE 6.5. Effects of Michaelis–Menton parameters on the outcome of competition between populations of microorganisms. (Reproduced from Veldcamp et al., 1983, with permission courtesy of the American Society for Microbiology).

out the importance of species-specific growth characteristics in determining the outcome of competition for limiting substrates.

The example just given is fairly simple. However, the modeling approach can be applied to much more complex questions concerning microbial ecology. For example, some bacteria are immotile, some are randomly motile, and some are chemotactic-that is, they are motile and are attracted by potential nutrients. The ecological question, therefore, is under what conditions these properties are advantageous and under what conditions are they disadvantageous. This is the type of question that can be addressed with ecological modeling (Kelly, Dapsis, and Lauffenburger, 1988).

While the model of Veldkamp et al. (1984) was based on a simple analytical model (equation 6.1), the approach taken by Kelly, Dapsis, and Lauffenburger (1988) to evaluate the effect of motility and chemotaxis on microbial competition was more complex. First, they wrote mass balance equations for the densities of two populations, c_1 and c_2, and substrate concentrations, s:

$$\frac{\partial c_i}{\partial t} = \frac{\partial J_{ci}}{\partial x} + G_i(s, c_i) \qquad i = 1,2$$

$$\frac{\partial s}{\partial t} = \frac{\partial J_s}{\partial x} - \Sigma U_i(s, c_i)$$

Then they used monad kinetics to describe the growth of each population

$$G_i = \left[\frac{k_i s}{(K_i + s)} - k_{di}\right] c_i$$

$$U_i = \frac{1}{Y_i} \frac{k_i s}{K_i + s} c_i$$

where G_i is the net growth rate of population i and U_i is the substrate uptake rate by population i. μ_{max}, s, K are the Monad parameters and are the same as given in equation 6.1. k_{di} is the specific death rate and Y_i is the cell yield for population i. It is assumed that nutrient transport is governed by a diffusive flux,

$$J_s = -D\frac{\partial s}{\partial x}$$

and that cell transport is variably chemotactic

$$J_{ci} = -\mu_i \frac{\partial c_i}{\partial x} + \chi_{0i} \frac{K_{Di}}{(K_{Di} + s)^2} \frac{\partial s}{\partial x} c_i$$

where J_{ci} is the local cell flux, c the local cell density, and X (chi) is the cell random motility coeficient. Boundary conditions for the simulation are given by

$$J_{ci} = 0 \qquad x = 0, L$$
$$J_s = 0 \qquad x = 0$$
$$s = S_O \qquad x = L$$

The first boundary condition states that cells are not transported into or out of the system. The second condition states that there is no diffusive flux of substrate out of the system. Finally, the substrate concentration at L is constant. This set of coupled partial differential equations was solved using a Galerkin finite-element scheme (Kelly, Dapsis, and Laufenberger, 1988).

Figure 6.6a shows the modeled effects of a randomly motile population (population 1) competing with an immotile population (population 2). Curiously, as the random motility of population 1 increases, it's success in competing with population 2 actually decreases. At ratios of random motility of less than 10^{-2}, the ratio of specific growth rate for population 2 and population 1 must be about 0.1 for population 1 to win the competition and become the predominant species. In other words, for a randomly motile population to compete successfully with an immotile population, it would have to have a growth rate about 10 times that of the immotile population. This general behavior has been observed experimentally (Lauffenburger, 1983). Thus, random motility may actually be a *disadvantage* relative to immotility under some conditions.

Figure 6.6b shows the results of another simulation made with this model. In this case, the survival of a chemotactic population (population 2) is compared with one that is randomly motile (population 1). As the chemotatic attraction of population 2 increases, the growth rate of population 1 must increase in proportion to the growth rate of population 2 in order for population 1 to win the competition. At equal growth rate ($K_2/K_1 = 1$), even slight chemotactic response gives population 2 an advantage. Thus, chemotaxis confers a definite competitive advantage to motile bacteria.

Figure 6.6c shows one more simulation. It compares the outcome of competition between immotile (population 1) and chemotactic (population 2) bacteria. In this case, it is clear that for the motile population to win the competition, relatively high values of chemotactic attraction are required. At low values of chemotactic attraction, the growth rate of a chemotactic organism must be at least 10 times that of the immotile organisms for the chemotactic bacteria to win. On the other hand, if the growth rates are the same ($K_2/K_1 = 1$), a high chemotactic attraction is required for the chemotactic population to win.

These simulations, as is true with any kind of mathematical modeling, require numerous simplifying assumptions. Indeed, many would argue that they are oversimplified and therefore not applicable to natural systems. However, it is interesting to compare the results of Kelly, Dapsis, and Lauffenburger (1988) with observations of subsurface bacteria. Kelly and associates suggest that immotile bacteria have a considerable advantage over randomly motile and chemotactic bacteria, particularly under conditions of scarce substrate availability. One would predict, therefore, that in substrate-limited subsurface environments, immotile (i.e. attached bacteria) would have a considerable competitive advantage over freeliving bacteria. It is widely observed that bacteria in subsurface environments tend to be sediment-bound rather than free-living (Wilson et al., 1983; Ghiorse and Balkwill, 1983; Hazen et al., 1991). Furthermore, it has been observed that as substrate availability increases, as in a plume of contaminated ground water, the proportion of free-living bacteria present increases (Harvey, Smith, and George, 1984). Therefore, the modeling analysis of Kelly, Dapsis, and Lauf-

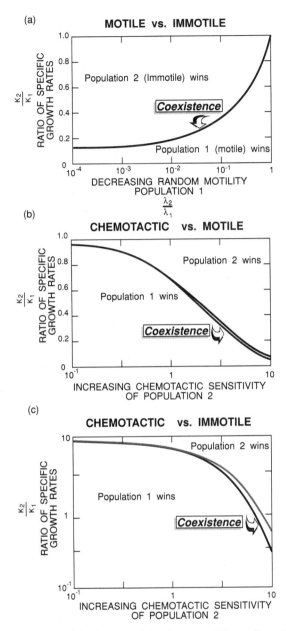

FIGURE 6.6. Ecological model of competition between (a) motile and immotile populations, (b) chemotactic and randomly motile populations, and (c) chemotactic and immotile populations. (Reproduced from Kelly, Dapsis, and Lauffenburger, 1988, with permission courtesy of Springer-Verlag, Inc.)

fenburger (1988) provides a theoretical framework within which some ecological features of subsurface environments may be understood. Ecological modeling, therefore, potentially is a useful tool that can be applied to subsurface environments.

6.3 MICROBIAL DIVERSITY AND NICHES IN AQUIFER SYSTEMS

A fundamental characteristic of aerobic aquifer systems is that they exhibit a high degree of morphological and physiological diversity in their bacteria. This was observed as early as the 1950's by German investigators (Wolters and Schwartz, 1956) using culture techniques. Subsequent studies, also primarily using culture techniques (Hirsch and Rades-Rohkohl, 1983; Balkwill and Ghiorse, 1985; Kolbel-Boelke, Anders, and Nehrkorn, 1988; Balkwill, Fredrickson, and Thomas, 1989), have shown that diversity among aerobic and facultatively anaerobic heterotrophic bacteria appears to be the rule rather than the exception. Similar culture studies of anaerobic aquifer systems have not been performed as extensively. However, the results of White et al. (1983) using biochemical marker techniques suggest that microbial diversity is also characteristic of anaerobic ground-water systems. In this section, the ecological implications of microbial diversity in ground-water systems is examined.

6.3.1 Measurement of Diversity

The term "diversity" describes heterogeneity within a system. As applied to ecology it refers to (1) the total number of species present in an ecosystem and (2) the number of individuals of each species present in that ecosystem. It is possible to define a "diversity index" based on these two quantities. The Shannon–Weaver index is the most widely used measure of diversity in ecology and consists of one term for describing total number of species (S) and a second term to describe the distribution of that species within the community ($Dist_s$). Total diversity (Div_{tot}) may thus be defined as

$$Div_{tot} = S + Dist_s \qquad (6.2)$$

When dealing with macroscopic organisms, such in as our earlier example of blackbirds, the information needed to quantify diversity may be obtained by direct observation. However, when dealing with microbes, individuals can only be identified using culture techniques. The selectivity of culturing microorganisms in the laboratory automatically introduces bias into measuring diversity. Even with the limitations imposed by the selectivity bias, it is evident that determining the number of individuals of each species present is not a practical undertaking.

Faced with these difficulties for quantifying diversity, Balkwill, Fredrickson, and Thomas (1989) used a multitiered approach for expressing the horizontal and vertical diversity of microorganisms in a ground-water system. In this study, sediments from three continuous core holes in the Atlantic Coastal Plain were collected from land surface to a maximum depth of 265 meters. Sediment samples from various depths, spanning at least eight different geologic units, were used to isolate microorganisms on a peptone-tryptone-yeast extract-glucose (PTYG) media and a 1% dilution of this media. The physiological characteristics of these isolates were then determined. These physiological characteristics included the presence or absence of particular enzyme activities such as nitrate reductase, and the ability to assimilate different carbon sources such as D-glucose. Of the 1,112 isolates obtained, 626 were physiologically distinct; that is, they differed by at least one characteristic from all the other isolates.

These data were then used to investigate how the physiological characteristics of the microorganisms varied spatially in the system. The first comparison made was between bacteria present in the soils (0 meters depth) with those present in deeper horizons (> 8 meters). In this case, the physiological data were presented as the percent of isolates testing positive for a particular capability. By expressing the data in this manner it became evident that soil bacteria at the sites exhibited different diversity trends than bacteria isolated from deeper horizons. Specifically, the soil bacteria were able to metabolize a wider range of substrates than the subsurface bacteria. For example, about 69% and 85% of the soil isolates were able to oxidize L-arabinose and D-mannose respectively whereas only 46% and 53% of the subsurface isolates were able to oxidize these compounds. Balkwill, Fredrickson, and Thomas (1989) suggested that this difference in metabolic flexibility reflected the more limited kinds of organic matter present in the subsurface than in the soil.

It is interesting, but not surprizing, that there were considerable differences between the physiological characteristics of isolates recovered on 1 % PTYG agar and the characteristics of isolates recovered on PTYG agar. For the subsurface isolates there were significant statistical differences for 16 of the 20 physiological tests between 1 % PTYG and PTYG isolates. This implies, as expected, that culture methods have an important impact on the organisms recovered. Because of this impact, Balkwill et al. (1989) were careful to point out the media-specific nature of the diversity estimates.

Given the data from the physiological tests, Balkwill, Fredrickson, and Thomas (1989) defined *diversity in specific horizons* as the number of physiologically distinct isolates divided by the total number of isolates tested. Note that this definition includes only the first term of div_{tot}, as defined by equation 6.2. This reflects the technical impracticality of determining how individual species are distributed in the sediments using culture techniques. Using this definition, the researchers compared diversity between geologic formations sampled from three different core holes, all of which were located within a radius of 15 kilometers (Table 6.4). It is evident from this data that each horizon exhibited a high degree of diversity. At depth 259 in core hole P24 [the Middendorf (MD) Formation], 23 of the 28 isolates differed by at least one of the physiological characteristics tested. Most of the other formations showed a similar degree of diversity. Diversity was equally apparent for isolates recovered on PTYG and 1% PTYG.

Another widely used technique for estimating microbial diversity in the subsurface is to recover microorganisms from pumped ground water (Hirsch and Rades-Rohkohl, 1983; Stetzenbach, Delley, and Sinclair, 1986; Pedersen and Ekendahl, 1990), count them, and then identify them using physiological criteria. For example, Pedersen and Edendahl, 1990 recovered four species of bacteria from a depth of 463 meters in a granitic aquifer in Sweden. The total plate count for this sample was 3.9×10^4 cells/ml. Of this total, 3.2×10^4 cells were identified as *Pseudomonas fluoresecens*. This data shows that most of the viable (i.e. culturable) cells present at that sampling depth were from a single species. That is to say, the observed diversity was fairly low.

TABLE 6.4 Microbial Diversity in Sediments of the Atlantic Coastal Plain as Determined by Balkwill, Fredrickson, and Thomas (1989)

Site P24			Site P28			Site P29		
	Diversity[a] on			Diversity on			Diversity on	
Depth (m)	PTYG	1 percent PTYG	Depth (m)	PTYG	1 percent PTYG	Depth (m)	PTYG	1 percent PTYG
0	13/13	8/8	0	17/21	11/12			
34	7/7	7/7	14	9/10	10/12	0	16/16	9/10
						8	25/28	9/9
45	13/16	7/7				29	26/34	10/10
58	17/19	10/10	31	8/9	15/15			
91	10/11	9/9	59	9/17	19/20	39	9/11	10/10
118	16/19	10/10	72	15/19	7/8	69	0/0	0/0
139	13/14	13/14						
145	6/7	3/3	112	14/15	15/15	94	9/12	10/10
180	15/22	8/9	115	0/0	1/1			
200	14/18	17/17	134	19/21	9/10	111	3/3	0/0
204	13/13	23/26	162	22/28	10/13	141	22/27	17/19
234	0/0	0/0						
244	11/13	14/17	180	12/13	15/15	151	10/17	13/16
255	0/0	0/0	182	11/11	5/5	176	15/15	10/10
259	23/28	10/12	191	16/19	7/9	181	0/0	4/4
265	9/10	2/2	203	11/14	13/13	187	19/22	12/15
			214	7/7	10/11	193	6/7	12/12
			216	10/11	6/6	200	11/12	8/9

[a] Diversity is presented as the number of physiologically distinct types per number of isolates tested. *Physiologically distinct types* are defined as those that differ in at least one (but usually several) of the 21 physiological characteristics assayed with the API Rapid-NFT kits.

Source: Reprinted courtesy of the American Society for Microbiology.

In a similar study, Stetzenbach, Delley,, and Sinclair (1986) identified bacteria isolated from ground water produced from deep wells in Tucson, Arizona. This data (Table 6.5) showed that more than 70 % of the total viable microorganisms present were members of the genus *Acinetobacter*. Almost 10 % of the microorganisms present were an unidentified pigmented organism. It is evident that, as in the example given by Pedersen and Ekendahl (1990), the diversity of the ground-water community is fairly low. It appears, therefore, that ground-water microbial communities are considerably less diverse than sediment-bound communities.

6.3.2 Niches and Sources of Microbial Diversity

In natural systems, habitats are seldom uniform; rather, there is considerable variation in the availability of nutrients and the suitability of living space. A *niche* is defined as the physical space and immediate environmental conditions within which an organism survives and reproduces. A *fundamental niche* is determined by the organism's genotype and is all of the conditions under which it may survive.

TABLE 6.5 Identification and Enumeration of Bacteria Isolated from Two Deep Wells in Tucson, Arizona

Organism Isolated	Well Number 1		Well Number 2	
	Number	Percent	Number	Percent
Acinetobacter sp.	183	70.4	95	37.4
PO*	25	9.6	95	14.2
Moraxella sp.	7	2.7	38	14.9
Flavobacterium sp.	9	3.4	29	11.4
Pseudomonas/Alcaligenes	1	0.4	3	1.2
Gram-positive rods	13	5.0	17	6.7
Gram-positive cocci	3	1.2	4	1.6
No growth on subculture	19	7.3	32	12.6
Total number of cfu	260	100	254	100
Mean cfu/ml	236		508	
Percent Gram-negative rods	86.5		79.1	

* PO = pigmented organism
Source: Data from Stetzenback, Delley, and Sinclair (1986). Reprinted courtesy of the National Water Well Association.

The actual niche occupied by an organism, however, is generally limited by competition with other organisms and is referred to as the *realized niche*.

There are a wide variety of niches available in saturated subsurface environments. Some of the most readily apparent are shown in Figure 6.7. There are large pores ($> 2\ \mu$), small pores ($<0.2\ \mu$), pores with open throats, pores with closed throats, grains composed of different minerals, and particles of organic matter that may also differ in composition. Differences in mineralogy alone provide a wide range in potential niches. For example, a microorganism growing on a microcline grain would have a potential source of potassium whereas a microorganism growing on a quartz grain would not. A microorganism growing on an organic-matter particle would have a source of solid carbon whereas one growing on a mineral grain would be limited to sources of dissolved or adsorbed organic carbon.

In general, one might expect a functional relationship between the number of exploitable niches in an ecosystem and the diversity of organisms. This depends, however, on how much each of the niches overlap and on the interaction between organisms. If there is significant overlap between niches and the organisms compete strongly, then a single organism may become dominant and the ecosystem may exhibit low diversity. Alternatively if competition is weak, there may be considerable diversity even if there is overlap of niches.

In ground-water systems, it would seem that some niches are largely isolated (i.e. a pore within a feldspar grain) and others have considerable overlap (a grain boundary between a quartz and feldspar grain). The observed high degree of diversity thus may represent the high number of available niches, the lack of competition, or both. It is likely that both of these factors enter in to the development of diversity in ground-water systems.

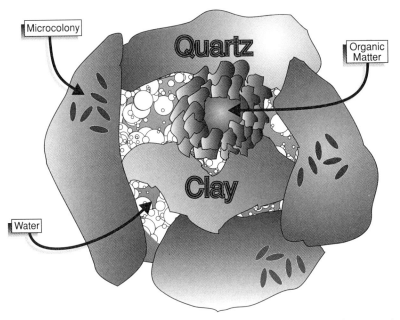

FIGURE 6.7. Niches available for exploitation by subsurface microorganisms.

One possible explanation for the observed low microbial diversity found in ground water pumped from aquifers, as opposed to the high diversity found in sediments, is that ground-water samples represent a single niche. This niche would be the pore-water present in the relatively large, open pores between sand-sized mineral grains. Water in these large pores is subject to less surface tension forces than water in small pores and is preferentially drawn into pumping wells. The large-pore niche, which would tend to be populated by free-living rather than attached microorganisms, would be expected to exhibit much less diversity than the sediment as a whole. This, in turn, could explain the lower diversity in ground water relative to sediments.

6.3.3 Stress and Microbial Diversity

In addition to responding to the availability of nutrients and the abundance of niches, microbial diversity also responds to environmental stresses. The impact of environmental stress on microbial diversity has been extensively studied in surface water systems. These studies show a measurable decrease in microbial diversity as the load of pollutant increases. For example, Larrick et al. (1981) examined the diversity of heterotrophic bacteria as influenced by effluents from fossil fuel power plants. Total bacterial counts at his stations, as measured by the number of colony-forming units (cfu), remained fairly constant with increasing stress from the effluents. However, the proportion of pigmented bacteria within the community correlated with physiochemical stresses. At an unpolluted reference site, the proportion of pigmented bacteria was about 13% of total cfu. At a heavily polluted ash basin, in contrast, pigmented bacteria made up almost 60 % of the cfu.

A decrease in diversity due to pollution stress is often, but not always, observed. For example, in a study of sewage effluent discharge to Lake Huron, Stevenson and Stoermer (1982) found high diversity of diatoms in areas receiving effluent. They hypothesized that species diversity is low in algal populations that either grow very rapidly (low stress) or in populations that grow very slowly (high stress). Intermediate growth rates, however, seem to encourage greater species diversity. Thus, species diversity could be greatest at a moderate level of pollution stress. This example points out that there is not always a simple relationship between species diversity and environmental stress.

Systematic studies of how diversity in subsurface microbial communities varies with stress are largely lacking. Based on analogy with surface water systems, however, several effects would be anticipated. In pristine systems, highly oligotrophic aquifers (high stress) might be expected to have lower diversity than moderately oligotrophic aquifers. Similarly, the onset of pollution due to chemical spills or migration of landfill leachates (high stress) might decrease species diversity in a moderately oligotrophic (low stress) aquifer. This decrease in species diversity may coincide with an actual increase in cell numbers as a few well-adapted strains become dominant. Because changes in microbial diversity may either decrease or enhance the bioremediation potential of a system, this is a topic that deserves detailed study.

6.4 POPULATION INTERACTIONS

The ecology of microbial communities is strongly affected by interactions between populations that make up the community. The term "population" in this context refers to organisms that are phylogenetically related to each other. In subsurface environments that lack a community of higher plants and animals, the most important interactions are those between microbial populations. Interactions between microbial populations are extremely varied. *Positive interactions* are those that enhance the abilities of populations to survive within a community. In some cases, development of positive interactions allows the exploitation of resources that would otherwise not be available. This, in turn, allows some microbes to survive in environments that would otherwise be closed to them. *Negative interactions* are those that are in some way detrimental to certain populations. Negative interactions act as negative feedback mechanisms that limit population densities. These feedback mechanisms, in turn, adjust populations to the resources of the habitat. Negative interactions also tend to preclude the invasion of an established community by allochthonous populations and tend to maintain community stability.

A classification of various possible interactions between two populations is shown in Table 6.6. As can be seen from this table, some interactions are beneficial to each population, some are beneficial to one population and detrimental to another, and some are detrimental to both. Other interactions have a neutral effect on one or both populations.

TABLE 6.6 Types of Interactions Between Two Populations and the Effects of Interaction on Each Population

Name of Interaction	Effect of Interaction	
	Population A	Population B
Neutralism	0	0
Commensalism	0	+
Synergism (protocooperation)	+	+
Mutualism (symbiosis)	+	+
Competition	−	−
Antagonism	− or +	−
Predation	+	−
Parasitism	+	−

0 = no effect.
+ = positive effect.
− = negative effect.
Source: Reprinted from Atlas (1984), with permission, courtesy of Macmillan Publishing Company.

6.4.1 Neutralism

The first kind of possible interaction is actually a lack of interaction. This is called *neutralism* (Table 6.6). Neutralism occurs if, for some reason, populations are physically separated from each other. Furthermore, neutralism is more likely to occur under conditions of low population density which increases the chances for lack of contact. The most common example of neutralism would be the coexistence of a living bacterial population with endospores. Because the bacteria cannot use resources needed by endospores (which need none) and do not prey on or otherwise degrade endospores, bacteria have no effect on the endospore population. Similarly, because the endospores do not use resources or attack bacteria, the endospore population does not affect the bacterial population.

In subsurface environments characterized by large numbers of potential niches and low population densities, neutralism is probably very common. For example, populations growing exclusively on dispersed microcline grains would have negligible interaction with populations growing exclusively on equally dispersed biotite grains. Similarly, populations inhabiting closed pore spaces would have virtually no interaction with those in interconnected pore spaces. There is no direct evidence for the presence (or absence) of neutralism in subsurface environments. This almost certainly reflects the lack of techniques for documenting such interactions.

6.4.2 Commensalism

In commensalism (Table 6.6), one population benefits from the interaction and the other population is unaffected. This type of interaction between bacterial populations probably is also very common in subsurface environments. The best, and probably most widespread, example of a commensal interaction in subsurface environments is consumption of oxygen. In and near outcrop areas of aquifers, oxygen is supplied to the system by vertically percolating recharge. As ground water flows downgradient, oxygen is systematically removed by aerobic and facultatively anaerobic bacteria. The removal of this oxygen by the aerobically respiring population creates the conditions required by obligately anaerobic bacteria. Thus, the anaerobic bacteria benefit by the presence of aerobes in aquifer systems. Since aerobes and anaerobes inhabit different niches, however, the aerobes are neither positively nor negatively affected by the anaerobic population.

In some commensal interactions, one microbial population physically alters the habitat in such a way as to allow a second population to exist. Another possible way this occurs in subsurface environments is the presence of boring bacteria. These bacteria bore though solid grains, presumably to obtain substrates for growth. It is possible, however, for these borings to be opportunistically colonized by other bacteria as well. In this case, the opportunistic bacteria benefit with no apparent corresponding benefit for the boring bacteria.

Another common commensal interaction is based on the production of specific growth factors. Some organisms produce specific compounds, vitamins or proteins for example, and excrete them to the surrounding environment. These compounds can be taken up and used by other organisms to support growth. As long as these compounds are produced in excess and the producing organism derives no particular benefit in their removal, this interaction is commensal. It is common for this relationship to develop in communities of free-living bacteria and algae in aquatic environments. While there have been no report of such relationships between subsurface microorganisms, this situation probably occurs quite commonly.

6.4.3 Synergism and Mutualism

Synergism (protocoorperation) and mutualism (symbiosis) refer to relationships that benefit both populations. The difference is that two synergistic populations, while benefiting from interaction, are capable of living without each other. Symbiotic populations, on the other hand, require interaction to sustain life. The distinction between synergism and symbiosis has not always been made. Symbiosis was originally used to describe any close relationship between populations. In present usage, however, the term *symbiosis* is restricted to interactions that are obligatory.

In microbial ecology, a classic example of synergism is the relationship between *Streptococcus faecalis* and *Lactobacillus arabinosus*. *L. arabinosus* requires phenylalanine, which is produced by *S. faecalis,* for growth. Similarly, *S. faecalis* requires folic acid for growth, and *L. arabinosus* happens to produce folic acid. These two microorganisms are quite capable of growing independently of each other, as long as these specific nutritional requirements are present in the media. However, if media are prepared that lacks both folic acid and phenylalanine *L. arabinosus* and *S. faecalis* can grow together in culture but they cannot grow separately.

Synergistic relationships are very important in the ecology of subsurface environments. The best example is the relationship between fermentative and respirative microorganisms in anaerobic ground-water systems. Fermentative bacteria are able to utilize directly complex organic matter, carbohydrates and lignins, present in subsurface environments to supply energy for growth. However, because fermentative metabolism cannot completely oxidize organic compounds (see section 3.6), fermentation products, such as hydrogen and acetate, still contain significant amounts of energy. Anaerobic respirative bacteria Fe(III) reducers, sulfate reducers, methanogens) are incapable of directly metabolizing carbohydrates or lignins, but are capable of metabolizing fermentation products. Thus, the presence of fermenting bacteria encourages the growth of respirative bacteria. Furthermore, since acetate and hydrogen are waste products, at least as far as the fermenters are concerned, their consumption by respirative bacteria encourages fermentation. Thus, both populations benefit.

The classic example of a symbiotic relationship between microorganisms is the growth of lichens. Lichens consist of photosynthetic algae or cyanobacteria and heterotrophic fungi. The algae fix atmospheric carbon dioxide and in some cases atmospheric nitrogen that provides the substrate for growth of the fungi. In return, the fungi produce organic acids that help dissolve rock material allowing for further growth of the algae. These symbiotes inhabit extremely cold and dry environments of the Arctic desert. Although not capable of surviving this extreme environment individually, together these two species can survive and propogate

Symbiosis has been closely studied by microbial ecologists for a number of reasons. One is that many ecologists feel that the formation of symbiotic relationships is a mechanism by which two populations unite to form new species. Many current theories of microbial evolution point to the structural similarities between organelles of eucaryotic cells (e.g. mitochondria, chloroplasts) and procaryotic microbes. The consensus is that eucaryotic cells developed through the formation of symbiotic relationships between procaryotic microorganisms.

6.4.4 Competition

Different microbial populations in particular environments often strive to utilize the same resource. In this case, the relationship between the populations becomes competitive. The resource that is the object of competition can include nutrients, moisture, space, or any other commodity that is necessary for life functions and is in scarce supply.

Competitive interactions tend to separate populations in ecosystems. In microbial ecology there are numerous well-documented examples of population separation by means of competitive interaction. Perhaps the classic example is the separation of methanogenic and sulfate reducing activity in aquatic sediments. It is observed ubiquitously in anaerobic marine sediments that a sulfate-reducing zone develops near the sediment-water interface and is underlain by a methanogenic zone (Froelich et al., 1979). For many years it was thought that this zonation reflected some direct inhibition of methanogenesis by sulfate-reducing bacteria. However, studies performed in M. J. Klug's laboratory at Michigan State University have shown that this separation reflects competition between sulfate reducers and methanogens for the same substrates (Lovley and Klug, 1983; Lovley and Klug, 1986).

In anaerobic sediments, both methanogenic and sulfate-reducing populations are supported by fermentation products such as acetate and hydrogen. These microorganisms thus compete directly for the same substrates. Because sulfate reducers have a higher affinity for acetate and hydrogen than methanogenic bacteria, sulfate reducers outcompete methanogens. Furthermore, because the sulfate reducers lower concentrations of these substrates below the threshold required by methanogens, methanogens are effectively excluded from the environment. This is termed *competitive exclusion*. Sulfate reducers continue to dominate the ecosystem until concentrations of dissolved sulfate are exhausted. Once sulfate reducers can no longer respire, concentrations of hydrogen and acetate increase to levels required by methanogens, and methanogenic bacteria become dominant.

It's interesting to note that methanogens may have both a commensal and competitive relationship with sulfate reducers. While methanogens cannot successfully compete with sulfate reducers for fermentation products, methanogens *depend* on the sulfate reducers to deplete pore water with respect to sulfate, thus providing the sulfate-free environment they require.

Competitive exclusion is very important in the ecology of subsurface environments and this importance is reflected in many commonly observed ground-water chemistry patterns. For example, one of the most frequently occurring ground-water quality problems is the presence of high concentrations of dissolved iron. Furthermore, it is often observed that these high concentrations of dissolved iron occur in discrete zones. The Middendorf aquifer of South Carolina is characterized by a 40-mile wide zone where concentrations of dissolved iron commonly exceed 0.3 mg/L (Fig. 6.8). Chapelle and Lovley (1992) have shown that this zonation of high-iron ground water reflects the competitive exclusion of sulfate-reducing and methanogenic bacteria by Fe(III)-reducing bacteria. Fe(III) reducers have a higher affinity for hydrogen, acetate, and formate than either sulfate reducers or methanogens. If suitable Fe(III) oxyhydroxides are present in the sediments, Fe(III) reducers lower concentrations of these substrates below thresholds required by sulfate reducers or methanogens, effectively excluding these communities.

Figure 6.9a shows that, in the high-iron zone where sediments contain abundant Fe(III) oxyhydroxides, concentrations of acetate and formate are significantly higher than downgradient of the low-iron zone (Fig. 6.9b) where sediments contain

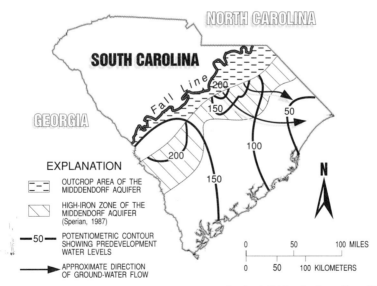

FIGURE 6.8. Zonation of high-iron ground water in the Middendorf aquifer. (Reproduced from Chapelle and Lovley, 1992, with permission courtesy of the National Water Well Association).

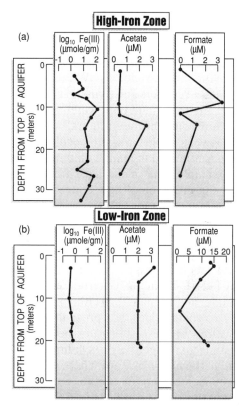

FIGURE 6.9. Concentrations of solid Fe(III) oxyhydroxides and dissolved pore water acetate and formate in (a) the high-iron zone and (b) the low-iron zone of the Middendorf aquifer. (Reproduced from Chapelle and Lovley, 1992, with permission courtesy of the National Water Well Association.)

little Fe(III) oxyhydroxides. This competition is reflected in the water chemistry as well. In the high-iron zone, hydrogen concentrations are in the 0.1 to 2.0 range characteristic of Fe(III) reduction. As Fe(III) oxyhydroxides become less abundant, concentrations of hydrogen increase to sulfate-reducing levels and dissolved iron concentrations decrease. This decrease reflects sulfide production by sulfate-reducing bacteria and subsequent precipitation of iron sulfides.

An interesting and-from a geoscientists viewpoint very important-aspect of this example is that the presence or absence of solid Fe(III) oxyhydroxides is largely controlled by geologic factors. In the updip portion of the Middendorf aquifer, the sediments are largely fluvial in origin, having been deposited in upper-delta plain environments, and therefore contain significant amounts of Fe(III) oxyhydroxides that occur as grain coatings. The lack of Fe(III) oxyhydroxides in downdip sediments reflects a change in depositional environments from upper-delta plain (above tidal influence) to lower-delta plain (marginal marine) and delta front (open marine) sediments. Fe(III) oxyhydroxide-coated clays typically flocculate as salinity increases and therefore are not as common in sediments characterized by marine influence.

This example shows that the outcome of microbial competition in aquifer systems reflects *both* microbiologic and geologic factors. From a microbiologic viewpoint, Fe(III) reduction excludes sulfate reduction in the high-iron zone because of the presence of Fe(III) oxyhydroxides in the sediments. However, it is the geological and sedimentological framework of the system that determines the distribution of Fe(III) oxyhydroxides. Thus, understanding the outcome of microbial competition in aquifer systems requires both a microbiologic and geologic perspective.

Similar patterns of competitive exclusion have been reported from contaminated aquifers as well. Beeman and Suflita (1987) studied a shallow water table aquifer in Oklahoma that had been contaminated by municipal landfill leachate. Most of the contaminated zone was characterized by high organic carbon concentrations and low sulfate concentrations. In these areas, methanogenesis was the predominant terminal electron-accepting process. In one localized area, however, the ground water contained significant concentrations of sulfate and sulfate reduction was the predominant process. Significantly, while methanogenic bacteria were more numerous than sulfate reducers in the methanogenic zone, sulfate reducers were present. Similarly, while sulfate reducers were more numerous than methanogens in the sulfate reducing zone, methanogens were also present. Thus, as was the case with Fe(III) reducers and sulfate-reducers in the Middendorf aquifer, it is apparent that competition inhibits the activity of microorganisms that are at a competitive disadvantage, but does not lead to their total exclusion.

6.4.5 Antagonism, Parasitism, and Predation

When one population produces a substance that inhibits another population, the interaction is termed *antagonistic*. The population that produces the inhibitory substance thus gains a competitive advantage over the other. The production of antibiotics by some strains of bacteria and fungi are thought by some microbial ecologists to be an example of antagonism in nature. Curiously enough, however, it has not been unequivocally shown through ecological studies that antibiotic production actually confers a competitive advantage in natural systems. One reason for this may simply be that virtually all studies of antibiotic production are performed by medical microbiologists who are not specifically concerned with ecological interactions in nature.

Studies showing the presence or absence of antagonistic interactions between microorganisms in ground-water systems have not been performed. It is notable, however, that numerous strains of bacteria characterized by antibiotic resistance have been isolated from subsurface environments (Fredrickson et al., 1988). It is possible that this resistance reflects antibiotic production by some species in the subsurface and thus may reflect antagonistic interactions. On the other hand, given the low population densities characteristic of subsurface environments, it's not clear that antibiotic production would confer much of a competitive advantage. As antibiotic production in different environments has been and will remain a topic of practical interest, studies on antagonistic relationships in subsurface environments are certainly warranted.

A *parasitic* relationship occurs when one population derives its nutritional requirements from hosts of another population. The parasitic population is clearly benefited by this relationship, generally to the detriment of the host population. Host-parasite interactions require long-term contact between the populations, and members of the parasitic population are generally much smaller than members of the host population.

The most common host-parasite interaction in microbial communities is between viruses and bacteria. Viruses are obligate parasites and are able to reproduce only inside the cells of higher organisms. Recent investigations have shown that a rather high percentage of planktonic marine bacteria harbor viruses. This shows, in turn, that parasitic interactions are rather common in that environment. Studies of viruses in aquifer systems have focused on the transport of human viruses rather than on viruses that parasitize indigenous bacteria. Thus, the extent to which virus parasitism exists in ground-water systems is largely unknown.

When one population actively consumes another population, the interaction is termed *predation*. The predatory population derives nutrients from this interaction, and the interaction is clearly beneficial. Conversely, such interaction is clearly detrimental to the prey.

The most common predator-prey relationship in microbial ecology is the interaction between protozoa and bacteria. Many protozoa actively feed, or graze as it is often termed, on bacterial populations. Early studies of the microbial ecology of subsurface environments noted a lack of protozoa in ground-water systems (Ghiorse and Balkwill, 1983), but subsequent studies have documented that protozoa are fairly common in aerobic aquifer systems (Sinclair and Ghiorse, 1989; Sinclair et al., 1990). Furthermore, there is evidence that protozoa actively graze on bacterial populations in some systems (Kinner et al., 1990). The observed higher numbers of protozoa in contaminated water table aquifers suggest that predation may act to limit bacterial degradation of contaminants in some systems. This important possibility needs to be explored in a variety of systems.

6.5 r AND K STRATEGIES IN MICROBIAL ECOSYSTEMS

Different populations in a microbial community employ different metabolic and reproductive strategies in their competition with each other. These different strategies allow different niches within a habitat to be successfully colonized. One scheme for classifying such strategies is to view them as falling along a gradient between pure r (high rate of reproduction) strategists and pure K (maximum carrying capacity) strategists (Fig. 6.10). In this scheme, r strategists rely on their high rates of reproduction to maintain themselves in the community. Because of their particular life style, r strategists tend to be characterized by catastrophic mortality events. The r strategists tend to have short life spans, high productivity-low efficiency utilization of resources, high migratory tendency, and variable population size. Conversely, the more resource-efficient K strategists are characterized by lower reproductive rates, longer life spans, low productivity-high efficiency utilization of resources, low migratory tendency, and stable population size. Clearly, most microorganisms are neither pure r nor pure K strategists but fall somewhere between the two end members (Fig. 6.10).

The possible effects of r or K strategies on species survival can be illustrated by considering the emergence of a new volcanic island in the Pacific Ocean. The

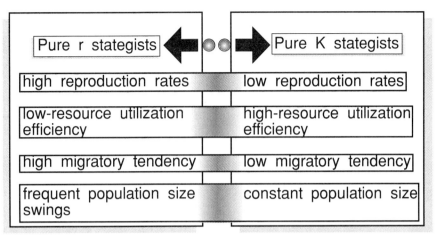

FIGURE 6.10. Attributes of pure r and K strategists.

creation of land obviously creates a new habitat that can potentially be exploited by plants and animals. In colonizing this new habitat, r strategists have an immediate advantage. Plants that produce highly mobile seeds (palm trees, for example) reach the new habitat first. Once on the island, the plants that can reproduce the fastest are the ones that can claim the most space. Since resources (space) is not limiting, there is little initial advantage to use available resources efficiently. Clearly, r strategists have the advantage over K strategists when colonizing a new habitat.

As time passes on the island, however, and as more and more plants come and colonize, all of the available niches become crowded. Shortages of resources begin to appear and there might be mass mortality events among the r strategists. The ability to reproduce quickly becomes less important and the ability to utilize resources efficiently becomes more important. Plants with K strategy physiological adaptations-adaptations that allow successful reproduction in space-limited conditions-have the advantage. Clearly, as time goes on K strategists will gradually occupy most of the available niches.

The concept of these two end-member strategies has been around a long time. Winogradsky, the great soil microbiologist of the early twentieth century, referred to some microbial species as being "zymogenous," or opportunistic. There are obvious similarities between the concept of zymogenous and r-selected microorganisms. Similarly, Winogradsky's concept of "autochthonous", or indigenous, microorganisms is analogous to K strategists.

6.5.1 r and K Strategies in the Aquifer Environment

At first glance, pristine subsurface environments would appear to be ideal for K-selected microorganisms, since there has been long-term stability in ambient conditions and nutrients are in limited supply. Indeed, available evidence suggests that K-selected microorganisms dominate in subsurface environments. For example, population densities appear to be fairly constant spatially (Balkwill, Fredrickson, and Thomas, 1989), reproduction rates are very low (Harvey and George,

1987), and separation of populations by competitive exclusion mechanisms are evident (Chapelle and Lovley, 1992). Furthermore, the presence of fairly diverse microbial communities (Balkwill, 1989) is consistent with the predominance of K strategies.

Interestingly, however, there is evidence that some microorganisms in pristine subsurface environments retain some r strategy characteristics. For example, in an experimental investigation Hirsch and Rades-Rohkohl (1990) found that sterilized sediments exposed to ground water were rapidly (within 12 weeks) colonized by bacteria. Furthermore, the diversity of the colonizing microorganisms was considerably less than the microbial diversity in the surrounding aquifer sediments. This evidence suggests that r strategists that are able to take advantage of a new habitat do exist in subsurface environments.

Ghiorse and Wilson (1988) suggest an intriguing hypothesis that may explain the long-term presence of r strategists in subsurface environments. Ghiorse and Wilson were concerned with explaining why subsurface bacteria appear to divert considerable carbon resources to synthesize exopolysaccharides—a diversion that appears wasteful of scarce resources. They suggest that organic carbon particles in aquifers serve to absorb and concentrate dissolved organic carbon transported in from the overlying soil zone. As organic carbon absorption continues, the carbon particle becomes capable of supporting microbial growth. If a microorganism comes in contact with this enriched particle, it is crucial that it bind strongly to the particle in order to exploit the resource fully. This neatly explains the presence of exopolysaccharides, which enable the microorganism to bind firmly to surfaces. After the organic resource is exhausted, the cells die or disperse, and the organic particle again begins to concentrate dissolved organic carbon. Ghiorse and Wilson (1988) thus envision a cycle of nutrient concentrations, colonization by microorganisms, exploitation, and finally dispersal (Fig. 6.11). Implicit in this cyclic model is the steady-state presence of r strategists capable of moving to and rapidly exploiting resources. Thus, the continued presence of r selected microorganisms in subsurface environments may not be as surprising as it might appear at first glance.

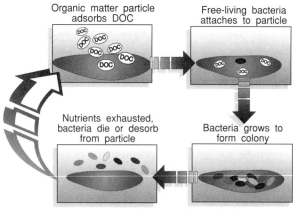

FIGURE 6.11. Cycle of microbial colonization and dispersal that may support populations of r strategists in nutrient-poor subsurface environments.

6.6 SUMMARY

The ecology of microorganisms in subsurface environments may be studied using a number of different techniques, including culture methods, direct observations, biochemical marker methods, activity measurements, geochemical methods, and ecological modeling. Each of these methods has advantages and disadvantages. Ultimately, the methods used for a particular problem must reflect the nature of the questions being asked.

Subsurface environments present a number of different niches that may be exploited by subsurface microorganisms. Ground water in relatively large pores, for example, may support a free-living population of bacteria. Alternatively, bacteria may attach themselves to a variety of aquifer materials including mineral grains and organic matter. The large number of potentially exploitable niches in aquifers apparently leads to highly diverse microbial populations in many ground-water systems.

Populations interact in a number of different ways in ground-water systems. Neutral interactions are neither positive nor negative for different populations. This may be the case for bacteria that depend on different carbon sources. In a commensal relationship, one population benefits from the interaction whereas the other is unaffected. In a synergistic relationship, such as between fermenting and respiring bacteria, both populations benefit. Symbiosis refers to a relationship in which interaction is a prerequisite for life. Relationships between populations can also be competitive, which is the case when each population exploits the same resource. Competitive relationships between different terminal electron-accepting microorganisms, each of which attempts to use the same carbon sources, are very common in subsurface environments. In particular, the competitive exclusion of sulfate reduction by Fe(III)-reducing bacteria leads to very distinctive zones of high-iron ground water in some systems.

Different microbial populations employ different metabolic and reproductive strategies in competing with each other. Pure r strategists rely on high rates of reproduction, while sacrificing efficient utilization of resources, to maintain themselves. Pure K strategists, on the other hand, rely on the efficient utilization of resources to maintain themselves and sacrifice high rates of reproduction. The low-nutrient conditions that are characteristic of subsurface environments appear to favor efficient K strategists in many systems. However, there is evidence that fast-colonizing r strategists are also present in many systems. This may reflect the way in which carbon is cycled in subsurface environments.

PART II

MICROBIAL PROCESSES IN PRISTINE GROUND-WATER SYSTEMS

CHAPTER 7

ABUNDANCE AND DITRIBUTION OF BACTERIA IN THE SUBSURFACE

The study of bacterial processes and how they affect ground-water geochemistry begins with consideration of the abundance and distribution of bacteria. That bacteria inhabit virtually all subsurface environments is not currently a matter of dispute. However, the types of bacteria present, their abundance, and the kinds of physiologic processes that they carry out differ widely from environment to environment.

In order to deal with bacterial abundance and distribution in subsurface environments, it is first necessary to draw some fundamental distinctions between the types of environments most commonly encountered. This requires some basis on which to classify subsurface environments.

7.1 CLASSIFICATION OF SUBSURFACE ENVIRONMENTS

One possible basis for classifying subsurface environments, and one that is often used informally, is one of depth. It seems reasonable that deep subsurface environments are fundamentally different from shallow subsurface environments. Closer inspection of this as a classification strategy, however, immediately points out several problems. First, "deep" and "shallow" are relative terms that have different meanings in different geologic settings. "Deep" to a soil scientist would mean depth in terms of meters, whereas to a petroleum geologist it would conjure up thoughts in terms of 10,000 meters. Aside from the semantic problems, the terms "deep" and "shallow" give no information concerning the hydrologic setting of the environment. A depth of 200 meters in the Tucson Basin of Arizona, for example, would be an aerobic water-table aquifer. In the Illinois Basin, on the other hand, a similar depth might intersect an accumulation of petroleum. Obviously, these two environments have little in common despite the fact that they occur at the same depth.

In describing microbial processes in subsurface environments, the hydrologic setting is probably the most important frame of reference. The hydrologic setting is what determines such important factors as the amount of communication with the surface, rates of ground-water movement, and transport of dissolved substrates. Thus, classifying subsurface environments based on their hydrologic attributes is logical when microbial processes are considered. Also, because the physics of ground-water flow does not vary between diverse hydrologic systems, such classification can be applied universally.

The most fundamental distinction between subsurface environments that may be drawn on the basis of hydrology is the difference between the saturated and unsaturated zones. The unsaturated zone is characterized by pore spaces that are incompletely filled with water. Pore space that is not filled with water is filled with gas. The significance of the unsaturated zone to hydrologists is that capillary forces between water and sediment particles prevent water from flowing to wells. Thus, wells completed in the unsaturated zone do not yield appreciable quantities of water. The amount of water present in the unsaturated zone varies widely and is highly sensitive to climatic factors.

Once the distinction between saturated and unsaturated hydrologic systems is made, the next step is to classify different types of saturated subsurface environments. Hydrologists have long recognized, on a qualitative basis, three categories of ground-water flow systems (Norvatov and Popov, 1961). These include an "upper zone" of active flow strongly influenced by local precipitation events, a "medium zone" of deeper flow only moderately affected by local precipitation events, and a "lower zone" of relatively stagnant water unaffected by local precipitation. This "zonation" was recognized entirely on an empirical basis and suffers from a vague notion of the importance of depth.

In the early 1960's J. Toth, a Canadian hydrologist, showed that the zonation of subsurface hydrologic systems follows logically from the physics of ground-water flow (Toth, 1963). By solving the steady-state ground-water flow equation

$$\partial^2\phi/\partial x^2 + \partial^2\phi/\partial z^2 = 0$$

for a hypothetical basin bounded on the sides by a ground-water divide and a valley bottom,

$$\partial\phi/\partial x = 0 \quad \text{at } x = 0 \text{ and } s \text{ for}$$
$$0 \le z \le z_0$$

bounded below by an impermeable layer,

$$\partial\phi/\partial z = 0 \quad \text{at } z = 0 \text{ for}$$
$$0 \le x \le s$$

and bounded above by an water-table surface that follows undulating topography.

$$\phi_i = g(z_0 + c'x + a' \sin b' x) \quad \text{at } z = z_0 \text{ for}$$
$$0 \le x \le s$$

Based on the governing equation and boundary conditions, Toth (1963) calculated lines of equal potential, termed *equipotential lines,* and from these deduced directions of ground-water flow. In this context, the term *flow system* is defined rigorously as "a set of flow lines in which any two flow lines adjacent at one point of the flow region remain adjacent throughout the whole region." Figure 7.1 illustrates this concept. In the upper part of the system, the flow lines cluster into "cells" in which water recharges the aquifer at topographic highs and discharges at topographic lows. Because the flow lines in these shallow cells are parallel, they constitute one individual flow system. Deeper in the system, a set of flow lines passes underneath the shallow cells. Again, because these flow lines are parallel, they constitute one individual flow system. Finally, those flow lines that pass through the deepest part of the system constitute a third flow system.

This type of reasoning shows that three distinctly different types of flow systems develop for a wide variety of topographic configurations (Fig. 7.1). These different types of flow systems have these general characteristics:

1. A local system has its recharge area at a topographic high and its discharge area at a topographic low that are located adjacent to each other.
2. An intermediate system is where recharge and discharge areas are separated by one or more topographic highs.
3. A regional system is where the recharge area occupies the water divide and the discharge area occurs at the bottom of the basin.

These three different hydrologic settings are similar to the "zonation" that had been recognized empirically by Norvatov and Popov (1961). For example, recharge rates are greatest in the local (i.e., "upper") systems and progressively less in the intermediate (i.e., "medium") and regional (i.e., "lower") systems. Thus, local systems will tend to respond quickly to local precipitation events whereas intermediate and regional systems respond less or not at all. Also, rates of ground-water flow are much more sluggish in regional than in local systems. Ground-water flow in regional (lower) systems may indeed appear to be "stagnant." There is, however, an important difference between Toth's classification and the empirical classification; the reference to depth is absent from the definition. In fact, Toth's definitions are independent of scale and depend only on the distribution of recharge and discharge to the hydrologic system. Thus, Toth's classification can be more universally applied than the empirical classification.

A major limitation of Toth's argument is that he assumes homogeneous and isotropic hydrologic properties within his hypothetical basins. While this was an obvious oversimplification, it was necessary in the early 1960's because only analytical techniques were available for obtaining solutions to the ground-water flow equation. The advent of numerical methods for solving ground-water flow problems in the late 1960's, however, showed that the development of well-defined local, intermediate, and regional flow systems also occurs in non-homogeneous and geometrically complex aquifer systems (Freeze and Witherspoon, 1967). This added complexity, however, tends to diffract flow lines so that flow in more

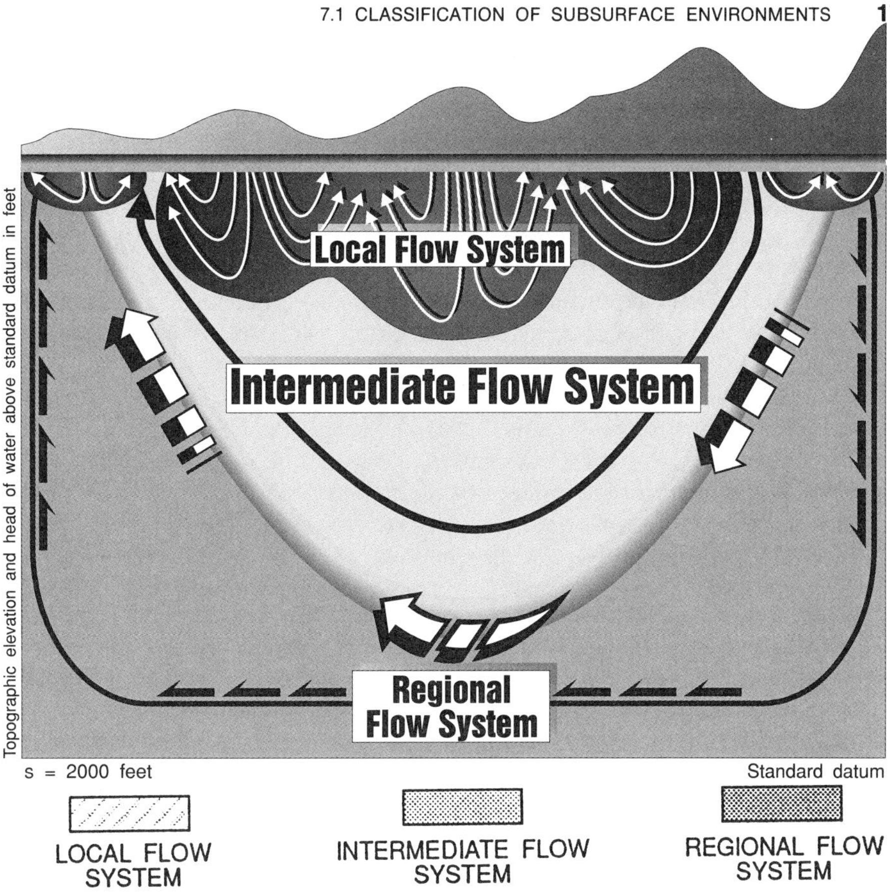

FIGURE 7.1. Delineation of local, intermediate, and regional ground-water flow systems. (Modified from Toth, 1963, with permission courtesy of the American Geophysical Union.)

permeable layers tends to become horizontal whereas flow in less permeable layers becomes largely vertical.

For example, given the common case in which material of low permeability is underlain by material of higher permeability, flow lines in local and intermediate regimes are mostly vertical and flow lines in the regional regime are almost exclusively horizontal. Similarly, if a zone of dipping high permeability material is overlain and underlain by lower permeability material, flow is concentrated in the high-permeability zone. While nonhomogeneity adds significant complexity to patterns of ground-water flow, this complexity occurs within the framework of the three hydrologic regimes identified by Toth (1963).

It is possible, therefore, to distinguish between four types of subsurface environments (Table 7.1). The first of these are unsaturated subsurface environments, which include most soil zones and are identified by the lack of water flow to wells. For saturated environments, a rigorous distinction between local, intermediate, and regional flow systems can be drawn. If water from a recharge area discharges into an adjacent stream, that subsurface environment is defined as "local." If water discharges into a river one or more basins removed from the recharge area,

178 ABUNDANCE AND DISTRIBUTION OF BACTERIA IN THE SUBSURFACE

it is defined as an intermediate subsurface environment. Finally, if water from a recharge area discharges only at the center of a basin, it is defined as a "regional" subsurface environment.

The classification of subsurface environments shown in Table 7.1 emphasizes the different degrees of connection with the surface. Obviously, the unsaturated zone will be most closely connected with the surface, with water entering the system with every major precipitation event. Local flow systems also may be closely connected with the surface and may be strongly influenced by precipitation. Intermediate and regional systems, on the other hand, are much more isolated from events on the surface. The varying degrees of connection with the surface will obviously influence the kinds of bacterial processes that may occur in different environments. For example, the unsaturated zone and local flow systems are often free to exchange gases with the atmosphere. Thus, a continuous source of oxygen is potentially available and aerobic bacteria may be able to continuously respire. In intermediate systems, however, the transport of dissolved oxygen is much more limited and anaerobic processes may predominate. Regional systems are, for practical purposes, entirely isolated from atmospheric oxygen and anaerobic bacteria ubiquitously predominate.

TABLE 7.1 Classifications and Hydrologic Attributes of Subsurface Environments

Connection with Surface Recharge	Flow Rates	Oxygen Status	Potential for Microbial Transport from Surface	Example
Unsaturated				
Extensive	Fast 0.5–5.0 ft/d	Usually aerobic	High	Soil zones
Saturated (Local)				
Extensive	Fast 0.5–5.0 ft/d	Often aerobic	High	Water table aquifer
Saturated (Intermediate)				
Small	Slow 0.5–5.0 ft/yr	Generally anaerobic	Moderate	Confined aquifer
Saturated (Regional)				
Virtually nonexistent	Almost stagnant	Anaerobic	Virtually nonexistent	Deep basin/ petroleum reservoirs

7.2 THE UNSATURATED ZONE

An important hydrologic feature of the unsaturated zone is that it contains both water and air in voids between sediment particles. More practically, water in the unsaturated zone is under negative hydraulic pressure—that is, less than atmospheric pressure—and, therefore, water cannot flow into well bores that penetrate it. The unsaturated zone is important from a hydrologic viewpoint because recharge to aquifers must first move through it. The rate that water moves through the unsaturated zone, therefore, directly controls the rate that an aquifer may be recharged. From a geochemical perspective, the unsaturated zone is important because gases are freely exchanged with water that ultimately recharges aquifers. Recharge water infiltrating through the unsaturated zone is readily charged with highly reactive gases, such as oxygen and carbon dioxide, due to the coexistence of the gas and aqueous phases. The carbon dioxide input is particularly important in studies of carbon isotopes in ground water (Rightmire and Hanshaw, 1973).

Microbial processes in the unsaturated zone are important for several reasons. First, bacterial processes such as oxygen consumption, carbon dioxide production, nitrification, and denitrification, will directly impact the chemistry of infiltrating water reaching the water table. Secondly, anthropogenic compounds such as petroleum hydrocarbons, herbicides, pesticides, and a host of others must pass through the unsaturated zone in order to impact ground-water quality. Because many of these compounds are biologically active, bacterial processes in the unsaturated zone will affect their fate and transport.

The unsaturated zone is frequently divided into three components (Fig. 7.2). The first of these is the *soil zone,* generally 1 or 2 meters thick, which contains

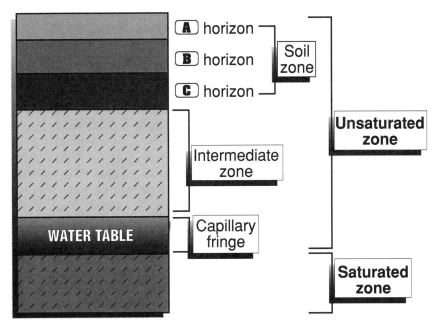

FIGURE 7.2. Three components of the unsaturated zone.

living roots and which supports plant growth. The porosity and permeability of the soil subzone is generally higher than that of underlying material. This underlying material, which varies in thickness from place to place, is often referred to as the *intermediate zone* (not to be confused with the "intermediate" zones of saturated flow systems) and consists of sediments or rocks that have not been exposed to extensive pedogenic (soil forming) processes. The boundary between the unsaturated zone and the saturated zone is termed the *capillary fringe*.

Capillarity, the process that forms the capillary fringe, results from two forces: the mutual attraction (cohesion) between water molecules and the molecular attraction between water and solid mineral surfaces. Because of the forces, water will rise into small tubes to varying levels above water in a container. Pores in sediments are often of capillary size and, therefore, water is pulled from the saturated water table into the unsaturated zone. As a general rule, finer-grained sediments have thicker capillary fringes (up to 1 meter) than sandy sediments (generally 1—10 centimeters).

The abundance and distribution of bacteria in the unsaturated zone, particularly in soils, has been intensely studied over the last 150 years (Alexander, 1977). Much of the impetus for this research has come from the recognition that microbial processes are directly involved in soil fertility and, therefore, agricultural productivity. For example, it had been known for hundreds of years that composting manure under aerobic conditions leads to the accumulation of nitrate, an important component of plant fertilizers. It was the soil microbiologist S. N. Winogradsky who showed that this process is brought about by bacteria oxidizing the ammonia present in manure to nitrates. In 1891, Winogradsky isolated in pure culture a bacterium that carried out this nitrification process. At about the same time, Winogradsky isolated an anaerobic bacterium of the genus *Clostridium* that was capable of nitrogen fixation: the process of converting atmospheric nitrogen into organic nitrogen. In 1888, the Dutch bacteriologist M.W. Beijerinck isolated a nitrogen-fixing bacteria from the roots of legumes, a microorganism later assigned to the genus *Rhizobium*. It's hard to overestimate the importance of these discoveries to modern agriculture, as they form the core of crop fertilization and crop rotation strategies.

The traditional association of soil microbiology with soil fertility and crop production has led to a curious lack of attention to microbial processes in sediments just below the root zone. In fact, the unsaturated zone between the root zone and the water table is probably the least studied of any subsurface environment. In many areas, there are only a few meters between the root zone and the water table, so this lack of attention is understandable. However, in some parts of the world, and particularly in arid climates, there may be hundreds of meters of unsaturated material below the root zone. Just from the standpoint of sheer volume, the deep unsaturated zone in these areas bears consideration.

7.2.1 The Unsaturated Zone as a Microbial Habitat

The unsaturated zone, by virtue of its proximity to photosynthetic organic carbon production at land surface, is the most biologically active and varied subsurface environment. By far the most biologically active part of the unsaturated zone is the soil.

There are five interactive factors that lead to the development of soil: climate, topography, parent material, time, and biologic processes. Soil microbiologists have long recognized that soil organisms play a significant role in the development of their own habitat. This is certainly true, to varying degrees, in other subsurface environments but is most easily observed in soils. Soil formation is initiated by chemical and physical weathering of rocks or sediments. The release of nutrients by these weathering processes allow colonization by algae and lichens. This initiation of primary production leads to the establishment of a heterotrophic bacterial population. In turn, these combined microbial processes speed the weathering processes by increasing the partial pressure of carbon dioxide and by releasing organic acids. Once a suitable combination of weathered rock debris and organic matter is achieved, plants are able to initiate growth. This, in turn, speeds organic carbon production, increases the activity of heterotrophic bacteria, and results in more efficient weathering processes. The development of soil, therefore, may be viewed as a combination of processes that are linked in positive feedback loops; that is, as each process becomes more efficient, the efficiency of the other processes is enhanced.

The net result of soil formation processes is the establishment of a stable, well-defined soil profile at the top of the unsaturated zone (Fig. 7.2). The formation of clay organic complexes results in the stabilization of clay, sand, and silt particles into aggregates. Aggregate formation is initiated when polysaccharides from roots and microorganisms combine with clays to form organic matter and mineral complexes. Microorganisms live in the pores of the soil aggregates, generally attached to particle surfaces. The porosity and size of pore throats in the soil are important structural features because they form living space and access to nutrients.

Moisture and Gas Content The coexistence of aqueous and gas phases in the unsaturated zone has important consequences to microbial processes. All microorganisms require water for cell maintenance and growth. The lack of water is, therefore, extremely detrimental to microbial activity. On the other hand, if pore space is completely filled with water, the transport of such metabolically important gases as oxygen is greatly decreased. Thus, too much or too little water will depress microbial processes.

The presence of a distinct gas phase is a feature that is unique to the unsaturated zone. The most abundant gases are those found in the atmosphere: nitrogen, oxygen, and carbon dioxide. In well-aerated sediments, the oxygen content is near that of the atmosphere ($\sim 20\%$) and the carbon dioxide content is less than 2%. However, high rates of bacterial activity can remove oxygen from the gas phase of the unsaturated zone and increase the carbon dioxide as high as 19%.

The vertical diffusion of gases is governed by Fick's law

$$q = D \, dC/dz$$

where q is the rate of diffusion (mmoles cm^{-2} sec^{-1}), D is the diffusion coefficient (cm2 sec^{-1}), C is the concentration of the gas (mmoles cm^{-3}), and z is depth. In the unsaturated zone, however, the diffusion of gases is considerably more complex than implied by Fick's law. For example, gaseous diffusion will depend upon the percentage of pore space actually filled with gas (as opposed to capillary water) and is subject to rapid changes with time. Also, biologically active gases such as oxygen or carbon dioxide may be consumed or produced as diffusion proceeds. A more accurate description of gaseous transport in the unsaturated zone incorporating these complexities was given by van Bavel, (1951) as

$$q = R + YP_g dC/dz$$

where q, D, and dC/dz are as given in Fick's law and where R is the production or consumption of a gas (mmole cm^{-3} sec^{-1}), Y is a factor that takes the tortuosity of the sediment into account (dimensionless), and P_g is the gas-filled porosity of the sediment (dimensionless).

A quantitative treatment of gas transport in the unsaturated zone of several sites in the western United States has been given by Thorstenson et al. (1983) and Wood and Petraitis (1984). These studies showed that oxygen concentrations decreased while carbon dioxide concentrations increased with depth, reflecting oxygen consumption by aerobic microorganisms and consequent carbon dioxide production. The unsaturated zone is extremely dynamic in terms of CO_2 production and diffusion. For example, the CO_2-depth profile for one site in North Dakota shows that in May, the pCO_2 is higher deep in the unsaturated zone than at land surface. Consequently, there is a diffusive flux of CO_2 from the unsaturated zone to the atmosphere. Later in the summer, however, CO_2 production in the root zone reverses the diffusive flux and there is a net transport of CO_2 from the soil zone downward toward the water table. Although year-round data from this site are not available to document the timing of the diffusive gradient reversal, it is clear that it occurs in the winter when root zone CO_2 production is depressed by cold weather. This site provides an excellent example of the dynamic nature of gas transport in the unsaturated zone.

7.2.2 Biomass Measurements in Soil Microbiology

Many of the methods commonly used in subsurface microbiology were developed by soil microbiologists. For this reason, a brief survey of some techniques commonly used in soil microbiology is of practical as well as historical interest.

Biomass is defined as the living part of soil organic matter, exclusive of roots and soil animals larger than 5,000 cubic micrometers in size. In soils, biomass consists of both bacteria and protozoa. Below the soil zone, however, biomass consists largely of bacterial cells. While protozoa are commonly found deeper in the unsaturated zone, their total biomass is generally small compared to that of the bacteria.

Measurements of biomass in soils has long been a preoccupation of soil microbiologists. The most important reason for this is the direct relationship between soil biomass and soil fertility. There are many techniques for measuring biomass, each of which has certain advantages and disadvantages.

Direct Microscopy Winogradsky, the father of soil microbiology, was the first investigator to successfully attempt direct microscopic examination of soils. The basic method developed by Winogradsky is simplicity itself: a microscope slide is placed into the ground, soil packed carefully around it, and left for several weeks.

After the incubation time has passed, the slide is removed, stained and the microorganisms that have adhered to the slide observed directly. This technique provides a qualitative view of microbial interaction in situ.

More recently, the use of the scanning electron microscope (SEM) has been used to observe microorganisms in soils. This technique works well for observing fungi and actinomycete structures, but is less useful for observing bacteria. This is because bacteria are difficult to unequivocally distinguish from small soil particles.

A common technique that was designed specifically for counting microorganisms is epifluorescent microscopy. This technique is especially notable because it has been widely adapted to enumerating bacteria in samples from the deep subsurface. A sediment sample is dispersed in a blender or shaker, sometimes with the help of a detergent (tween 80), and a small volume of the suspension is filtered onto a membrane. The membrane is then stained with a fluorescent dye, and bacteria are enumerated with an epifluorescent microscope. This type of microscope illuminates the stained sample with ultraviolet (UV) light, causing the fluorescent dye to emit visible light that can be observed by the microscopist. Acridine orange is a widely used fluorescent dye. Acridine orange binds to nucleic acids, RNA and DNA, and the resulting complex fluoresces at a characteristic wavelength of UV light. Fluorescein isothiocyanate (FITC) binds to sulfydrl groups of proteins in cell walls and is sometimes used. After enumeration of the stained sample, the number of bacteria in the sample is calculated from the volume of sample filtered, the area of the filter, and the area of filter actually enumerated.

There are significant practical problems with obtaining reproducible counts of bacteria using the epifluorescent method. It is impossible, for example, to be sure that all bacteria have been desorbed from sediment grains by the shaking process. Also, acridine orange is a cationic dye and binds to negatively-charged clay mineral surfaces. Thus, there is generally a large amount of interference in counting samples containing large amounts of clay. Despite these problems, direct counting procedures have been widely used.

Chemical Techniques The most widely used chemical technique for determining soil biomass is the *soil fumigation method*. Chloroform ($CHCl_3$) under pressure is forced through the soil in order to completely saturate it. The chloroform destroys the cell membrane of living cells allowing the cytoplasm to leak into the soil. After the fumigation has lysed the bacteria, the soil is inoculated with a small amount of soil containing living microorganisms, and the mixture incubated for 10 days along with untreated controls. Biomass is then calculated as

$$B_c = (CO_{2\,tr} - CO_{2\,un})/K_c$$

where $CO_{2\,tr}$ is the amount of carbon dioxide evolved from the chloroform-treated sediment, $CO_{2\,un}$ is the carbon dioxide evolved from the untreated soil, and K_c is the percentage of biomass carbon mineralized to carbon dioxide (often assumed to be 0.41).

A chemical technique that has been used to measure biomass in estuarine and marine environments is to quantify the abundance of phospholipids (Gillan, 1983; White, 1983). This method has been applied to biomass measurements in soils (Federle et al., 1986) and in other subsurface environments (White et al., 1983).

The phospholipid technique takes advantage of the fact that bacterial membranes (as well as the cell membranes of some eucaryotes) contain phospholipids and that these compounds are unstable once the cell dies. Thus, the phospholipid content of sediments is directly proportional to biomass. Phospholipids can be extracted from sediments using chloroform, hydrolized into their various fatty acid components, and quantified using gas chromatography. Given the average amount of phospholipid present in bacterial cells (generally on the order of 50 mmoles/g bacteria), phospholipid content can be converted to bacterial numbers (Gehron and White, 1983).

A significant advantage of the phospholipid method over other measures of biomass is that the kinds of fatty acids liberated from sediment samples gives some information about microbial diversity. This reflects the unique phospholipid content of each bacterial strain. Also, because some phospholipids are found only in procaryotic organisms and some are found only in eucaryotic organisms, the method can help distinguish the biomass component attributable to each class. More importantly, in deep subsurface sediments where eucaryotes are rare, the method gives information on bacterial diversity that cannot be gathered using culture techniques.

Another chemical technique for estimating biomass in soils is by ATP assay. In this method, the cells in a soil sample are lysed, often by chloroform fumigations as discussed earlier, and the ATP content is determined. This determination is based on the reaction

$$\text{D-Luciferin} + \text{ATP} + \text{Luciferase} + O_2 \rightarrow \text{AMP} + CO_2 + \text{Luciferase} + \text{light}$$

The light released by this process can be measured by a photometer and related to ATP content by comparison to standard curves. This technique is widely known as the "firefly" assay, because firefly tails are a ready source of D-Luciferin and Luciferase.

Activity Measurements The essence of heterotrophic microbial metabolism is the oxidation of organic carbon compounds to carbon dioxide, with the microorganisms using the energy obtained for cell maintenance and growth. Thus, measurement of carbon dioxide production from soils has been widely used as an indicator of microbial activity. As with methods for estimating biomass, this method has been widely adapted for activity measurements of subsurface environments other than the soil horizon (Chapelle et al., 1988). The most straightforward method of estimating carbon dioxide production from soils is to place a sample in a closed vial and monitor CO_2 production over time. Carbon dioxide is most conveniently quantified by gas chromatography equipped with a thermal conductivity detector (McMahon, Williams, and Morris, 1990). However, CO_2 can also be stripped from incubation vials by a stream of nitrogen gas, and trapped in a NaOH solution. The carbonate ions generated by the trapping procedure may then quantified by titration with $BaCl_2$ and precipitation of $BaCO_3$.

The accuracy and sensitivity of quantifying respiratory activity by measuring CO_2 production can be significantly enhanced by use of radiotracers. For example, amending a soil sample with ^{14}C-labeled glucose and quantifying the rate of $^{14}CO_2$ evolved is a powerful tool for measuring metabolic rates in soils. This radiotracer technique is particularly useful for determining relative decomposition rates of different organic compounds by indigenous microorganisms. For example, numerous radiotracer studies have shown that aliphatic compounds degrade much faster than phenolic compounds. This has been interpreted as reflecting the relative difficulty that soil microorganisms encounter in degrading compounds containing relatively stable benzene rings.

A significant problem with both the CO_2 and $^{14}CO_2$ methods is that they tend to overestimate respiration rates. This may reflect soil disturbance due to sampling procedures, enhancement of activity by addition of artificial nutrients, or a host of other possibilities. For this reason, these procedures are most effective when used to determine the activities of different soils relative to each other. In this way, sampling and experimental protocols can be standardized, and the relative activities of soil samples determined fairly accurately.

Other Measures of Biomass and Activity A large amount of work has been invested into alternative methods of measuring biomass and microbial activity. One of these methods, measuring the activities of specific enzymes in soils, has been widely applied. Because microorganisms catalyze a wide variety of metabolic functions enzymatically, the activity of a particular enzyme should reflect the rate of metabolism. Some of the enzyme activities used for this purpose include oxidoreductases such as catalase, dehydrogenase, glucose oxidase, and oxidase, as well as such hydrolases as amylase and cellulase.

Dehydrogenases, enzymes used by microbes to transfer hydrogen ions, have been widely used as a direct measure of metabolic activity. Dehydrogenase activity is measured by determining rates of 2,3,5-triphenyl tetrazolium chloride reduction to triphenyl formazan in soil under anaerobic conditions. The triphenyl formazan can then be extracted and measured spectrophotometrically. Significantly, a positive correlation between dehydrogenase activity and respiration rates determined by CO_2 or $^{14}CO_2$ production is not commonly observed. This has been interpreted as reflecting the inactive state of much of the ambient soil microflora.

One variation of the enzyme activity method is to measure the products of an enzymatic process such as hydrolosis, that may be carried out by a wide variety of enzymes. For example, fluorescein diacetate (FDA) is hydrolyzed by many different enzymes such as lipases, proteases, and esterases. Thus, the rate of FDA hydrolysis in soil slurries is a qualitative measure of a broad spectrum of enzymatic activity and, thus, bacterial activity (Schnurer and Rosswall, 1982). The FDA hydrolosis technique has been applied to a number of soils. A notable example was given by Federle et al. (1986) who showed that microbial activity was present well below the soil zone and pointed out that this activity could well affect the transport of contaminants. These authors also showed that FDA activity correlated well with the abundance of phospholipids, a measure of biomass.

Another method long used in soil microbiology as a measure of microbial activity is heat output. This was first proposed by Hesselink van Suchtelen in 1931 and significant advances in the technique, termed microcalorimetry, have been made recently in applying this concept to soils (Sparling, 1981).

Given the wide variety of methods available for measuring microbial activity in soils, it is reasonable to wonder how well these techniques agree with each other. In fact, the methods do not always agree as well as one might expect. One such comparison was given by Sparling (1981), who showed that heat output consistently correlated with respiration, ATP content, amylase activity, and biomass. Curiously, however, heat output did not correlate well with dehydrogenase activity. In addition, there was little correlation between the various activity measurements and soil properties such as pH, carbon content, or nitrogen content.

The lack of consistently strong correlations of the various measures of soil biomass and microbial activity troubled microbiologists for many years. Recently, however, a consensus has emerged that no one technique is capable of giving the "answer" as to soil biomass or microbial activity. Each of these techniques is suited for answering particular questions. Thus, the technique selected generally depends more on the question being asked than on the "accuracy" of the technique. If, for example, the question deals with relative respiration rates between the soil zone and sediments deeper in the unsaturated zone, CO_2 or $^{14}CO_2$ production rates are suitable techniques. If, on the other hand, the question is how much of the soil organic matter is tied up in microbial biomass, the chloroform fumigation or ATP techniques may be more appropriate. Finally, if the question deals with the diversity of the microbial community, the phospholipid technique perhaps is most appropriate. This generality applies to biomass and microbial activity measurements in all subsurface environments. The measurement technique largely depends on the question being asked.

7.2.3 Distribution of Bacteria in the Unsaturated Zone

The unsaturated zone is inhabited by bacteria, viruses, actinomycetes, cyanobacteria (formerly called blue-green algae), fungi, and algae as well as by microscopic arthropods. Bacteria, however, are the most numerous of the microorganisms in this environment and their metabolism is the most broadly based. For these reasons, bacterial processes probably have the largest impact on the chemistry of infiltrating ground water. No attempt will be made here to discuss the distribution of microorganisms other than the bacteria.

The Soil Zone A great deal is known about the types and distribution of bacteria in soils. Much less is known about bacterial abundance and distribution in the unsaturated zone below the soil. Early in the century, it was observed that total numbers of bacteria decreased sharply with depth in the soil (Waksman, 1932). Because of this, it was easy for soil microbiologists to conclude that bacteria ceased to exist in appreciable numbers below the soil zone. This reasoning, while inaccurate, was logical based solely on the distribution of bacteria in the soil zone. For example, if one fits an exponentially decreasing function to a bacterial profile like that of Wakman (1932), it is found that the numbers of bacteria approach zero at a depth of only a few meters. Based on this type of reasoning, it's easy to see why some soil microbiologists earlier in this century dismissed the possibility that bacteria could live deep in the unsaturated and saturated zones.

The Intermediate Unsaturated Zone While studies of bacteria in the saturated portion of the subsurface have become fairly common in the last 20 years, relatively little attention has been paid to the unsaturated zone. A notable exception to this was the Lula, Oklahoma study (Balkwill and Ghiorse, 1985). These investigators carefully noted the saturation status of samples taken from the subsurface and reported them as being from the unsaturated zone, from the interface of the saturated and unsaturated zones (i.e., the capillary fringe), or from the saturated zone. Total counts of bacterial cells (determined by acridine-orange direct count) were in the range of 10^6 to 10^7 cells per gram of dry sediment (Table 7.2). The highest counts were found just below the soil zone and were somewhat lower in the capillary fringe. Interestingly, there was no significant change of bacterial numbers from the capillary fringe into the saturated zone.

Balkwill and Ghiorse (1985) also reported the total numbers of distinct colony types observed on several types of growth media (table 7.2). The number of distinct colony types observed growing on dilute PTYG (peptone, trypticase, yeast extract, glucose) agar declined noticeably with depth. In contrast, samples plated onto SSA (surface soil extract agar) media showed no noticeable decrease in numbers of colony types with depth. While it's tempting (and probably inappropriate) to overinterpret these data, they do suggest that bacteria from deeper in the unsaturated and saturated zones grew most efficiently on carbon sources readily available from indigenous sources.

The types of bacteria identified by Balkwill and Ghiorse (1985) from the unsaturated and saturated zones were dominated by isolates of the genera *Pseudomonas* and *Arthrobacter*. Significantly, members of these genera are typically observed to be the dominant types of bacteria present in soils. It is possible that the types of bacteria present in the unsaturated zone and shallow saturated zone largely reflect bacterial transport from the overlying soil zones. However, data are presently insufficient to decide this question.

The Deep Unsaturated Zone Significant areas of the earth are characterized by arid or semiarid climates. These areas often exhibit unsaturated zones that are tens or even hundreds of meters thick. The microbiology of these deep unsaturated zones have not been extensively characterized. However, one such study has been reported by Colwell (1989) that focused on unsaturated sediments from the high desert of Idaho.

In the Idaho desert study, surface soils exhibited direct counts of 10^6 cells/g whereas unsaturated sediments from a depth of 70 meters showed direct counts of about 10^5 cells/g. However, viable counts of bacteria made by plating sediment slurries onto PTYG agar showed a much more dramatic decrease with depth. Surface soils showed about 10^5 cells/g whereas unsaturated sediments from a depth of 70 meters showed viable counts of 50 and 21 cells/g, respectively. This finding contrasts markedly with studies of saturated sediments in which there is typically only about two orders of magnitude difference between total and viable counts of bacteria. This, in turn, suggests that the bacteria in these deep sediments are under considerable stress.

One unusual finding reported by Colwell (1989) is that of 32 strains of aerobic, heterotrophic bacteria recovered from these deep unsaturated sediments, 84 % were gram positive. This again contrasts with studies of saturated subsurface sediments where gram-negative strains typically predominate. It is possible that a major component of the stress in this environment is due to the lack of mois-

TABLE 7.2 Total and Viable Cell Counts of Bacteria in Aquifer Sediment Samples

Depth (m)	Saturation status[a]	Moisture content (%)	Incubation atm[b]	AODC [no. of cells (SD) g (dry wt) 1]	Viable counts [CFU (SD) g (dry wt)$^{-1}$] on the following media[c]:		
					PTYG agar	Dilute PTYG agar	SSA
1.2	U	13.31	AE	$6.8(4.9) \times 10^6$	$3.4(0.9) \times 10^4$	$1.9(0.4) \times 10^5$	$1.3(0.2) \times 10^5$
1.2	U	13.31	AN	ND	$2.1(0.6) \times 10^3$	$1.2(0.5) \times 10^4$	$5.0(1.9) \times 10^3$
1.2	U	14.35	AE	$9.8(1.3) \times 10^6$	$3.0(0.6) \times 10^3$	$3.1(0.8) \times 10^4$	$2.1(0.5) \times 10^4$
1.2	U	ND	AE	$6.8(0.9) \times 10^6$	ND	ND	ND
3.1	I	18.26	AE	$3.4(2.6) \times 10^6$	$2.0(0.5) \times 10^4$	$2.6(0.2) \times 10^6$	$2.9(0.6) \times 10^6$
3.1	I	18.26	AN	ND	$3.0(1.0) \times 10^2$	$4.0(2.0) \times 10^4$	$5.0(1.0) \times 10^3$
3.1	I	17.94	AE	$3.7(1.0) \times 10^6$	$1.2(0.3) \times 10^3$	$6.0(0.2) \times 10^3$	$7.4(2.1) \times 10^3$
3.1	I	ND	AE	$4.6(3.4) \times 10^6$	ND	ND	ND
2.8	I	17.2	AE	$2.9(3.1) \times 10^6$	$3.3(0.3) \times 10^5$	$6.1(0.7) \times 10^5$	$1.1(0.1) \times 10^6$
3.1	I	25.0	AE	$4.8(2.3) \times 10^6$	$5.7(0.6) \times 10^4$	$5.8(0.9) \times 10^4$	$6.5(0.9) \times 10^4$
4.9	S	19.35	AE	$6.8(4.3) \times 10^6$	$2.6(0.7) \times 10^3$	$3.5(0.1) \times 10^6$	$4.1(0.2) \times 10^6$
4.9	S	19.35	AN	ND	$2.0(2.0) \times 10^2$	$7.0(5.0) \times 10^2$	$<10^2$
4.9	S	23.60	AE	$3.8(1.1) \times 10^6$	$8.4(1.1) \times 10^4$	$7.7(1.1) \times 10^5$	$3.8(1.6) \times 10^5$
4.9	S	ND	AE	$4.8(1.9) \times 10^6$	ND	ND	ND
4.5	S	19.3	AE	$5.0(3.8) \times 10^6$	$2.4(0.4) \times 10^4$	$3.0(1.0) \times 10^5$	$4.4(0.3) \times 10^5$
4.4	S	13.6	AE	$3.8(3.0) \times 10^6$	ND	ND	ND
4.7	S	19.3	AE	$4.0(2.0) \times 10^6$	$2.0(1.5) \times 10^3$	$4.7(2.6) \times 10^4$	$2.0(2.0) \times 10^5$
6.3	S	19.9	AE	$9.3(2.9) \times 10^6$	$6.4(1.5) \times 10^4$	$6.8(0.8) \times 10^5$	$1.1(0.3) \times 10^6$
5.5	S	20.46	AE	$5.2(3.1) \times 10^6$	$6.3(1.0) \times 10^2$	$3.1(1.0) \times 10^6$	$2.5(1.0) \times 10^6$

[a] U, Unsaturated sample from above water table: S, Saturated sample from within water table: I, sample from interface between saturated and unsaturated zones.
[b] AE, Aerobic: AN, anaerobic (N_2).
[c] Average of triplicate spread plates.

Source: After Balkwill and Ghionse, 1985. Reprinted courtesy of the American Society for Microbiology.

ture. If this is the case, the relatively thick peptidoglycan layer in the cell wall of gram-positive bacteria may confer upon them a competitive advantage over gram-negative bacteria. In any case, the observed presence of nearly atmospheric levels of gaseous oxygen in these deep unsaturated sediments certainly reflects the extremely low indigenous activity of microorganisms in this environment.

7.3 LOCAL FLOW SYSTEMS

A local system of ground water flow is defined as having its recharge area at a topographic high and its discharge area at an adjacent topographic low (Fig. 7.1). From a practical perspective, local flow systems often correspond to shallow water table aquifers that are recharged directly from precipitation and that provide base flow to adjacent streams or rivers. Local flow systems are hydrologically very active. In humid, temperate climates, like those of the eastern United States where annual rainfall reaches 130 to 150 centimeters, as much as 25 to 50 centimeters of water may recharge local flow systems. For example, in a regional simulation of ground-water flow of the Atlantic Coastal Plain physiographic province, an estimated 90-98% of recharge to the water table was subsequently discharged to surface water bodies in adjacent topographic lows. In areas of such high recharge and discharge, rates of ground-water flow may be as high as a meter per day.

The depth that local flow systems penetrate into the subsurface is largely dependent upon topographic relief. For example, Toth (1963) shows that for the case of a homogeneous aquifer, the depth of the local flow system varies directly with topographic relief. In one example similar to that shown in Figure 7.1, Toth showed that if topographic relief was about 15 meters, then the local flow system would penetrate about 150 meters into the subsurface. If, on the other hand, topographic relief was increased to 61 meters and all other parameters were held constant, then the local flow system would penetrate all the way to the impermeable lower boundary.

This is an extremely important result, and it illustrates why depth is such a poor predictor of the type of flow system predominating at any site. For high-relief topographies, local flow systems may penetrate hundreds of meters into the subsurface. For topographic settings with little relief, local flow systems may penetrate only a few tens of meters into the subsurface.

The other important variable that influences the depth of local flow systems is subsurface permeability variation (Freeze and Witherspoon, 1967). In the case of a low-permeability ($K = 1$) surface material underlain by a high-permeability ($K = 100$) material, Freeze and Witherspoon showed that the local flow system penetrated about 75 meters into the subsurface. If, on the other hand, there was a layer of intermediate permeability ($K = 10$), the local flow system penetrated only about 25 meters into the subsurface. In general, high-permeability sediments underlying low-permeability sediments tend to restrict the depth to which local flow systems penetrate. This situation tends to carry water past adjacent recharge and discharge areas so that intermediate flow systems occupy a greater thickness of the section relative to local flow systems.

7.3.1 Local Flow Systems as a Microbial Habitat

Local flow systems are very active hydrologically. This is to say that water from precipitation events quickly passes through the unsaturated zone, recharges local flow systems, and is readily discharged to surface streams or transpired by phreatophytic plants that tap the saturated zone. It's not uncommon for the water table of local flow systems to rise or fall several meters over time periods of only a week or two. It is generally considered that residence times of water in local flow systems are on the order of days to years.

The dynamic nature of local flow systems has important consequences as to the microbial habitat that is generated. Fast rates of recharge and discharge mean that local flow systems are relatively open to sources of nutrients from the surface or from the unsaturated zone. For example, dissolved organic carbon (DOC) generated in the unsaturated zone may be quickly transported into local flow systems. Figure 7.3 shows a conceptual model of DOC production, transport through the unsaturated zone, and delivery to the water table (Thurman, 1985). DOC that is transported into local flow systems, while often consisting of high-molecular weight compounds that are not easily assimilated by microorganisms, may provide an important source of nutrients to bacterial populations.

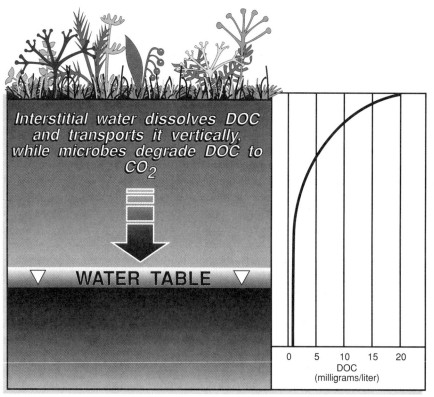

FIGURE 7.3. Model of DOC production and transport to local flow systems. (Modified from Thurman, 1985, with permission courtesy of Martinus Nijhoff/Dr. W. Junk Publishers).

Fast rates of recharge and the relatively free exchange of gases within the unsaturated zone encourage the transport of dissolved oxygen to local flow systems. Because aerobic bacteria obtain more energy per mole of organic material oxidized than bacteria using other electron acceptors (such as ferric iron or sulfate), local flow systems are often dominated by aerobic bacteria. Such aerobic activity can be sustained only if there is a steady flux of oxygen into the system, and this requires the close connection with the unsaturated zone characteristic of local flow systems.

Recharge to local flow systems that passes through the unsaturated zone may also facilitate the transport of dissolved nitrogen species in the form of nitrate or ammonia. Because nitrogen fixation, ammonification, and nitrification are processes that are characteristic of the soil zone, recharge moving through soils to local flow systems is a potential source of organic nitrogen. Because nitrogen is required for protein synthesis, it is an important nutrient requirement for bacterial growth. In many subsurface environments, nitrogen is the principal limiting factor for bacterial growth. In this respect, however, local flow systems have less limitation than intermediate or regional flow systems.

The proximity to land surface and the dynamic nature of ground-water movement in local flow systems makes them relatively accessible to chemical pollutants from human activities. These pollutants come from a variety of sources and include a variety of compounds. From the standpoint of a microbial habitat, such pollution is often a boon for bacterial growth, as it may provide carbon (gasoline, sewage wastes) or nitrogen (fertilizers, septic effluents) sources, each of which is often limiting in subsurface environments. Although precise figures are difficult to compile, it is probable that as much as 80% or 90% of all documented cases of ground water pollution occur in local flow systems. Thus, our experience with the microbiology of ground water pollution and bioremediation strategies (Chapters 12 and 13) is largely confined to local flow systems.

7.3.2 Distribution of Bacteria in Local Flow Systems

Systematic studies on the distribution of bacteria in shallow water table aquifers, herein termed *local flow systems*, are relatively new to subsurface microbiology. Whereas the distribution of bacteria in unsaturated sediments (the soil zone) and in regional flow systems (petroleum reservoirs) has been studied since the early 20th century, studies on local flow systems began only in about 1980 (Ghiorse and Wilson, 1988, p. 126). To a large degree, this neglect simply reflects a lack of pertinent questions concerning microbial processes in local flow systems. In the case of the unsaturated zone (soils), the obvious connection between soil fertility and the abundance of soil microbes presented investigators with a wide range of important questions. Addressing these questions has spurred research in soil microbiology for 100 years. In the case of regional flow systems (largely petroleum reservoirs), it was recognized in the 1920s that bacterial processes affected petroleum production. This led to many studies on bacterial corrosion of well casings, biofilm plugging of reservoir rocks, and metabolism of hydrocarbons (Davis, 1967).

However, there was no similar impetus for studying microbial distributions in shallow water table aquifers. The increasing awareness that some human activities contaminated water table aquifers to the point of their having to be abandoned as sources of potable water provided this impetus beginning in the early 1970's. It was widely thought, for example, that simple dilution by uncontaminated water was the most practical method of renovating contaminated water supplies. Since many surface water bodies are completely turned over in a matter of days or weeks, the "dilution solution" was deemed a workable remediation strategy. By this criteria, however, ground-water reservoirs could not be salvaged once contamination took place. Even local flow systems are recharged very slowly relative to surface water bodies, and the conventional wisdom was simply to write off contaminated ground water as a loss.

This conventional wisdom did not stand up to closer examination. It became a common observation in the consulting industry, for example, that the size of plumes resulting from gasoline spills often tended to shrink over time. How could such observations be explained? To microbiologists, an obvious question to ask was whether microbial metabolism could be involved. This question was systematically attacked by a group of researchers brought together by the U.S. Environmental Protection Agency in Ada, Oklahoma (Dunlap and McNabb, 1973, McNabb and Dunlap, 1975, Wilson et al., 1983; Ghiorse and Balkwill, 1983). These investigators showed not only the presence of bacteria in shallow aquifers but also that the bacteria were capable of degrading a variety of organic pollutants (Wilson et al., 1983). This realization rapidly changed the conventional wisdom that pollution in ground water reservoirs was irreversible and led to an explosion of interest in subsurface microbiology.

The distribution of bacteria indigenous to a local flow system near Lula, Oklahoma, as measured by direct microscopy as well as aerobic plate count methods, is shown in Table 7.2 (Balkwill and Ghiorse, 1985). The direct counts of bacteria were all in the 10^6 cells per gram (dry weight) and were remarkably consistent. The plate counts, on the other hand, varied widely. The lowest counts were found on nutrient-rich PTYG media. Higher counts, performed on duplicate sediment samples, were found using dilute (1:20 dilution) PTYG media and surface soil extract (SSE) media. The fact that higher counts were consistently found using dilute media was interpreted by Balkwill and Ghiorse (1985) as reflecting the acclimation of subsurface bacteria to low-nutrient conditions. This acclimation apparently tends to lower the organisms' tolerance to high-nutrient conditions and, hence, the lower counts on the rich media. Anaerobic plate counts were generally one to four orders of magnitude less than aerobic counts.

Just as interesting was the demonstration that the total counts were consistently higher than the plate counts (Table 7.2). Two possible explanations for this observation include (1) the presence of a large number of inactive organisms or (2) the inability of some of the organisms present to grow effectively on the media offered. Balkwill and Ghiorse (1985) tended to lean toward the second possibility, because plate counts on SSE media consistently approached (within one order of magnitude) the direct counts. However, that particular question may never be resolved adequately.

Perhaps the most extensive characterization of bacterial distributions in local flow systems has been at the Savannah River Site (SRS) in South Carolina (Sinclair and Ghiorse, 1989; Balkwill, 1989, Fredrickson et al., 1989). The geology of this site is extremely complex and represents a series of overlapping deltas that have been deposited since Late Cretaceous Time (~ 80 MY). The hydrology of the site is similarly complex. Figure 7.4 shows a generalized stratigraphic column together with counts of bacteria and eucaryotic organisms at one location (site P 24) at SRS (Sinclair and Ghiorse, 1989). Local topographic relief shows elevation changes of from 25 to 50 meters between ridges and stream bottoms. This relatively high relief, combined with the high permeability of the surface sediments allows the local flow systems to penetrate deeply into the subsurface at this site. At site P24 (Fig. 7.4), the local flow system penetrates about 75 meters into the subsurface and grades into an intermediate system below the low-permeability Williamsburg formation.

Within the different geologic horizons in the local flow system, counts of bacteria

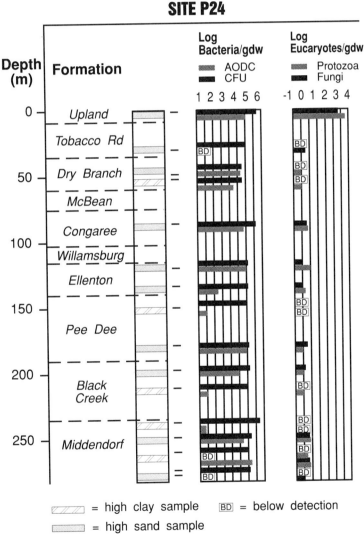

FIGURE 7.4. Counts of bacteria, protozoa, and fungi in deep coastal plain sediments. (From Sinclair and Ghiorse, 1989, reprinted with permission courtesy of Taylor and Francis Publishers.)

by direct microscopy (AODC) are all in the 10^6–10^8 cells per gram range. As was observed at the Lula, Oklahoma site, plate counts of bacteria showed much more variation than direct counts. Interestingly, these sediments contained a significant population of eucaryotic organisms, with numbers ranging from below detection to as much as 10^3 cells per gram of sediment. There was little variation of eucaryotes with depth and no measureable change between the local and intermediate portions of the flow system. The most consistent predictor of the numbers of eucaryotes was clay content with high-clay samples containing the least numbers of cells. While present throughout the column at the SRS, numbers of eucaryotes were always from 3 to 5 orders of magnitude less than bacterial numbers. These are the most comprehensive data yet compiled showing that bacterial numbers far exceed numbers of eucaryotes in subsurface sediments below the soil zone.

A major finding of the microbial characterization at the SRS was that numbers of bacteria correlated inversely with the amount of clay in the sediments (Fig. 7.4). Fredrickson et al. (1989) attribute this to the lower hydraulic conductivity of clay-rich sediments. However, these authors also point out that the pH of pore water in clays was lower then in the sands and that this might also contribute to the observed trend.

Not only were bacterial numbers lowest in the clay horizons, but bacteria activities were correspondingly low (Phelps et al., 1989a). For example, clayey samples from borehole P24 oxidized ^{14}C-labeled acetate to $^{14}CO_2$ at rates of 100-500 disintegrations per minute (dpm) per day whereas sandy sediments ranged from 10,000-50,000 dpm/day. Not only were rates of substrate oxidation greater in sandy sediments, but other measures of bacterial activity, such as ^{14}C acetate incorporation into lipids and [^{3}H]thymidine incorporation into DNA, were also consistent with this finding.

The aquifers underlying the SRS-both those in the local flow system above the Ellenton Formation and those of the deeper intermediate flow system-are largely aerobic with dissolved oxygen concentrations varying between 1 and 6 mg/L. In spite of this, anaerobic bacteria are part of the ambient microflora (Jones, Beeman, and Suflita, 1989). Numbers of sulfate-reducing bacteria ranged from undetectable to as high as 10^4 cells per gram of sediment. While these numbers are much lower than numbers of aerobic bacteria, it does show a significant presence. Similarly, there was methane production from a number of zones. The presence of anaerobic bacteria in a predominantly aerobic aquifer is not entirely unexpected. Aerobic soils, for example, often contain significant populations of anaerobic bacteria. This is often interpreted as reflecting anaerobic microenvironments in the sediment where the diffusion of oxygen is insufficient to overcome oxygen consumption. The actual activity of anaerobic bacteria in the local flow system of the SRS is probably fairly low. Concentrations of methane in the ground water are generally less than 0.01 micromolar, and sulfate concentrations show no decreasing trend along aquifer flow paths. The documented presence of anaerobic bacteria in these sediments, however, confirms the similarity of local flow systems with sediments from the soil zone.

The distribution and abundance of bacteria have been studied in local flow systems developed in other geologic settings. For example, in the Midwest, rivers with beds incised into Paleozoic bedrock often have narrow (1-10 km) alluvial aquifers associated with them. These aquifers are termed *buried-valley aquifers*, since they represent Pleistocene river channels that have subsequently been filled with sediment. In a survey of microbial populations in Kansas, Sinclair et al. (1990) have shown that many of the microbial characteristics found in coastal plain sediments of the SRS also occur in buried valley aquifer sediments. For example, total counts of bacteria range between 10^6 and 10^8 cells per gram of sediment, which is the range found at the SRS. Also, counts of eucaryotes were much lower than those of bacteria. Finally, a similar trend of decreasing bacterial abundance with increasing clay content was found (Table 7.3).

The statistical correlating matrix (Table 7.3) prepared by Sinclair et al. (1990) demonstrated several other patterns. For example, there was a strong ($r = 0.64$) positive correlation of number of colony types (NCT) with the total number of colony-forming units (CFU). This suggests that those zones characterized by the greatest number of bacteria also exhibit the greatest diversity of bacterial types. This trend was also evident in the correlation between NCT and most probable number of protozoa (MPNP), suggesting that protozoa are most abundant where bacteria are most abundant. From an ecological point of view, these observations make intuitive sense. One would expect that, given a favorable environment, several different types of microbes, both procaryotes and eucaryotes, should thrive. Intuition, however, is notoriously unreliable, and it is comforting to see these trends confirmed by hard data.

An interesting feature of the aquifers studied by Sinclair et al. (1990) is the presence of an apparent redox boundary, indicated by a change of sediment color from tan (oxidized) to gray (reduced). There was no noticeable change in total number of bacteria across this boundary but there was a drop in the number of viable bacteria as evaluated by plate count enumeration. One possible explanation for this is that a higher percentage of cells present in the gray sediments were strict anaerobes. They would, therefore, be countable by direct microscopy, but not countable by aerobic plate counts. Many local flow systems are well-oxygenated and, thus, the predominant types of bacteria are aerobic. However, anaerobic zones are not uncommon and these conditions can be expected to change the kinds of bacterial processes that are carried out in situ.

7.4 INTERMEDIATE FLOW SYSTEMS

The major characteristic of an intermediate flow system is that one or more topographic low areas separate recharge areas from discharge areas (Fig. 7.1). In practice, intermediate flow systems generally correspond to confined aquifer systems of moderate (less than 300 m) depth. Because intermediate flow systems produce a high percentage of the ground water used for human consumption, animal consumption, and irrigation in the United States, they are economically very important.

Intermediate flow systems differ from local flow systems primarily in the extent of connection with the surface and with the unsaturated zone. Because of this relative lack of connection, recharge rates are much lower than in most local flow systems. For example, digital modeling studies of Atlantic Coastal Plain aquifers show that average recharge rates to the confined part of the system (i.e. the intermediate flow system) are typically in the 1-3 centimeter per year range (Leahy

TABLE 7.3 Correlation Matrix Showing r Values of Paired Variables Measured by Sinclair et al., 1990

	Sand	Clay	MPNP	NPT	CFU	NCT	AODC	NH$_4$+	TOC	pH
Sand	1.00									
Clay	-0.80[b]	1.00								
MPNP	0.03	-0.06	1.00							
NPT	0.20	-0.12	0.72[b]	1.00						
CFU	0.40	-0.44[a]	0.21	0.53[a]	1.00					
NCT	-0.10	-0.15	0.49[a]	0.58[b]	0.64[b]	1.00				
AODC	0.60[b]	-0.63[b]	0.02	0.05	-0.01	-0.20	1.00			
NH$_4$+	-0.68[b]	0.80[b]	-0.14	-0.12	0.12	0.02	-0.79	1.00		
TOC	-0.35	0.29	-0.04	-0.26	-0.65[b]	-0.41	0.31	0.02	1.00	
pH	-0.62[b]	0.44	-0.01	-0.10	-0.10	0.23	-0.37	0.38	-0.03	1.00

Sand = % sand in sediment; Clay = % clay in sediment; MPNP = Most probable number of protozoa; NPT = Number of protozoan types in MPN enumeration plates; CFU = Numbers of colony forming units of viable bacteria; NCT = Number of colony types on CFU enumeration plates; AODC = Acridine orange direct count of total numbers of bacteria; NH$_4$+ = mg NH$_4$+ per kg sediment; TOC = mg C per g sediment; pH = pH of sediment slurry in distilled water.

[a] Significant at the 5% level.
[b] Significant at the 1% level.

Source: Reprinted courtesy of the National Water Well Association.

and Martin, 1986). This contrasts with recharge rates to shallow local flow systems of 30 to 50 centimeter per year. In particularly humid climates, characterized by permeable aquifer sediments such as the Floridan aquifer system of the southeastern United States, recharge rates to intermediate flow systems are somewhat higher and are in the 10 cm/yr range (Bush and Johnston, 1986). Recharge rates to the High Plains aquifer systems in the Central USA, where the climate is arid, are as low as 0.1 cm/yr (Weeks, 1986).

Partly because of these low recharge rates, intermediate flow systems are characterized by relatively slow rates of ground-water flow. Typical rates of ground water flow in intermediate flow systems are on the order of 0.1 to 1.0 meter per year.

7.4.1 Intermediate Flow Systems as a Microbial Habitat

The low rates of recharge and ground-water flow characteristic of intermediate flow systems create a microbial habitat that is very different from that of local flow systems. Available evidence suggests that the delivery of DOC from the soil zone to local flow systems is fairly low. This being the case, delivery of DOC from surface soils to intermediate flow systems is probably negligible. Because of this, microbial populations in intermediate flow systems are largely dependent upon organic material present in the sediments as a primary carbon source. These carbon materials are often refractory in nature, and this severely limits the metabolic and growth potential of microbes.

For example, McMahon (1990) showed that sedimentary organic material present in the Black Creek aquifer of South Carolina consists largely of complex aromatic and aliphatic polymers. The ^{13}C nuclear magnetic resonance (NMR) spectra of such materials show that carbohydrate carbon is largely absent, having been removed by prior microbial processes. McMahon (1990) also showed, however, that indigenous bacteria are able to metabolize this residual organic carbon, concentrating on the aliphatic fraction. Evidence for this was provided by evaluating the NMR spectra of organic material cored from the subsurface before and after a long-term incubation. In one instance, the NMR spectra of organic carbon cored from a depth of 58.5 meters showed two large peaks: one corresponding to aliphatic carbon compounds and the other to aromatic compounds. The aromatic : aliphatic ratio (R) of the peak areas for the freshly cored material was 1.28. However, after being incubated in the lab for one year, a procedure that allowed indigenous bacteria to metabolize the material, the ratio increased to 1.70. This indicates that indigenous bacteria are capable of metabolizing the refractory carbon compounds available in the aquifers and that they preferentially oxidize the aliphatic portion. Because this pool of available carbon is so refractory, intermediate flow systems are typically very oligotrophic (nutrient-poor) environments.

Due to these oligotrophic conditions, rates of bacterial metabolism in intermediate flow systems are slower than in local flow systems. Rates of bacterial metabolism as measured by CO_2 production in a local flow system underlying Hilton Head Island, South Carolina, were in the 10^{-2} mmoles per liter per year range (Chapelle et al., 1988). In contrast, rates of bacterial metabolism in the intermediate flow systems of the Black Creek and Middendorf aquifers were three orders of magnitude slower, in the 10^{-5} millimole per liter per year range (Chapelle and Lovley, 1990).

In addition to being strongly carbon-limited, microbial processes in intermediate flow systems are limited by the availability of organic nitrogen. Delivery of dissolved nitrogen compounds to intermediate flow systems from soil water is limited for the same reasons that delivery of DOC is limited. Not surprisingly, therefore, concentrations of nitrogen species in ground water of intermediate flow systems are generally less than 0.1 milligrams per liter.

7.4.2 Distribution of Bacteria in Intermediate Flow Systems

The presence of bacteria in intermediate flow systems, has been recognized for much of the twentieth century (Gurevich, 1962). However, data documenting the occurrence and distribution of bacteria in particular hydrologic systems has not generally been available until the 1980's.

Early work showing the presence of bacteria in intermediate flow systems concentrated on documenting the presence of bacteria in water produced from the aquifers. For example, Leenheer et al. (1976) showed that methanogenic bacteria were present in ground water produced from coastal plain aquifers of North Carolina. Godsy (1980), using an antibiotic disc method, showed the presence of methanogenic bacteria in water produced from the Floridan aquifer.

Studies of bacterial distributions in local flow systems had clearly shown that a large percentage of indigenous bacteria were attached to sediment particles rather than living free in interstitial water. However, by the late 1980's there were no systematic studies of bacteria in sediments of intermediate flow systems. The reason for this lack of attention, as was the case with local flow systems prior to 1970, was that there were no pressing scientific questions that required such investigations. In the case of local flow systems, this impetus was supplied by concerns about ground-water pollution. However, because of the low recharge rates, chemical contamination of intermediate flow systems was a rare occurrence. Thus, microbial studies of intermediate flow systems were not mandated by concern about ground-water pollution.

The primary motivation for studying bacterial processes in intermediate flow systems came from another, somewhat unlikely, source. In 1983, techniques for studying the ground-water geochemistry of large-scale aquifer systems was codified in a series of papers by L.N. Plummer, D.L. Parkhurst, and D.C. Thorstenson (1983) of the U.S. Geological Survey. This series of techniques, loosely termed "geochemical modeling," provided geochemists with a means of quantifying the amounts of minerals and gases that reacted along flowpaths of individual aquifers. As geochemical modeling studies progressed in many different aquifer systems, it quickly became apparent that production of CO_2 gas in aquifers was one of the most important processes affecting ground-water chemistry (Thorstenson, Fisher, and Croft, 1979; Plummer and Back, 1980; Chapelle, 1983). The problem in 1980 was that there was no known process capable of generating this CO_2 in deeply-buried sediments below the soil zone. This led Chapelle and Knobel (1985) to resurrect a hypothesis originally put forth by Cedarstrom (1946) that the metabolism of bacteria in these confined aquifer systems was the ultimate source of the observed incoming CO_2.

Cedarstrom's hypothesis was strengthened when Chapelle et al. (1987) enumerated total and viable bacteria in a 200 meter deep core hole in southern Maryland. This core hole penetrated the Aquia and Patapsco aquifers, both of which function as intermediate flow systems in that part of Maryland. This aquifer system was much different from the local flow systems that had been studied previously. The most important difference was that ground water in both the Aquia and Patapsco aquifers is anaerobic. Not surprisingly, counts of total bacteria were in the 10^6-10^5 cell per gram range, about one order of magnitude less than found in aerobic aquifers. Furthermore, the viable cell counts were very low, on the order of 10^4 cell/gram, which is near the detection limit of the direct viable count procedure used for the measurements. Most significantly, the presence of viable sulfate reducing bacteria in the sediments, as had been postulated forty years earlier by Cedarstrom (1946), was confirmed.

The primary significance of this work was to show that bacterial metabolism was a plausible source of CO^2 to ground water. As often happens, however, these data raised more questions than they answered. For example: How diverse were the bacterial communities? What were the interactions between different types of bacteria? Were there bacteria present that were in any way unusual? Did bacterial numbers and diversity vary with sediment properties?

At present, these questions have been considered only in a few intermediate aquifer systems. One of the most intensely studied intermediate aquifer systems is the Upper Cretaceous section of South Carolina. These studies, performed at the Savannah River Site (SRS) in South Carolina, has been discussed previously in reference to local flow systems. The three deepest units present at the SRS, the Pee Dee, Black Creek, and Middendorf aquifers, largely function as intermediate flow systems. Unlike many intermediate aquifer systems, these aquifers are aerobic, and, therefore, they are somewhat unusual. However, most aquifer systems exhibit some unusual characteristics. A "typical" intermediate flow system probably does not exist.

Figure 7.4 shows the distribution of bacteria and eucaryotes in the intermediate part of the flow system, consisting of the Pee Dee, Black Creek, and Middendorf aquifers, in a core hole drilled at the SRS in South Carolina (Sinclair and Ghiorse, 1989). As was found in the shallower sediments comprising local flow system at the SRS, the most consistent predictor of microbial abundance was sediment texture. Sands generally contained the greatest numbers of microbes and clays contained the lowest numbers of microbes. Significantly, the diversity of bacteria also correlated with sediment type with the greatest diversity being found in the sandiest sediments.

Not surprisingly, the aerobic activity of bacterial populations behaved in a similar manner. Hicks and Fredrickson (1989) investigated the metabolism of ^{14}C-labeled acetate, 4-methooxybenzoic acid, and phenol in cored sediments over a 21-day incubation period. Acetate and phenol were metabolized relatively easily in the sandy sediments whereas 4-methooxybenzoate was metabolized less readily. In general, the highest rates of mineralization were observed in the sandy sediments and the lowest in the clays for each compound.

The dependence of bacterial activity on the clay content of sediments is a general observation and is found in other intermediate flow systems with quite different geochemical properties. For example, in the anaerobic portion of the

Black Creek and Middendorf aquifers, rates of ^{14}C-labeled acetate and glucose oxidation were orders of magnitude higher in sandy aquifer sediments than in clayey confining bed sediments (Chapelle and Lovely, 1990). There are probably many mechanisms that contribute to this effect, but one important factor is the small size of pore throats in clays relative to sands. Figure 7.5 shows that the average pore throat diameter in clays, as measured by mercury injection, is less than 0.05 microns. In sands, on the other hand, average pore throat diameters are much greater and are in the 2-20 micron range. Because bacteria generally have diameters from 0.1 to 1.0 microns, the small pore throats of clays must greatly restrict the ability of bacteria to move about and reproduce effectively. Similarly, helium injection into sediments shows that the interconnected porosity of sands is in the 90-95% range whereas in clays it is much lower. This would tend to restrict the mobility of nutrients to active cells as well as restrict the movement of waste products away from cells. Again, this would tend to depress bacterial activity in clays relative to sands. Geochemical effects, such as lower pH of confining bed pore water, may also contribute to this effect (Fredrickson et al., 1989).

While data on the abundance and distribution of bacteria in intermediate flow

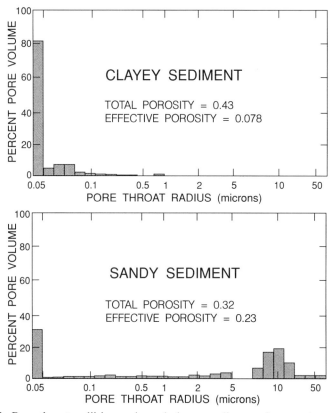

FIGURE 7.5. Pore throat radii in sandy and clayey sediments from an intermediate flow system. The smaller radii characteristic of clays may depress bacterial mobility and activity. (Modified from Chapelle and Lovley, 1990, with permission courtesy of the American Society for Microbiology.)

systems are too sparse to allow comparisons between systems in a meaningful way, data on the activity of bacterial processes are available for three different systems. Figure 7.6 shows the ranges of bacterial activity, as measured by rates of CO_2 production, for the Madison aquifer system of the western United States (Plummer et al., 1990), the Patapsco aquifer system of southern Maryland (Chapelle et al., 1987), and the Black Creek-Middendorf-Cape Fear aquifer system of South Carolina. The notable feature of these data is how similar they are, with CO_2 production rates in the 10^{-4} to 10^{-6} mmole per liter per year range. Furthermore, the ranges of CO_2 production are considerably lower than in deep-sea sediments. In fact, the only known environment with similarly low rates of CO_2 production are the nutrient-deprived waters of deep ocean basins. These data serve to illustrate that, while bacteria and bacterial activity are evident in many intermediate flow systems, overall rates are extremely low. In fact, these aquifer systems are among the most oligotrophic microbial environments that have ever been described.

There is evidence that some intermediate flow systems lack bacterial activity entirely. For example, dissolved oxygen in concentrations of 6-8 milligrams per liter have been reported in deep (400 meters) alluvial-basin aquifers of the Southwestern USA (Winograd and Robertson, 1982). This oxygenated water has been dated at 20,000 to 30,000 years old, which implies that rates of bacterial oxygen uptake are less than 10^{-7} mmoles per liter per year. In those samples where dissolved oxygen concentrations were still at atmospheric saturation, bacterial respiration rates must be virtually zero. The reasons for this lack of bacterial activity are not difficult to understand. These sediments were deposited under arid

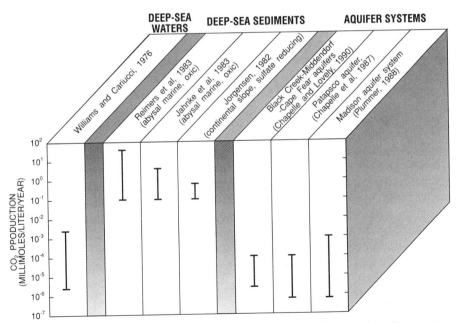

FIGURE 7.6. Ranges of bacterial CO_2 production in three intermediate flow systems. (Modified from Chapelle and Lovley, 1990, with permission courtesy of the American Society for Microbiology.)

conditions and the amount of organic carbon present in the sediments is extremely low. The lack of a suitable organic carbon substrate evidently precludes the metabolism or growth of bacteria. However, studies on the distribution and abundance of bacteria in this system are not available to confirm or refute this.

7.4.3 Microbial Processes in Confining Beds

While it is generally observed that microbial activity in clayey confining beds is much less than in more permeable aquifers, activity is often not completely absent. Furthermore, this activity can have important impacts on the chemistry of water in adjacent aquifer sediments in intermediate flow systems.

The first systematic evaluation of how microbial processes in confining beds affects ground water chemistry was given by McMahon and Chapelle (1991). This study pointed out some unusual patterns in the concentrations of organic acids in aquifer and confining-bed pore water. Specifically, confining-bed pore water contained fairly high concentrations of acetate and formate (50-70 micromolar) compared with the fairly low concentrations found in the pore water of sandy aquifer sediments (1-10 micromolar). Furthermore, the clayey confining bed sediments showed much greater hydrogen (H_2) production than sandy aquifer sediments. Since hydrogen, acetate and formate are central to anaerobic food chains, these data were exactly opposite of what one might have expected given reported low rates of microbial activity in clayey sediments (Phelps et al., 1989a, Chapelle and Lovley, 1990).

The reasons for the observed accumulation of organic acids in confining beds relative to aquifer pore water were clarified by means of radiotracer experiments. When confining bed sediments were amended with ^{14}C-labeled glucose, a fermentable substrate, it was observed that most of the ^{14}C accumulated in formate and acetate rather than being oxidized to CO_2. In sandy aquifer sediments, however, all of the ^{14}C was oxidized rapidly to CO_2. In anaerobic sediments, complete oxidation of fermentable substrates such as glucose requires the symbiotic action of fermentative and respirative bacteria. Fermentative bacteria partially oxidize the substrate with the production of organic acids and hydrogen. These simpler compounds are then utilized by respirative bacteria. The observed accumulation of organic acids, therefore, indicated that fermentation outpaced respiration in clayey confining-bed sediments.

This finding has important implications for the microbial cycling of carbon in ground-water systems. The organic acid concentration differences between confining beds and aquifers appears to drive a net diffusive flux of organic carbon to aquifers. This, in turn, provides a constant supply of simple organic substrates for respirative bacteria in the aquifers.

The model proposed by McMahon and Chapelle (1991) is shown in Figure 7.7. Complex organic carbon material, including both aliphatic and aromatic compounds, in confining beds are subjected to slow continuous microbial fermentation. Respiration in confining beds, for reasons not well-understood, is inhibited leading to the accumulation of organic acids in the pore water. These organic acids then slowly diffuse to aquifers where they are consumed by respirative bacteria such as sulfate reducers (SRB) or Fe(III)-reducers (FRB). This model is important

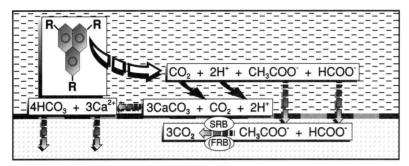

FIGURE 7.7. Model linking the fermentation of complex organic matter in confining beds and the production of acetate (CH_3COO^-) and formate ($HCOO^-$) with sulfate-reducing (SRB) or Fe(III)-reducing (FRB) respiration in aquifers.

because it is the first proposed mechanism that effectively links the large pool of organic carbon often present in confining beds with microbial processes in aquifers.

This model may also explain the observation that overall rates of microbial metabolism in geologically diverse intermediate flow systems are apparently so similar (Fig. 7.6). If diffusion is the rate-limiting step for delivery of substrates to aquifers in intermediate flow systems, it is to be expected that overall rates of metabolism should also be similar. Although this is an interesting hypothesis, data are presently insufficient to come to a firm conclusion as to it's general applicability.

7.5 REGIONAL FLOW SYSTEMS

A regional flow system is defined as having its recharge area at a ground-water divide and its discharge area at the bottom of the regional basin. A regional flow system is characterized by long flowpaths and extremely slow rates of ground-water flow (Fig. 7.1). Because of the long residence times of ground water in regional systems, the water is often highly mineralized and unsuitable for human consumption, animal consumption, or irrigation. Also, the great depth of many regional flow systems (often greater than 300 meters) makes water-well drilling prohibitively expensive. For these two reasons, regional flow systems are not often suitable for supplies of potable water.

The deep sedimentary basin in which regional flow systems develop are, however, very important economically because they are often sites of hydrocarbon accumulation. It was recognized very early on that bacterial processes, such as sulfate reduction and methanogenesis, had important impacts on the production of oil and gas from wells. In addition, it was soon recognized that bacterial processes may have been involved in petrogenesis, the processes that convert buried organic matter into petroleum. For these and other reasons, studies of bacterial processes in regional flow systems were undertaken early in the twentieth century and have continued to the present.

7.5.1 Early Observations from Petroleum Reservoirs

The beginning of the petroleum industry in the United States is usually traced back to 1859 when Drake drilled his first oil well in Pennsylvania. Throughout much of the nineteenth and early twentieth centuries, prospecting for petroleum reservoirs and the production of oil and gas were decidedly unscientific enterprises. Perhaps the classic example was the discovery of the Spindletop oil field near Beaumont, Texas, at the turn of the century. A gentleman named P. Higgins decided, on the basis of some remarkably poor geologic mapping, that a large pool of oil was present under Spindletop. In fact, Higgins got almost everything wrong about the geology except this: There really was oil under Spindletop!

The first oil wells drilled at Spindletop were spectacular gushers. It was common for oil to spurt hundreds of feet into the air once the "cap rock," or upper confining bed to the reservoir, was breeched. Spindletop set off a wild oil rush, and literally thousands of wells were drilled in a short period of time. Beaumont became a boom town, and oil was produced frantically in an effort to make as much money as quickly as possible. Unfortunately this led to massive overpumping. Salt water moved into the oil-bearing strata and within a few years the reservoir was ruined.

The experience at Spindletop and other fields made a deep impression on the fledgling petroleum industry. It gradually became clear that the best way to produce consistent profits in this high-risk business was by knowing as much as possible. The major oil companies that emerged from these early years began to invest in scientific studies to determine the origin, transport, and accumulation of petroleum in the subsurface. These studies soon led to consideration of microbial processes in regional flow systems.

The single largest problem for a petroleum prospector is, obviously, how to locate accumulations of oil. To this end, enormous effort was invested in understanding the geology, hydrology, and geochemistry of known petroleum reservoirs in order to identify leads for finding new reservoirs. One such lead was that brines associated with petroleum tended to be characterized by low concentrations of sulfate and high concentrations of sulfides (Rogers, 1917). The practical importance of this was that the simple analysis of water for sulfate and sulfide was a possible indicator of the presence or absence of nearby petroleum accumulations. This observation by itself, however, was unsatisfactory until its cause could be identified. This was accomplished by Bastin (1926) when he showed that sulfate-reducing bacteria were present in oil-field brines and that the activity of these microorganisms could account for the observed phenomenon.

One example of how this information was useful in petroleum prospecting is the case of the Carrizo Sand in Texas (Davis, 1967). The Carrizo is a confined aquifer system that grades from a local flow system where it crops out at land surface to a regional flow system in downdip areas (Fig. 7.8). Throughout its regional extent, Carrizo aquifer ground water contains 12-52 milligrams per liter of sulfate except in the vicinity of a fault-controlled accumulation of petroleum (Fig. 7.8). Near the petroleum (well no. 6), sulfate concentrations are less than 1 milligrams per liter. In this case, the lack of sulfate in ground-water is a clear indication of a nearby petroleum accumulation. Davis (1967), as did Bastin, was able to relate this phenomenon to the presence of sulfate-reducing microorganisms,

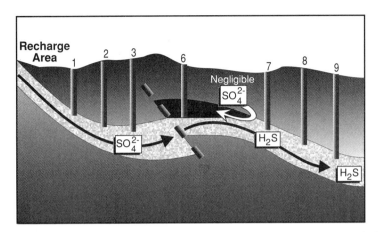

FIGURE 7.8. Depletion of sulfate near a fault-controlled petroleum accumulation in the Carrizo Sand. (Modified from Davis, 1967, with permission courtesy of Elsevier Publishing Company).

which were present in ground water at concentrations from 10^3 to 10^4 cells per liter. Davis also noted the presence of facultative anaerobic bacteria.

There is much literature concerning the presence of microorganisms and microbial processes associated with petroleum reservoirs. A good deal of this work was performed by Russian scientists (Kuznetsov, 1950; Ekzertsev, 1951; Ekzertsev and Kuznetsov, 1954; Al'tovskii, 1962). This tradition has been continued and extended in more recent years (Belyaev and Ivanov, 1983).

7.5.2 Distribution of Bacteria in Regional Flow Systems

Sampling sediments from regional flow systems presents extreme technological problems. While auger drilling rigs are easily adapted for aseptically sampling local flow systems, and mud-rotary rigs can be used for aseptically sampling local and intermediate flow systems, aseptically sampling sediments from deep regional flow systems is much more difficult. One of the most obvious problems is cost. Drilling into regional flow systems is expensive under any circumstances and coring sediments is prohibitive. Although cores are routinely taken in such environments, they are usually taken for stratigraphic control or for analysis of organic carbon content. In addition, procedures for obtaining cores that are not contaminated by drilling fluids are difficult to apply. For these reasons, microorganisms in regional flow systems are most commonly recovered from ground water.

Culture Techniques As in local and intermediate flow systems, various culture techniques have been used to investigate the presence and kinds of microorganisms in regional flow systems. The work of Bastin (1926) and Davis (1967) are fairly typical of this approach. More recent examples of how the distribution of microorganisms in regional flow systems were given by Dockins et al. (1980) and Olson et al., 1981 who documented the presence of methanogenic and sulfate-reducing bacteria in water produced from the Madison aquifer system of Montana. The work of Dockins, Olson, and their associates is particularly notable in that the investigators showed that the bacteria grew at the temperatures of the aquifers (~ 70°C) but would not grow at room temperature (20° C). This showed that the bacteria were acclimated to the specialized conditions of the aquifer and ruled out the possibility that the observed bacteria were contaminants from the near surface.

Because of the slow movement of fluids in regional aquifers and because of their isolation from the surface, their has been considerable interest in using them as repositories of hazardous and radioactive wastes. Due to the potential for microorganisms to affect waste mobility, evaluations of such flow systems as waste repositories have led to microbiologic studies. One such study was described by Pedersen and Ekendahl (1990). These investigators showed that ground water produced from a deep (860 meters) granitic aquifer contained on the order of 10^5 cells per milliliter total bacteria of which between 10^2–10^3 were cultureable under anaerobic conditions. Sulfate-reducing and methanogenic bacteria were part of the ambient microflora as well as members of the genera *Shewanella* and *Pseudomonas*.

Pedersen and Ekendahl (1990) were microbiologists; they approached the problem of microorganisms in the deep granitic terrain using classic microbiologic techniques. When they were confronted with the problem of determining which microbial processes were active in situ, however, it was clear to them that the mere presence of certain microorganisms was equivocal. Consequently, these investigators, as had the earlier petroleum geologists, turned to evidence provided by ground-water chemistry. They noted that sulfate (and sulfide) was present in the ground water and that sulfate reducers largely exclude the activity of methanogens. The conclusion, therefore, was that sulfate reduction was the most likely predominant process in situ.

Geochemical Methods There are obvious problems with relying solely on culture methods to study the distribution of microorganisms in regional flow systems. In addition to the problems of media selectivity and determining which processes are most active in situ, there is the problem of how representative microorganisms in ground water are of the sediment-bound population. Davis (1967) recognized this problem explicitly as did Pedersen and Ekendahl (1990). In both cases, it was suggested that ground-water geochemistry gave important information as to in situ microbial processes.

In practice, much of what is known about the distribution of microbial processes in regional flow systems has been deduced from ground-water chemistry data. The Carrizo sand (Fig. 7.8) is one example of this that uses fairly simple water chemistry patterns (lack of sulfate, presence of sulfide) to describe distribution of microbial processes. The development of geochemical modeling to deduce operative geochemical processes (including microbial activity) has made this procedure more quantitative.

One of the first examples of this was given by Thorstenson, Fisher, and Croft (1979). The system under study was the Fox Hills-Basal Hell Creek aquifer. This aquifer ranges from being a local flow system near the recharge area, an intermediate flow system downgradient of the recharge area, to a regional flow system in the most downgradient parts of the system. Using ground-water chemistry data and a computer program for making equilibrium chemistry calculations, Thorstenson, Fisher, and Croft (1979) identified sulfate reduction as the predominant microbial process in the intermediate portion of the flow system which graded to methanogenesis as the predominant process in the regional portion of the flow system. This paper showed that geochemical modeling was one of the best available methods for deducing the distribution of microbial processes in inaccessible regional flow systems. This topic is discussed in greater detail in Chapter 10.

7.7 SUMMARY

In dealing with microbial processes in subsurface environments, it is first necessary to classify the kinds of environments that are most commonly encountered. In this chapter, a classification is given that divides subsurface environments on the basis of their hydrologic attributes and that is independent of depth. The two broad environments recognized are (1) the unsaturated zone characterized by the presence of both gas and water in sediment interstices, and (2) the saturated zone where water fills all available interstices. The saturated zone is further subdivided into local, intermediate, and regional flow systems. These divisions are made on the basis of quantitative considerations of ground-water flow in closed basins as described by Toth (1963). Local systems are characterized by abundant local recharge, fast ground-water flow velocities (meters per day), and are generally aerobic. Intermediate systems are not directly affected by transient precipitation events, have slower flow velocities (meters per year), and are generally anaerobic. Regional systems are largely isolated from surface influences, often have very sluggish flow velocities (meters per century), and are ubiquitously anaerobic.

These different subsurface environments are characterized by different microbial processes. The unsaturated zone and local flow systems are dominated by aerobic heterotrophic bacteria that are present in numbers of between 10^6–10^8 cells/gram sediment. These systems also may contain measurable numbers of eucaryotic microorganisms. Intermediate flow systems often contain very diverse populations of anaerobic and facultatively anaerobic microorganisms in numbers ranging from 10^4–10^6 cells/gram sediment. Regional flow systems often contain microorganisms as reflected in their presence in ground-water. However, little data is available on the numbers or kinds of microorganisms present in sediments. Geochemical modeling is a primary method for studying microbial processes in regional flow systems.

CHAPTER 8

MICROBIOLOGIC SAMPLING OF SUBSURFACE ENVIRONMENTS

Subsurface environments, whether they are shallow local flow systems or deep regional flow systems, share a common characteristic. This is that they are relatively inaccessible and therefore very difficult to sample. Simply obtaining materials from subsurface environments, which may consist of unconsolidated sediments or fully lithified rocks, is difficult under most circumstances. However, if the additional requirement is made that these samples must be free of chemical and biological contamination induced by the sampling process, this becomes an even more formidable undertaking.

Early work on the microbiology of deep subsurface environments, such as the isolation of sulfate-reducing bacteria by Bastin (1926), relied on water produced from oil or water wells as a sample source. This is certainly the easiest and most direct method for sampling the deep subsurface. Unfortunately, there are many potential problems with such a method, the most serious of which is that the drilling process, for either water wells or oil wells, can potentially contaminate the environment under study. Drilling fluids and grouting materials contain substantial quantities of microorganisms, and these can "inoculate" subsurface environments during drilling and well construction. When investigators are sampling wells, therefore, they may find it difficult to judge whether the microorganisms recovered are indigenous or introduced by the drilling process.

A second problem with using wells for sampling the subsurface is that the construction of a well introduces a new "environment" that did not exist prior to drilling. This new environment includes the borehole, the gravel pack, the casing and screen, and the grout. Since all of this material is artificially introduced, it may substantially alter conditions in the vicinity of a well. This alteration of the physical conditions may significantly affect the microbial processes near the well. Perhaps the best example of this is the potential introduction of oxygen to the subsurface. Well bores, being open to the surface, may transport oxygen into previously anaerobic environments. This, in turn, may support growth of aerobic bacteria that are not active under undisturbed conditions. Clearly, the possibility of such alteration must be carefully accessed when using wells for microbial sampling.

A third problem with using water from wells for microbiologic analysis is that most microorganisms in the subsurface tend to be sediment-bound. Thus, the kinds and diversity of free-living microorganisms may be significantly different from those attached to sediment surfaces. This possibility, together with the other problems associated with sampling well water, introduces considerable uncertainty into the data obtained.

For these reasons, modern studies of microbial processes in subsurface environments have tended to emphasize direct sampling of subsurface materials. Significant advances have been made since the early 1970's in techniques for obtaining uncompromised samples of unconsolidated sediments from subsurface environments for microbiologic analysis. These sampling techniques have probably been more important in the investigation of microbial processes in subsurface environments than any other factor. Much less attention has been paid, however, to sampling consolidated rock materials. In this chapter, different methods of sampling subsurface environments for microorganisms and microbial processes are outlined.

8.1 SAMPLING THE UNSATURATED ZONE

The unsaturated zone, as was discussed in Chapter 7, is characterized by a negative hydraulic pressure; that is, water in the unsaturated zone tends to be bound to sediments by capillary forces and does not flow readily to wells. This characteristic has a major impact on methods for sampling sediments from the unsaturated zone. The relative lack of water pore pressure means that the sediment framework is supported by grain-to-grain contacts rather than by hydrostatic pressure. This greatly increases the cohesiveness of the sediment and makes it much easier to sample.

A sampling system for subsurface sediments must simultaneously solve three problems. First, and most obviously, the system must be able to bore a hole from the surface to the horizon of interest. Second, the system must be capable of removing sediments from the hole as it is bored. This sediment retrieval may or may not be coupled to the sampling process. Third, the system must be able to maintain the sides of the borehole so that it does not collapse. The cohesive nature of sediments in the unsaturated zone allows each of these problems to be solved with relatively simple technology.

8.1.1 Hand Augering

By far the simplest method of obtaining sediment samples from the subsurface is by the use of a hand auger. A typical hand auger is equipped with a simple blade system that enables the tool to cut into unconsolidated sediments. The head of the auger also is designed to capture and to hold sediments so that they can be retrieved. The sediments that are retrieved serve to deepen the hole and become the sample material as well. Thus, for hand-augering systems, there is no distinction between hole construction and sample recovery. The stem of the hand auger is equipped with a handle that enables the head of the auger to be twisted into the ground. Once the head of the auger is twisted into the ground, the handle serves as a means of pulling the head out of the hole. The handle of the auger, and thus the length of the auger stem, can be extended by additional lengths of rod.

The usual procedure for obtaining sediment samples by means of a hand auger generally consists of three steps. First, the auger is twisted into the ground the approximate length of the auger head. This is usually about 6 or 8 inches. Second, the sample is retrieved by pulling the head of the auger back to land surface. Third, the sediment recovered is subsampled by means that are appropriate to the problem at hand. If the sediments are to be used for microbiologic analyses, subsampling is accomplished using flame-sterilized instruments and the subsamples placed into appropriate sterilized containers.

It is generally possible to sample subsurface sediments to a depth of about 20 or 30 feet using hand augers. Beyond 20 feet or so, the weight of the augers, as well as the difficulty in twisting them into the ground and retrieving them quickly becomes limiting. It is often difficult to hand auger below the water table. The increased pore pressure of the water tends to separate the sediment grains, thus decreasing the sediment cohesiveness. Sand-sized sediments will usually collapse once the auger is removed so that it becomes impossible to deepen the hole. Even in clayey sediments that are more cohesive, pulling the auger out of the sediments tends to collapse the hole. In general, hand augers are not suitable for routine sampling of sediments below the water table.

The advantage of hand augering is the technical ease of obtained subsurface sediments and its applicability to a number of unsaturated subsurface environments. However, because of the limited depth obtainable with this method, the inability to penetrate hard or indurated sediments, and the difficulty of sampling saturated sediments, its applicability to many subsurface environments is greatly reduced.

8.1.2 Air Drilling and Coring

In arid regions of the western United States, hydrologic systems are often characterized by extensive unsaturated zones that may be hundreds of feet thick. Because of the unsaturated nature of these materials, even unconsolidated sediments tend to be relatively cohesive and therefore do not require technology for holding a bored hole open. For these unsaturated zone sediments, the use of a core barrel for sample recovery together with compressed air for removing cuttings from the hole is an effective sampling method.

In air drilling and coring, a large compressor is used to force air down the drill pipe, out the bit, and up the outside of the borehole. As the core barrel cuts the subsurface materials, the air serves to blow the cuttings out of the hole as well as to cool the core barrel. Often, a small amount of water or surfactant is injected into the airstream to control dust and to aid with cooling the bit. However, use of these materials increases the possibility of contamination of the recovered sediments.

An example of the use of air coring for recovering unsaturated sediments for microbiologic analysis has been given by Colwell (1989). In this case, air coring was considered to be the sampling method of choice because it minimized the possibility of sample contamination by circulating fluids. To reduce the possibility of airborne microorganisms compromising the sampling procedure, the air being circulated into the bore hole was first filtered through a 0.3 micron filter. Because water or surfactants were not used to help cool the core barrel, drilling had to proceed fairly slowly in order to avoid overheating the core barrel. Clearly, if the core barrel were allowed to become too hot, the microbiologic integrity of the samples could be compromised. The main disadvantage of this method, therefore, is that sample recovery must proceed fairly slowly. The advantage of this method is that the probability of sample contamination by the coring process is fairly low.

8.2 SAMPLING LOCAL FLOW SYSTEMS

Water-saturated sediments, which by definition occur below the water table, are much less cohesive than unsaturated sediments. When a hole is bored into saturated unconsolidated sediments, therefore, there is much more tendency for the hole to collapse. A major technical problem in sampling saturated sediments is to be able to hold a bore hole open. There are numerous solutions to this technical problem with saturated sediments. By far the most common technique in relatively shallow water table aquifers (local flow systems) is the use of auger drilling combined with split-spoon and shelby-tube core sampling.

8.2.1 Split-spoon Sampling

By far the most widely used method for sampling sediments from relatively shallow subsurface environments is by means of hollow-stem auger drilling combined with split-spoon coring. The split-spoon technique was developed for use by civil engineers in foundation studies for buildings, bridges, and roads. The dual purpose of this technique was to provide estimates of the in situ strength of the subsurface materials and to provide samples for laboratory testing. Because of the ease of using this technique in relatively shallow systems, however, it has been widely adapted for ground-water applications. In turn, this technique has been widely adapted for microbiologic sampling of subsurface environments.

A typical system for hollow-stem coring includes hollow auger flights, a plug for closing off the hole in the auger flights, a cutter bit on the auger flights, a pilot bit on the plug, and rods that are coupled to the plug and pilot bit. Both the pilot and the auger cutter bits are designed to penetrate unconsolidated sediments.

The usual procedure for obtaining sediment samples with an auger system is to drill to the desired horizon and then remove the plug by pulling the rods. The plug is then replaced with a split spoon. A typical split spoon consists of two halves of an 18-inch cylinder 1 1/2 inches in diameter that are machined to fit together. The lower part of the spoon is threaded to accommodate a bit. This bit is generally fitted with a core retainer that is designed to keep the core from slipping out of the barrel during sample retrieval. The upper part of the spoon is threaded for attachment of the drilling rods used to lower and retrieve the spoon. When the spoon is at the bottom of the hole, it is driven into the ground ahead of the auger flight. The most common method for driving the spoon is by means of a 140-pound hammer that is repeatedly dropped from a height of 36 inches. The number of "blows," or number of hammer drops, needed to drive the spoon 6 inches is related to the foundation strength of the material being cored and is typically recorded in engineering applications of this method.

Once the split spoon has been driven the 18 inches of its length, the sediment-containing spoon is removed from the hole, generally by upward tapping with the hammer. Once the spoon has been retrieved, it is removed from the string of drilling rods. The bit of the spoon is removed, and the spoon "broken" in half by gentle tapping. Excessive pounding to separate the spoon halves can disturb the core. Once the spoon is broken in two, the sediment recovered is accessible for whatever sampling procedure is to be used. The plug is then recoupled to the drilling rods, inserted into the auger flights, and drilling to the next sample depth can commence. This sampling procedure is then repeated.

The main advantage of the hollow-stem auger method of drilling is that the auger flights remain in place during sampling thus preventing collapse of the bore hole. Also, because the screw action of the auger flights is what removes cuttings from the bore hole, circulating fluids are not required. This greatly decreases the potential for contamination of the sampled sediments.

The principal disadvantage of the split-spoon method is that the depth of sampling is greatly limited. Because of their great weight, the number of auger flights that can be manipulated by the drilling rig is limited. A typical drilling rig designed for hollow-stem augering has a depth limitation of about 100 feet. If more auger flights are added, the rig would be incapable of holding the added weight and the drill string would be lost. Another disadvantage of this method is that the diameter of recovered sediment samples is typically limited. A sample diameter of 1 1/2 inches, the diameter of core recovered by a standard split spoon, is the minimum size for obtaining sediments for microbiologic analysis. In clayey-textured sediments that are fairly cohesive, this sample size is not a significant problem. In some coarse sandy sediments, however, the lack of cohesion often makes it difficult to obtain or to manipulate uncompromised sediments. In such sediments, or for applications where more sample material is needed, other drilling techniques must be used.

8.2.2 Push-tube (Shelby Tube) Sampling Methods

For many applications, particularly involving aseptic techniques for microbiologic analysis, the limited sample diameter provided by split-spoon sampling is insufficient. An alternative is provided by push-tube sampling methods. These methods are notable from a historical perspective because they were among the first used to obtain subsurface sediments that were uncontaminated by surface-sediment microorganisms (Dunlap et al., 1977).

An example of this kind of aseptic sampling procedure was described by Wilson et al. (1983). A diagram of the sampling apparatus is shown in Figure 8.1a. It consists of a core retainer, a core barrel, and an adapter, much like a split spoon apparatus. In this system, however, the core barrel consists of a single tube that cannot be broken in order to retrieve the sample. Instead, the sample is removed from the core barrel by a hydraulic cylinder. During sample extrusion, the outside of the sediment sample is removed by a paring device (Fig. 8.1b). Thus, the sample

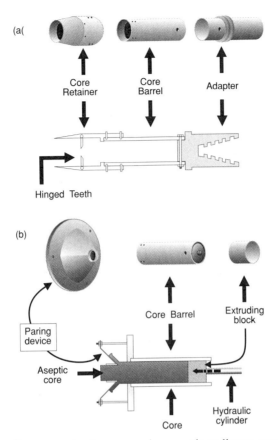

FIGURE 8.1. Sampling apparatus for recovering aseptic sediment samples using hollow-stem auger drilling. (a) Diagram of core barrel and retainer, and (b) core extrusion and paring device. (From Wilson et al., 1983, reprinted courtesy of the National Water Well Association).

can be extruded directly into a sterile container which reduces the possibility of contamination.

In order to obtain samples using this type of coring device, the hole is augered to the desired depth for sampling, and the auger flights are removed from the hole. Next, the core barrel is lowered to the bottom of the hole with drilling rods, and the barrel is pushed into the sediment by the drilling rig. It is also possible, and sometimes necessary, to drive the sample tube into the sediment with a hammer apparatus such as that used in split-spoon sampling. Once the tube has been driven into the sediment, it is retrieved with the drilling rig and brought to the surface for subsequent processing.

A *shelby tube* is a specialized sampler often used for soil science or foundation engineering applications. The dimensions of a shelby tube are 3 inches in diameter by about 20 inches long. The purpose of a shelby tube is to obtain sediments in an undisturbed condition. The wall of a shelby tube is relatively thin, requiring the tube to be pushed rather than pounded into the sediment. Once it is retrieved, the ends of the tube are often sealed by dipping them into melted wax or with special caps, and the entire tube is returned to the laboratory for testing. This method is designed primarily for clayey or silty sediments and is not suitable for recovering sandy sediments.

An important limitation of the push-tube method for obtaining subsurface sediments is that the auger flights must be removed prior to sampling. This, in turn, requires the sediments be sufficiently cohesive so as to prevent the bore hole from caving in as the flights are withdrawn. This is possible in some, but not all, locations.

An ingenious modification of this method, which circumvents possible bore hole-caving problems, has been described by Leach et al. (1989). In this method, the plug at the head of the auger flights is replaced by a set of telescoping plates that can open outward (i.e. down hole) but cannot open inward (allowing sediment into to the hollow stem of the auger). Using this system, the hole can be augered to the desired depth, and the augers lifted a few inches to allow the telescoping plates to open. Next, the sample tube is lowered down the hole without removing the augers. The sample tube pushes through the plates and can then be pushed into the sediment for sample recovery. After sample retrieval, the augers are pulled from the hole and the telescoping plates are reset. The augers are then returned to the hole and drilled to the depth of the next sample. This method is more technically difficult and consumes more time than split-spoon sampling, but it allows recovery of larger-diameter samples while solving the hole collapse problem.

8.2.3 Aseptic Technique with Split-spoon and Push-tube Sampling

The goal of sediment sampling for microbiologic analysis is the recovery of materials that are representative of the subsurface environment under study. In order to accomplish this, the most obvious requirement is that the samples not be contaminated by microorganisms from other horizons. Such possible contamination needs to be carefully assessed in each individual application.

Tool Contamination There are numerous possible sources of contamination when hollow-stem auger drilling is used for sampling. First, there is possible contamination from the sampling tool itself. This possibility has been emphasized by Dunlap et al. (1977) and Wilson et al. (1983), who stress that the tools must be sterilized prior to sampling. Wilson et al. (1983) favor flame sterilization, in which the tool is rinsed with alcohol and then ignited until the alcohol is completely burned off. This is probably the most convenient method of sterilizing sampling tools and is common in field applications.

Another method of sterilizing sampling tools is by means of compressed steam. It is common practice, when drilling sites characterized by chemical contamination, for drillers to steam-clean auger flights and sampling tools in order to prevent cross-contamination between holes. These steam cleaners operate at temperatures in excess of 100°C and are capable of sterilizing the tools as well cleansing them of chemical contamination. Because the steam is delivered at atmospheric pressure, there is some doubt that bacterial spores can be inactivated. However, this possibility has not been systematically investigated.

Down-hole Contamination Another possible mode of sample contamination is from down-hole contamination. This is to say that materials that originated from the surface or from the side of the bore hole fall to the bottom of the hole and are recovered by the sampling tool. Such contamination is often readily apparent upon sample recovery by the presence of loose sediment at the top of the sample container.

Down-hole contamination presents two technical problems. The first being that the top portion of the recovered core is not representative of the sample interval of interest. The generally accepted solution to this problem is to discard the upper 4 or 5 inches of a recovered core as a matter of standard procedure (Wilson et al., 1983). The second technical problem is that, because the sampling tool must pass through a layer of down-hole contamination before undisturbed sediments are reached, the tool itself becomes recontaminated. Thus, even if the tool had been properly sterilized at the surface, it would under most circumstances not be sterile during actual sampling.

The accepted method for dealing with this potential contamination is to assume, as a matter of course, that the outside of a sediment is contaminated. This outside portion of the core must be removed (Fig. 8.1), or the inside of the core must be sampled separately with sterilized instruments.

8.3 SAMPLING INTERMEDIATE AND REGIONAL FLOW SYSTEMS

The technology involved with split-spoon or push-tube sampling is relatively simple and straight-forward. However, it is rarely feasible to use these methods for sampling sediments deeper than about 100 feet. It is possible, with special equipment, for auger rigs to penetrate as deep as 250 feet. However, this is rare and seldom done in practice. When researchers are dealing with these deeper systems, which generally include intermediate and regional ground-water systems, more sophisticated technology is required. This technology introduces several more problems related to aseptic technique which must be carefully evaluated.

8.3.1 Mud Rotary Drilling

In mud rotary drilling, the hole is bored by a rotating bit, as was the case with auger drilling. However, the twin problems of removing cuttings and stabilizing the hole (i.e. preventing the hole from collapsing) are solved by the use of drilling fluids. Drilling fluids are continually circulated in the hole to remove cuttings and to apply pressure to the sides of the bore hole in order to keep it from collapsing. This technique has been widely used in microbiologic investigations of deep subsurface environments and considerable effort has been invested in solving the sample-integrity problems encountered with this technique.

A typical mud-rotary drilling rig has a *mast* (also called a boom), which can be raised for drilling and lowered for traveling by means of *hydraulic cylinders*. The hole is cut by a bit which has ports to allow drilling fluid to be pumped into the hole. The drilling fluid is pumped into the drilling rods through the *water swivel* (since the rods are rotating) with a specially-designed *mud pump*. The fluid travels down through the rods, out of the bit, and back up the hole. As the fluid moves up the hole, cuttings are carried out of the hole. The cuttings are then separated from the fluid, and the fluid is reused.

Mud rotary drilling is most commonly used to construct a bore hole for later construction into a well. However, mud rotary technology can also be used for coring applications. For subsurface microbiology applications, mud rotary coring is the most common method for obtaining samples of relatively deep sediments (100-3000 feet). This method has been extensively used in studying the microbiology of coastal plain sediments (Chapelle et al., 1988; Plelps et al., 1989b; Fredrickson et al., 1991).

8.3.2 Drilling Fluids

Drilling fluids are generally referred to as "mud" by drillers and geologists. This term, while being a fairly accurate description, gives little hint as to the technological sophistication involved. Drilling fluid is a marvel of applied chemistry and has undergone extensive research and development since its invention in the nineteenth century.

Drilling fluids are designed to accomplish four main functions:

1. *To remove cuttings from the hole during drilling.* After cuttings are generated by the drill bit, it is critical to remove them immediately from the hole. If the cuttings are allowed to settle around the drill bit, they will prevent the bit from turning freely. Ideally, cuttings are immediately entrained by the fluid stream and carried out of the hole.
2. *To keep the borehole from collapsing.* When sediments, particularly sandy sediments, are saturated by water, they have little cohesive strength. Under most conditions, a bore hole drilled into sandy sediments will not remain open without some sort of support. Drilling fluid provides this support by virtue of the pressure it exerts on the bore hole wall. If the pressure exerted by the drilling fluid is greater than the pressure in the formation, the bore hole will not collapse. A bore hole that is prevented from collapsing with drilling fluid is referred to as a "stable" hole.
3. *To cool and lubricate the bit.* Drilling bits, due to friction with the sediments being drilled, generate considerable amounts of heat. Drilling fluids prevent a buildup of heat in the drilling tools by lubricating the bit, thus minimizing heat generation, and by transferring heat away from the bit.
4. *To control fluid loss into the formations being penetrated.* When drilling in highly permeable formations, drilling fluids will tend to migrate from the bore hole into the formation. Drilling fluids are designed to control this fluid loss by generating a *filter cake* on the outside of the bore hole. This filter cake is made primarily from low-permeability clays that inhibit migration of fluid out of the bore hole.

Each of these functions is addressed by one or a combination of drilling fluid properties. The principal properties of water-based drilling fluids are listed in Table 8.1.

8.3 SAMPLING INTERMEDIATE AND REGIONAL FLOW SYSTEMS

TABLE 8.1 Principal Properties of Water-Based Drilling Fluids

1. Density (weight)	4. Gel strength
2. Viscosity	5. Fluid-loss-control effectiveness
3. Yield point	6. Lubricity (lubrication capacity)

Source: Reprinted courtesy of Johnson Division, St. Paul, Minnesota.

Density The *density* of a drilling fluid is defined as the weight per unit volume. Thus, it is common practice to use the terms "density" and "weight" interchangeably. Density is usually expressed in terms of pounds per gallon (lbs/gal) or pounds per cubic foot (lbs/ft^3).

The pressure exerted by drilling fluid on the walls of the bore hole is a function of density and the length of the fluid column. In order to prevent the bore hole from collapsing, the pressure of the drilling fluid must be greater than the pore pressure of the formation being drilled. Thus, if very unstable sediments are penetrated near land surface, the drilling fluid must be denser in order to prevent hole collapse than if those same sediments were penetrated at greater depth. It is considered good practice for the drilling fluid pressure to exceed formation pore pressure by about 5 pounds per square foot in order to maintain a stable hole. The hydrostatic pressure (H_p) exerted by drilling fluid equals the product of the fluid density (F_d) and the height of the fluid column (F_h)

$$H_p = F_d \times F_h$$

where H_p is in lbs/ft^2, F_d is in lb/ft^3, and F_h is in feet. Dense drilling fluids are also needed for drilling formations with relatively high pore pressures.

The density of drilling fluids is adjusted by the use of clay and soluble salt additives. Bentonites have a specific gravity of between 2.6 and 2.7, which often provides sufficient density for drilling shallow (<500 feet) bore holes. If unusual pore pressures are encountered or if deeper formations are being drilled, it is often necessary to use fluid densities greater than can be obtained with just clay additives.

The most important high-density additive used in drilling fluids is barite (barium sulfate) which has a specific gravity of from 4.2 to 4.35. By varying the amount of barite added to the drilling fluid, it is possible to obtain fairly high density. When drilling through formations with high hydrostatic pressure, progressively more barite is added to the drilling fluid. When drilling through formations with low hydrostatic pressure, which includes most shallow formations, little or no barite is added to the mud. Other soluble salts such as sodium chloride and calcium chloride are also used to adjust drilling fluid density. Figure 8.2 shows the density ranges obtainable using water with different additives.

Viscosity The *viscosity* of a drilling fluid refers to the sheer stress that it can exert on surfaces or particles. The ability of drilling fluids to lift cuttings from the hole is a function of fluid viscosity and fluid velocity. Curiously, viscosity and density (described above) are not functionally related to each other and are measured in different units.

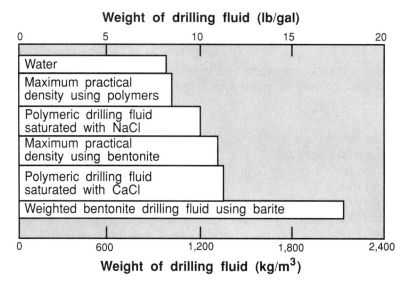

FIGURE 8.2. Range of densities obtainable with water-based drilling fluids. (Reprinted courtesy of Johnson Filtration Systems, St. Paul, Minnesota).

In the field, the viscosity of drilling fluid is measured with a Marsh funnel. This is a funnel that carries a standard volume and has a standard-sized opening at the bottom. Viscosity is expressed as the number of seconds required for the fluid to drain from the funnel. A Marsh viscosity reading of 35—40 seconds is sufficient to remove most fine sands from the hole. Coarser sediments, such as coarse gravels, may require a Marsh viscosity of 75—85 seconds.

The viscosity of most drilling fluids is adjusted by the addition of clay minerals. In fact, the early practice of adding clay minerals to drilling fluids used for drilling oil wells is where the term "drilling mud" originated. When clay minerals such as montmorillonite are added to water, they tend to swell as much as ten-fold. This swelling leads to the clay particles occupying more area and, therefore, to a more viscous fluid. Nonswelling clays (such as kaolinite) can be used to adjust viscosity of drilling fluids as well. However, swelling clays are much more compact when they are dry and are more convenient to work with. Bentonite, which is the rock name for the swelling clay montmorillonite, is a basic component of most drilling fluids and is added primarily to increase viscosity.

Another method of varying the viscosity of drilling fluids is by addition of synthetic polymers such as polysaccharides. Plain cornstarch was the first such additive used in this manner. In more recent years, very sophisticated polysaccharides have been developed for use in drilling fluids. The structure of one such polymer, which is called guar gum, is shown in Figure 8.3. This polymer consists of two kinds of sugar molecules, D-mannose and D-galactose. The D-mannose molecules are linked together to form a chain and the D-galactose molecules are linked to, on average, every other D-mannose molecule.

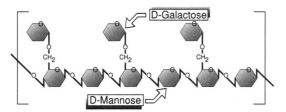

FIGURE 8.3. Molecular structure of guar gum, an example of a polymer used to enhance properties of drilling fluids. (Modified with permission courtesy of Johnson Division, St. Paul, Minnesota).

One of the advantages of using guar gum as a viscosity-enhancing additive is that the shear strength is a function of fluid velocity. When a colloidal suspension of guar gum is at rest, the polysaccharide chains tend to coil and become entangled with each other thus increasing viscosity. When shear stress is increased by pumping the suspension, the coils are stretched out and untangled, thus decreasing viscosity. This property enables the driller to stop mud circulation without allowing sediment particles entrained in the drilling fluid to settle out and possibly trap the string of drilling tools in the bore hole.

The use of organic polymers as drilling fluid additives when drilling or coring for microbiologic applications must be carefully considered. Most of these polymers are biodegradable: in fact, many of them are designed to be biodegradable, and this property may significantly alter conditions in the hydrologic systems being drilled. For example, if there is substantial penetration of the drilling fluid into a formation, biologic activity will, in most cases, be stimulated. This stimulation may alter conditions in the formation to the point that they are no longer representative of the natural system. In general, biodegradable drilling fluids should be avoided when drilling for microbiological investigations.

Yield Point Water is referred to as a *Newtonian fluid* because it deforms in direct proportion to an applied stress. Drilling fluids that contain clay minerals and other additives, however, tend to behave as plastics; that is they do not deform until a significant amount of stress is applied. The *yield point* of a drilling fluid is the pressure at which the pump is able to move the drilling fluid and the fluid begins to behave as a Newtonian fluid. Below the yield point, the viscosity of the drilling fluid changes with increasing stress. Above the yield point, the drilling fluid viscosity does not increase with increasing stress.

Gel Strength Gel strength is a measure of a drilling fluid's ability to support particles when the fluid is not moving. In clay-based drilling fluids, gel strength is provided by the electrical interaction of the clay particles. The flat surfaces of clay particles typically are positively charged whereas the edges are negatively charged. When the fluid is at rest, the attraction between clay particles provided by these charge differences provides its *gel strength*. Gel strength can also be enhanced by means of polymer additives, such as guar gum.

The gel strength of a drilling fluid is maximized if the clay particles and polymers are well-dispersed, allowing them to align themselves according to their electrical and or chemical properties. For this reason, a good driller takes care to mix clays and polymers thoroughly in the fluid prior to use.

Fluid Loss Control An important function of drilling fluid additives is the prevention of fluid migration into the formations being drilled. Because fluid pressure in the bore hole must exceed the hydrostatic pressure in the formation to prevent borehole collapse, the tendency is for drilling fluids to migrate into the formation. This is undesirable for two reasons. First of all, drilling fluids are expensive and their loss during drilling is a significant economic drain on the driller's operation. Second, the presence of substantial quantities of drilling fluid in a water-producing formation can degrade water-quality, rendering any well useless for water-chemistry studies or for water supplies.

A combination of clay minerals and organic additives is generally used to control fluid loss during drilling operations. When the drill bit penetrates a formation, some amount of drilling fluid flows into the formation due to the pressure difference between the bore hole and the formation. However, the clay minerals and or organic polymers, or both, that are present in the drilling fluid tend to be trapped in the pore throats of the formation materials. As this process continues, a *filter cake* is deposited on the bore hole surface that becomes less and less permeable to fluid penetration. Eventually fluid penetration virtually ceases.

Clay minerals added to drilling fluids in order to increase viscosity confer significant mud cake-forming properties. Several organic polymers, such as guar gum (Fig. 8.3) are also effective in preventing fluid loss. One of the advantages of using organic polymers to prevent fluid loss is that they can be subsequently removed by chemical or biologic action. For example, in order to develop a well to maximum capacity, it is obviously desirable to remove the mud cake from the bore hole. Organic polymers can be removed by addition of oxidizing agents (chlorine) or by natural biological activity. Clay mud cakes, on the other hand, require mechanical methods such as hole-scrubbing techniques to effect complete removal.

Lubricity One function of drilling fluid is to lubricate and cool the bit during drilling operations. The flat structure of clay minerals used in drilling fluids confers excellent lubricating properties. When physically pressed together, clay minerals tend to align themselves along their flat positively charges surfaces. Because the positive charges repel each other, there is little tendency for them to stick together, and the clay surfaces slide easily past each other. The lubricating characteristics of drilling fluids maximizes the efficiency of drilling and minimizes downhole heat build-up.

8.3.3 Mud Rotary Coring

The technology that has been most extensively utilized for microbiologic sampling of intermediate and regional flow systems is mud rotary coring. This coring technology was developed parallel to mud rotary drilling methods and utilizes many of the same tools and drilling fluids. However, there are significant philosophical differences between drilling and coring technologies. The primary goal of drilling technology is the production of a bore hole that can be used for constructing a well. The primary goal of coring, however, is the recovery of sediment samples that have been disturbed as little as possible, and the bore hole itself is generally of less importance. The net effect of these differences is that coring technology has become a narrowly focused specialty in the drilling trades that is practiced by relatively few individuals. It is important for the subsurface microbiologist to recognize that successful coring operations require a coring specialist and cannot be carried out by just any drilling contractor.

Equipment Used for Mud Rotary Coring The essential features present on a coring rig include a boom for raising and lowering the drill string, a rotary table for turning the drilling tools, and a mud pump for circulating drilling fluid. In all of these particulars, a coring rig is virtually identical to any mud-rotary rig. The primary differences are in the tools that are used.

The basic tools involved in wire-line coring are shown in Figure 8.4. The *outer barrel* (Fig. 8.4a) of the coring system attaches directly to the drilling rods and carries a drilling bit designed to cut semiconsolidated sediments. The outer barrel rotates with the drilling rods and actually cuts the bore hole. The bit on the outer barrel has ports, through which drilling fluid is circulated. The *inner barrel* (Fig. 8.4b) fits into the outer barrel and is raised and lowered by the wire-line. The function of the inner barrel is to trap the sediments as the outer barrel drills downward. When lowered downhole, the inner barrel locks into the outer barrel. However, the inner barrel is supported by a swivel equipped with bearings. This enables the outer barrel and bit to rotate while the inner barrel remains stationary. This, in turn, minimizes stress on the sediments in the inner barrel and helps avoid core disturbance. The top of the inner barrel is equipped with a latching apparatus. This latch allows the retrieval tool on the wire-line to attach to the inner barrel. This attaching mechanism simultaneously unlocks the inner barrel from the outer barrel so that it can be retrieved and brought to the surface.

The inner barrel does not actually cut the formation and is not equipped with a bit. The end of the inner barrel, however, is coupled to what is termed a *shoe* (Fig. 8.4b). The shoe is designed to allow the core into the inner barrel during drilling, but to prevent the core from falling out as the inner barrel is retrieved. There are two separate design features of the shoe that can be employed to prevent core loss during retrieval. The shoe itself is slightly tapered so that it is narrowest at its tip. Inside the shoe, a split metal sleeve is fit that also is tapered. During

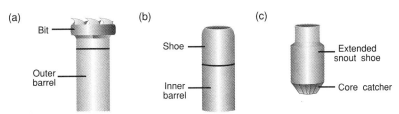

FIGURE 8.4. Tools used for mud rotary coring: (a) outer barrel and cutting bit, (b) inner barrel and shoe, and (c) extended snout shoe and core catcher.

coring, this sleeve is pushed upward where the shoe is widest, allowing the sleeve to expand slightly so that cored sediments can move easily past it. However, when the inner barrel is retrieved, the sleeve is pushed down in the tapered shoe, and the sleeve contracts around the core and holds it in the core barrel.

The shoe may also be equipped with a metal or a plastic core catcher (Fig. 8.4c). This core catcher is identical in function to catchers used during split-spoon coring, allowing the sediment into the core barrel, but discouraging it from falling back out. Some coring systems have been modified to use both the contracting sleeve and plastic core catcher simultaneously. This can greatly increase the efficiency of core recovery.

Under most drilling conditions, the shoe fits slightly behind the drilling bit on the outer barrel. However, when drilling through coarse sandy materials, drilling fluid pumped out of the drilling bit may tend to wash the sediments away before they can be trapped in the inner barrel. The Christensen system solves this problem by having a series of shoes of different length. For most coring situations, the short shoe (Fig. 8.4b) that fits behind the drilling bit is used. For sandier sediments, a shoe that protrudes slightly ahead of the bit is used (Fig. 8.4c). For very coarse sediments, an even longer shoe may be utilized. This system greatly increases sediment recovery when drilling sandy sediments. Because the sandy sediments typically comprise the aquifer portion of subsurface environments, this system has greatly aided investigations of aquifer materials.

After a core has been drilled, the inner barrel is retrieved by the wire line and the shoe is removed from the inner barrel. Once the shoe is removed, the core is extruded from the inner barrel. For sandy sediments that do not expand significantly when removed from the ground, simply removing the shoe often allows the core to slide out of the inner barrel. Other sediments, notably clays, tend to expand and become tightly trapped inside the inner barrel. In these cases, the core must be extruded using a rod, or in some cases, hydraulic pressure.

The Role of the Driller It is difficult for the casual observer to readily appreciate the skill involved in mud rotary coring. When watching a coring operation, an observer sees the physical work of raising and lowering the inner barrel, of adding rods to the drilling string, and manipulating the rig's controls. These physical activities are by far the easiest and least important aspects of the driller's job. Far more important are things that are not immediately apparent.

When coring a particular "run," as a length of core barrel is called, the driller continuously adjusts three variables:

1. Down-hole pressure on the drilling bit.
2. Drilling fluid pressure at drill bit.
3. Rate of drill bit rotation.

The driller must exert enough down hole pressure on the bit to enable it to cut though the sediments. However, if too much pressure is exerted, the sediments will be cut faster than they can be removed by the drilling fluid and the bore hole will become clogged. Similarly, the driller must provide enough drilling fluid

8.3 SAMPLING INTERMEDIATE AND REGIONAL FLOW SYSTEMS

pressure at the drill bit to remove cuttings effectively. If the pressure is too great, however, the drilling fluid will wash away the sediments before they are pushed into the core barrel. In this case, there will be no sediment recovery at all. Clearly, the driller must adjust the pressure of the drilling bit and drilling fluid so as to maximize core recovery and coring rate. Typically, subtle adjustments need to be made continuously in order to obtain the best results. This, in turn, requires considerable skill on the part of the driller.

The third factor that the driller varies is the rotation speed of the drill bit. In general, the size of the drill shavings cut by the drill bit is proportional to the relative hardness of the materials being drilled. Hard materials are not cut as easily and therefore produce smaller shavings. Thus, in order to drill hard materials, it is efficient to increase the rate of bit rotation in order to increase the rate of drilling. Softer materials, of course, can be efficiently drilled with progressively lower rotation rates.

Each of these three factors—drill bit pressure, drilling fluid pressure, and rotation rate are interrelated during drilling. It is much like having one equation with three variables. Once drill bit pressure and the rotation rate are fixed, there is an optimum drilling fluid pressure that must be used in order to maintain optimum drilling rate with optimum recovery. Deviations from this optimum will result in poorer core recovery, slower drilling progress, or both. Furthermore, because the properties of the material being drilled often changes rapidly, the optimum drilling parameters also must be constantly changed.

When watching an experienced driller work, it is possible for the observer to follow some of the constant adjustments as they are being made. When the driller glances at the return flow of drilling fluid, he is judging the efficiency with which the fluid is removing the cuttings. Having established that he is getting return flow, he might then look at the drilling fluid pressure gauge. If that pressure is fairly high in order to maintain fluid flow he might ease some of the pressure on the drilling bit slightly. This will allow the fluid pressure to drop and keep the fluid from washing the core away. This adjustment is made by moving a lever perhaps one quarter of an inch. This whole process, which might be critical to core recovery, generally takes a matter of seconds and is repeated over and over again. Many drillers do not like to be spoken to while they drilling. This is due to the tremendous concentration required by the job.

It is important for the geoscientist and microbiologist working with cored sediments to be at least cursorily familiar with the drilling problems involved in core recovery. Many of the same factors that enter into successful core recovery also are involved in potential core contamination by drilling fluids. For example, many drillers tend to keep as much drilling fluid pressure at the drilling bit as possible. This is because if the fluid jets become clogged with sediments, the entire string of rods must be removed in order to clean them. For a drilling string that is 1,000 feet long, this could easily become a one or two day job. However, high fluid pressures also increase the possibility that drilling fluids will penetrate into the core and contaminate them. When informed of the special requirements of microbiologic sampling-that is samples that are not contaminated with drilling fluids-most drillers will adjust their techniques to accommodate the more exacting standards. Communicating these special requirements, however, is much more effective if the scientists understand the many technical problems facing the drillers.

Drilling Fluid Technology and Coring Proper application of drilling fluid technology is critical to successful coring operations. In simple bore hole drilling, drilling fluids are mixed to maximize drilling rates and to minimize use of expensive drilling fluid additives. Thus, a bore hole driller may choose to work with a less viscous drilling fluid and make up for cutting removal efficiency by increasing the rate of fluid circulation in the hole. As we have already seen, however, the coring driller often does not have this extra flexibility, as excessive fluid velocity inhibits core recovery.

Drillers engaged in core recovery often try and maximize drilling fluid viscosity in order to minimize the drilling fluid pressure needed for efficient cuttings return. This practice also has the effect of minimizing fluid penetration into cored sediments and is thus consistent with the needs of microbiologic sampling.

One technical problem encountered in coring that is not encountered in bore hole drilling is when working in sediments-generally clean sands-that lack cohesive strength. Such sediments may tend to clump together when pushed into the inner barrel creating a "sand lock." This sand lock prevents any other sediments from entering the core barrel and usually results in lost recovery. One method of dealing with this problem is the addition of organic polymers to the drilling fluid. One such additive often used in coring has the tradename "Quik Trol" and is manufactured by Baroid. Quik Trol is an organic polymer that forms a thin coat on the core as it enters the inner barrel. This coat provides just enough cohesion so as to minimize sand lock problems and increase core recovery.

8.4 DRILLING FLUID CONTAMINATION OF CORED SEDIMENTS

The use of drilling fluids in coring operations introduces a significant source of possible chemical and biologic contamination in recovered sediments. The possibility of such contamination must be carefully evaluated when using cored sediments for microbiological or pore-water chemistry investigations.

8.4.1 Down-hole Saturation Contamination

There are several ways in which drilling fluids can penetrate and contaminate cores during coring operations. First, if the pressure of the drilling fluid at the bit face is excessive relative to the permeability of the formation, the fluid will saturate sediments several inches below the bit. When this drilling fluid-saturated sediment is subsequently recovered by the inner barrel, the entire sample is always contaminated. This type of contamination, which is referred to as *down-hole saturation* contamination, is very common when coring sandy or gravelly material.

There are several techniques that an experienced driller can use to minimize or eliminate down-hole saturation contamination. The most important technique, and one that a good coring driller practices as a matter of course, is to minimize the drilling fluid pressure at the bit. When coring silty or clayey sediments, this practice is often sufficient to prevent down-hole saturation.

When drilling permeable sandy sediments, however, even minimal drilling fluid pressure can result in down-hole saturation contamination. The most obvious, and also most dangerous, drilling technique for avoiding down-hole saturation in sandy sediments is simply to stop fluid circulation during coring. The lack of circulation fluid will greatly reduce the possibility of contamination. However, the lack of circulation fluid makes it easy for the fluid jets on the coring bit to become clogged. Once clogged, fluid cannot be circulated and the entire string of drilling rods must be pulled in order to unclog the bit. Also, if cuttings are not removed from around the outer barrel, it may become trapped. For these reasons, halting circulation is a last resort that drillers prefer to avoid. However, this method can be used to sample a particularly important sandy horizon if no other option is available.

A more technologically advanced technique for minimizing down-hole saturation contamination is the use of the Christensen extended snout shoe (Fig. 8.4c). By extending the shoe of the inner barrel as much as 2 inches below the drilling bit, it is possible to recover sediments below the zone of downhole saturation. By combining the proper length of drilling shoe and by minimizing drilling fluid pressure, the driller can recover uncontaminated sediments even in very coarse sandy sediments. It is not an exaggeration to say that the development of the extended shoe design is the most important advance in drilling technology to modern studies of microbial processes in deep hydrologic systems.

8.4.2 Core Seepage Contamination

A second kind of contamination is called *core seepage*. When cored material is pushed into the inner barrel, some drilling fluid is present on the outside of the core and can seep into the center of the core. This type of contamination can become particularly severe if the driller maintains an incorrect ratio between fluid pressure and drill stem pressure. When the driller is coring sandy material, excessive fluid pressure relative to drill stem pressure results in the core being slightly smaller than the diameter of the inner barrel. This, in turn, allows excessive quantities of drilling fluid in with the core and encourages drilling fluid seepage into the core.

Core seepage occurs to some extent with all cores recovered by mud rotary drilling, and the outside portion of any core must be considered contaminated. The extent to which such contamination occurs, however, varies widely. The degree of core contamination is largely dependent upon sediment texture. Sandy sediments often experience severe seepage contamination during coring, whereas clayey sediments generally experience much less seepage contamination. The reason for this difference is related to the distribution of porosity in each sediment type. In sandy sediments, typically 70-80% of the total porosity is interconnected. This allows fluids, including drilling fluid to move through it with relative ease. In clayey sediments, on the other hand, less than 10% of the total porosity is interconnected, which retards penetration of drilling fluids.

8.4.3 Core Fracture Contamination

Another common way for cored sediments to become contaminated by drilling fluids is by core fracturing. It is fairly common for cored sediments to come under considerable compressional stress as they are pushed into the inner barrel. If some portion of the sediment has less compressional strength than another portion, the sediment can fracture. Once the core is fractured, drilling fluid from the outside of the core can seep inward and contaminate the inner portion of the core. Core-fracture is the most common way in which clayey sediments can become contaminated.

Cores can become fractured by means of extensional stress as well as by compressional stress. When the inner barrel containing a freshly recovered core is lifted, extensional stress can develop along the length of the core. As before, if a particular portion of the sediment is relatively weak, a fracture can develop and drilling fluid seepage along the fracture trace can occur.

In both the case of compressional and extensional core fracturing, the problem can be minimized by careful application of proper drilling technique. To minimize compressional stresses on the core during drilling, the down-hole pressure on the drilling string and the drilling fluid pressure must be properly balanced. If the drilling fluid pressure is insufficient, the core will be of slightly greater diameter and will increase the compressional stress as it is pushed into the core barrel. A good driller always examines each core as it is retrieved in order to make sure that string pressure and fluid pressure are properly balanced.

Extensional core fracturing can also be minimized by good drilling technique. At the end of a core run, the sediment in the core barrel is still attached to sediments at the bottom of the hole. The driller can effect a clean break of the sediment at the bit by alternately spinning and stopping the drill string. If this break is made successfully, there will be less extensional stress applied to the core as the inner barrel is retrieved.

8.4.4 Evaluating Drilling Fluid Contamination

Using cores recovered by mud rotary techniques for biologic or pore-water chemistry studies requires that the extent of drilling fluid contamination be quantitatively evaluated. As all of the cores are contaminated to some extent, the problem is to determine the extent of contamination and then to judge whether the cores are usable for the purpose at hand.

A number of methods have been used over the years to evaluate drilling fluid contamination. All of these methods involve the utilization of some kind of tracer to track contamination. The characteristics of a good tracer include:

1. It is a substance that is present or can be added to the drilling fluid but that is not present in large amounts in the sediments being cored.
2. It is present or can be added in sufficient quantity so as to be easily recognizable.
3. It can be readily analyzed with adequate sensitivity.

There are at least four classes of tracers that have been used in subsurface microbiologic investigations. These include:

1. The drilling fluid itself
2. Chemical additives to the drilling fluid
3. Particulate tracers
4. Biologic tracers

Each of these methods has particular advantages and disadvantages, and each has been utilized in subsurface microbiological studies with varying degrees of success.

Drilling Fluid as a Tracer Drilling fluid has a carefully controlled chemical composition and, under many circumstances, makes a perfectly good tracer. For example, drilling fluids that are weighted with barite as an additive typically contain between 10 and 100 mg/L dissolved barium. Because barium is generally present in sediment pore waters at concentrations of less than 10 ug/L, barium can be used as a tracer.

An example of using barium as a drilling fluid tracer was given by Chapelle and Lovley (1990). The formations being drilled were under relatively high pressure and the driller weighted the drilling fluid with barite to about 12 lbs/gal. After the cores were recovered, they were shipped to the laboratory for processing. In the laboratory, the cores were broken in half and one of the faces was sampled along a traverse from the outside obviously contaminated portion of the core to the inside. Sediment samples of 1 ml were mixed with 9 mls of deionized water, shaken, and the supernate analyzed for barium concentrations using plasma atomic emission spectroscopy. The results were normalized to mg/L of pore fluid using measured moisture content. The results (Fig. 8.5) showed that some cores contained measureable barium concentrations into the center of the core, which probably reflected down hole saturation contamination. Other sediments only showed the presence of barium on the outside of the core, reflecting a small degree of core seepage contamination.

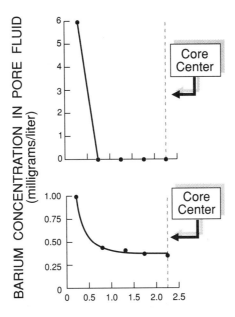

FIGURE 8.5. Concentrations of barium, an additive of drilling fluid, as a tracer of core contamination: (a) contamination present on outside of core but absent in core center, and (b) contamination present throughout the core. (Reprinted from Chapelle and Lovley, 1990, with permission courtesy of the American Society for Microbiology).

In addition to soluble cations present in drilling fluids, soluble anions may also be used. Chapelle and McMahon (1991), in a study of the pore-water chemistry of aquifer and confining bed sediments, measured sulfate concentrations in sediment pore-waters. The drilling fluid used for the coring operation contained about 5 mM dissolved sulfate. Analysis of ground water from a nearby production showed that sandy aquifer sediments contained between 0.05 and 0.1 mM sulfate. The pore water of uncontaminated (as determined using a particulate tracer) sandy sediments showed sulfate concentrations in the 0.05-0.1 mM range-identical to concentrations in well water. The pore water of contaminated sediments, on the other hand, contained as much as 2 or 3 mM dissolved sulfate. Thus, the sulfate in the drilling fluid could have been used as a tracer for contamination.

The principal problem with using dissolved constituents of drilling fluids as tracers is that analysis tends to be fairly cumbersome. In the case of the barium tracer, the samples had to be returned to the laboratory before contamination could be evaluated. In the case of sulfate, pore water was collected from the sediments in the field, but analysis was performed in the laboratory. It is highly desirable, in terms of both time and effort spent, to be able to screen the cores for contamination in the field. Analysis of cations—usually by means of atomic adsorption or emission spectroscopy—in the field is not generally feasible. Similarly, analysis of anions, usually by ion chromatography, is also not feasible in the field.

Chemical Additive Tracers A number of chemicals may be added to drilling fluid to serve as tracers. Many of these substances have been adapted from other tracer technologies. For example, rhodamine dye, an organic compound that absorbs light at one wavelength (~ 550 nM) and emits light at another (~ 590 nM), can be added to the drilling fluid as a tracer. This method has been used by Phelps et al. (1989b) in microbiologic studies of deep coastal plain sediments.

The advantages of rhodamine dye are that it does not occur naturally, it can be easily detected with a portable fluorometer, and it is relatively sensitive. Rhodamine dye in water can be detected in concentrations of as low as 100 parts per trillion. The principal disadvantage of rhodamine is that it is degraded by contact with chlorine and chloride-solutes common in drilling fluids-and that turbidity interferes to some degree with analysis. Also, rhodamine is strongly adsorbed by some organic materials. Another disadvantage is that because it alters the color of the drilling fluid, drillers dislike working with it. Drillers keep track of how much formation clay is suspended in drilling fluid by monitoring color changes over time. If these color changes are masked by use of a dye, it makes the driller's job more difficult. Drillers can, however, work within this if use of rhodamine is judged to be necessary.

Fluorescein is another fluorescent dye that has been used as a tracer for drilling fluid contamination. Pedersen and Ekendahl (1990) described the use of fluorescein in a study of bacteria in a granitic aquifer. When drilling hard rock formations such as granites, heavy drilling fluids are not needed to hold open the bore hole. In the case described by Pedersen and Ekendahl (1990) water produced from a nearby well was used as the primary drilling fluid. Prior to use in drilling, fluorescein was added to the ground water. During later microbiologic sampling of the wells, the amount of fluorescein present was quantified and the approximate percentage of drilling water contamination assessed. It was found that very low levels, as low as 0.1 %, drilling water contamination was measureable using this technique.

In addition to fluorescent tracers, other chemical tracers can be used. Phelps et al. (1989b) and Garland, Murphy, and McFadden (1991) added potassium bromide to drilling fluids as a tracer. Bromide is present in only trace amounts in the pore fluids of most aquifer systems and can readily flag drilling fluid contamination.

Another class of chemical tracer that has been applied to sampling deep subsurface sediments are perfluorocarbons (PFTs). PFTs are fully fluorinated alkyl substituted cycloalkanes. They are chemically inert, nontoxic, and are detectable in very low concentrations ($\sim$ 1 picoliter/gram of sample) using gas chromatography with an electron capture detector. PFTs are widely used in atmospheric tracing studies and as tracers in oil and gas reservoirs for evaluation of enhanced recovery potential. PFTs have been used for tracing drilling fluid contamination in deep coastal plain sediments (Senum and Dietz, 1991). In this application, differently substituted PFTs were used to separate potential contamination due to drilling of a preliminary test hole from contamination due to the coring operation. These results, which showed some cross-contamination of the two bore holes, indicated that contamination from one bore hole can extend several meters away from a bore hole. The sensitivity of this method, combined with the flexibility provided by using different PFTs for different parts of the drilling operation shows that it is a potentially useful tool for contamination tracing.

Particulate Tracers In addition to solutes, it is also possible to use particulates as tracers of drilling fluid contamination. One such particulate tracer that has been used in microbiologic investigations are micron-sized polystyrene spheres impregnated with a fluorescent dye. These microspheres, which are manufactured by Polysciences, Inc., come in a number of sizes, chemical characteristics, and fluorescent properties. The sizes available range from 0.1 to 10 micrometers in diameter. Importantly, the polystyrene from which they are made has a density of about 1.05, or very near that of fresh water. Because the spheres are much less dense than sediment particles, they are easily separated for microscopic examination. The microspheres are manufactured with a neutral, positive, or negative surface charges and are available with a variety of fluorescent properties.

Column experiments with sandy sediments have shown that the adherence of microspheres to sediment surfaces is a function of size and surface charge. The smaller spheres tend to adhere to sediment particles more frequently than larger spheres. Furthermore, since the surface of most silicate minerals grains have a net negative charge, negatively charged (carboxylated) microspheres adhere less frequently than those that are neutral or positively-charged. Microspheres can be easily detected with epifluorescent microscopy. With this microscopic technique, the sample is illuminated with fluorescent light. Those objects that emit visible light when exited with fluorescent light are then readily observed by the microscopist.

The manufacturer supplies the microspheres in 10-ml vials. The total number of spheres per 10-ml vial is dependent on sphere diameter, the weight percent of solids per ml, and the density of the solids. For 1-micron diameter spheres, a 10-ml vial contains 4.5×10^{11} spheres. The final concentration of microspheres in the drilling fluid is dependent on the volume of drilling fluid being used and the effectiveness of mixing it with the mud. An even distribution of microspheres can be obtained by having the driller add microspheres as drilling fluid is mixed. Typically, a 100 foot bore hole (6 inch diameter) uses about 400 gallons of drilling fluid. Each next 100 feet of bore hole requires approximately an additional 100 gallons. Thus, for a 200 foot bore hole, addition and complete mixing of one 10-ml vial results in a final microsphere concentration of about 10^7 spheres/ml of drilling fluid. Experience has shown that a concentration of 10^5 to 10^6 microspheres/ml drilling fluid is sufficient to trace contamination of cored materials.

The use of fluorescent microspheres as a tracer for drilling fluid contamination of cored sediments has been described by Chapelle and McMahon (1991). For this method, the epifluorescent microscope was set up at the drill site and microspheres were added to the drilling fluid for a final concentration of 10^5 spheres/ml. After a core had been recovered, it was broken in half to expose a fresh sediment surface. One of the sediment surfaces was sampled along a traverse from the outside to the middle of the core. Approximately one ml of sediment was placed in 9 mls of distilled water and shaken. When the larger sediment particles had settled, one ml of the suspension was filtered onto a nucleopore membrane. The membrane was placed directly on a microscope slide, and the microspheres enumerated by scanning about 10 % it's total area.

The principal advantage of the microsphere method is that cores can be screened for contamination within minutes of core recovery. This allows contaminated sediments to be quickly identified so that valuable time is not spent processing them.

Biologic Tracers Drilling fluids normally contain large numbers ($\sim 10^8$ cells per ml) of microorganisms. Because many of these microorganisms are different than the natural flora of deep subsurface sediments, they can be used as a tracer. One application of this general technique was given by Chapelle et al. (1988), who observed that the bacterium Gluconobacter was present in the drilling fluid but not in the sediments being cored. Thus, the presence or absence of this microorganism was an indication of the presence or absence of drilling fluid contamination.

Another example of this kind of tracer was described by Beeman and Suflita (1988). These investigators noted that drilling fluids contained large numbers of coliform bacteria (bacteria that normally inhabit the intestines of animals), whereas subsurface sediments did not. Thus, the presence or absence of coliform bacteria could serve as a "serendipitous" tracer for drilling fluid contamination.

The principal drawback for using microorganisms as tracers is that it is difficult to use them quantitatively. Although they are often useful qualitative indicators of the simple presence or absence of contamination, it is difficult to determine the sensitivity of the method. For quantitative purposes, chemical or particulate tracers are more suitable.

8.5 SAMPLING GROUND WATER FOR MICROORGANISMS

Sampling ground water for microorganisms is fraught with numerous potential contamination problems. Drilling fluid used in constructing bore holes for wells contain substantial numbers of microorganisms. The drilling process, therefore, effectively "inoculates" any aquifer system being drilled. It may well be that the majority of bacteria introduced by drilling die off with time in their new environment; however, this has never been studied quantitatively. It must be assumed that some microorganisms recovered from well water may represent introduced, rather than indigenous, species.

For some applications, however, it is just as important to consider introduced microorganisms as well as indigenous ones. For example, microorganisms introduced by drilling procedures into wells to be used for water supply are probably of more concern from a health standpoint than indigenous species. For some applications, therefore, sampling ground water may be the procedure of choice.

The procedures used to sample ground water for microorganisms are generally fairly simple. Olson et al. (1981), in a study of methanogenic and sulfate-reducing bacteria in the Madison aquifer of Montana, allowed the completed well to purge for 8 hours prior to sampling. A sterile piece of rubber tubing was then inserted into the discharge pipe, and this tube used to fill the sterile sample bottles.

Hirsch and Rades-Rohkohl (1983) used a more sophisticated setup to sample ground water from a shallow aquifer in Schleswig-Holstein, Germany. For sampling fully penetrating wells, these investigators sterilized a submersible pump and its attached tubing with 70% ethanol alcohol and lowered it into the wells to be sampled. Three casing volumes were then pumped from the well and discarded in order to flush out traces of ethanol and to remove free standing water from the well bore. Ground water was then collected in a sterile 500 ml sampling flask equipped with a screw cap for subsequent plate count and isolation procedures. These procedures included spread-plating 0.1 ml of ground water onto triplicate plates of solid media, followed by standard streak plate isolation methods.

In addition to sampling fully penetrating wells in the shallow aquifer, Hirsch and Rades-Rohkohl (1983) sampled a series of multilevel screened wells. As before, about three well-volumes of ground water were removed from each well prior to sampling. Plate counts of this water that had stood in the well bores showed higher numbers of bacteria than present were in water pumped directly from the aquifer. Water was lifted from the wells by suction into a desiccator and a 500 ml sterile sampling flask was filled with water for microbiologic analysis.

Most strategies for sampling ground water from wells for microbiologic analysis are similar to those employed by Olson et al. (1981) and Hirsch and Rades-Rohkohl (1983). It is assumed that free-standing water in the bore hole is not representative of aquifer water and so at least 3 well volumes-and often tens of well volumes-are evacuated prior to sampling. After well development, the ground water samples are collected in appropriate presterilized vessels. Similar techniques have been employed in more recent years by Stetzenbach, Delley, and Sinclair (1986) in the Tucson Basin and by Pedersen and Ekendahl (1990) in a deep granitic aquifer in southeastern Sweden.

8.6 SUMMARY

Subsurface sediments can be sampled for microbiologic analysis using a number of methods. The unsaturated zone is most conveniently sampled using a hand auger to a depth of about 20 feet. For relatively thick unsaturated sediments, air drilling combines ease of sample recovery with relatively low risk of sample contamination. Local flow systems, which are usually relatively shallow, are most easily sampled using hollow-stem augering combined with split-spoon or push-tube methods. Hollow-stem augering is the method of choice for sample depths that are less than about 100 feet.

Intermediate and regional flow systems, being deeper, often cannot be sampled using auger rigs. In these cases, mud rotary coring has been most often employed for sampling. Mud rotary coring is technologically sophisticated and uses a number of highly specialized drilling tools and drilling fluid additives. While coring, the driller constantly adjusts drill stem pressure, fluid pressure, and drill bit rotation. The efficiency of core recovery depends in large measure on the skill of the driller and the degree to which these three variables are optimized. Coring efficiency also may be enhanced by the proper use of specialized drilling fluid additives.

The use of drilling fluid in mud rotary coring often results in contamination of the core. There are three common ways for cored sediments to be contaminated by drilling fluids. One is down-hole saturation, where drilling fluid saturates the sediments ahead of the core barrel. Another is core seepage, where drilling fluid seeps into the core while it is in the inner barrel. Fracturing of the core once it has entered the inner barrel will also contaminate cored sediments, particularly relatively hard clays or indurated sediments.

Because of the possibility of core contamination by drilling fluids, it is necessary for each core used for microbiologic analysis to be screened for contamination. To this end, a number of different types of tracers can be used. Some components of drilling fluids, such as barium or sulfate from barite additives, are potential tracers. It is also possible to use microorganisms that are present in the drilling fluid but not in subsurface sediments, such as colliform bacteria, as tracers. Often, however, the best indicators of contamination are chemical or particulate tracers added to drilling fluids. Rhodamine dye or bromide are examples of chemicals used as chemical tracers, and fluorescent microspheres are examples of particulate tracers.

The microbiology of subsurface environments can also be sampled by producing ground water from properly constructed wells. Although it is fairly easy to preclude water contamination after it has left the well, it is difficult to rule out sample contamination by microorganisms present in drilling fluids or microorganisms growing gratuitously in the well bore. Furthermore, the similarity between water borne microbial communities and sediment-bound communities is problematic. Nevertheless, these techniques are suitable for some applications.

CHAPTER 9

BIOGEOCHEMICAL CYCLING IN GROUND-WATER SYSTEMS

The primary goal of ground-water geochemistry is to understand the processes governing the distribution of chemical species in ground water. If ground-water systems were sterile environments that lacked any sort of microbial life, this goal could be attained solely from fundamental principles of mineral equilibria and chemical kinetics. However, available evidence suggests that sterile ground-water systems are the exception rather than the rule. It is important, therefore, to consider how the biochemistry of microorganisms affects the occurrence and movement of chemical species in these systems.

The term *biogeochemical cycling* refers to the alternate storage and release of chemical energy as an element moves through the biosphere. Because the chemical characteristics of each element are different, the pathways that cycle them through the biosphere are also different. Biogeochemical cycling in ground-water systems has a significant impact on the speciation and mobility of biologically active elements.

The concept of biogeochemical cycles comes from ecological theory. Ecosystems can be viewed at many different levels depending on the questions being asked. For example, the relationships between primary producers (plants), grazers, predators, and decomposers are often depicted ecologically (Fig. 9.1a), which illustrates the interactions between different populations. At another level, however, relationships between populations can be expressed as the flow of energy and chemicals through the ecosystem (Fig. 9.1b). In this view, primary producers store energy and biochemically active elements in a form which can be utilized by consumers. The cycle is completed when a specialized group of consumers, the degraders, return the elements to a form that can again be stored by the producers.

Many of the chemical solutes important in ground-water systems—oxygen, organic and inorganic carbon species, sulfur species, nitrogen species, and iron species—move through microbial food chains in ways analogous to those shown in Figure 9.1b. For this reason, understanding the nature of biogeochemical cycles

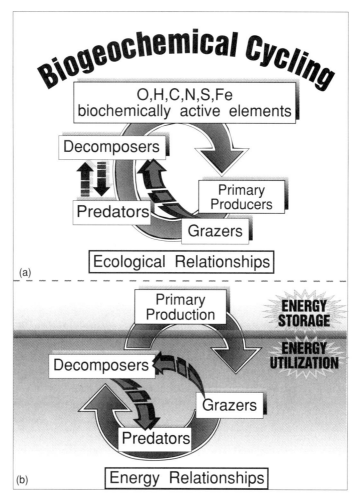

FIGURE 9.1. Biogeochemical cycling in terms of (a) ecological relationships and (b) energy relationships.

is fundamental to understanding the distribution of chemical species in microbially active ground-water systems. However, because of the unique nature of subsurface ecosystems, the manner in which microbial elemental cycling is expressed is often much different than observed in other sedimentary environments. In this chapter, the biogeochemical cycling of oxygen, carbon, nitrogen, iron, and sulfur in ground-water systems—and some effects of this cycling on ground-water geochemistry—are considered.

9.1 THE OXYGEN CYCLE

Molecular oxygen is present in the earth's atmosphere solely because of photosynthetic production of reduced carbon compounds from carbon dioxide by green (chlorophyll-bearing) plants and microorganisms. Prior to the evolution of photosynthetic algae (roughly 2 billion years B.C.) the earth's atmosphere was highly reducing. The overall stoichiometry of oxygen production via photosynthesis is

$$CO_2 + H_2O \xrightarrow{\text{Sunlight}} CH_2O + O_2$$

where CH_2O represents reduced organic carbon present in carbohydrates. When this reaction proceeds from left to right, it is highly endergonic and requires energy from sunlight in order to proceed. Molecular oxygen produced from this reaction subsequently accumulates in the atmosphere and in surface and ground water. The solar energy that drives this reaction and that is stored as chemical energy can be subsequently recovered and utilized by respiring micro- and macroorganisms:

$$CH_2O + O_2 \rightarrow CO_2 + H_2O + \text{Energy}$$

On a global scale, the production of oxygen by photosynthesis is exactly balanced by oxygen consumption by respiring organisms, as shown in Figure 9.2. This large-scale balance between oxygen production and consumption is referred to as the *global oxygen cycle*.

9.1.1 Oxygen Cycling in Ground-Water Systems

The balance between oxygen production via photosynthesis and oxygen consumption via respiration is maintained only on a global scale. On more local scales, it is common for imbalances to develop between oxygen production and consumption. For example, in surface water bodies in which photosynthetic algae are active, it is possible for the rate of oxygen production to exceed respiration during daylight hours resulting in a net accumulation of dissolved oxygen in the water. During the night, however, the activity of respiring heterotrophic bacteria outpaces photosynthesis and oxygen concentrations decrease. The push-pull relationship between oxygen production and consumption on a local scale largely determines the oxygen status of a particular environment.

In many environments, such as ground-water systems and many aquatic sediments, sunlight is excluded and photosynthetic production of oxygen cannot occur. This effectively truncates the oxygen-producing component of the oxygen cycle

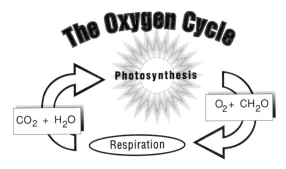

FIGURE 9.2. The oxygen cycle.

while leaving the oxygen-consuming component intact. In these cases, the oxygen status of an environment is determined by rates of oxygen transport from the atmosphere and by rates of respiration. If rates of oxygen transport into the system exceed rates of endogenous respiration, oxygen will accumulate and the system will be aerobic. On the other hand, if respiration rates exceed the flux of oxygen, the system will become anaerobic.

The flux of oxygen to all sedimentary environments is severely limited by its low solubility in water. At 20°C, only about 9 mg/L oxygen can dissolve in water. This low solubility affects different systems in different ways. In surface-water systems such as rivers or lakes, water generally moves fairly rapidly and aerobic conditions are common. In aquatic sediments or ground-water systems, however, rates of water movement are much slower. This decreases the available flux of oxygen to these environments and anaerobic conditions are more common.

From a water quality standpoint, dissolved oxygen is one of the most important geochemical constituents of ground water. Aerobic ground water generally lacks such undesirable constituents as dissolved organic carbon, dissolved Fe(II), sulfides, or methane. For this reason, oxygenated ground water is generally desirable for human and animal consumption.

The distribution of dissolved oxygen in ground-water systems generally follows a predictable pattern. The distribution of dissolved oxygen in the Patuxent aquifer near Baltimore, Maryland (Chapelle and Kean, 1985) is a good example of this (Fig. 9.3). Where the aquifer crops out-and hence where it functions as a local flow system-the ground water contains measureable concentrations of dissolved oxygen. As water flows downgradient and enters the intermediate flow system, however, concentrations of dissolved oxygen decrease below measureable levels.

The pattern of dissolved oxygen changes shown in Figure 9.3 reflects relative rates of dissolved oxygen transport to and oxygen consumption in each aquifer. Local flow systems closely connected to the atmosphere tend to be oxygenated. In these local flow systems, the flux of oxygen from the atmosphere exceeds rates of respiration in the aquifer and hence oxygen accumulates in solution. As ground water moves away from local flow systems and enters confined intermediate systems, the flux of dissolved oxygen is typically less than oxygen consumption, and concentrations decrease.

Aquifer systems vary widely in their rates of respiration and, therefore, in their rates of oxygen consumption. Some organic-rich aquifers, even where they function as shallow local flow systems, are entirely anaerobic. For example, the Hawthorne aquifer of South Carolina (Chapelle et al., 1988) becomes anaerobic within a few meters of the water-table. The rate of oxygen respiration (Ox_{re}) in this system can be calculated from the equation

$$Ox_{re} = \Delta O_2 \cdot R/L$$

where ΔO_2 is the change in oxygen concentrations, R is the average rate of ground-water flow, and L is the length of the flowpath segment. Given the average downward flow rate (R) of 0.23 m/yr, a dissolved oxygen concentration change of 5 mg/L (0.16 mM), and a flowpath length of 6 meters, the rate of oxygen respiration in this system is calculated as being on the order of 10^{-2} mM/L/yr. In contrast, calculated rates of respirative oxygen consumption in the Patuxent aquifer (Fig.

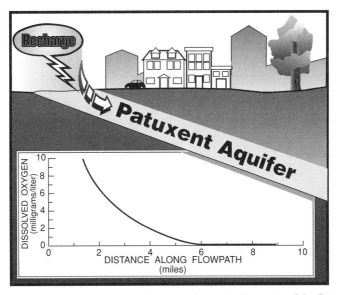

FIGURE 9.3. Consumption of dissolved oxygen along the flowpath of the Patuxent aquifer near Baltimore, Maryland.

9.3), given an average flow rate of 5 m/yr, a dissolved oxygen change of 10 mg/L (0.32 mM), and a flowpath length of 6 miles (9,654 m), indicates a much slower oxygen consumption rate on the order of 10^{-4} mM/L/yr. It is clear that respirative oxygen consumption rates vary widely between different ground-water systems.

When rates of dissolved oxygen consumption are relatively low, it is possible for low concentrations of oxygen to persist over wide areas of aquifer systems. For many years, analytical methods have been inadequate for studying these low-oxygen aquifers. Standard methods for measuring dissolved oxygen concentrations in ground water, the Winkler titration and oxygen-sensitive membrane electrodes, have detection limits of about 0.1 mg/L ($\sim 3 \mu M$) dissolved oxygen. This has limited the study of aquifers with fairly low (3.0–0.2 μM) dissolved oxygen concentrations.

A study by White, Peterson, and Solbau (1990) introduced a more sensitive oxygen-measuring technique based on the oxidation of rhodazine-D. This study showed that oxygen could be reliably measured in concentrations as low as 0.2 μM (0.006 mg/L). Furthermore, it showed that the presence of selenate, an oxidized selenium compound that is relatively mobile as well as toxic, correlated with the presence of low dissolved oxygen concentrations in a shallow aquifer. It is probable that application of low detection level techniques for dissolved oxygen will reveal a variety of aquifer systems characterized by low but measurable levels of dissolved oxygen.

Some aquifer systems exhibit virtually no oxygen consumption, even when ground water enters the deeper intermediate and regional flow systems. For example, deep sedimentary basins of the American Southwest, in which ground water is on the order of 10,000 years old, in some cases have oxygen concentrations in the 7–8 mg/L range and thus still near atmospheric saturation (Winograd and Robertson, 1882). In this case, the aquifer consists of alluvial material that is virtually devoid of organic material or other reduced compounds. Consequently, there is little or no consumption of oxygen. This aquifer system is a rare example of a ground-water system lacking significant microbial respiration.

The oxygen cycle illustrates a general principle of biochemical cycling in subsurface environments-that the cycles become truncated at one or more places. In the case of oxygen, the truncation of the cycle is due to the lack of energy input in the form of photosynthesis. This, in turn, simply results in the establishment of anaerobic conditions in many subsurface environments. Other elemental cycles may also be truncated by the lack of energy input resulting in the *accumulation* of particular chemical species. An example of this behavior is the carbon cycle, which is closely intertwined with the oxygen cycle. This will be considered next.

9.2 THE CARBON CYCLE

The carbon atom, with its ability to be stable in a number of different oxidation states (-4 to $+4$) and its tendency to form stable covalent bonds, is very efficient at storing and releasing energy. The ability of carbon to absorb (solar) energy by forming reduced organic compounds, and then release this (chemical) energy through oxidation reactions is the chemical basis for all life on earth. The biochemical mechanisms and pathways that control the flow of energy through oxidation-reduction reactions involving carbon are known collectively as the *global carbon cycle*.

A schematic diagram of the global carbon cycle is shown in Figure 9.4. The central compound in this cycle is carbon dioxide (CO_2) present either in the atmosphere as a gas or in water as dissolved inorganic carbon species (CO_2, HCO_3^-, $CO_3^=$). In terrestrial and near-surface marine environments, sunlight is available and CO_2 is reduced to carbohydrates via photosynthesis. The reduction of oxidized carbon in CO_2 (oxidation state $+4$) to organic carbon present in carbohydrates (oxidation state zero) releases free oxygen and results in aerobic conditions. Much of this organic carbon is aerobically re-oxidized via the respiration of plants, animals, and microorganisms and returned to the atmosphere or ocean as CO_2. A significant quantity of this oxidized carbon in marine environments is sequestered by calcium carbonate-precipitating marine organisms.

Certain types of anaerobic bacteria, notably the green sulfur bacteria, the purple sulfur bacteria, and nonsulfur purple bacteria are also capable of photosynthesis. The biochemistry of this photosynthesis differs markedly from that of green plants and algae in that molecular oxygen is not produced. Thus, in anaerobic aquatic environments exposed to sunlight, bacterial photosynthesis contributes to the reduction of CO_2 to carbohydrates and other organic compounds but without the production of oxygen.

A large amount of organic carbon produced by plant photosynthesis is cycled back to CO_2 by means of anaerobic oxidation. In anaerobic oxidation, fermentative bacteria incompletely oxidize organic carbon with the production of organic acids, alcohols, and molecular hydrogen. These simple reduced compounds are then completely oxidized by anaerobic respirative bacteria that use mineral electron acceptors such as Mn(IV), Fe(III), and sulfate. The methanogenic bacteria pro-

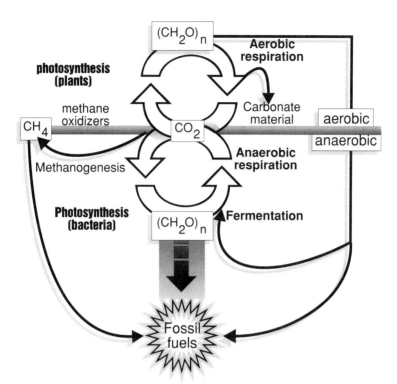

FIGURE 9.4. The global carbon cycle.

duce methane (CH_4) under anaerobic conditions. Much of this methane subsequently diffuses into aerobic environments where oxidation to CO_2 takes place. A portion of this CH_4 becomes trapped in geologic structures and accumulates. Where aerobic and anaerobic oxidation of organic carbon is incomplete, organic carbon may accumulate over geologic time to form coal, petroleum, and other fossil fuels.

The observed rise in atmospheric CO_2 levels over the last century reflects a number of human perturbations of the global carbon cycle. In particular, the rapid burning of fossil fuels has oxidized and released large amounts of carbon from storage within the earth. This, in turn, has increased the available pool of CO_2. Because of the complexity of the carbon cycle, however, the effects of this increased pool of atmospheric CO_2 remains controversial. Inspection of Figure 9.4 suggests, for example, that increased availability of CO_2 may stimulate photosynthetic production of organic carbon. This process would act as a negative feedback mechanism for CO_2 accumulation in the atmosphere. However, such increased stimulation of photosynthesis cannot occur if the world's forests are progressively removed by an expanding human population and if oceanic algal populations are depleted by pollution. Possible political solutions to this potential problem will rest on, among other things, a correct interpretation of the global carbon cycle and how it responds to perturbation. This topic is and will remain a major scientific challenge.

9.2.1 The Integrated Carbon, Oxygen, and Hydrogen Cycles

Although it is a convenience to consider each elemental cycle individually, it is clear that these cycles are closely intertwined. It is not possible, for example, to discuss the production of molecular oxygen (Fig. 9.2) without also considering the production of organic carbon from CO_2 (Fig. 9.4). Furthermore, the hydrogen cycle, which has not been discussed separately, is also intertwined with the carbon and oxygen cycle.

A schematic representation of the integrated carbon, oxygen, and hydrogen cycles is shown in Figure 9.5. These cycles are driven by energy provided in sunlight. Photosynthesis reduces carbon in CO_2 with the oxidation of oxygen present in water. Under aerobic conditions, much of this reduced carbon is cycled back to CO_2, with oxygen being reduced to water. Under anaerobic conditions, organic carbon compounds may also be oxidized, but these oxidations proceed in stepwise fashion. The oxidation is initiated by fermentative microorganisms that

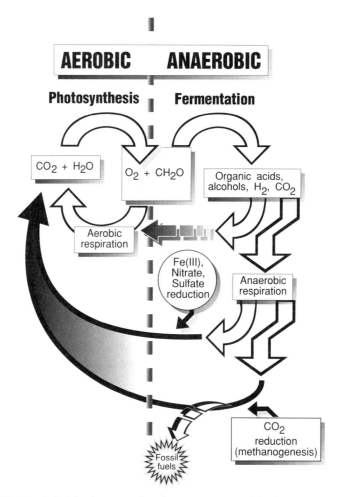

FIGURE 9.5. The integrated carbon, oxygen, and hydrogen cycles.

partially oxidize the carbon to CO_2 with the production of reduced compounds such as organic acids, alcohols, and molecular hydrogen (H_2). These reduced compounds can then be completely oxidized anaerobically with Fe(III), sulfate, or carbon (methanogenesis) acting as terminal electron acceptors. Under aerobic conditions, these reduced compounds are cycled directly back to CO_2 and water.

From a microbiologic point of view, the nature of these integrated cycles includes a number of important ecological implications. For example, it is clear that the cycling of carbon under anaerobic conditions requires a cooperative food chain in order to achieve complete oxidation. From a geologic point of view, it's interesting to note the steps necessary for the production of biogenic methane gas. First of all, if mineral oxidants such as Fe(III) or sulfate are available, there will be little production of methane. Only in the absence of these mineral oxidants is organic carbon oxidation coupled to the production of methane. This methane may go into storage, creating a pool of potential fossil fuel, or may be reoxidized under aerobic conditions. Judging from the relatively small pool of methane trapped in geological structures relative to the scale of the global carbon cycle, it is clear that aerobic reoxidation is by far the most common fate of biogenic methane gas.

9.2.2 Carbon Cycling in Ground-Water Systems

The carbon, oxygen, and hydrogen cycles are driven by two pools of energy (Fig. 9.5). The photosynthetic fixation of organic carbon, which initiates the cycle, is driven exclusively by solar energy. In turn, the oxidation of organic carbon back to CO_2 is driven by chemical energy stored in reduced carbon and hydrogen compounds. A fundamental characteristic of ground-water systems removed from active photosynthesis is that these cycles are effectively truncated. In the case of the oxygen cycle, the truncation of the cycle leads to depletion of oxygen in many ground-water systems. In the case of the carbon cycle, however, truncation would be expected to lead to the *accumulation* of CO_2 and CO_2-derived carbonate species in ground water.

The accumulation of dissolved inorganic carbon species associated with intermediate and regional flow systems was one of the first chemical attributes of ground water that was systematically investigated. Effervescent spring waters have been known to humanity since prehistoric times. These waters were closely studied over the years, primarily because of their purported medicinal properties, but also out of curiosity about their effervescence. During the early nineteenth century, when rudimentary techniques of analytical chemistry became available, it became fashionable for physicians to perform chemical analyses of spring waters in order to discover the chemical basis of their observed medicinal properties.

One particularly well-documented case of this was recorded by a physician named John H. Steel (1782-1838). Steel had practiced medicine in the area of Saratoga Springs in New York State for many years. As such, he became familiar with the medical uses of water from several springs. Being very much a man of the Age of Enlightenment, he resolved to chemically analyze the spring waters and thereby discover the causes of the medicinal effects. Steel (1838) describes the water of one particular spring, Congress Spring, in this manner:

> The surface of the well is constantly agitated by the escape of gas in fine bubbles, giving the appearance of simmering, not unlike that which water exhibits just before the process of violent ebullition (boiling) takes place. When first dipped, the water is remarkably sparkling, and were it not for the constant escape of gas in innumerable fine points, it would be perfectly transparent.

Steel went on to describe some effects of the gas issuing from the spring on animals:

> The respiration of all breathing animals is immediately affected by coming in close contact with the surface of this fountain. The gas which issues from it is immediately fatal to the lives of animals which happen to be immersed in it, and even fishes and frogs survive but a short time when placed in it.

Steel observed that this gas, when driven out of the water by heating and collected in an inflatable bladder, was absorbed by "potash of soda" (potassium hydroxide), thus deflating the bladder. This showed that the gas was predominantly "carbonic acid gas" or carbon dioxide. This was a result that explained the observed effects on animals (it suffocated them) and neatly explained the effervescence. Deducing that the gas was carbon dioxide was an accomplishment of which Steel was greatly proud. Incidentally, the water also contained relatively high concentrations of magnesium sulfate, which probably also explains the cathartic medicinal properties of the water.

Showing that an accumulation of carbon dioxide explained the effervescent properties of this (and countless other) artesian waters did not explain the source of the CO_2. Steel clearly recognized this problem. In the last paragraph of his book, he writes:

> The production of the unexampled quantity of carbonic acid gas, and the medium through which the other articles are principally retained in solution, is yet, and probably will remain, a subject of mere speculation. The low and regular temperature of the water seems to forbid the idea that it is the effect of subterranean heat, as many have supposed. Its production is therefore truly unaccountable.

The observed accumulation of CO_2 and other dissolved carbonate species, however, may be regarded as a manifestation of a truncated carbon cycle in ground water systems. In turn, the nature of this cycling is dependent on the properties of the ground-water flow system.

Local Flow Systems Local flow systems are characterized by relatively close interaction with the surface and surface processes. This close connection has a marked impact on carbon cycling in local flow systems which distinguishes them from intermediate or regional flow systems.

Figure 9.6 shows a schematic diagram of how carbon is cycled in local flow systems. Photosynthetic production of organic carbon at land surface begins the cycle. The decomposition of organic matter near land surface, a designation that includes the soil zone, releases some CO_2 as well as dissolved organic carbon (DOC) compounds and particulate organic carbon (POC). Much of the CO_2 is immediately released to the atmosphere. A small portion of CO_2 as well as some

9.2 THE CARBON CYCLE

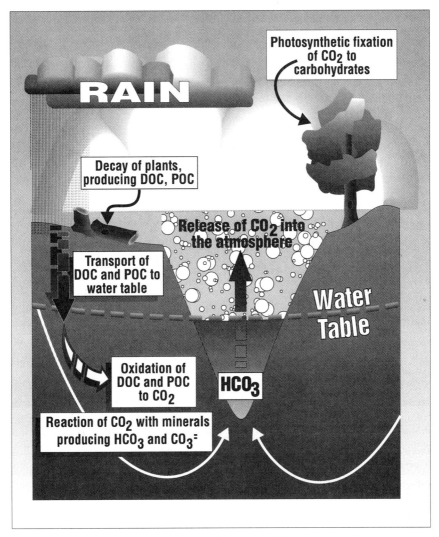

FIGURE 9.6. Carbon cycling in local flow systems.

DOC and POC is carried to the water table and into the local flow system by vertically percolating recharge. Much of the DOC and POC transported in this manner are subsequently oxidized to CO_2. However, because the aquifer is under unconfined, water-table conditions, CO_2 in excess of water solubility is readily exchanged with soil gases and ultimately the atmosphere. Most of the CO_2 generated by these processes is transported by ground-water flow to an adjacent recharge area where it seeps to the surface and escapes to the atmosphere.

While many variations on this general theme are possible, the basic outline applies to most local flow systems. The outcome of these cycling processes is that there is little net accumulation of CO_2 or CO_2-derived species in ground water. Rather, the close connection of local flow systems with the surface, and therefore with the atmosphere, insures rapid and complete cycling of carbon. Carbon cycling in local flow systems closely resembles the kinds of processes that occur in soils, unsaturated zones, and other near-surface environments.

Intermediate and Regional Flow Systems The carbon cycle in intermediate and regional flow systems tends to be much different from that in local flow systems. One difference is the relative importance of organic carbon from surface decomposition of organic matter. The ultimate source of organic carbon is, of course, photosynthesis. However, organic carbon originally deposited with the sediments tends to be much more important in intermediate and regional flow systems. This reflects the relatively slow recharge rates from the surface. Because recharge rates are generally slow, DOC or POC fluxes from surface sources also tend to be low.

Some time after deposition, often millions of years, a ground-water flow system is established and slow oxidation of the organic carbon continues. In or near the outcrop areas of the aquifer, oxidation of organic matter may occur aerobically. However, with the consumption of dissolved oxygen, oxidation under anaerobic conditions quickly becomes the dominant process. Oxidation is not confined to the sandy aquifers exclusively, but also occurs in clayey confining beds. It is rare that anaerobic oxidation proceeds only via fermentation. Rather, fermentation and oxidation by mineral electron acceptors tend to be coupled. Because of the presence of confining beds, there is little connection between aquifers and the surface. Therefore, CO_2 and other dissolved inorganic carbon (DIC) species generated by these anaerobic oxidation processes accumulate in solution.

The progress of carbon cycling in intermediate and regional flow systems can have several different outcomes. If the ground water remains isolated from the surface by means of structural traps or other geologic features, then much of the DIC may ultimately be precipitated as carbonate cements. This effectively removes that carbon from the carbon cycle for long periods of time. A more common outcome, however, is the eventual discharge of ground water to the surface by means of springs or by diffuse upward discharge. This type of discharge accounts for the occurrence of mineral springs, such as those at Saratoga Springs, and countless other artesian seeps. The occurrence of CO_2-charged waters at these springs is thus an inevitable manifestation of carbon cycling in these subsurface environments.

On a large scale, both in terms of time and space, it is clear that complete truncation of the carbon cycle (and thus the oxygen and hydrogen cycles) in intermediate and regional ground-water systems is rare. Even dissolved carbonate species from organic carbon oxidation that become sequestered as carbonate cements will eventually return to the cycle when the sediments are uplifted and eroded. The concept of cycle truncation, therefore, is useful only when we consider the time frame of human life and history. This is, however, a time frame that is of special importance and interest.

9.3 THE NITROGEN CYCLE

While the carbon, oxygen, and hydrogen cycles involve both macroorganisms and microorganisms in producing and decomposing organic compounds, the nitrogen cycle involves principally microorganisms, as well as some abiotic processes. Nitrogen in the atmosphere is present primarily as N_2 gas which is a particularly stable compound. Under ambient atmospheric conditions, oxidized (NO_3) or reduced (NH_3) nitrogen compounds are almost never produced from N_2. However, such compounds are produced nonbiologically during electrical storms as well as in the higher atmosphere where photochemical processes are important. These nitrogen compounds are then carried to the earth's surface by precipitation. At the earth's surface, however, microbially-mediated *nitrogen fixation*-the production of organic nitrogen compounds from N_2-and microbial nitrogen oxidation and reduction are the sole nitrogen-cycling processes.

A schematic representation of the nitrogen cycle is shown in Figure 9.7. Molecular nitrogen gas is fixed aerobically by a number of microorganisms including *Azotobacter*, *Rhizobium*, and many species of *Cyanobacteria*. Anaerobically, N_2 is fixed by members of the genus *Clostridium*.

The production of organic nitrogen compounds, such as ammonia, from molecular nitrogen requires a relatively high amount of energy input by microorganisms. The overall reaction of ammonia formation is approximately

$$N_2 + 6H + 16ATP \rightarrow 2NH_3 + 16\ ADP + 16\ P$$

Thus, a cell must spend about eight ATP molecules in order to synthesize a single molecule of ammonia. From this it is clear that microorganisms can only afford to fix nitrogen actively if a ready supply of energy, either from photosynthesis (in the *Cyanobacteria*), symbiotic plants (as in *Rhizobium*), or from organic carbon (as in *Clostridia*). The high "cost" of nitrogen fixation is a crucial factor in nitrogen cycling in ground-water systems.

The crucial enzyme system involved in nitrogen fixation is *nitrogenase,* a complex molecule consisting of an iron-molybdenum protein and two iron proteins.

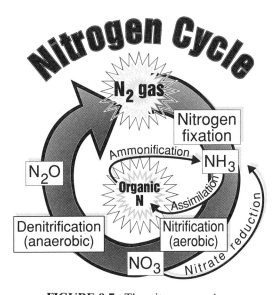

FIGURE 9.7. The nitrogen cycle.

The general mode of nitrogenase action is for the two iron proteins to reduce the iron-molybdenum protein to a low enough state so that the triple bond of the N_2 molecule may be broken and hydrogens attached to the nitrogen atoms:

$$N_2 + 3H_2 \rightarrow 2NH_3$$

Once the nitrogen cycle is initiated by N_2 fixation, the nitrogen cycle can continue (Fig. 9.7). Ammonia (NH_3) produced by N_2 fixation may be *assimilated* into proteins by microorganisms and by some macroorganisms such as plants. Upon the death of these organisms, the nitrogen present in proteins may be returned to ammonia via *deamination* processes.

Under aerobic conditions, ammonia may be oxidized to nitrite (NO_2^-) by specialized nitrifying bacteria such as *Nitrosomonas*. Nitrite is then further oxidized to nitrate ($NO_3^=$) by other nitrifying bacteria such as *Nitrobacter*. These combined *nitrification* processes are the basis of the production of nitrate fertilizers from human and animal wastes-a procedure that has been used by farmers for millennia. Once complete oxidation of nitrogen is accomplished, the nitrate may be *assimilated* by a variety of microorganisms and incorporated as NH_2 groups in proteins either by plants or by microorganisms. Note that the assimilation into proteins involves the reduction of nitrate. This *assimilative nitrate reduction* production is a process carried out by a large number of microorganisms.

In addition to being available as a component in protein synthesis, the nitrate ion can function as an electron acceptor in bacterial metabolism. Under anaerobic conditions, nitrate is the most thermodynamically favored electron acceptor and is used by a number of facultatively anaerobic bacteria including many *Pseudomonas* species. Nitrate reduction coupled to respirative energy production is referred to as *denitrification*, in order to distinguish it from assimilative nitrate reduction. Denitrification occurs in a number of steps, each of which is catalyzed by different microorganisms, in the order $NO_3 \rightarrow NO_2 \rightarrow N_2O \rightarrow N_2$.

As was the case with the oxygen and carbon cycles, the nitrogen cycle is driven primarily by solar energy. The manner in which some microorganisms gain access to the solar energy, however, is interesting and very important for agricultural applications. Members of the bacterial genus *Rhizobium* have evolved a symbiotic relationship with several plants, most notably legumes (soybeans and related crops). In *symbiotic nitrogen fixation*, *Rhizobium* infects the root hairs of the host plant, forming nodules. Within these nodules, the plant supplies sugars to the bacteria. This energy supply gives the bacteria the wherewithal to produce the large amounts of ATP needed to fix nitrogen. This fixed nitrogen is then made available to the plant, which uses it for protein synthesis and growth. It is this symbiotic relationship that is responsible for the relatively high protein content of soybeans and other legumes.

In some cases, the symbiotic relationship between the plant and the microorganism has reached the molecular level. For example, *Rhizobium* is an obligate aerobe, but the activity of nitrogenase is irreversibly destroyed by exposure to oxygen. A specialized protein called *leghemoglobin* controls access of oxygen to the nodule in order to allow respiration and nitrogen fixation to continue simultaneously. In order to construct leghemoglobin, *Rhizobium* carries the gene for one portion of the protein molecule whereas the plant carries the gene for another portion. Thus, nitrogen fixation requires both organisms. This is a classic example of true symbiosis, in which each partner requires the other in order to carry out critical life functions.

9.3.1 Nitrogen Cycling in Ground-Water Systems

The behavior of nitrogen species in ground-water systems is strongly affected by three key aspects of the nitrogen cycle. First, fixation of organic nitrogen from N_2 requires a tremendous input of energy. In oligotrophic ground-water systems, this necessarily places limits on the rate and amount of nitrogen fixation that takes place. Second, the oxidation of organic nitrogen to nitrate (nitrification) takes place only under aerobic conditions. Third, denitrification takes place only under anaerobic conditions. These aspects insure that nitrate accumulates in ground water under oxygenated conditions and that nitrate is depleted under anaerobic conditions.

In the case of the oxygen and carbon, we saw that the lack of solar energy input in ground-water systems effectively truncated the cycles resulting in depletion of oxygen and accumulation of carbonate species. In the case of the nitrogen, however, the behavior differs depending on the oxygen status of the ground-water system. Under aerobic conditions, nitrification may take place, but on the opposite side of the cycle, denitrification cannot take place. Thus, under aerobic conditions, the cycle truncates after nitrification and nitrate accumulates in solution. Under anaerobic conditions, the reverse is true: nitrate can be converted to N_2 via denitrification, but nitrification is blocked. This generally results in depletion of fixed nitrogen species in anaerobic ground-water systems.

The occurrence of nitrate in ground water is a significant public health issue that has received considerable attention over the years. Excessive intake of nitrate in infants has been linked to methemoglobinemia, often referred to as the "blue baby syndrome," in which oxygen in the bloodstream is depleted. Furthermore, intake of nitrate has been linked to rates of mortality from gastric cancer. Because of these health concerns, the processes that influence the mobility of nitrate and nitrogen species in ground water have been closely studied.

As a practical matter, the cycling of nitrogen in ground-water systems is strongly influenced by human activities. Nitrogen fertilizers are widely used in agricultural practice, and the application of these often affects nitrogen cycling in particular systems. Similarly, organic nitrogen is present in a number of waste products, notably sewage effluents, and discharge of these affects nitrogen cycling. It is convenient, therefore, when considering nitrogen cycling in ground-water systems, to distinguish between different kinds of human impacts on ground-water systems. In human-impacted ground-water systems, the principal sources of nitrogen are (1) animal excrement and manure, (2) sewage effluents, (3) nitrogen fertilizers, and (4) municipal wastes.

Animal Excrement and Manures Many studies have documented that animal excrement can affect nitrate concentrations in local flow systems. For example, Robertson (1979) showed that nitrate concentrations in a shallow water table aquifer in Delaware tended to be associated with confined feeding operations of poultry or cattle. Furthermore, Robertson (1979) showed that there was a general decrease in nitrate concentrations with well depths.

These observations can be readily interpreted in the context of the nitrogen cycle. Animal manure contains organic nitrogen in the form of ammonia. Under the aerobic conditions at land surface, ammonia is converted to nitrate via nitrifying bacteria. This nitrate is soluble and readily leaches into the shallow ground-water aquifer. This aquifer tends to be aerobic near the water table surface, and so nitrate accumulates in solution. However, in deeper portions of the aquifer, where anaerobic conditions are more common, nitrate concentrations decrease because of denitrification. The accumulation of nitrate in this aquifer is therefore primarily dependent on the delivery of nitrate and the presence or absence of oxygen.

The relationship between the delivery of nitrate to local flow systems and the application of manure to fields has also been widely studied. In one such study in eastern Pennsylvania, Gerhart (1985) showed that concentrations of nitrate in a shallow aquifer mirrored recharge events. As water levels in the well rose, reflecting recharge from rainfall events, concentrations of nitrate increased significantly. In general, nitrate concentrations in the ground water were higher during periods of recharge than between periods of recharge.

This behavior reflects the two-step nature of nitrate delivery to shallow ground-water systems. First, the ammonia present in manure must be oxidized to nitrate by nitrifying bacteria under aerobic conditions, often at or near land surface. Second, nitrate produced by this reaction must be delivered to the ground-water system by means of vertically percolating recharge. If the aquifer being recharged is aerobic, nitrogen cycling truncates at this point and nitrate accumulates in ground water. If the aquifer grades from aerobic to anaerobic, as commonly occurs along aquifer flowpaths, the nitrogen cycle continues through denitrification and nitrate does not accumulate in solution. Again, the nitrogen cycle truncates at this point-this time because of the lack of nitrogen fixation.

Sewage Effluents Another important source of nitrogen to ground water in human-impacted local flow systems is by the discharge of sewage effluents. The problems of ground-water quality generated by sewage effluents have been widely studied worldwide. In the United States, one place where this problem has been studied particularly closely is on Long Island, New York.

Long Island is underlain by coastal plain sediments of the Magothy aquifer which is in turn overlain by coarse-grained sediments of glacial origin. The coarse, organic matter-poor glacial sediments generally contain dissolved oxygen at or near atmospheric saturation. This allows nitrate to accumulate in solution. Much of Long Island has been dependent on septic drains for disposal of sewage effluent. This has, in some places, resulted in significant accumulations of nitrate in ground water.

The influence of septic systems on nitrate accumulations in ground water of Long Island was studied by Katz, Lindner, and Ragone (1980). Katz and coworkers made a statistical comparison of nitrate concentrations in unsewered residential areas, where septic tanks were the predominant method of sewage disposal, and sewered areas where sewage was collected for treatment. Surprisingly, there was no significant difference found in mean nitrate concentrations between sewered and unsewered areas over a 15 year period. The data, however, did indicate some differences. For example, the unsewered areas were much more likely to have nitrate concentrations greater than 20 mg/L than the sewered areas. Furthermore, although there was little difference in nitrate concentrations, concentrations of ammonium were significantly lower in the sewered area relative to the unsewered area.

These differences can be interpreted in the context of nitrogen cycling in this ground-water system. For example, a major source of dissolved nitrogen from sewage is the deamination of organic matter to produce ammonia and ammonium ions. In the unsewered areas, therefore, a major portion of the nitrogen cycle appears to be deamination followed by nitrification. In the sewered areas, in contrast, deamination was much less important. Katz, Lindner, and Ragone (1980) suggested that sewage effluents were a more important source of dissolved nitrogen in the unsewered areas than in the sewered areas. Furthermore, it was suggested that the lack of significant nitrate differences between the two areas reflected applications of nitrate fertilizers.

Perhaps the most comprehensive study of microbial denitrification associated with sewage effluents in local flow systems was described by Smith and Duff (1988). These investigators measured rates of denitrification in a 4-km long sewage effluent plume on Cape Cod, Massachusetts. Near the source of the effluent, nitrate concentrations exceeded 50 mg/L. As water in the plume moved downgradient, however, concentrations of nitrate decreased below detectable levels.

Smith and Duff (1988) obtained samples of aquifer sediments along the hydrologic gradient and measured rates of denitrification using the acetylene block technique. In this technique the final step of denitrification, conversion of N_2O to N_2, is blocked by adding acetylene to sediment slurries. The accumulation of N_2O gas in the slurry vial thus gives an estimate of the overall rate of denitrification. Using this technique, Smith and Duff found that rates of denitrification decreased as nitrate concentrations decreased along the flowpath. Furthermore, rates of denitrification in the plume were greater than in the pristine portion of the aquifer. Denitrification of nitrate to N_2, rather than nitrate reduction to ammonia (Fig. 9.7) was the predominant reductive process in this system.

Perhaps the most interesting finding of this study was that denitrifying activity in the contaminated plume was not limited by the availability of nitrate. Rather, the limiting factor to denitrification rates was the availability of a suitable organic carbon substrate. In anaerobic aquifers in which measurable concentrations of nitrate persist for any distance along the flowpath, denitrification must be limited by factors other than nitrate availability. In such cases, carbon limitation is often responsible for the observed persistence of nitrate.

Nitrogen Fertilizers Nitrogen fertilizers are a significant source of nitrate contamination of ground water. Nitrogen fertilizers may be composed of a number of nitrogen compounds including anhydrous NH_3, NH_4NO_3, $(NH_4)_2HP_2O_5$, and urea. In general, nitrogen delivered to the soil as NH_3 or NH_4^+ is retained better than nitrogen delivered as nitrate. Ammonium, being a positively charged ion, tends to adsorb to the negatively charged silicate minerals present in soils. The negatively charged nitrate ion, however, is repelled from mineral-grain surfaces and moves freely with percolating pore water.

The impacts of agricultural nitrogen applications on the chemical composition of ground water have been extensively studied. In the United Kingdom, agricultural nitrogen applications to fields overlying a chalk aquifer have been studied by Gray and Morgan-Jones (1980). This study documented the accumulation of nitrate in three well fields tapping the chalk aquifer over a 40-year period. One of these well fields in particular showed a close correlation between agricultural activity and nitrate concentrations. In the early 1940's, when fertilizer applications were lowered because of World War II, concentrations of nitrate decreased. However, over the next 30 years, as agricultural activity progressively increased, concentrations of nitrate in the ground water increased dramatically. These accumulations reflected the familiar pattern of nitrification in the soil zone, followed by nitrate leaching into and accumulating in, an aerobic local flow system.

Nitrate contamination of ground water due to fertilization practices is an important water-quality issue in the Midwest. Corn, an important crop in the Midwest, requires large amounts of nitrogen for yields to be economically viable. It is common practice for corn growers to apply as much as 180 lbs N/acre/yr in order to obtain sufficient yields. Nitrogen applied in this manner is rapidly depleted from soil both by plant uptake and by nitrification followed by leaching of nitrate into ground water.

In one study of the impacts of such practices on ground-water quality, Spalding et al., 1978 showed that zones of nitrate contamination in ground water were closely correlated with coarse-textured, well-drained soils. This reflected rates of nitrification, rates of leaching, and the occurrence of oxygenated conditions in the water table aquifer. This general pattern has been observed in many locations and is characteristic of nitrate contamination of ground water.

Municipal Wastes In addition to animal excrements, sewage effluents, and agricultural fertilizers, municipal wastes that have been disposed in landfills may be important sources of nitrogen contamination to ground water. An example of such nitrogen contamination is given by Baedecker and Back (1979) in their description of a municipal landfill in Delaware. The source of the nitrogen species in the ground water was organic nitrogen in the refuse. Near the landfill, ground water was anaerobic and most of the nitrogen was present as reduced organic nitrogen and ammonium. However, as the contaminated water moved downgradient and mixed with the ambient oxygenated ground water, concentrations of reduced nitrogen decreased and concentrations of nitrate increased. Baedecker and Back (1979) described this behavior using a nitrogen index, defined as the ratio of reduced nitrogen species (determined by Kjeldahl digestion) to nitrate. As ground water became progressively more oxygenated, the nitrogen index decreased. This reflected the progressive nitrification of reduced nitrogen with the production of nitrate.

Since anaerobic conditions are more likely in and near landfills of municipal waste, the pattern described by Baedecker and Back (1979) is probably the most common. However, patterns of nitrogen cycling are very sensitive to the availability of oxygen and other patterns may occur. A very different pattern of nitrogen cycling was described near a municipal landfill and waste-water holding cells in North Dakota (Bulger, Kehew, and Nelson, 1989). This study showed that nitrate was the primary nitrogen species emanating from the landfill under oxidizing conditions. However, once the nitrate plume intersected a plume of high dissolved organic carbon, which was anaerobic, the nitrate was progressively reduced to NH_4^+. This pattern indicates that denitrification, with the conversion of NO_3 to N_2, is not the only pathway of nitrate reduction found in ground-water systems. In this case, nitrate is apparently reduced to NH_4^+ as the plume becomes anaerobic.

Distinguishing Sources of Nitrogen Contamination Because of the numerous sources of nitrogen contamination in ground-water systems, distinguishing whether contamination originated from agricultural fertilizers, animal wastes, or sewage effluents is often difficult. One method for distinguishing different sources of nitrogen contamination that has been widely applied is by the use of stable nitrogen isotopes.

Nitrogen occurs as two stable isotopes: ^{15}N and ^{14}N. Ratios of these isotopes are commonly referred to a standard and reported as $\delta^{15}N$. Ranges of $\delta^{15}N$ for nitrogen from different sources are shown in Figure 9.8. Clearly, there is a wide difference between nitrogen present in fertilizers (-4 to $+2$ per mil) and nitrogen present in animal (and human) wastes ($+9$ to $+18$ per mil). All things being equal, therefore, it should be possible to distinguish fertilizer nitrogen from animal waste nitrogen based solely on $\delta^{15}N$.

However, nitrogen isotopes are subject to a number of fractionation processes in the nitrogen cycle. For example, some ammonia nitrogen applied to soils is subsequently volatilized, resulting in enrichment in ^{15}N. In addition, denitrification also results in a ^{15}N enrichment as microorganisms tend to discriminate against the heavier isotope. These twin processes tend to enrich nitrate with ^{15}N after fertilizer application. For example, Gormly and Spalding (1979) showed that there is an inverse correlation between δ-^{15}N and nitrate concentrations in ground water. This reflects the fact that ground water with lower concentrations of nitrate undergoes more denitrification than the high-nitrate water. However, volatilization and denitrification do not appear to increase $\delta^{15}N$ values above $+10$ per mil. As such nitrogen isotopes may be used to distinguish between animal waste nitrogen contamination and fertilizer nitrogen contamination.

This technique has been applied on Long Island, New York, in order to distinguish between animal waste contamination and fertilizer contamination. For example, Flipse and Bonner (1985) showed that the $\delta^{15}N$ values of nitrate under a heavily fertilized potato field and under a golf course were in the range of $+6.5$ to $+6.2$. While these values were significantly heavier than the fertilizers being applied ($-.6$ to $+.2$), they were lighter than the $+10$ values characteristic of animal wastes. Thus, while fractionation processes altered the isotopic signatures, the technique was still useful for distinguishing between fertilizer contamination and animal waste contamination.

Nitrate Accumulation Due to Dry-Land Farming Practices Although nitrogen species in ground water are commonly associated with contamination events, natural processes also contribute to delivery of nitrogen to ground water. In the soil zone, organic nitrogen produced by microbial nitrogen fixation is cycled through deamination, nitrification, and denitrification processes. Because organic nitrogen is often a limiting nutrient for both plant and microbe growth, it is unusual for ground water in pristine local flow systems to accumulate more than 1 or 2 mg/L

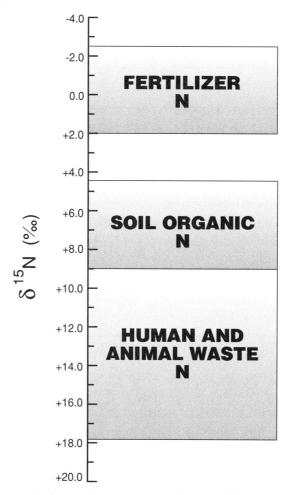

FIGURE 9.8. Range of nitrogen isotope composition for different sources of dissolved nitrogen in ground water.

dissolved nitrate. However, under some conditions, natural processes can contribute substantial nitrate concentrations to ground water.

One example of this was given by Kreitler and Jones (1975). In 1968, several cattle in Runnels County, Texas, died of anoxia from drinking water produced from the water table aquifer. Upon analysis of the water, it was found to contain between 100 and 1,000 mg/L nitrate. This was surprising in that the use of nitrogen fertilizers for agriculture was minimal. What could be the source of the observed nitrate?

One possibility was that animal wastes were the principal source. In order to test this possibility, Kreitler and Jones (1975) examined the $\delta^{15}N$ of nitrate in the ground water. Surprisingly, the observed values were generally less than +10 per mil, indicating that animal wastes were not the principal source. These values, however, did generally match values of soil nitrogen that were present naturally. Prior to 1900, this area was covered with buffalo grass, the roots of which form a symbiotic relationship with nitrogen-fixing bacteria much like legumes. These soils, therefore, were naturally high in reduced nitrogen compounds. The advent of dryland farming practices after 1900, particularly with increased oxygen delivery to the newly plowed soils, resulted in the oxidation of soil nitrogen and subsequent leaching to the water table. This leaching was greatly accelerated by terracing the fields in order to capture and retain more moisture. The intriguing result, therefore, was the production of a nitrate contamination problem without human or animal application of nitrogen.

This is an excellent example of why considering elemental cycling processes is important for understanding some ground-water chemistry problems. The accumulation of biochemically active elements in ground water *always* reflects the perturbation or truncation of one or more elemental cycles. These may be due to natural causes, such as nitrate accumulating due to truncation of denitrification in an aerobic aquifer. On the other hand, the perturbation may reflect human influences, such as adding a nitrogen source (fertilizer) where no source existed before. In either case, the key to understanding the observed phenomenon is to consider the nature of the nitrogen cycle.

9.4 THE IRON CYCLE

Iron in the environment exists predominantly in either the reduced ferrous (Fe(II)) or oxidized ferric (Fe(III)) form. Ferrous iron is relatively soluble in water and is therefore quite mobile. Ferric iron, on the other hand, tends to form insoluble Fe(III) oxyhydroxides and is therefore relatively immobile. The iron cycle in most environments involves alternate reduction (mobilization) of ferric iron followed by oxidation (immobilization) of ferrous iron. Most of these reactions are microbially catalyzed under natural conditions.

A schematic diagram of the iron cycle is shown in Figure 9.9. Under anaerobic conditions, Fe(III) oxyhydroxides are reduced by a variety of microorganisms that use Fe(III) as a terminal electron acceptor. Curiously, the role of microorganisms in iron reduction has only recently been extensively studied. Prior to 1985, most geochemists and microbiologists considered Fe(III) reduction to be an abiologic reaction initiated by "reducing conditions." However, with the isolation and characterization of Fe(III)-reducing microorganisms in the late 1980's it became clear that this important component of the iron cycle is largely mediated by microbial processes (Lovley, Phillips, and Lonergan, 1991). Several microorganisms, including strain GS-15 and *Shewanella putrifaciens,* are now known to carry out this metabolic function.

Much of the Fe(II) generated by Fe(III) reduction remains sediment-bound, being sequestered in Fe(II)-bearing minerals such as illite or magnetite. A small percentage (1-5%) of the Fe(II) accumulates in solution, however, and moves in response to diffusion gradients or with flowing water. If dissolved Fe(II) is transported to an aerobic environment, it is subsequently oxidized to Fe(III) oxyhydroxides. This oxidation is exergonic and can take place spontaneously without microbial mediation. However, several types of microorganisms have evolved mechanisms to harness this available energy. These include representatives of the genera *Gallionella, Thiobacillus,* and *Leptothrix*. Because spontaneous oxidation of Fe(II) occurs much less rapidly at acidic pH, many of these microor-

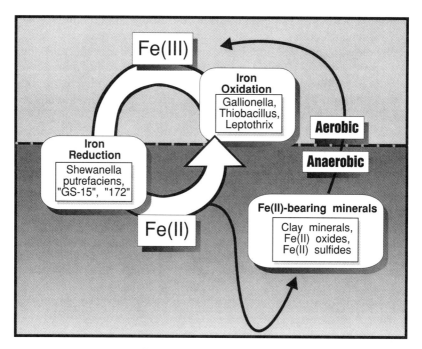

FIGURE 9.9. The iron cycle.

ganisms are specifically adapted to acidic environments. *Thiobacillus,* for example, has been widely described in streams impacted by acid mine drainage. There is relatively little energy to be had from oxidizing Fe(II), and consequently, Fe(II)-oxidizing microorganisms need to process large amounts of Fe(II)-and thus produce large amounts of Fe(III) oxyhydroxides-in order to provide enough energy for growth and reproduction.

The microbially mediated mobilization of Fe(II) and the subsequent oxidation to Fe(III) is responsible for many economically important accumulations of iron ore. *Bog iron ores,* for example, occur widely at the margins of swamps and bogs throughout the world. These accumulations of Fe(III) oxyhydroxides result from the mobilization of Fe(II) within the bog, migration of Fe(II) to the aerobic margins of the bog, and subsequent precipitation. Historically, bog iron ores were extremely important in the development of iron technology. The fact that iron ores existed in close proximity to peat, a ready source of fuel, was the basis for many early iron-working technologies such as those in England and Ireland.

9.4.1 Iron Cycling in Aquatic Sediments

Iron is required as a nutrient by most organisms, including microorganisms, plants, and animals. Iron is a critical component of many enzyme systems involved in metabolism and energy storage. For these uses, however, iron is generally needed in only trace quantities. The great importance of iron cycling in aquatic sediments is related more to the availability of phosphorous and other trace elements than to the availability of iron.

Fe(III) oxyhydroxides are very efficient at adsorbing a variety of chemical species including trace metals and phosphate ions. The growth of photosynthetic algae and cyanobacteria, which form the base of the food chain in many aquatic systems, are typically limited by the availability of phosphate. When decaying organic matter accumulates at the bottom of an aquatic environment, a considerable amount of the phosphate released is promptly adsorbed by Fe(III) oxyhydroxides in the sediment and thereby removed from the food chain. In the aerobic portion of the aquatic sediment, Fe(III) oxyhydroxides are stable and phosphate ions remain adsorbed. However, when the sediments become anaerobic, Fe(III) oxyhydroxides are reduced with the remobilization of adsorbed phosphate. Fe(II), phosphate, and other previously adsorbed trace elements diffuse into the aerobic zone, and a portion of the phosphate diffuses back into the water column. The cycling of iron between the oxidized and reduced forms provides a mechanism for returning phosphate to the food chain.

While Fe(II) is much more soluble than Fe(III) in aquatic sediments, much of the Fe(II) produced by iron reduction does not enter the aqueous phase; rather, a considerable portion, perhaps as much as 98%, is sequestered in the form of iron-rich clays such as illite, Fe(II)-bearing oxides such as magnetite or ilmenite, or Fe(II) sulfides such as pyrite. In these forms, iron can be removed from biological iron cycling for long periods of time. However, upon reexposure to aerobic conditions, Fe(II)-bearing minerals are rapidly oxidized with or without microbial mediation, and the iron is cycled back to Fe(III) (Fig. 9.9).

9.4.2 Iron Cycling in Ground-Water Systems

In terms of world wide economic impact, the biologic cycling of iron is one of the most important processes in ground-water systems. High concentrations of dissolved iron (Fe(II)) is by far the most common water quality problem associated with ground water. While dissolved iron is not toxic to humans or animals, the red staining of clothing and plumbing fixtures that it causes is highly undesirable. In addition, iron imparts an unpleasant taste to water that, while harmless, gives the impression that it is somehow contaminated. Finally, the oxidation of high-iron water as it is applied to cultivated fields can lead to low-pH ferric hydroxide-rich soils that may severely damage agricultural productivity. Treating ground water to remove Fe(II) from municipal, agricultural, and domestic wells is a multi-million dollar a year business throughout the world.

It has long been recognized that microorganisms play an important role in the oxidation of Fe(II) associated with ground water. However, the role of microorganisms in the other side of the iron cycle, that of Fe(III) reduction, has only recently been worked out. As discussed previously, the first microorganism known to carry out this function was isolated in 1987 (Lovley and Phillips, 1988). Two years later, similar techniques were used to isolate and characterize an Fe(III)-reducing microorganism from a ground-water environment (Lovley, Chapelle, and Phillips, 1989). This strain, informally named "172" (the microorganism was isolated from a depth of 172 feet), could perform many of the same biochemical functions as GS-15. However, "172" was clearly a different microorganism. Morphologically, "172" is a short, blunt rod whereas GS-15 is an elongated rod. Furthermore, the two microorganisms exhibit different optimum temperatures for growth.

The isolation of strain "172" was fortuitous and the manner in which it came about has implications for the cycling of iron in subsurface environments. The upper 150 feet of sediment in the core hole from which "172" was isolated was predominantly sand of a uniform reddish-brown color, reflecting the presence of ferric oxyhydroxides. Nearby wells screened in these brown sands produced ground water containing 2-6 mg/L of dissolved oxygen and little dissolved iron. Below about 180 feet the sediment remained predominantly sand, but was a grayish-white color reflecting the lack of ferric oxyhydroxides. Wells screened in these gray sands produced anaerobic water that often contained between 0.3 and 1.0 mg/L of dissolved iron. The cored interval between 170 and 180 feet was neither uniformly reddish-brown or grayish-white; rather, it was a mottled mixture of red and white that also contained streaks of purple sediment. These purple streaks appeared to be zones where dissolved Fe(II) had been oxidized from solution and thereby concentrated in the sediment. In addition, this mottled interval clearly occurred at the interface between the aerobic and anaerobic zones of this aquifer system.

As Fe(III)-reducing microorganisms had never been recovered from subsurface environments before, there were no precedents available for determining which horizons to sample. Luckily, however, a sample was taken at 172 feet that turned out to contain Fe(III)-reducing microorganisms. Samples in the overlying aerobic zone and in the underlying anaerobic but Fe(III)-lacking sediments did not yield viable Fe(III) reducers. Apparently, the "172" microorganism required both anaerobic conditions *and* the presence of grain-coating Fe(III) oxyhydroxides in order to maintain viability. The presence of purple Fe(III) precipitates implied that Fe(III) reduction and Fe(II) oxidation occurred in close proximity to each other. Interestingly, strain "172" exhibited much more tolerance to oxygen (although it would not reduce Fe(III) in the presence of oxygen) in laboratory studies than did the strict anaerobe GS-15.

These observations suggest that the mottled zone is actively involved in iron cycling on a small scale. Under anaerobic conditions, grain-coating Fe(III) oxyhydroxides are reduced and Fe(II) is mobilized. Some of this Fe(II) evidently enters the anaerobic portion of the ground-water system and remains in solution. This would explain the presence of high-iron concentrations in the deeper wells. Evidently, however, some of the Fe(II) comes into contact with dissolved oxygen and is reprecipitated as Fe(III).

Wells and Iron-Oxidizing Bacteria If one were to query a roomful of water well field operators as to what was the single most important microbial problem that they faced, the answer would overwhelmingly be "iron bacteria." When used in this context, the term "iron bacteria" refers to those species of aerobic bacteria that oxidize ferrous iron as a source of energy.

There are several reasons why iron-oxidizing bacteria are so troublesome for well operators, and these are related to their physiology. For one thing, aerobic oxidation of Fe(II) yields a relatively small amount of energy. For the reaction

$$Fe^{2+} + H^+ + 1/4 O_2 \rightarrow Fe^{3+} + 1/2 \, H_2O$$

only about 40 kcal of free energy is available per mole of reactant. In contrast, complete oxidation of hydrogen sulfide yields about 160 kcal. So, in order for iron-oxidizers to maintain their cellular structure and to accumulate enough energy for reproduction, they must oxidize a large amount of Fe(II). Fe(III) is relatively insoluble and tends to form oxides and hydroxides

$$Fe^{3+} + 3H_2O \rightarrow Fe(OH)_3 + 3H^+$$

The combination of iron oxidizers producing an insoluble waste product and their tendency to produce a lot of it, tends to clog wells and water lines rather efficiently.

In addition to these physiologic properties, the ecological niche filled by iron oxidizers works against well field operators. Iron oxidizers are adapted to living at the interface of anaerobic environments, where dissolved Fe(II) is available, and aerobic environments, where dissolved oxygen is available. When a well is drilled into an anaerobic aquifer, a aerobic-anaerobic interface is automatically created. This, in turn, creates the perfect environment for iron oxidation.

Examples of the havoc that iron oxidizers can wreak on a municipal well system are numerous. One particularly interesting case occurred in Florence, South Carolina. The ground water, which was pumped from the Middendorf and Cape Fear aquifers, generally contains 1 or 2 mg/L of Fe(II). However, because the oil-lubricated turbine well pumps used to pump the water tended to minimize introduction of oxygen from the air, for years iron oxidizing bacteria were only a minor nuisance. However, in the 1980's, the state environmental protection agency decreed that oil-lubricated pumps could no longer be used (they often leaked oil into wells, raising fears of organic chemical contamination) and that instead, water-lubricated pumps were to be used. When not in operation, these pumps were lubricated by a constant stream of tap water running down the well. This, of course, injected a constant stream of oxygen into an anaerobic aquifer—the perfect environment for iron-oxidizing bacteria. The predictable result was a series of massive well-clogging problems. These problems were only solved when the city gave up on water-lubricated turbine pumps and installed submersible pumps.

Well clogging is not the most insidious aspect of iron-oxidizing bacteria, however. In pipes containing oxygenated water, iron oxidizers may form colonies. The inside of the colonies are typically anaerobic. The difference of electrical potential between the aerobic and anaerobic zones encourages the flow of electrons which in turn corrodes the iron pipe. This kind of cathodic corrosion, mediated by iron-oxidizing bacteria, is the source of millions of dollars worth of damage per year. Well operators go to extraordinary lengths to minimize this damage. Periodic chlorination of well pipes, or other antibacterial measures, are the most common damage control methods.

High Iron Concentrations in Ground Water By far the most economically important aspect of iron cycling in ground-water systems is the production of high iron concentrations in ground water. It is often observed that high-iron ground water occurs in distinct zones (Langmuir, 1969; Back and Barnes, 1965). As was the case with other biologically active elements, the accumulation of dissolved iron in ground water represents a truncation of the iron cycle.

When Fe(III) reduction occurs at the interface of an aerobic and anaerobic zone, such as in aquatic sediments, there is the possibility that some Fe(II) produced will be cycled back to Fe(III) oxyhydroxides. In many aquifers, however, Fe(III) reduction occurs where molecular oxygen is absent and there is no possibility of reoxidation. In these cases, dissolved Fe(II) may accumulate in solution, causing high-iron concentrations in ground water. The truncation of the iron cycle due to the lack of iron reoxidation is an important mechanism leading to the accumulation of iron in ground water.

9.5 THE SULFUR CYCLE

Sulfur occurs in three common oxidation states in nature. These are the -2 state, in which sulfur occurs as sulfides ($S^=$), the 0 state in which sulfur occurs as elemental sulfur (S^o), and the $+4$ state in which sulfur occurs as sulfate ions ($SO_4^=$). As in the case of carbon, nitrogen, and iron, this variety of oxidation states allows chemical energy to alternately be stored and released. The series of sulfur-based energy storage and energy releasing processes is known as the *sulfur cycle*.

The major microbially mediated processes involved in the sulfur cycle are shown schematically in Figure 9.10. Sulfur in the -2 state is an integral component of many proteins. Most microorganisms are therefore capable of reducing sulfate to sulfide in order to supply the necessary reduced sulfur. This process is termed *assimilatory* sulfate reduction. Interestingly, assimilatory sulfate reduction is carried out under both aerobic and anaerobic conditions. The reduced sulfur component of proteins is returned to the sulfur cycle by means of *putrifaction*, in which sulfur is released from decaying tissue in the form of hydrogen sulfide (H_2S).

Hydrogen sulfide is also produced in anaerobic environments by means of *dissimilatory* sulfate reduction. In this process, sulfate acts as the terminal electron acceptor in the oxidation of organic carbon or elemental hydrogen. Dissimilatory sulfate reduction is extremely important in the global cycling of carbon. In marine and estuarine environments, sulfate reduction is often the most important process leading to the oxidation of plant debris and the consequent production of carbon dioxide.

The fate of hydrogen sulfides produced by assimilatory and dissimilatory sulfate reduction is extremely complex. Some portion of the H_2S reacts with metals, especially iron, in order to form sulfide minerals. Such mineral formation temporarily removes sulfur from the cycling process. Some of the sulfide is cycled back to sulfate as it diffuses into aerobic zones. The oxidation of H_2S to sulfate under aerobic conditions is mediated by a number of microorganisms, notably *Thiobacillus*. *Thiobacillus* is capable of using H_2S as its sole energy source during growth and reproduction.

It has recently been shown that much H_2S produced under anaerobic conditions is also reoxidized anaerobically (Jorgensen, 1990). The anaerobic oxidation of sulfide appears to be coupled to the reduction of Fe(III) compounds present in the sediments, and thiosulfate ($S_2O_3^=$), a partially oxidized intermediate compound, is produced. Thiosulfate is then *disproportioned* to either sulfate or sulfide according to the stoichiometry

$$S_2O_3^= + H_2O \rightarrow SO_4^= + HS- + H^+$$

The term "disproportioned" refers to the fact that the thiosulfate anion contains

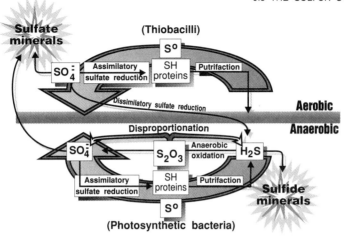

FIGURE 9.10. The sulfur cycle.

a sulfur atom in the oxidized (+4) state, as well as one in the reduced (−2) state. Thus, the decomposition of thiosulfate to sulfate and sulfide does not involve an oxidation or a reduction step. This "thiosulfate shunt" for the anaerobic oxidation of sulfide appears to be a significant process for the cycling of sulfur in aquatic sediments.

The thiosulfate shunt is not the only pathway by which sulfides may be anaerobically oxidized. Most photosynthetic cyanobacteria and algae use H_2O as an electron donor, with the consequent production of molecular oxygen (O_2). The anaerobic photosynthetic sulfur bacteria, which includes the green and purple sulfur bacteria, use H_2S as an electron donor. The oxidation of H_2S by these microorganisms is another anaerobic mechanism by which reduced sulfur is cycled back to sulfate.

In the same manner that reduced sulfur can be sequestered and temporarily removed from the sulfur cycle by precipitation of sulfide minerals, sulfate may also form mineral deposits. Because of the relatively high solubility of most sulfate minerals, significant sulfate mineralization occurs only in evaporative basins. However, the precipitation of sulfate minerals in these hot, dry environments is responsible for the temporary removal of much sulfate from the sulfur cycle. The exposure and weathering of both sulfate and sulfide minerals during tectonic events eventually returns most of this sequestered material to the sulfur cycle.

9.5.1 Sulfur Cycling in Ground-Water Systems

The behavior of sulfur in ground-water systems depends in large part on the presence or absence of sulfur-bearing minerals. Many clastic aquifers lack sulfate minerals such as gypsum or anhydrite. In these cases, sulfur cycling reflects reactions involving sulfide minerals and pools of dissolved sulfate available in pore water. As was the case with oxygen, carbon, and nitrogen, the behavior of sulfur in ground-water systems often reflects the truncation of the cycle due to lack of energy input. However, in the case of sulfur, the processes leading to these truncations are more varied.

Sulfide Oxidation in an Oxygenated Aquifer

An example of sulfur accumulation due to cycle truncation was given by Chapelle and Kean, (1985). In the outcrop area of the Patuxent aquifer near Baltimore, Maryland, ground water is oxygenated (Fig. 9.3). This aquifer, however, contains pyrite as a minor constituent, and the oxidation of pyrite has two apparent effects as illustrated by the overall stoichiometry

$$FeS_2 + 15/4\ O_2 + 7/2\ H_2O \rightarrow Fe(OH)_3 + 4H^+ + 2\ SO_4^=$$

First, concentrations of sulfate increase along aquifer flowpaths. Second, the pH of the ground water decreases due to the release of H^+ ions. Thus, when the sulfur cycle is truncated after sulfide oxidation, the accumulation of sulfate in solution is observed in some systems.

In the Patuxent aquifer, the sulfur cycle truncates after sulfide oxidation because of interaction with the iron cycle. After dissolved oxygen is consumed, iron reduction becomes the predominant microbial process as indicated by the rapid increase in iron concentrations. Because Fe(III)-reduction precludes active sulfate reduction, sulfate is not immediately recycled and sulfate accumulates in solution. One would expect, however, that once Fe(III) reduction ceased, sulfate reduction activity would increase and the sulfur cycle would continue. This interplay between the iron and sulfur cycles occurs widely in ground-water systems and is one more mechanism leading to the (temporary) truncation of the sulfur cycle.

Sulfate Reduction in Sulfate Mineral-Free Aquifers

A wide variety of aquifer systems lack significant mineral sources of sulfate. Clastic sediments deposited in humid or temperate fluvial and marine environments generally lack such mineralization. Despite the lack of a mineral source of sulfate, however, sulfur cycling is often important to the water chemistry of these ground-water systems.

The presence or absence of sulfate reduction in clastic coastal plain aquifers is one of the oldest debates in ground-water chemistry. Cedarstrom (1946) proposed that sulfate reduction was the primary oxidation process driving the accumulation of bicarbonate in coastal-plain aquifer ground water. As Foster (1950) correctly pointed out, however, there were no observed changes in sulfate concentrations along aquifer flowpaths. In both recharge and discharge areas, sulfate concentrations remained in the 10-20 mg/L range. There was a brief heated debate between supporters of each position in the early 1950's, and then, due largely to the lack of any new ideas for resolving the issue, interest in the problem waned.

As often happens in science, however, new data from an unexpected source cast the dilemma in a new light. In 1984, as part of the U.S. Geological Survey's RASA (Regional Aquifer Systems Analysis) program, an effort was made to locate more accurately the salt-water/fresh-water interface in the coastal plain of the eastern United States. As finishing and developing wells was very expensive, the hydrologist in charge of the project, Harold Meisler, decided to squeeze cored sediments and to analyze the pore water for chlorides and other ions. As it turned out, the analytical procedure-ion chromatography-was capable of quantifying sulfate together with chloride, and so sulfate was analyzed as well.

The results (Trapp et al., 1984) were puzzling. Although the aquifers contained relatively low concentrations of sulfate (20-30 mg/L), the confining beds contained relatively high concentrations (200-600 mg/L). This suggested that diffusion of sulfate from confining beds to aquifers was possible. Could this be the source of sulfate for sulfate reduction? The data collected by Meisler were intriguing, but by themselves were not conclusive. Beginning in 1987, a more comprehensive study of confining bed pore water chemistry was conducted in the New Jersey Coastal Plain by Pucci and Owens (1989). They confirmed the presence of relatively high concentrations of sulfate in some but not all confining beds. Specifically, confining beds of marine origin contained up to 600 mg/L of dissolved sulfate whereas confining beds of non-marine origin contained low concentrations (less than 50 mg/L). Similar results were subsequently observed by Chapelle and Lovley (1990) in coastal plain sediments of South Carolina.

By now it was clear that coastal plain sediments contained a previously unknown pool of sulfate that was present in the fine-grained confining beds. Now the question was how was this sulfate involved in sulfur cycling in these ground-water systems? This question was addressed by Chapelle and McMahon (1991) who repeated the observation that confining beds contained higher concentrations of sulfate than aquifers. In addition, it was shown that while viable sulfate-reducing bacteria were readily recovered from aquifer sediments using culture techniques, such bacteria were not recovered from confining bed sediments. Evidently, sulfate-reducing activity was depressed in confining beds relative to aquifers, possibly explaining the observed higher concentrations of sulfate. More important, it was shown that the diffusive flux of sulfate into the aquifers implied by the concentration differences balanced the observed rate of microbial CO_2 production in this system. Forty-five years after Cedarstrom (1946) proposed that sulfate reduction accounted for the observed CO_2 production in these coastal plain aquifers, the mechanisms involved in the process were finally worked out.

One unresolved issue is the source of the observed sulfate. Because pore-water concentrations of chloride were relatively low ($\sim$ 20 mg/L), it did not seem to be residual seawater. In any case, if diffusion were ongoing in this system, seawater sulfate would be exhausted within just a few million years. Pucci and Owens (1989) pointed out that the confining beds contained large amounts of sulfide minerals that could be oxidized by Fe(III)-bearing minerals. This process of slow oxidation could maintain sulfate concentrations as long as Fe(III) minerals were present. Interestingly, Jorgensen (1990) suggested that Fe(III) oxidation was an important mechanism in the anaerobic oxidation of sulfides generated by sulfate reduction in aquatic sediments. While the ultimate source of sulfate in confining beds is presently still unknown, it is possible that it is an important component of the sulfur cycle in some aquifer systems.

The essential features of sulfur cycling in sulfate mineral-free, coastal-plain aquifers involves a complex interplay between aquifers and their confining beds (Chapelle and McMahon, 1991). Confining beds contain relatively high concentrations of sulfate but lack substantial sulfate-reducing activity. In the aquifers, on the other hand, there is active sulfate-reduction. The concentration gradient that develops allows sulfate to diffuse to aquifers. Thus, even though there is ongoing sulfate reduction in the aquifers, there is little overall change in sulfate concentrations along aquifer flowpaths.

Sulfides produced by sulfate reduction are rapidly immobilized by reaction with Fe(III)-bearing minerals, particularly glauconite, and so there is little accumulation of dissolved sulfides. The overall result is the net transport of sulfur from fine-grained confining beds to coarse-grained aquifers and the accumulation of reduced metal sulfides in the aquifers. Concurrently, there is a net accumulation of dissolved inorganic carbon in the aquifers.

Sulfate Reduction in Sulfate Mineral-Bearing Aquifers Many aquifer systems contain a mineral source of sulfate, and this markedly affects observed patterns of sulfur cycling. One of the best documented examples of sulfur cycling in this type of aquifer is the Floridan aquifer of central Florida (Rightmire et al., 1974; Plummer, 1977).

The Floridan aquifer consists of limestones and dolostones of the Avon Park, Ocala, Suwannee, and Tampa Formations. Some of these limestones were deposited under shallow marine conditions in a hot, arid climate. These conditions resulted in the accumulation of significant amounts of gypsum during carbonate deposition, particularly in the Avon Park Formation. This gypsum, in turn, is a ready source of sulfate to ground water associated with the Floridan aquifer.

This distinctive mineralogy is reflected in the water chemistry of the aquifer. As one would expect, dissolution of dolomite and gypsum results in increasing concentrations of calcium, magnesium, bicarbonate, and sulfate. In its recharge areas, the Floridan aquifer is oxygenated and sulfate reduction is not an important process. Downgradient, however, the aquifer becomes anaerobic and sulfate reduction becomes the predominant microbial process. This is evident from the relatively high concentrations of hydrogen sulfide present in the ground water ($\sim$ 1–2 mg/L). It is not evident, however, from any change in sulfate concentration along the flowpath. This occurs because the flux of sulfate to the aquifer via gypsum dissolution (1.970 mmol of gypsum per liter of ground water between Wauchula and Arcadia) is much greater than the rate of sulfate reduction (0.225 mmol sulfate reduced per liter of water).

A better indicator for the presence of active sulfate reduction in this aquifer system is the isotopic composition of dissolved sulfate. Gypsum that has precipitated from seawater typically has a $\delta^{34}S$ of about +22 per mil. Maritime rainfall, on the other hand, has a $\delta^{34}S$ of about +10 per mil. Thus, if the only processes affecting the isotopic composition of sulfate in this system were input of sulfate from rainwater and from dissolution of gypsum, $\delta^{34}S$ values would track along a mixing curve (Rightmire et al., 1974). In the upgradient aerobic portion of the aquifer where sulfides are absent, observed values of $\delta^{34}S$ of the sulfate did scatter in the vicinity of this mixing curve. In the anaerobic downgradient portion of the aquifer, where the water contains sulfides, the $\delta^{34}S$ values were consistently more positive than predicted by the mixing curve. Rightmire et al. (1974) interpreted this as reflecting isotopic fractionation during sulfate reduction. Sulfate-reducing microorganisms actively discriminate against the heavier stable isotope of sulfur (^{34}S), preferring the more reactive light isotope (^{32}S). Thus, sulfate reduction occurring along the aquifer flowpath will actively enrich the available pool of sulfate in ^{34}S, resulting in more positive $\delta^{34}S$ values.

The essential features of sulfur cycling in the Floridan aquifer, therefore, include production of dissolved sulfate via dissolution of gypsum. In the aerobic portion of the aquifer, sulfate accumulates without significant sulfate reduction. In the anaerobic portion of the aquifer, sulfate reduction becomes an important process.

However, because the rate of gypsum dissolution outpaces the rate of sulfate reduction by at least one order of magnitude, this process is not evident from concentration changes of sulfate. Sulfate reduction is evident, however, from the accumulation of dissolved sulfides and by the observed fractionation of sulfate stable isotopes.

From a water-quality standpoint, one important way that sulfur cycling in the carbonate Floridan aquifer differs from sulfur cycling in the clastic Black Creek aquifer is in the availability of iron. In the Black Creek aquifer, iron is present in a number of mineral phases and is available in excess to react with and precipitate pyrite. There is therefore no observed accumulation of dissolved hydrogen sulfide, which imparts the characteristic "rotten egg" smell, in Black Creek aquifer water. In the carbonate Floridan aquifer, where clastic iron-bearing minerals are much less abundant, sulfide production outpaces pyrite formation and significant quantities of sulfide accumulate in the water. Thus, even though sulfate reduction dominates sulfur cycling in both systems, the sulfide water quality problem associated with it is only present in the Floridan aquifer.

9.6 SUMMARY

Biologically active elements-oxygen, hydrogen, carbon, nitrogen, iron, and sulfur- are cycled continuously between oxidized and reduced states in the biosphere. These cycles are driven by the input of energy from the sun. Because subsurface environments are isolated from solar energy, many elemental cycles are truncated. In some cases, such as the oxygen cycle, this truncation results in the depletion of oxygen in ground water. In other cases, such as the carbon cycle, this truncation results in the accumulation of dissolved inorganic carbon species in solution. The truncation of biogeochemical cycling is one of the unique features of microbial processes in subsurface environments and has a significant impact on observed ground-water chemistry.

Consideration of biogeochemical cycling in ground-water geochemistry is useful primarily in that it gives a qualitative framework for understanding of the behavior of biochemically active elements. It is difficult, however, to use this framework in a quantitative manner. As we have seen, these cycles do not act independently of each other. For example, the oxygen and hydrogen cycles are intimately connected with the carbon cycle, and all three interact with the nitrogen and sulfur cycles. To make things even more difficult, mineral sources of inorganic carbon and sulfur often overshadow biologic sources in many systems. Thus, while useful in describing overall behavior, the concepts of elemental cycling are an inadequate basis for a quantitative treatment of ground water chemistry. Such quantitative treatments, which integrate both biological and non-biological processes, are considered in the next chapter.

CHAPTER 10

GEOCHEMICAL MODELING AS A TOOL FOR STUDYING MICROBIAL PROCESSES IN GROUND-WATER SYSTEMS

Microbial processes are difficult to study in ground-water systems for a number of reasons, but certainly the greatest difficulty is the inaccessibility of subsurface environments. As we have seen in Chapter 8, sampling subsurface environments directly for microorganisms is a difficult, technology-intensive, and expensive endeavor. However, it is possible to obtain much information about microbial processes using indirect methods. In aquatic sediment microbiology, for example, it is common practice to use pore water chemistry as an indicator for the zonation of microbial processes (Froelich et al., 1979). In the same manner, ground-water chemistry data can be used to deduce a great deal about the zonation, extent, and rates of microbial processes in ground-water systems.

There are several advantages to using ground-water chemistry as an indicator of microbial processes. As we have seen, obtaining uncompromised samples of deeply buried aquifer materials is fairly difficult. Ground-water samples, on the other hand, are relatively easy to obtain. Furthermore, because wells are generally available over a large area in most ground-water systems used for water supply, information on the areal extent of microbial processes is more readily available. Finally, techniques for quantitatively evaluating chemical processes in ground-water systems, including both microbially mediated processes and abiotic mineral equilibria processes, have been worked out in great detail (Plummer, Parkhurst, and Thorstenson, 1983). This methodology for deducing operative chemical and microbial processes from ground-water chemistry data is termed *geochemical modeling*.

Historically, geochemical modeling can be traced to the 1960's and the work of Garrels and Mackenzie (1967); Helgeson (1968), Helgeson, Garrels, and Mackenzie (1969), Helgeson et al. (1970), and Truesdell and Jones (1974). At that time it had been established that ground-water chemistry reflected the dissolution of gases and minerals that occurred as ground-water flowed along aquifer flowpaths (Back, 1966). Garrels, Mackenzie, Helgeson, and their students from Northwestern University took the next logical step which was to use the principles of conservation of mass (i.e., mass balance) and mineral equilibria to quantify the *amounts* of minerals dissolving to account for observed water chemistry. Garrels and Mackenzie (1967) showed, for example, that the composition of spring water in a granitic terrain could be accounted for by weathering a given amount of plagioclase, orthoclase, and biotite to kaolinite. Furthermore, they showed that the extent of reaction was constrained by the solubility characteristics of each mineral and its stable weathering product. This realization was critical because it implied that the amount of each mineral dissolving was not random, but rather followed predictable pathways based upon thermodynamic principles.

It is no accident that quantitative techniques in ground-water chemistry developed in the late 1960's. As we shall see, the calculations involved with this process lead to a large amount of tedious algebra and arithmetic. Before computer technology became widely available in the late 1960's, many of the calculations involved were simply not feasible. With the advent of computer technology, however, geochemists were quick to apply it in performing chemical speciation and mineral equilibria calculations (Truesdell and Jones, 1974).

The geochemists who pioneered the techniques of geochemical modeling were largely interested in inorganic chemical processes, such as mineral equilibria. From a historical viewpoint, therefore, geochemical modeling was not conceived as a method for studing microbial processes. It was, however, conceived as a method for deducing the operative chemical processes in aquifer systems. As we shall see, one of the important contributions of geochemical modeling has been to show explicitly the importance of microbial processes on ground-water chemistry. Furthermore, these techniques make it possible to evaluate the ecology of microorganisms in subsurface environments in ways that are impractical using any other methods.

10.1 CONSIDERATIONS IN GEOCHEMICAL MODELING

The purpose of constructing geochemical models is to deduce and quantify the chemical and biologic processes that modify ground-water chemistry in a hydrologic system. There are two quite distinct methods available to accomplish this goal. First, there is the *inverse method,* in which available hydrologic and groundwater chemistry data are used to deduce the operative geochemical processes in a particular hydrologic system. Second, there is the *forward method,* in which a priori predictions of water chemistry are made based on assumed geochemical processes and thermodynamic constraints (Plummer, Parkhurst, and Thorstenson, 1983).

In practice, the inverse method of geochemical modeling provides the most information about a given hydrologic system because it is constrained by real-world data. The forward method has the advantage of predicting details of thermodynamically valid reaction paths for a given system. But without data to constrain them, such models may predict water compositions and mass transfers that are far from reality. The greatest use of geochemical modeling as applied to subsurface microbiology is in the deducing of operative microbial processes from ground-water chemistry data. Thus, it is the inverse method that is most widely applicable.

266 GEOCHEMICAL MODELING AS A TOOL FOR STUDYING MICROBIAL PROCESSES

10.1.1 Hydrologic Considerations

The basic strategy of inverse geochemical modeling is to evaluate how ground-water chemistry changes as it moves through a hydrologic system. In practice, this is accomplished by determining directions of ground-water flow in an aquifer system and by selecting water analyses at various points along individual flowpaths. For example, Figure 10.1 shows a water level map of a hypothetical flow system with four wells A, B, C, and D. Wells A and B are oriented along a single flow line, and, therefore, the water chemistry changes between them will be representative of the chemical and biological processes occurring between these two points in the hydrologic system. Similarly, wells A, B, and C are oriented along a flow line. It's clear, however, that there is no hydrologic relationship between wells B and D. Therefore, to perform a geochemical modeling analysis of water chemistry differences between wells B and D would be meaningless.

Hydrologic considerations place considerable limitations on the application of geochemical modeling. First of all, if the only analyses of ground water available do not fall along a particular flowline, then the technique is not applicable. Secondly,

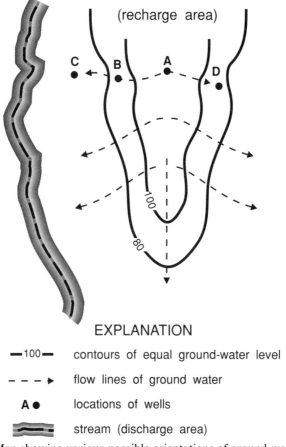

FIGURE 10.1. Map showing various possible orientations of ground-water flowpaths and wells.

because water-level maps are approximations and are usually based on incomplete data, there is always the possibility of pairing analyses of ground water inappropriately. For these reasons, great care must be taken in evaluating the hydrology of an aquifer system prior and during any application of geochemical modeling.

Once the hydrology of a ground-water system has been properly evaluated, the selection of wells to be used in a geochemical modeling analysis depends on the questions being asked. For example, if the chemical processes in the recharge area of Figure 10.1 are of primary interest, then the geochemical modeling effort might focus on the chemical difference between rainfall and well A. On the other hand, if the chemical processes occurring near the discharge area are of primary interest, than wells B and C would give the most relevant chemical data. Matching ground-water sampling locations to the questions being asked is an important component of geochemical modeling.

10.1.2 Mineralogic Considerations

In constructing a geochemical model of any ground-water system, a detailed knowledge of the minerals that comprise the aquifer matrix is necessary. The following questions are particularly relevant (Plummer, 1984).

1. What minerals are present, and how abundant are they?
2. How does mineral abundance, including trace mineralogy, vary spatially in the system?
3. How does mineral composition, including elemental substitutions and exchangeable ions, vary spatially?
4. What is the isotopic composition of the minerals present, and how does this vary spatially?
5. How does mineralogy vary with respect to the direction of ground-water flow?
6. Using petrographic or electron microscopy, is there evidence that any of the minerals present are secondary?

The importance of such mineralogic information is largely self-evident. Clearly, if one is going to construct a geochemical model that requires calcite to dissolve in order to explain observed water chemistry changes, the presence of calcite in the aquifer must be adequately documented. Furthermore, mineralogic observation of secondary minerals or mineral coatings can be critical in refining a geochemical model. For example, if there is clear evidence of secondary pyrite cementation, this places constraints on the possible number of geochemical models that may be constructed. First, it requires consideration of sulfate reduction as an operative process. Second, it requires that some portion of the sulfide generated by sulfate reduction not accumulate in solution, but be transferred to pyrite.

10.1.3 Thermodynamic Considerations

The presence or absence of particular minerals obviously places constraints on the kinds of rock–water interactions possible in any given system. Beyond the simple presence or absence of minerals, however, is the more subtle question of mineral solubility. If a mineral is present in an aquifer but is insoluble, it is illogical to consider it as a source of dissolved material in a geochemical model. Questions of mineral solubility may be addressed using thermodynamic considerations.

The *saturation index* (SI) of a particular mineral in the presence of an aqueous solution is defined as

$$SI = \log IAP/K$$

where IAP is the ion activity product and K is the equilibrium constant for a particular reaction. Take, for example, the reaction

$$Ca^{2+} + CO_3^= = CaCO_3$$

The IAP of this reaction is given by

$$[aCa^{2+}]\,[aCO_3^=]$$

where a denotes activity and the equilibrium constant is given by

$$[aCA^{2+}]\,[aCO_3^=]/[aCaCO_2] = K$$

At equilibrium and assuming that the solid phase $CaCO_3$ has unit activity (i.e., an activity of 1), it is clear that

$$IAP = K$$

so that

$$IAP/K = 1$$

and

$$\log IAP/K = 0$$

Thus, an SI of zero implies that a mineral is at thermodynamic equilibrium with the solution. An SI <0 implies that the solution is undersaturated with respect to $CaCO_3$ and that, therefore, that mineral may dissolve. Similarly, an SI >0 implies that the solution is oversaturated and that the mineral may precipitate.

Mineral solubility data are useful in constraining the kinds of reactions that may be used in constructing geochemical models. In the simplest case, if a particular ground water is oversaturated with respect to calcite, it is illogical to invoke calcite dissolution as a possible reaction contributing dissolved material to ground water. On the other hand, if the ground water is undersaturated with calcite, it is illogical to invoke calcite precipitation as an operative chemical process.

There are probably no ground-water systems that are in overall chemical equilibrium with all of the minerals present in the aquifer matrix. This is largely due to the presence of chemical and biologic reactions that proceed irreversibly. Plummer (1984) gave an example of how such irreversible reactions may impact chemical processes in a particular system. Consider a water in equilibrium with calcite, siderite, geothite, pyrite, and gypsum at 25°C and 1 atmosphere of total pressure. The phase rule, which governs the number of phases that may coexist with a particular number of chemical components may be written as

$$p = C - F + 2$$

where P is the number of phases, C is the number of chemical components, and F the number of degrees of freedom. As the temperature and pressure of this system are fixed, the number of degrees of freedom equals 2 and the equation simplifys to

$$P = C$$

This relation is known as the *mineralogic phase rule* and simply states that in a system of fixed temperature and pressure, as most ground-water systems are, the number of phases must equal the number of chemical components at equilibrium. For this system, therefore, the phases and components are as follows:

Phases	Components
Calcite ($CaCO_3$)	$CaCO_3$
Siderite ($FeCO_3$)	$FeCO_3$
Geothite (FeOOH)	Fe_2O_3
Pyrite (FeS_2)	H_2S
Gypsum ($CaSO_4 \cdot 2H_2O$)	H_2SO_4
Aqueous phase	H_2O

Because temperature and pressure are fixed in this system, the activities of all six components, and all other aqueous species derived from these components, are also fixed. This is therefore referred to as an *invarient* system.

If a particular aquifer were composed only of these minerals and components, then, once equilibrium was reached, there would be no net dissolution or precipitation of minerals and the ground-water chemistry would not change composition. However, if an irreversible process were present, such as the oxidation of organic matter (microbial respiration), then dissolution and precipitation reactions *would* occur. This is illustrated in Figure 10.2a, which shows total mass transfer (dissolution and precipitation) of these minerals as a function of organic carbon (CH_2O) addition. As CH_2O is added to the system, CH_2O is oxidized with the reduction of ferric iron in goethite and sulfate from gypsum. Calcium ions from gypsum dissolution combine with carbonate ions from the oxidation of CH_2O to precipitate calcite. Similarly, ferrous iron, sulfide, and carbonate ions drive the precipitation of pyrite and siderite.

This example illustrates perhaps the most common interaction of microbial processes with thermodynamic constraints in ground-water systems. The reduction of sulfate and ferric iron coupled to the oxidation of organic carbon is, at these temperatures and pressures, always mediated by microorganisms. Thus, this is a particularly clear example of how microbial processes can drive the dissolution or

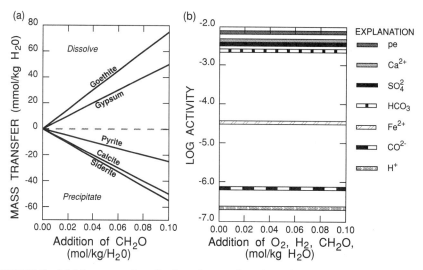

FIGURE 10.2. (a) Mass transfer of minerals as a function of CH_2O oxidized, and (b) net change in water composition due to mass transfer. Note that the lack of water composition change does not indicate a lack of mass transfer. (Modified from Plummer, 1984, with permission courtesy of the National Water Well Association.)

precipitation of minerals in an aquifer system. In order to identify and quantify these microbial processes, however, it is necessary to consider the thermodynamic properties of the minerals involved.

This example illustrates another commonly observed phenomenon in groundwater systems. Note that, as the oxidation of organic matter drives the simultaneous dissolution and precipitation of mineral phases, *there is no net change in water composition* (Fig. 10.2b). In other words, if one were simply to measure water composition along an aquifer flowpath in which these processes were operative, one would have no indication that any microbiological or geochemical processes were in fact occurring. In terms of inverse geochemical modeling, therefore, additional data would be needed to solve the problem. These additional data could be in the form of petrographic observations (evidence of calcite and pyrite precipitation, for example) or isotopic evidence, if the carbon isotopic composition of CH_2O is distinct from that of calcite.

10.1.4 Kinetic Factors

The mineralogy of an aquifer determines which phases are available for reaction, and thermodynamics determines whether these phases may dissolve or precipitate. Thermodynamics, however, gives no information as to the *rate* at which such processes actually occur.

Mineral-water reaction rates are strongly influenced by mechanisms of reaction. The dissolution of many minerals—calcite for example—is a surface phenomenon. Only those calcium and carbonate ions on the surface of the crystal are available to be transported to the aqueous phase. In this case, therefore, the dissolution rate is strongly dependent on the surface area of the dissolving crystal. Also, the transport of calcium and carbonate ions from the crystal surface may be inhibited if trace metals or organic compounds are present on the crystal surface. This inhibition becomes more and more pronounced as equilibrium is approached. The practical effect of these complicating factors is that it is not presently feasible to predict water chemistry changes in aquifer systems based solely upon kinetic considerations.

While prediction of chemical kinetics is impractical, it is possible to determine rates of reactions in ground-water systems using inverse geochemical modeling. This is accomplished by combining the calculated mass transfer of a particular process along an aquifer flowpath with the travel time of ground water. In one such study, Plummer and Back (1980) used ^{14}C age-dating techniques to estimate rates of ground-water movement in the carbonate Floridan and Madison aquifers. Next, they used mass balance calculations to estimate amounts of dolomite and gypsum dissolution, and calcite precipitation along particular flowpath segments in each aquifer. By combining these estimates, they were able to determine that dolomite and gypsum were dissolving at a rate of about 10^{-4} mmol/L/yr and that calcite was precipitating at about the same rate.

These same techniques can be applied to rates of microbial processes as well as to rates of precipitation and dissolution reactions. As with mineral dissolution rates, it is not presently feasible to make a priori estimates of microbial processes in natural systems. Furthermore, traditional methods of determining rates of microbial processes are difficult to apply to subsurface environments. Thus, one of the most potentially useful applications of geochemical modeling in subsurface microbiology is for estimating rates of microbial processes.

10.2 METHODOLOGY OF GEOCHEMICAL MODELING

Geochemical modeling is a way of thinking about ground-water chemistry data in order to deduce and to quantify the operative chemical and biologic processes in aquifer systems. It is, therefore, a mental process that is driven exclusively by the questions at hand. Because questions necessarily vary from system to system, there is considerable variation in the kinds of geochemical models that have been developed. All of these models, however, are based on three kinds of calculations: (1) speciation calculations, (2) mineral equilibrium calculations, and (3) material balance calculations.

10.2.1 Speciation Calculations

When a water sample is analyzed for it's chemical composition, only a small fraction of the chemical species present are actually quantified. For example, if a water sample contains both carbonate and sulfate ions, calcium in solution will be present not only in the familiar ionized form:

$$Ca^{2+}$$

but also as charged and uncharged ion pairs:

$$CaOH^+$$
$$CaHCO_3^+$$
$$CaCO_3^o$$
$$CaSO_4^o$$

Thus, a chemical analysis of water does not give the concentration of Ca^{2+} directly but rather gives the sum of all calcium ions and ion pairs present in solution

$$Ca_{tot} = Ca^{2+} + CaOH^+ + CaHCO_3^+ + CaCO_3^o + CaSO_4^o$$

For calculation of mineral stabilities—calcite, for example—it is necessary to know the concentration of Ca^{2+} explicitly and not just Ca_{tot}. One possible way of attacking this problem is simply to analyze for each dissolved species. However, in practice, the sheer number of dissolved species that would need to be considered renders this impractical for most purposes. The method that is most commonly used is to build an equilibrium aqueous model of the water sample, and *calculate* the distribution of chemical species using the model.

The Aqueous Model An aqueous model is a mathematical representation of a water sample at chemical equilibrium. Aqueous models are built from equations describing the thermodynamic activity of aqueous species, the activity of water, effects of temperature, and exchange of electrons (redox reactions).

In real solutions, there are electrical interactions between charged ions that have the effect of decreasing apparent concentrations of dissolved species. This effect is formally defined by the equation

$$a_i = \gamma_i m_i,$$

in which a_i is the activity, γ_i is the *activity coefficient*, and m_i is the actual molar concentration of the ith species. The activity coefficient is generally less than 1 and is dependent upon the charge of the ion and on the concentration of other charged ions in solution. For relatively dilute solutions (< 0.10 molar solution of NaCl), the *Debye–Hüchel* equation is used:

$$\log \gamma = - \frac{A z^2 \sqrt{I}}{1 + Ba\sqrt{I}}$$

where γ is the activity coefficient, A and B are constants related to the properties of the solvent (water, in this case), z is the ionic charge, a is a constant related to the effective diameter of the ion, and I is the ionic strength. The constants A and B have been determined experimentally (they are equal to 0.5085 and 0.3281 respectively, for water): values for a are known experimentally, and therefore the activity coefficient can be calculated directly from a knowledge of the solution's ionic strength. The ionic strength may be calculated from analytical data with the equation

$$I = \Sigma(m_i z_i^2/2)$$

where m is the molar concentration of the ith species and z is the charge of the ith species.

At high ionic strengths (> 0.1 molar), the Debye–Huckel equation is a poor predictor of activity coefficients. For these systems, the aqueous model must incorporate a more sophisticated approach. One such approach, based on the equations of Pitzer (1973), is capable of estimating activity coefficients up to 20 molar.

The temperature dependence of solute behavior in the aqueous model may be handled in a number of ways. The great majority of equilibrium constants and free-energy data have been determined at 25°C, and data at other temperatures may be entirely lacking. In this case, the temperature dependance of the equilibrium constant K may be estimated from enthalpy ($^\wedge$H) data using the equation

$$\log K = \log K_{Tr} - \frac{\Delta H_{Tr}}{2.3R}\left(\frac{1}{T} - \frac{1}{Tr}\right)$$

In other cases, where experimental data are available at other temperatures, the temperature dependence may be calculated from an empirical equation of the form

$$\log K = A + BT + C/T + D \log T.$$

The aqueous model may also include pressure-dependent corrections for equilibrium constants. However, most chemical processes of importance to solution chemistry show little pressure dependency up to several hundred atmospheres. Thus, pressure dependency is generally neglected for many applications.

In addition to activities for individual solutes, the aqueous model must assign an activity to water. For relatively dilute water, this activity may be calculated from the approximate relation

$$a\text{H}_2\text{O} = 1 - 0.017\Sigma m_i.$$

where sum m_i is the molar sum of all anions, cations, and neutral species. For sum m_i greater than about 1 molar, this approximation breaks down and other approaches must be taken.

Speciation Calculations with the Aqueous Model Once the framework of the aqueous model is established, it is possible to calculate the distribution of dissolved species at chemical equilibrium. Returning to the example of dissolved calcium, the equilibrium speciation in a particular water sample can be described by the following set of equations:

$$K_1 = \frac{a\text{CaOH}^+}{a\text{Ca}^{+2}\, a\text{OH}^-}$$

$$K_2 = \frac{a\text{CaHCO}_3^+}{a\text{Ca}^{+2}\, a\text{HCO}_3^-}$$

$$K_3 = \frac{a\text{CaCO}_3^0}{a\text{Ca}^{+2}\, a\text{CO}_3^{-2}}$$

and

$$K_4 = \frac{a\text{CaSO}_4^o}{a\text{CA}^{+2}\, a\text{SO}_4^{-2}}$$

where a denotes the activity of each dissolved species. Because the activity of each species can be related to their concentrations by means of activity coefficients, the concentrations of each species may be written as follows

$$m\text{CaOH}^+ = \frac{K_1 a\text{OH}^-\, m\text{Ca}^{+2}\, \gamma\text{Ca}^{+2}}{\gamma\text{CaOH}^+}$$

$$m\text{CaHCO}_3^+ = \frac{K_2\, a\text{HCO}_3^-\, m\text{Ca}^{+2}\, \gamma\text{Ca}^{+2}}{\gamma\text{CaHCO}_3^{+2}}$$

$$m\text{CaCO}_3^o = \frac{K_3 a\text{CO}_3^{-2}\, m\text{Ca}^{+2}\, \gamma\text{Ca}^{+2}}{\gamma\text{CaCO}_3^o}$$

and

$$m\text{CaSO}_4^o = \frac{K_4 a\text{SO}_4^{-2} m\text{Ca}^{+2}\, \gamma\text{Ca}^{+2}}{\gamma\text{CaSO}_4^o}$$

Combining these equilibrium equations with the mass balance equation

$$\text{Ca}_{tot} = m\text{Ca}^{2+} + m\text{CaOH}^+ + m\text{CaHCO}_3^+ + m\text{CaCO}_3^o + m\text{CaSO}_4^o$$

yields five equations with five variables that are easily solved. This, in turn, gives the equilibrium distribution of all calcium ions and complexes present in the water sample. Although it is possible to solve the set of five equations in this example by hand, it must be remembered that models of real systems typically consider as many as 200 or 300 dissolved species and, consequently, 200 or 300 simultaneous equations. As a practical matter, therefore, a computer is required to compute ionic strength, to calculate the activity coefficients, and to solve for the equilibrium distribution of dissolved species.

Oxidation-Reduction Reactions In addition to considering ion complex equilibria, as shown in the previous example, the aqueous model must be capable of considering oxidation-reduction reactions. Redox reactions, such as

$$\text{Fe}^{2+} = \text{Fe}^{3+} + e^-$$

can be treated like any other equilibrium reaction with a conventional equilibrium constant

$$K = (a\text{Fe}^{3+})\,(ae^-)/(a\text{Fe}^{2+})$$

where ae^- is defined as the electron activity, or pE

$$ae^- = 10^{-pE}$$

This definition is controversial since free electrons do not, for all practical purposes, exist in aqueous solution at 25°C (Thorstenson, 1984). However, the concept is commonly used in constructing aqueous models. For example, given a measured value of oxidation-reduction potential (*Eh*), the *pE* of an aqueous system may be calculated from the relation

$$pE = Eh/(2.303RT/F)$$

where R, T, and F are the universal gas constant, the temperature in degrees Kelvin, and the Faraday constant, respectively. Given a value for electron activity, the distribution of redox sensitive species is calculated in the same manner as before. For the Fe^{2+}—Fe^{3+} problem, the equilibrium expression

$$K = (aFe^{3+})(ae^-)/(aFe^{2+})$$

is combined with the mass balance expression

$$Fe_{tot} = Fe^{3+} + Fe^{2+}$$

where Fe_{tot} is known from chemical analysis of the water. This gives two equations in two unknowns that can then be solved simultaneously for the distribution of the two dissolved iron species.

The concept of *pE* greatly simplifies construction of an aqueous model. However, *pE* is not a physically measurable quantity, since there are no free aqueous electrons. *Eh* measurements made with platinum electrodes reflect reaction of redox couples on the electrode surface. For some redox couples that react quickly and reversibly on a platinum electrode, such as the Fe^{3+}/Fe^{2+} couple, measured *Eh*s may closely approach the actual redox potential of a system. Many redox couples, notably the O_2/H_2O couple do not react at all on an electrode surface. In this case, measured Ehs have no physical meaning.

Perhaps the largest problem with using pE is the observed lack of thermodynamic equilibrium in most water samples. For example, Lindberg and Runnels (1984) showed that Ehs, as measured by platinum electrode, and numerous other redox pairs, including the O_2/H_2O, Fe^{3+}/Fe^{2+}, $SO_4^=/H_2S$, and CO_2/CH_4 couples, showed little agreement with each other. Lindberg and Runnels (1984) interpreted this lack of agreement as reflecting the substantial lack of thermodynamic equilibrium in ground-water systems. This, in turn, calls into question the basic premise of using equilibrium considerations in constructing aqueous models. However, at present there is no alternative procedure that has proven more useful. The difficulty in adequately defining a "redox potential" for a given water sample is a serious problem in constructing aqueous models.

Computer Programs for Making Speciation Calculations It should be clear by now that making speciation calculations concerning the hundreds of possible dissolved species by hand is a practical impossibility. However, there are a large number of computer programs for making these calculations. In 1983, there were at least 50 different such programs available, all of which differed to a greater or smaller degree in the details of their aqueous models. For example, if a program is to be applied primarily to speciation in saline lakes, then the activity of water in the model must be handled differently than if the program is to be applied to relatively fresh waters. Similarly, if a program is to be used for hydrothermal systems, the aqueous model might be handled differently than one for use at relatively low temperatures. However, no single program has been developed that is capable of treating the wide range of environmental problems to which speciation calculations may be applied.

Nordstrom et al. (1979) compared the results of the various codes when applied to the same problem and highlighted some (mostly minor) differences. In general, the selection of a code for making geochemical calculations depends upon the problem at hand. The model of Harvie, Moller, and Weare (1984) is designed for thermodynamic calculations of high-ionic strength saline solutions. The aqueous model in this program is based on the equations of Pitzer (1973) for aqueous electrolyte solutions. For calculations at geothermal temperatures (up to 300°C), the programs SOLMNEQ (Kharaka and Barnes, 1973) and EQ3NR (Wolery, 1983) have been widely applied. For speciation calculations in low-temperature, low-salinity water characteristic of local and intermediate flow systems, the program WATEQF (Plummer, Jones, and Truesdell, 1976) is widely used. WATEQF was designed to be applied to carbonate equilibria problems, which are commonly encountered in ground-water systems, and is most suitable for that purpose.

10.2.2 Mineral Equilibrium Calculations

Once the aqueous model has been defined and the distribution of aqueous species calculated based upon chemical analysis of a given water sample, it is possible to calculate the saturation state of a wide variety of minerals. We saw earlier that the saturation state of any particular mineral can be defined in terms of the saturation index (SI):

$$SI = -\log IAP/K$$

where IAP is the ion activity product from a given water sample and K is the equilibrium constant for the reaction. Application of an aqueous model to a particular water sample gives the distribution of dissolved species, which allows calculation of the IAP for any particular reaction. Given the IAP, the saturation index for a mineral is easily calculated from a library of equilibrium constants. If a particular water sample is undersaturated with respect to a given mineral, the SI is negative. If, on the other hand, the water sample is oversaturated with respect to a mineral, the SI will be positive. Most speciation programs, such as WATEQF or SOLMNEQ also calculate saturation indices for a wide range of minerals.

Determining the saturation state of minerals with respect to particular water samples is generally the purpose of constructing an aqueous model. Clearly, the distribution of complex ions in a water sample is only of minor interest. The saturation state of a water sample with respect to minerals, however, is of great practical importance. For example, consider the case of a petroleum engineer designing a facility for separating petroleum from oil field brines. If the process of bringing a saturated brine to the surface, with subsequent cooling, results in the precipitation of halite (NaCl), then the separation facility must be designed to deal with salt encrustation. If, on the other hand, cooling of the brine will not result in halite precipitation, then the design may be greatly simplified. Aqueous models are widely used to address this type of problem in chemical engineering.

In ground-water chemistry, the saturation state of particular mineral phases is often of primary interest. If, for example, calcite is present in an aquifer and the ground water is undersaturated with respect to calcite, then dissolution of calcite is possible as ground water flows along the hydrologic gradient. Conversely, if calcite is oversaturated, than precipitation of this mineral is possible.

The utility of the saturation state to predict possible dissolution or precipitation of a mineral is often self-evident. However, there are numerous subtleties that commonly occur. For example, it is fairly common for different minerals with the same composition to coexist in an aquifer. Calcite and aragonite have the same chemical composition ($CaCO_3$), but, because the ions are arranged differently in the two minerals, they have slightly different solubilities. The equilibrium constant for calcite (containing 2 mole % Mg) is log $K_c = -8.492$, whereas the equilibrium constant for aragonite is log $K_a = -8.336$ and is slightly higher. Thus, if a water sample is saturated with respect to aragonite, it would be slightly oversaturated with respect to calcite.

An example of how this affects saturation indices in real systems was given by Plummer (1984) in a study of the Madison aquifer in Wyoming and Montana. It was observed that whereas the SI for aragonite was very close to saturation in most of the water samples, calcite was slightly oversaturated. In fact, dissolution of aragonite (as well as dissolution of gypsum and oxidation of CH_2O) drives the precipitation of calcite in this system. This pattern of aragonite dissolution driving precipitation of calcite is very common in ground-water systems. This is of great practical significance as the precipitation of calcite cement may destroy porosity and decrease the yield potential of an aquifer.

10.2.3 Material Balance Calculations

It is commonly observed that the composition of ground water changes as it flows downgradient in most aquifer systems. It is intuitively obvious that the net change in ground-water composition must reflect the dissolution or precipitation of solid material and gases present in the aquifer. Stated more precisely, the chemical evolution of ground water along a flowpath must conform to material balance constraints, including conservation of mass, electrons, and isotopes.

Mass Balance A general statement of the principle of conservation of mass as applied to ground-water systems is

Initial water composition + Reactant phases

= Final water composition − Product phases

This general principle may be formulated precisely by the equation (Plummer, Parkhurst, and Thorstenson, 1983):

$$\{\sum_{p=1}^{P} \alpha_p b_{p,k} = \Delta m_{T,k}\} \quad k = 1, J$$

which states that the change in total moles of element ($\Delta m_{k,\text{tot}}$) along a flowpath is equal to the sum of all the sources and sinks for element k. Furthermore, this equation states that these relationships hold for 1 through j elements. The sources and sinks may include dissolution, precipitation, microbial degradation, gas transfer, and so forth, of P phases (minerals, gases) along the flowpath where α is the number of moles reacting and b is the stoichiometric coefficient for the element in the Pth phase. Because all phases are electrically neutral, and because all changes in water composition are assumed to come from the phases, this treatment implicitly requires that charge balance in the final solution is maintained.

For applications of inverse geochemical modeling, $\Delta m_{\text{tot}},k$ values are derived from analytical data using the equations of the form

$$\Delta m_{\text{tot},k} = m_k \text{ (initial)} - m_k \text{ (final)}$$

where m_k (initial) is the molar concentration of k in the upgradient water and m_k (final) is the molar concentration of k in the downgradient water.

In theory, mass balance equations of this kind can be written for each element present in the phases. Thus, if there are j elements, one could write j mass balance equations. In practice, however, it is not feasible to include a mass balance on oxygen and hydrogen, because it is analytically impossible to determine the total masses of these elements in aqueous solutions. Thus, of a possible j number of equations in j unknowns, mass balance considerations actually yield only $j - 2$ equations in j unknowns and additional constraints are required in order to obtain a unique solution.

One constraint that may be applied is to assume that the mass of water (and therefore the mass of hydrogen and oxygen) remains constant along a flowpath. For relatively dilute solutions (less than 1 molar) this assumption introduces very little error. For geochemical modeling applications to potable ground water, the assumption of constant water mass may be reasonably applied.

For additional constraints in formulating these problems, it is possible to consider factors such as electron and isotope balance.

Electron Balance Because hydrated electrons effectively do not exist in aqueous solution (Thorstenson, 1984), any electrons transferred in a system must be conserved among the dissolved species and the mineral phases. Thus, it is possible to write electron balance equations in the same manner as mass balance equations

$$\sum_{p=1}^{P} \mu_p \alpha_p = \Delta RS$$

where μ_p is the operational valence of the pth phase, α_p is the mass transfer of the pth phase, and ΔRS is the change in redox state of the dissolved constituents along a flowpath segment. In practice, ΔRS is calculated from analytical data using the equation

$$\Delta RS = \sum_{i=1}^{\blacktriangleleft} v_i m_i \text{(final)} - \sum_{i=1}^{\blacktriangleleft} v_i m_i \text{(initial)}$$

Note that ΔRS includes the redox state of *all* dissolved constituents present in the water.

The conventions for assigning redox states to particular species (v_i) and phases have been established by Plummer, Parkhurst, and Thorstenson (1983), and follow these rules:

1. The redox state of aqueous redox species equals their formal elemental valence.
2. For redox complexes, the redox state equals the sum of the elemental valences.
3. For non-redox species, the redox state equals zero.
4. The redox state of H and O in aqueous species, including OH^- and H^+, equals zero.
5. The redox state of H_2 equals -2.0 and the redox state of O_2 equals $+4.0$.

According to these rules, the value for $v_{Fe^{2+}}$ is $+2.0$ and $v_{Fe^{3+}}$ is $+3.0$ (rule 1), the value for $v_{SO_4^=}$ is $+6.0$ (rules 1 and 4), the value for $vFe_{SO_4}^\circ$ is $+8.0$ (rules 2 and 4), the value for vNa^+ is 0.0 (rule 3), and the value for vH_2O is 0.0 (rule 4). Rule 5 is entirely self-explanatory.

Isotope Balance Because many of the elements that make up mineral phases in nature are composed of more than one stable isotope, and because the stable isotope composition of different mineral phases are often distinct, isotope balance equations can often be included with mass balance equations when constructing geochemical models.

The elements for which isotopic data are most commonly available in groundwater systems are carbon and sulfur. Carbon compounds and minerals consist of two stable isotopes ^{13}C and ^{12}C. The ratio (R) of these isotopes are commonly related to a standard of marine carbonate material (the Pee Dee Belemnite, or PDB) and reported in the standard del notation where

$$\delta^{13}C = [R_{sample}/R_{std} - 1] \times 1,000$$

Because the ratio of sample and standard are multiplied by 1000 (Since ^{13}C is much rarer than ^{12}C), the ratio are referred to in units of "per mil". In the same manner, sulfur compounds and minerals consist of the stable isotopes ^{34}S and ^{32}S and the ratios of these isotopes are commonly reported in the same del notation, $\delta^{34}S$.

If the only geochemical reactions occurring in a system are dissolution, then isotope balance is exactly analogous to mass balance. For the case of sulfur isotope balance, the equation is

$$\sum_{p=1}^{P} \alpha_p b_{p,s} \delta^{34}S_p = \Delta m_T^{34}S$$

where $b_{p,s}$ is the stoichiometric coefficient of sulfur in the pth phase and $del^{34}S_p$ is the sulfur isotopic composition of the pth phase and where $\Delta m_{tot}{}^{34}S$ is given by the equation

$$\Delta m_T{}^{34}S = (m_{T,s}\delta^{34}S_T)_{final} - (m_{T,s}\delta^{34}S_T)_{initial}$$

For example, if both pyrite (FeS_2) and gypsum ($CaSO_4$) were dissolving along a flowpath, the isotope balance equation would be

$$\alpha_{gypsum}\delta^{34}S_{gypsum} + 2\alpha_{pyrite}\delta^{34}S_{pyrite} = \Delta m_T{}^{34}S$$

Obviously, however, precipitation reactions as well as dissolution reactions occur in natural systems. Because mineral precipitation generally results in isotopic fractionation, these fractionations must be considered. In the case of sulfur isotopes, there is a very large fractionation of about 55 per mil that occurs during sulfate reduction. On the other hand, the fractionation between dissolved sulfide and sulfide in pyrite is fairly small and on the order of 1 or 2 per mil. In this case, a linear isotope balance is an excellent approximation. For example, if gypsum (+22 per mil) were dissolving, the sulfate reduced, and pyrite (−33 per mil) were precipitated, the linear isotope balance equation would be

$$\alpha_{gypsum}{}^{22} - \alpha_{pyrite}{}^{66} = \Delta m_{S,tot}{}^{34}S$$

The fractionation of carbon isotopes cannot be treated in this simple fashion. Unlike the large fractionation that occurs during sulfate reduction, fractionations encountered in precipitating carbonate minerals are much smaller, often on the order of 1 per mil. These fractionations introduce nonlinearity into isotope balances that must be treated using a more sophisticated approach. Specifically, the isotopic composition of carbon precipitated in calcite and the remaining carbon isotopic composition of ground water depends on the nonlinear fractionation between precipitating calcite and dissolved carbonate, and linear isotope dilution from carbon sources. Equations for describing carbon isotope evolution for ground water systems have been worked out in great detail by Wigley et al. (1978, 1979).

Computer Programs for Making Material Balance Calculations Simple material balance problems, such as those considering four or five components and phases and therefore having four or five simultaneous equations, are easily solved by hand. However, the algebra involved quickly becomes tedious and computer programs for solving the equations have been written. Perhaps the most widely used is the program BALANCE (Parkhurst, Plummer, and Thorstenson, 1982). This program solves whatever set of simultaneous equations, including mass balance, electron balance, and linear isotope balance, as specified by the user. It is not capable of handling nonlinear isotope balance problems such as arise with carbon isotopes in dissolution and precipitation (Rayleigh distillation) of carbonate minerals.

10.2.4 Logic of Geochemical Modeling

In sections 10.2.2 and 10.3.3, we have seen how water-chemistry data can be used (1) to evaluate the saturation state of water with respect to various mineral phases, and (2) to formulate material balance equations based on the conservation of mass, electrons, and isotopes. These two procedures may be applied in a logical hierarchy in order to address particular geologic or microbiologic questions.

A schematic diagram of the logical hierarchy most commonly followed in geochemical modeling (Plummer, 1984) is shown in Figure 10.3. The first judgment that must be made concerns the availability of water chemistry data. If water chemistry data are available and of sufficient quality to perform speciation calculations (i.e., the data are "saturation sufficient"), then inverse geochemical modeling may proceed. If, on the other hand, analytical data are not saturation sufficient, the only recourse available is to make theoretical reaction-path calculations based upon thermodynamic constraints. This "forward method" of geochemical modeling gives no new information concerning geochemical or microbial processes in a given system and has little application in subsurface microbiology.

Given water chemistry data, however, inverse geochemical modeling can continue. First, the water chemistry data is used to perform mineral equilibrium calculations. Second, a set of "plausible phases", P, are identified for the aquifer

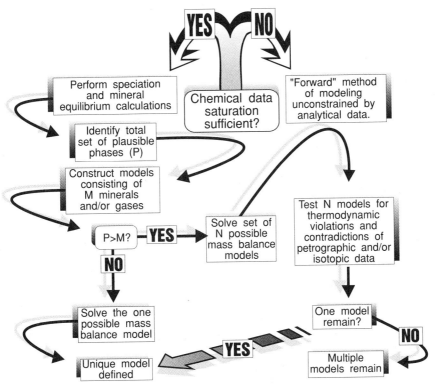

FIGURE 10.3. Logical hierarchy followed in geochemical modeling. (Modified from Plummer, 1984, with permission courtesy of the National Water Well Association.)

from geologic and petrographic information. For example, if an aquifer matrix consists entirely of quartz and calcite, quartz and calcite are the only possible plausible phases that can be identified. In real aquifers, however, the number of major, minor, and trace minerals present—and therefore the number of plausible phases—commonly varies between 5 and 30.

Once the number of plausible phases present in an aquifer is determined, a test is made to see if a unique algebraic solution to the material balance problem is possible. For a system containing J elements, it is possible to write $P = J - 2$ material balance equations if electron balance (i.e., redox reactions) is not involved, or $P = J - 1$ equations if electron balance is considered. If the number of material balance equations is equal to or less than the number of plausible phases, then a unique algebraic solution can be found. On the other hand, if the number of plausible phases is greater than the number of material balance equations (as commonly occurs), multiple algebraic solutions are possible.

If a unique material balance solution is possible, the problem can be solved to yield the mass transfer coefficients (α) for each mineral dissolving or precipitating. If a unique solution does not exist, a set of possible models (N) is identified and solved. These models are then tested for thermodynamic consistency, contradictions of petrographic observations, and consistency with isotopic data.

For example, if the speciation and mineral equilibrium calculations indicate that ground-water is oversaturated with respect to calcite, then any material balance model that requires calcite *dissolution* violates equilibrium considerations. This model can then be eliminated from the set of N possible material balance models. Similarly, if there is strong petrographic evidence that calcite is precipitating in an aquifer (such as the presence of calcite overgrowths), then any model requiring the dissolution of calcite can be eliminated on the basis of petrographic observation.

This process of eliminating possible material balance models continues until either one or several models are left that cannot be eliminated. If only one model remains, then a unique solution has been found. Typically, however, multiple models remain that cannot logically be eliminated from consideration without acquisition of additional data. It sometimes happens that *all* of the possible models that have been formulated are eliminated. This is an obvious indication that one has failed to include a key feature (i.e. a plausible phase) in the models that have been constructed.

Formulating Hypotheses of Microbial Processes The goal of geochemical modeling, especially as applied to identifying microbial processes in subsurface environments, is not necessarily simply to obtain a unique model of a system. The very process of iterating through the logic of Figure 10.3 is an effective way of formulating hypotheses concerning a particular aquifer system that are testable using microbiologic techniques.

One example of this was given by Chapelle and McMahon (1991). The aquifer being studied, the Black Creek aquifer, exhibited large increases of dissolved inorganic carbon species along the hydrologic gradient. However, there were not proportional decreases in oxygen concentrations, increases in dissolved iron (indicating Fe(III) reduction), decreases in sulfate (indicating sulfate reduction), or increases in methane (indicating methanogenesis). Given these water chemistry data, there was no possible combination of plausible phases that could satisfy constraints of mass balance and electron balance.

When such apparent logical contradictions are encountered, the conclusion can reasonably be drawn that important processes in the system are being overlooked. Application of a microbiologic technique, using concentrations of dissolved hydrogen as an indicator of the predominant terminal electron accepting processes, indicated that sulfate reduction was the predominant electron-accepting process in this system. Further investigation showed that while sulfate concentrations in aquifer water were low, there were large concentrations of sulfate in the pore water of adjacent confining beds. In this case, it turned out that sulfate diffusing from confining beds was driving sulfate reduction in this system. Furthermore, a geochemical model could be constructed of this process that was consistent with available mineralogic and water-chemistry data (McMahon and Chapelle, 1991). Without the logical iteration made possible by quantitative geochemical modeling, the importance of a microbial process, sulfate reduction, in this system could not have been deduced.

10.3 GEOCHEMICAL MODELING AND MICROBIAL PROCESSES IN AQUIFER SYSTEMS

Prior to about 1975, the material balance approach to studying ground-water chemistry was largely applied in a qualitative fashion. For many problems, this qualitative approach was quite useful in resolving questions concerning the kinds of chemical processes that impacted ground-water chemistry. When the material balance approach was used in a quantitative, mathematically rigorous basis, it had the effect of forcing geochemists to consider the nature of the reactions they were applying to geochemical models. This, in turn, began to focus attention on microbial mechanisms driving observed ground-water chemistry changes.

There are numerous examples of this in the literature. Perhaps the best documented and the most relevant to ground-water chemistry studies was the identification of microbial metabolism as a principal mechanism in the production of carbon dioxide, the reduction of sulfate, and production of methane in deeply buried aquifers. It had long been qualitatively known that carbon dioxide was being produced in some aquifer systems from sedimentary organic material (Cedarstrom, 1946; Foster, 1950). However, the mechanisms involved with carbon dioxide production remained obscure. Successive applications of geochemical modeling during the late 1970's and 1980's were crucial in showing the importance of microbial processes in CO_2 production and, in consequence, the chemical evolution of ground water.

10.3.1 CO_2 Production in the Floridan Aquifer

One of the first applications of quantitative geochemical modeling was made on the Floridan aquifer of central Florida (Plummer, 1977). Several attributes of this particular system made it a convenient aquifer for study. First of all, because ground-water is the most important source of water for municipal use in the State of Florida, the hydrology of the Floridan aquifer was well documented. The aquifer is recharged via atmospheric precipitation in the topographically high central part of the State near Polk City. Water flows downgradient from the recharge area into deeper parts of the flow system, eventually discharging into the Gulf of Mexico. Several towns located along this flowpath, including Fort Meade, Wauchula, and Arcadia, used the water for municipal supply so that ground water could be systematically sampled at various parts of the flow system. Second, the geology of the Floridan had also been closely studied. In particular, the Floridan had been cored at a number of places and the mineralogy of the aquifer matrix carefully documented.

As water flows along the hydrologic gradient of the Floridan aquifer, concentrations of dissolved solids progressively increase. This reflects increases in dissolved bicarbonate, calcium, magnesium, and sulfate along the flowpath from Polk City to Arcadia (Table 10.1). The Floridan aquifer is a carbonate system composed mainly of calcite, dolomite, and gypsum. If, in addition to these solid phases, a source of CO_2 is considered, then the number of phases (calcite, dolomite, gypsum, CO_2) equals the number of dissolved components (HCO_3, Ca^{2+}, Mg^+, $SO_4^=$) and a unique mass balance for the system is given by the equations:

$$\alpha_{Cal} = \frac{(2-y)(\Delta m_{Ca}^T - \Delta m_{SO4}^T) - y\Delta m_{Mg}^T}{2x - y}$$

$$\alpha_{Dol} = \frac{x\Delta m_{Mg}^T - (1-x)(\Delta m_{Ca}^T - \Delta m_{SO4}^T)}{2x - y}$$

$$\alpha_{CO2} = \Delta m_{CO2}^T - \Delta m_{Mg}^T - (\Delta m_{Ca}^T - \Delta m_{SO4}^T)$$

$$\alpha_{Gyp} = \Delta m_{SO4}^T$$

where values for $\Delta^m Ca$, $\Delta^m Mg$, $\delta^m CO_2$, and $\Delta^m SO_4$ are calculated from Table 10.1.

The results of this mass balance model (Table 10.2) showed several interesting features. For example, the chemical evolution of the water appears largely driven by the dissolution of gypsum. The production of dissolved Ca^{2+} from gypsum dissolution, together with the production of carbonate from CO_2 and dolomite dissolution, drives the precipitation of low-magnesium calcite. Furthermore, the amounts of dolomite dissolving and calcite precipitating were sensitive to the assumed composition of dolomite. This so-called dedolomitization reaction has subsequently been found in a number of hydrologic systems (Plummer and Back, 1980).

The most important result, however, was the behavior of CO_2 in this system. In the upgradient recharge area, water percolating through the soil zone was expected to provide a source of CO_2. Consistent with this hypothesis, the mass balance for water moving between Polk City and Fort Meade required dissolution of 0.264 mmol/L of CO_2. Unexpectedly, however, addition of CO_2 to the ground

10.3 GEOCHEMICAL MODELING AND MICROBIAL PROCESSES IN AQUIFER SYSTEMS

TABLE 10.1 Chemical and Carbon Isotope Composition of Selected Waters from the Floridan Aquifer

| Well | pH | Ca, mmol/l | Mg, mmol/l | CO_2, mmol/l | SO_4, mmol/l | Log P_{CO_2},* atm | Saturation Index | | | d13C,0/00 |
							SI_C	SI_D	SI_G	
Polk City	8.00	0.84	0.23	2.14	0.025	−2.89	0.17	−0.07	−3.10	−11.4
Fort Meade	7.75	1.45	0.70	2.77	0.739	−2.55	0.23	0.31	−1.53	−10.8
Wauchula	7.69	1.65	1.19	2.87	1.614	−2.48	0.17	0.37	−1.21	−8.5
Arcadia	7.44	2.65	2.47	3.63	3.584	−2.15	0.14	0.43	−0.81	−8.3

Source: From Plummer, 1977. Reprinted courtesy of American Geophysical Union.

water in the more downgradient areas between Fort Meade and Wauchula, and between Wauchula and Arcadia required dissolution of 0.285 and 0.450 mmol/L CO_2. In the more upgradient part of this system between Fort Meade and Wauchula, these results could be rationalized somewhat because downward leakage of water from the soil zone was possible. In the most downgradient portion of the aquifer, however, leakage could not be reasonably invoked as a CO_2 source. The only reasonable source of carbonate species was sulfate reduction:

$$2CH_2O + SO_4^= \rightarrow H_2S + 2HCO_3^-$$

This hypothesis was supported by the relatively high concentrations of dissolved sulfide (~ 1.5 mg/L) in Arcadia ground water (Rightmire et al., 1974).

The significance of these results is that taking a quantitative mass balance approach in this system forced the investigator to consider sulfate reduction, that is, a microbial process, as a source of CO_2. The next logical step, incorporating microbial-mediated redox reactions directly into the geochemical modeling process, was soon taken.

TABLE 10.2 Summary of Mass Balance Reaction Coefficients for the Floridan Aquifer

		Values of aj. [a]mmol/l					
Reaction	Path[b]	$CaMg(CO_3)_2$	$Ca_{1.08}Mg_{0.92}(CO_3)_2$	$Ca_{0.95}Mg_{0.05}CO_3$	$CaSO_4\ 2H_2O$	CO_2	$Ca_{0.98}Mg_{0.02}CO_2$
1	PC-FM	0.482			0.714	0.264	−0.598
2	PC-FM		0.526		0.714	0.264	−0.686
3	PC-FM			15.423	0.714	0.264	−15.057
4	FM-W	0.514			0.875	0.285	−1.214
5	FM-W		0.561		0.875	0.285	−1.307
6	FM-W			16.457	0.875	0.285	−16.642
7	W-A	1.327			1.970	0.450	−2.344
8	W-A		1.448		1.970	0.450	−2.585

[a] Reactants are positive, and products are negative.
[b] PC stands for Polk City, FM for Fort Mead, W for Wauchula, and A for Arcadia.
Source: From Plummer, 1977. Reprinted courtesy of the American Geophysical Union.

10.3.2 Sulfate Reduction and Methanogenesis in the Fox Hills–Basal Hell Creek Aquifer

In the late 1970's, it was clear that consideration of ground-water chemistry required a quantitative treatment of redox processes. To this end, a study was initiated of redox reactions in the Fox Hills–Basal Hell Creek aquifer (from now on referred to as the Fox Hills aquifer) in parts of North and South Dakota. The Fox Hills aquifer was in many ways an excellent system to study redox reactions. It is recharged by atmospheric precipitation in it's outcrop area. However, because the aquifer contains abundant organic matter, water quickly becomes anaerobic along the hydrologic gradient. The original research plan was to use electrode measurements of redox potential in order to map the progression of redox processes. As it turned out, however, the measurements of redox potential were not amenable to quantitative treatment. Far more was learned by means of quantitative techniques of geochemical modeling and by comparing results to observed water chemistry data (Thorstenson, Fisher, and Croft, 1979).

The methodology used in the geochemical modeling effort for the Fox Hills aquifer differed somewhat from that used by Plummer (1977). In the Floridan aquifer, the number of plausible phases could reasonably be assumed to equal the number of components in solution (within the uncertainty of the dolomite composition). The mineralogy of the Fox Hills aquifer, a complex deltaic unit, was much more diverse. This diversity insured that the number of plausible phases was much greater than the number of dissolved components in the ground water, rendering a unique algebraic solution impossible.

Because of the mineralogic complexity, the Fox Hills modeling was based on the computer program PHREDX, an early unpublished version of the program PHREEQE (Parkhurst, Thorstenson, and Plummer, 1980). In this approach, equilibrium was assumed with respect to selected minerals (i.e. plausible phases). Next, a given amount of reactant (CO_2, for example) was added to the solution to perturb the equilibrium condition. Finally, the solution was allowed to return to equilibrium conditions. This procedure has come to be known as "reaction path modeling," since it shows the pathways taken by a solution as it evolves and is primarily used for forward geochemical modeling. In the Fox Hills study, however, the calculated results of the modeling were compared to measured ground water compositions. Thus, this approach is conceptually identical to the material balance approach of inverse geochemical modeling. The main difference is that the possible number of mass balance models (N, Fig. 10.3) remains unevaluated.

The first step in the modeling process was to evaluate the saturation status of various minerals present in the system using the speciation program WATEQF. This indicated that the water was saturated with respect to calcite and dolomite throughout the extent of the aquifer. This, in turn, indicated that it was reasonable to specify equilibrium with respect to these minerals in the PHREEQE calculations.

A major problem, however, was how to deal with the abundant aluminum silicate clay minerals present in the aquifer. Clay minerals commonly exhibit a wide variety of compositions, many of which are metastable (i.e., a transient form between unstable and stable compositions). Because concentrations of dissolved silica were fairly low and because there was no evidence of silica precipitation, it was assumed that dissolution-precipitation reactions of silicate minerals had little impact on water chemistry changes. However, because clay minerals are an efficient medium for cation exchange reactions, they could not be completely disregarded.

Given the cation exchange reaction

$$Ca^{2+} + Na_2 \cdot Clay \leftrightarrow 2Na^+ + Ca \cdot Clay$$

the equilibrium expression between the adsorbed and aqueous sodium and calcium ions may be written

$$K_{ex} X Na^2 / XCa = aNa^+ / aCa^{2+}$$

Because the composition of clays in the system was not known, this expression could not be used directly. However, assuming that the composition of clays remained constant, a provisional constant equal to

$$K_{ex}' = K_{ex} XNa_2 / XCa$$

can be defined so that the equation is dependent only on the composition of the ground water

$$K_{ex}' = aNa^+ / aCa^{2=}$$

Because the activity ratio of sodium and calcium is easily calculated using a speciation program such as WATEQF, this approach allowed consideration of cation exchange in the system. This is important, because, when cation exchange occurs simultaneously with calcite dissolution-precipitation reactions

$$CaCO_3 + H^+ + 2Na_{clay} = Ca_{clay} + 2Na^+ + HCO_3^-$$

pH depends on both the exchange reaction and on calcite dissolution. Values of K_{ex}' derived from water chemistry data ranged from 40 to 5, and these values were used in the modeling effort.

The results of the modeling are shown in Figure 10.4. In the recharge area of the aquifer, it was reasonable to assume a source of CO_2 directly from soil gas. Note that if just calcite-dolomite equilibrium were specified, calculated concentrations of bicarbonate did not agree at all with measured values. However, if the cation exchange equilibria were included, most of the observed bicarbonate concentrations would be bracketed by the curves defined by K_{ex}' values between 40 and 5.

Downgradient of the recharge area in the "transition zone," the aquifer was clearly closed to an atmospheric or soil-gas source of CO_2. However, CO_2 addition was evidently required as bicarbonate concentrations continued to increase. Furthermore, the water chemistry data showed that concentrations of sulfate decreased from about 2.7 mmolar to less than 0.2 mmolar. Based on these data, the only reasonable interpretation was that microbial sulfate reduction was the primary CO_2-producing process. However, because there was little observed dissolved

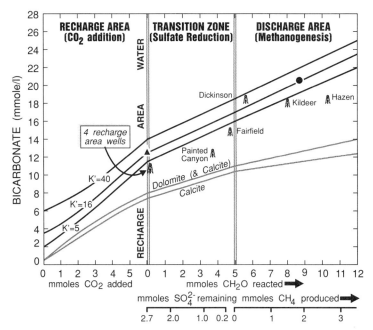

FIGURE 10.4. Geochemical modeling results of the Fox Hills–Basal Hell Creek aquifer. (Modified from Thorstensen, Fisher, and Croft, 1979, with permission courtesy of the American Geophysical Union.)

sulfide present, sulfate reduction was apparently accompanied by a sulfide-consuming reaction, such as pyrite precipitation:

$$15CH_2O + 2Fe_2O_3 + 8SO_4^= + H_2CO_3 \rightarrow 4FeS_2 + 16HCO_3 + 8H_2O$$

This irreversible reaction formed the basis for the simulation of the water chemistry in the transition zone. Using this equation, the bicarbonate concentrations of aquifer water could be reproduced as a function of sulfate consumed and organic matter (CH_2O) reacted (Fig. 10.4).

Downgradient of the transition zone, in the "discharge area," so called because water discharges from the aquifer via upward leakage, concentrations of bicarbonate continue to rise. As before, this indicated a source of CO_2 to the ground water. However, as sulfate had been removed from the water in the transition zone, another CO_2-producing process had to be occurring. As before, the only reasonable source of CO_2 was microbially mediated decomposition of organic material. In the absence of sulfate, methanogenesis was the most logical terminal electron accepting process. It was assumed that the overall decomposition reaction could be represented by the stoichiometry

$$2CH_2O \rightarrow CH_4 + CO_2$$

Again, reasonable agreement was obtained between simulated and measured bicarbonate concentrations (Fig. 10.4), as well as with measured pH and methane concentrations.

The study of the Fox Hills aquifer (Thorstenson, Fisher, and Croft, 1979) was significant for a number of reasons. First of all, it was one of the first geochemical modeling efforts that explicitly incorporated microbial processes. Furthermore, the study was one of first to show that the succession of redox processes (sulfate reduction → methanogenesis) was identical to the succession observed in aquatic sediments. The only differences were of scale. The sulfate-reducing zone in aquatic sediments is typically on the order of meters, whereas in the Fox Hills aquifer the sulfate reducing zone is on the order of 100 km. Finally, the study is an excellent example of how microbial processes (CO_2 production) interact with abiotic mineral-water reactions (cation exchange) in driving ground-water chemistry changes in hydrologic systems.

10.3.3 Sulfate Reduction in the Madison Aquifer

Early applications of quantitative geochemical modeling to the Floridan aquifer clearly indicated that microbially-mediated sulfate reduction was an important geochemical process (Plummer, 1977). In subsequent applications of these techniques, microbially-mediated electron transfer considerations were incorporated directly into the material balances. This treatment allowed the overall geochemical effects of microbial processes to be considered in geochemical models.

Certainly the most comprehensive application of these techniques to a predominantly sulfate-reducing system was made in the Madison aquifer of Wyoming and Montana (Plummer, 1984; Plummer et al., 1990). The Madison aquifer is a lithologically complex carbonate unit. The mineralogy is predominantly a mixture of calcite, dolomite, and anhydrite, with 3 to 4 percent accessory minerals including sedimentary organic matter, illite, smectite, kaolinite, quartz, feldspar, chalcedony, hematite, pyrite, halite, and sylvite. Based on these mineralogic data, trends in water composition, and computed saturation indices throughout the Madison, calcite ($CaCO_3$), dolomite ($CaMg(CO_3)$), gypsum/anhydrite ($CaSO_4$), organic matter (CH_2O), carbon dioxide (CO_2), ferric hydroxide ($FeOOH$), pyrite (FeS_2), Ca-Na exchange, halite ($NaCl$), and sylvite (KCl) were identified as plausible phases.

Note that organic matter is included as a plausible phase. This is so that microbially-mediated electron transfer reactions can be considered. Specifically, organic carbon of valance zero (CH_2O) is considered as an electron donor, and sulfate and ferric hydroxide are considered as electron acceptors. Sulfate reduction was considered to be the predominant terminal electron-accepting process in this system for a number of reasons. First of all, sulfate concentrations in the ground water were sufficient (one to 20 mmolar) to preclude significant methanogenesis. Secondly, dissolved sulfides were present in much of the ground water produced from this unit. Finally, sulfate-reducing microorganisms had been recovered in ground water produced from the Madison aquifer (Dockins et al., 1980).

For the ten plausible phases listed above, ten equations are needed in order to calculate the mass transfer coefficients (α) for each phase. Eight of these equations are mass balance equations:

$$\alpha_{calcite} + 2\alpha_{dolomite} + \alpha CH_2O + \alpha CO_2 = \Delta m_{T,C}$$

$$\alpha_{gypsum} + 2\alpha_{pyrite} = \Delta m_{T,S}$$

$$\alpha_{calcite} + \alpha_{dolomite} + \alpha_{gypsum} - \alpha_{exchange} = \Delta m_{T,Ca}$$

$$\alpha_{dolomite} = \Delta m_{T,Mg}$$

$$\alpha_{halite} + 2\alpha_{exchange} = \Delta m_{T,Na}$$

$$\alpha_{sylvite} = \Delta m_{T,K}$$

$$\alpha_{halite} + \alpha_{sylvite} = \Delta m_{T,Cl}$$

$$\alpha_{FeOOH} + \alpha_{pyrite} = \Delta m_{T,Fe}$$

One is an electron-balance equation:

$$4\alpha_{calcite} + 8\alpha_{dolomite} + 6\alpha_{gypsum} + 4\alpha_{CO_2} + 3\alpha_{FeOOH} = \Delta RS$$

And one is a sulfur isotope balance equation:

$$\alpha_{gypsum}\delta^{34}S_{gypsum} + 2\alpha_{pyrite}\delta^{34}S_{pyrite} = \Delta m_T{}^{34}S$$

These equations provide the ten independent equations required to solve for the unknown mass transfer coefficients, α_p. The algebraic solution, in terms of mass transfer for each phase, may be written

$$\alpha_{gypsum} = \frac{\Delta m_T{}^{34}S - \delta^{34}S_{pyrite}\Delta m_{T,S}}{(\delta^{34}S_{gypsum} - \delta^{34}S_{pyrite})}$$

$$\alpha_{pyrite} = (\Delta m_{T,S} - \alpha_{gypsum})/2$$

$$\alpha_{KCl} = \Delta m_{T,K}$$

$$\alpha_{NaCl} = \Delta m_{T,Cl} - \alpha_{KCl}$$

$$\alpha_{exchange} = (\Delta m_{T,Na} - \alpha_{NaCl})/2$$

$$\alpha_{dolomite} = \Delta m_{T,Mg}$$

$$\alpha_{FeOOH} = \Delta m_{T,Fe} - \alpha_{pyrite}$$

$$\alpha_{calcite} = \Delta m_{T,C} + \alpha_{exchange} - \alpha_{gypsum} - \alpha_{dolomite}$$

$$\alpha_{CO_2 gas} = \frac{(\Delta RS - 3\alpha_{FeOOH} - 4\alpha_{calcite} - 8\alpha_{dolomite} - 6\alpha_{gyp}}{4}$$

$$\alpha_{CH_2O} = \Delta m_{T,C} - \alpha_{CO_2 gas} - 2\alpha_{dolomite} - \alpha_{calcite}$$

In order to solve these equations, water chemistry data from at least two points along the flowpath are required. One set of data for the Madison aquifer, for water characteristic of the recharge area and the "Mysee well" located about 150 miles downgradient of the recharge area, is listed in Table 10.3.

TABLE 10.3 Average Composition of Recharge Water and the Mysse Well

	Recharge Water			Mysse Well	
Constituent	mg/l	mmol/kg H_2O	mmol/kg H_2O	mg/l	mmol/kg H_2O
Calcium	48.0	1.20	1.01	450.00	11.28
Magnesium	24.5	1.01	1.27	110.00	4.54
Sodium	0.5	0.02	0.02	730.00	31.89
Potassium	0.85	0.02	0.02	99.00	2.54
Iron	0.06	0.001	0.001	0.02	.0004
Chloride	0.6	0.02	0.02	630.00	17.85
Sulfate	15.0	0.16	0.16	1900.00	19.86
Alkalinity	245.0	—	—	320.00	—
Total inorganic Carbon	—	4.30	4.37	—	6.87
Redox state	—	18.14	—	—	146.13
H_sS	—	—	—	8.80	.259
Tritium	27.6		27.60	—	—
$\delta^{34}S_{SO_4^{2-}}(0/00)$	9.73		9.73	16.30	16.30
$\delta^{34}S_{H_2S}(0/00)$	—		—	−22.09	−22.09
$\delta^{13}C(0/00)$	−6.99		6.99	−2.34	−2.34
$^{14}C(\%)$	33.1		33.10	0.80	0.80
T°C	9.9		9.90	63.00	63.00
pH	7.55		7.67	6.61	6.61
SI calcite	−0.05		0.00		.29
SI dolmite	−0.27		0.00		.51
SI gypsum	−2.47		−2.55		−.15
log P_{CO_2}	−2.26		−2.36		−.94
Charge balance (meq/kgH_2O)		0.11	0.11		3.38

Source: From Plummer, 1984. Reprinted courtesy of the National Water Well Association.

Having formulated material balance equations and obtained appropriate data to solve the equations is not the end of the geochemical modeling process. At this point, several judgments need to be made. For example, inspection of the equations show that the value for αCO_2 is a function of the del^{34}S value of gypsum. Furthermore, in the deep confined portion of the aquifer the only plausible CO_2 source is from CH_2O oxidation and not from atmospheric or soil gases. Thus, the solution of these equations should give a value of αCO_2 that is near zero if it acturately describes the system. For these reasons, the del^{34}S value of gypsum may be treated as a variable in order to force values of αCO_2 to near zero. Given these criteria, the net mass transfer of the system is calculated to be

Recharge water + 10.15 gypsum + 3.54 dolomite + 0.87 CH_2O
    + 15.31 NaCl + 2.52KCl + 0.09 FeOOH + 8.28 Na_2Ex → 5.33 calcite
    + 0.09 pyrite + 8.28 CaEx + Mysse water

where the reaction coefficients are in mmol/L water and Ex denotes ion exchange sites.

This is only one of the many plausible solutions to the material balance equations depending on the composition of plausible phases and on the criteria selected for mass transfer of CO_2. For example, the possibility of both calcium and magnesium exchanging for sodium might be considered. Assuming that the selectivities of calcium and magnesium (i.e. attraction for exchange sites) are equal, the exchange of calcium and magnesium for sodium will be in proportion to the calcium-magnesium ratio in the ground water. Using the calcium-magnesium ratio at the Mysee Well indicates that the overall exchange reaction may be written

$$0.71\ Ca^{2+} + 0.29\ Mg^{2+} + Na_2EX \rightarrow Ca_{0.71}Mg_{0.29}Ex + 2Na^+$$

The net mass transfer calculated with this set of assumptions is

Recharge water + 20.15 gypsum + 5.91 dolomite + 0.87 CH_2O
+ 15.31 NaCl + 2.52 KCl + 0.09 FeOOH + 8.28 Na_2 Ex →
10.09 calcite
+ 0.09 pyrite + 8.28 $Ca_{.71}Mg_{.29}Ex$ + Mysee water

Note that the reaction coefficients for gypsum, NaCl, KCl, and FeOOH dissolution are unchanged. However, the amount of dolomite dissolving and calcite precipitating increase by 67% and 89% respectively.

Some effects of alternative criteria used to constrain material balance between recharge water and the Mysee well are tabulated in Table 10.4. In case 1, the del^{34}S of gypsum is assumed to be 11.8, only Ca-Na exchange is considered, and all gas produced from CH_2O oxidation is carbon dioxide (i.e. no methanogenesis). Case 1 requires the oxidation of considerable quantities of CH_2O (5.00 mmol/L) which in turn requires the "precipitation" of CO_2. In other words, if this model accurately simulates the ground-water chemistry, CO_2 outgassing of the ground-water should be observed. As CO_2 outgassing was not observed at the Mysee well, case 1 can be ruled out as an accurate representation of the ongoing reactions. Similarly, cases 2, 3, and 4, which test the effects of various cation exchange reactions and methanogenesis, may be ruled out using the same criteria. Only case 5, where the del^{34}S of gypsum is assumed to be 15.5, is the mass transfer of carbon dioxide gas close to zero. Interestingly, two measured values of $\delta^{34}S$ for gypsum in the vicinity of the Mysee well were 16.7 and 17.2, which are relatively close to the modeled value. Carbon isotopes were used by Plummer et al. (1990) as an additional criteria for refining the geochemical models of the Madison aquifer.

From a microbiologic point of view, the geochemical modeling exercise of the Madison aquifer demonstrates several points. First of all, by far the most important reactions in this system are not microbially mediated at all. Rather they are the inorganic dissolution of gypsum, halite, and sylvite. Dissolution of these salts accounts for about 75 % of the total mass transfer in this system. In contrast, microbially mediated oxidation of CH_2O accounts for less than 2% of the total mass transfer in this system. When one is studying microbial processes in ground-water systems, it is important not to lose sight of the fact that inorganic processes are also important.

TABLE 10.4 Summary of Modeling Alternatives Applied to the Mysse Flowing Well: Modeling Parameters

Case	$\delta 34S$,[a] 0/00	Ion Exchange X_{Mg}[b]	Proportion of Carbon Dioxide in Gas[c]	Methane Fractionation,[d] 0/00	$\delta 13C$, 0/00 Organic matter	Dolmite
1	11.8	0.0	1.0	—	−25.0	+2.0
2	11.8	1.0	1.0	—	−25.0	+2.0
3	11.8	0.0	0.8	−60.	−25.0	+2.0
4	11.8	1.0	0.8	−60.	−25.0	+2.0
5	15.5	0.0	1.0	—	−25.0	+2.0
6	15.5	0.0	1.0	—	−25.0	+2.0
7	15.5	0.0	1.0	—	−25.0	+4.0

Parts per thousand (0/00).
[a] Dissolving anhydrite.
[b] X_{Mg} is the fraction of magnesium/sodium ion exchange: 0.0 is pure calcium/sodium ion exchange and 1.0 is pure magnesium/sodium ion exchange.
[c] Proportion of carbon dioxide gas in a carbon dioxide methane mixture: 1.0 is pure carbon dioxide gas and 0.0 is pure methane gas.
[d] Modeled carbon isotopic fractionation of methane relative to the carbon isotopic composition of the dissolved inorganic carbon.
Source: From Plummer et al., 1990. Reprinted courtesy of the American Geophysical Union.

This having been said, however, the relatively small molar amounts of CH_2O oxidized by microbial processes in the Madison aquifer are nevertheless important in determining water chemistry changes. This is particularly true with respect to concentrations of dissolved carbonate species and equilibrium with respect to carbonate minerals. For example, the water is oversaturated with respect to calcite throughout the system and calcite progressively precipitates as water moves along regional flowpaths (table 10.4). The supersaturation of ground water with respect to calcite occurs because of dissolution of aragonite and dolomite. However, the ground-water is also very close to saturation with these minerals. It is the irreversible oxidation of organic matter via microbial metabolism and consequent production of CO_2 that drives aragonite dissolution in this system. This, in turn, drives calcite precipitation. Thus, even though microbial processes account for a relatively small portion of overall mass transfer in this system, they serve to drive the principal chemical processes. This is a general principle and occurs in many ground-water systems.

Application to Microbiological Studies The preceding example shows that geochemical modeling can be an important tool in the study of microbial processes in ground-water systems. Not only does geochemical modeling require water chemistry data that are more easily obtained than asceptically cored sediments, but the results are more representative of the overall system. Although geochemical modeling cannot replace microbiological evaluation of subsurface environments, it certainly can be used to answer particular questions. Perhaps the most promising approach to studying ground-water systems is to use geochemical modeling to

294 GEOCHEMICAL MODELING AS A TOOL FOR STUDYING MICROBIAL PROCESSES

formulate particular hypotheses concerning operative microbial processes. These hypotheses can then be systematically tested using microbiologic techniques.

10.4 SUMMARY

In this chapter we have given an overview of both the mathematical and philosophical basis of geochemical modeling. The first step in geochemical modeling is the application of an aqueous model (e.g., WATEQF or MINTEQ) that calculates the equilibrium distribution of dissolved species given a particular water analysis. Given the calculated distribution of species, it is possible to calculate the saturation state of particular mineral phases. These calculations allow judgments to be made concerning the possible dissolution-precipitation reactions of various minerals. Given a detailed knowledge of aquifer mineralogy, and appropriate water-chemistry data, it is possible to formulate precise mass balance, electron balance, and isotope balance equations describing material transfer in the system. When these equations are solved, generally with the help of a computer program such as BALANCE, the amounts of each mineral dissolving or precipitating along a particular flowpath can be calculated.

Such calculations are used to identify the important operative chemical processes, both biological and non-biological in an aquifer system. Specifically, geochemical modeling can be used to identify microbial processes such as sulfate reduction or methanogenesis as well as inorganic dissolution-precipitation reactions. In addition, geochemical modeling combined with hydrologic or radiometric dating of ground water is probably the most accurate method for estimating rates of microbial metabolism in aquifer systems. As such, these techniques can give important information about microbial processes in many subsurface environments.

PART III

MICROBIAL PROCESSES
IN CONTAMINATED
GROUND-WATER SYSTEMS

CHAPTER 11

MICROBIAL ACCLIMATION TO GROUND-WATER CONTAMINATION

A topic that never fails to arouse interest at professional meetings of microbiologists is the presence of microorganisms in what are called "extreme environments." A handy definition of an *extreme environment* is one that is either so cold or so hot or so saline or so oligotrophic or so chemically contaminated that it strains human credulity that life can exist at all. Microorganisms have been found living in boiling water emanating from volcanic terrains, on the snowpack of Arctic and Antarctic wastelands, in brines three or four times more concentrated than seawater, in the bottom waters of deep ocean basins, and in sediments contaminated with the most noxious, poisonous, xenobiotic chemicals imaginable. As a general rule, most microbiologists have ceased expressing amazement at the ability of microbes to adapt themselves to difficult circumstances. Rather, understanding the biochemistry of microbial acclimation to stressed environments has become an active field of investigation.

Understanding the mechanisms of microbial acclimation to various stresses is interesting in its own right, but it also has very practical applications to pollution control technology. Numerous studies of soils, sediments, and aquifer systems have shown that indigenous microorganisms are capable of acclimating to a variety of chemical stresses imposed by human activity. In the process of acclimating themselves to these stresses, microorganisms often accelerate the natural degradation of the chemicals involved. Thus, understanding the acclimation response is basic to the utilization of microorganisms in bioremediation technology.

In this chapter, aspects of microbial acclimation to chemical stresses in ground-water systems are considered. The first step in this discussion is to consider different manifestations of microbial acclimation to chemical stresses. The second step is to discuss the various mechanisms that lead to microbial acclimation. As we shall see, there is a wide variety of biochemical and genetic mechanisms that enable microorganisms to acclimate themselves to various conditions. The final step is to give some examples of microbial acclimation in aquifer systems, and to consider how such acclimation may affect the efficiency of bioremediation efforts.

11.1 MICROBIAL RESPONSE TO ENVIRONMENTAL CHANGES

When microorganisms collected from just about any natural environment are returned to the laboratory and placed in enrichment media, there is a characteristic lag time before they begin to grow and reproduce actively. The generally accepted reason for this phenomenon is that the conditions encountered in the enrichment media are considerably different from those in the environment to which the microorganisms are accustomed. Even if a microbiologist were careful to provide a carbon source, electron acceptor, trace nutrients, and a physical environment similar to that of the natural environment, there would inevitably be some differences. For example, the enrichment media may contain higher concentrations of carbon sources than are present in the natural environment. The temperature, ionic strength, pH, or one of a dozen other environmental factors may be different from what the microorganisms are used to. Presented with these differences, microorganisms typically require an adjustment period to acclimate themselves to the new conditions. This period of acclimation is the reason for the observed lag time before active growth commences.

The extent of the lag time is dependent upon a number of factors. If the conditions offered are reasonably similar to those in the environment, the lag time generally is fairly short. On the other hand, if there are significant differences between the media and the environment, the lag time generally is more extended. If the differences are too extreme, the microorganisms will be unable to acclimate themselves and will become inactive or eventually die.

Once acclimation to laboratory conditions has been achieved, however, microbial growth becomes exponential and continues until nutrients are exhausted or a buildup of toxic by-products limits growth. When acclimated microorganisms are reinoculated onto fresh media, growth continues at an exponential rate with no lag time. Thus, one indication that microorganisms have acclimated themselves to particular conditions is the absence of a lag time.

In many ways, ground-water contamination is analogous to the case of microorganisms acclimating themselves to a particular growth media under laboratory conditions. In both cases, conditions for growth are substantially changed and microorganisms must adapt themselves to the new conditions. Also, many kinds of contamination offer potential substrates for microorganisms to grow on, although these substrates are often much different from those encountered in pristine aquifers. If this analogy held, one would predict that microorganisms subjected to contamination in subsurface environments would behave in at least two ways. First, when the contaminant was introduced, there should be a lag time as microbial populations adjust themselves to the new conditions. Second, one would expect that microorganisms that were acclimated to the contamination would utilize the contaminants more efficiently than microorganisms that are were not acclimated to the contamination.

In fact, this behavior is widely observed in studies of aquatic sediments and contaminated ground-water systems. An example of this observed behavior was given by Madsen et al. (1991) in a study of a shallow water table aquifer contaminated by coal tar derivatives. In sediments cored at the water table (WT) and in the shallow saturated (SS), zones of the pristine part of the aquifer, there was a pronounced lag of ten or twelve days before mineralization of radiolabled p-hydroxybenzoate was apparent (Fig. 11.1). In sediments cored from within the contaminated zone, on the other hand, mineralization of p-hydroxybenzoate occurred with no lag at all (Fig. 11.1).

This behavior, which has been observed in a number of studies, has important implications for bioremediation of contamination in ground-water systems. First of all, it is evident that rates of biodegradation will be a function of the acclimation of indigenous microorganisms to the contaminated conditions. In turn, this implies that biodegradation rates will change over time as acclimation proceeds. Finally, it is evident that acclimation has the potential for substantially increasing biodegradation rates. Thus, technologies for improving the acclimation of indigenous microorganisms to specific compounds are potentially powerful tools for remediating contaminated environments. For these reasons, understanding the metabolic, physiologic, and genetic mechanisms involved in microbial acclimation is especially important.

11.2 MECHANISMS OF ACCLIMATION

Microbial cells are capable of adapting themselves to a wide variety of environmental conditions. For example, bacteria regulate cellular water content by adjusting the properties of their cell walls. Within limits, bacteria can reduce the osmotic flux of water into or out of the cell by adjusting the membrane properties of the cell wall. In similar fashion, bacterial cells can adjust to different pH and temperature conditions.

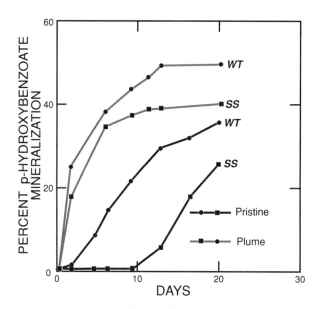

FIGURE 11.1. Evidence for a lag time in the mineralization of PHB by subsurface bacteria in pristine sediments and the lack of a lag time in sediments contaminated by PHBs. (Modified from Madsen et al., 1991, with permission courtesy of the American Association for the Advancement of Science.)

In addition to adapting to various environmental conditions, microorganisms are capable of adjusting their metabolism to new substrates. From the viewpoint of microorganisms present in ground-water systems, contamination events often introduce previously unavailable substrates. It is generally recognized that microbial populations adapt to new substrates by means of three distinct mechanisms: (1) induction of specific enzymes not present (or present at low levels) before exposure, (2) selection of new metabolic capabilities produced by genetic changes, and (3) an increase in the number of organisms able to metabolize newly available substrates. The third type of change generally follows one of the first two.

11.2.1 Induction

The process of *induction* has been discussed previously in the context of bacterial metabolism (Chapter 4). Induction refers to the series of processes by which microbial cells produce enzymes specific to a particular substrate only when that substrate is actually present. One of the best studied examples of induction in bacteria is that of the lactose operon in *E. coli*. Metabolism of lactose requires the presence of three enzymes: beta-galactosidase, galactoside permease, and thiogalactoside transacelylase. Beta-galactosidase cleaves the disaccharide lactose into the monosaccharides galactose and glucose, and galactoside permease is required to transport lactose across the cell membrane. The role of thiogalactoside transacelylase is still unclear.

The induction of the enzymes needed to metabolize lactose involves the activation of specific regions of the bacterial genome (Fig. 11.2). The lactose operon consists of a promoter region (P), a regulatory gene (R) that codes for a repressor protein, and a operator region (O) that occurs between the promoter and the genes coding for the lactose enzymes. When lactose is not present in the cell (Fig. 11.2a), the regulatory gene is expressed and a repressor protein binds to the promoter region of the lactose operon. This effectively blocks transcription of the Lac genes that code for lactose-metabolizing enzymes. However, when lactose is present (Fig. 11.2b), a derivative of lactose, allolactose, acts as an inducer by binding to the repressor protein. This inactivates the repressor and allows the synthesis of the three enzymes needed to metabolize lactose.

Induction of Hydrocarbon-Degrading Enzymes The lactose operon is a particularly clear example of how induction can lead to the acclimation of microorganisms to specific substrates. These concepts are important in the subsurface microbiology of contaminated sites since many contaminant-degrading enzyme systems are inducible. There are numerous examples of this in the literature, but one particularly good example is the induction of aliphatic hydrocarbon-degrading enzyme systems.

The lactose operon is present on the chromosomal DNA of *E. coli*. However, it is commonly observed that enzyme systems for degrading xenobiotic compounds are found in extra-chromosomal DNA (plasmids). Well-known examples of this are the OCT and TOL plasmids of *Pseudomonas* which code for the degradation of n-octane and toluene respectively.

The OCT plasmid of *Pseudomonas putida* has been shown to be inducible by a wide range of straight-chained compounds. Induction promotes the synthesis of

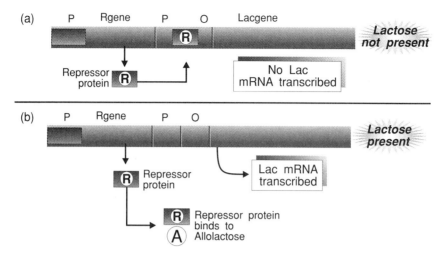

FIGURE 11.2. (a) Repression of LAC genes in the absence of lactose and (b) expression of LAC genes in the presence of lactose.

two enzymes, alkane hydroxylase and alcohol dehydrogenase, which are involved in alkane degradation. In the case of the lactose operon, only one compound (allolactose) served as the inducer. In contrast, alkane hydroxylase and alcohol dehydrogenase are induced by a wide variety of straight-chained compounds. Any straight-chained compound in the C_6–C_{10} range will induce the synthesis of these enzymes. Synthesis occurs even if carbonyl, hydroxyl, or methyl groups are present on the terminal carbon of one side of the chain. Interestingly, these enzymes may be induced by compounds against which the enzymes have no activity. For example, dicyclopropyl ketone and dicyclopropyl methanol will induce alkane hydroxylase activity, even though they are not degraded by the enzyme. These compounds are referred to as *gratuitous inducers*. Conversely, some compounds, such as undecane will *not* induce enzyme production but are degraded if the enzymes are produced gratuitously.

11.2.2 Catabolite Repression

Induction allows microorganisms to synthesize degradative enzymes if a particular substrate becomes available. However, the mere availability of a substrate does not necessarily mean that its utilization is in the microorganism's best interest. For example, if two usable substrates, A and B, are present but more energy is available from substrate B, clearly the microorganism is better off suppressing A-degrading enzymes and producing B-degrading enzymes. In fact, microorganisms have evolved mechanisms for distinguishing between available substrates. One of these mechanisms is termed *catabolite repression*.

In 1942 Jacques Monod discovered the phenomenon called *diauxie,* or the ability of certain microorganisms to utilize one substrate present in growth media preferentially to another substrate. The classic example of this is the sequential utilization of glucose and lactose. When a glucose-lactose growth media is inoculated with *E. coli,* there is a brief lag time as the cells acclimate themselves to the new conditions. After growth commences, growth is supported only by glucose oxidation with lactose not being used at all. After the glucose is exhausted, there is another brief lag as the cells acclimate themselves to lactose utilization, followed by growth at the expense of lactose. The phenomenon of diauxie typically produces a *biphasic growth curve*.

It took many years to elucidate the mechanism by which microorganisms accomplish diauxie. In the case of glucose-lactose utilization, it turns out that glucose utilization greatly decreases the intracellular concentration of cyclic adenosine monophosphate (AMP). When glucose is used up, intracellular concentrations of cyclic AMP increase, allowing the formation of a complex called the cyclic AMP repressor protein (CRP). The CRP complex binds to the promoter region of the lac operon, which cannot be activated in the absence of the CRP complex. Once the CRP complex is in place, allolactose is able to induce production of lactose-degrading enzymes (Fig. 11.2) with subsequent utilization of lactose as a substrate. Because low levels of cyclic AMP result from glucose catabolism, this mechanism has come to be called catabolite repression.

Catabolite Repression of Xenobiotic Oxidation There is some evidence to suggest that catabolite repression of potential organic contaminants may occur under some conditions. As is the case with glucose-lactose, microorganisms can obtain more energy from glucose oxidation than from oxidation of most xenobiotics. It is therefore in their interest to oxidize xenobiotics only in the absence of glucose as a substrate.

Van Eyk and Bartels (1968) showed that n-alkane oxidation in *Pseudomonas aeruginosa,* presumably conferred by the OCT plasmid, could be suppressed by addition of glucose or malate to the growth media. There is also some evidence that the mechanism of the catabolite repression is similar to that of glucose repression of lactose oxidation. Dalhoff and Rehm (1976) showed that the repression could be reversed by adding cyclic AMP plus an inducer substrate to the growth media.

The presence of catabolite repression in microbial populations is of particular importance to possible strategies of bioremediation. Although it is important to stimulate growth of natural populations in order to facilitate degradation, it is equally important not to add nutrients that might repress degradation of organic contaminants.

11.2.3 Genetic Mutations

The dual mechanisms of induction and catabolic repression provide microorganisms with powerful tools for acclimating their metabolism to different conditions. Clearly, however, microorganisms have evolved these pathways because they were advantageous in helping them deal with the environmental conditions they encountered. Since enzymatic mechanisms for degrading particular substrates have developed in the past, it reasonable to expect that microorganisms might evolve new mechanisms for degrading xenobiotic compounds. What genetic changes are potentially involved in the evolution of xenobiotic degradative capacity?

There is an interesting literature on the genetic mechanisms involved in producing new metabolic capacity in microorganisms (Clarke, 1984). For example, Betz and Clarke (1972) showed that point mutations produced changes in substrate specificity of *Pseudomonas aeruginosa* amidase. In this case, these changes enabled the mutant strains to grow on amides that the wild-type strain was unable to use. Furthermore, sequential mutations were shown to convert what had been solely an acetamidase to a phenylacetamidase. These and other experiments have shown that the specificity of an enzyme can be changed by point mutation of the genome, and that these point mutations can sometimes improve the ability of an enzyme to degrade particular substances.

In addition to point mutations, the metabolic capabilities of bacteria can be significantly altered by deletions and additions of DNA via transposable genetic sequences (transposons). For example, in the process of constructing a 4-chlorobenzoate degrading strain, Jeenes et al. (1982) found that by inserting the TOL plasmid into a particular *Pseudomonas* strain, the plasmid lost a sequence of 39 kilobases. This particular strain gained the ability to degrade 4-chlorobenzoate but lost the ability to degrade m-toluate. These authors suggest that transposable genetic elements are an important source of variation in the kinds of substrates that microoganisms can degrade.

These are just a couple of examples of the tremendous capacity for change inherent in the bacterial genome. It is reasonable to assume that point mutations, transposon deletions, and transposon insertions occur as a matter of course in microorganisms. Most of the time, such genetic changes result in a decrease in metabolic efficiency. However, if a microorganism is subjected to novel substrates, there is a finite chance that a fortuitous genetic change will increase its efficiency. Thus, it is entirely possible for microorganisms to develop the ability to degrade even the most xenobiotic compounds. However, the importance of these mechanisms in stimulating biodegradation of xenobiotic compounds in ground-water systems has yet to be conclusively demonstrated.

11.2.4 Acclimation to Available Electron Acceptors

Pristine water table aquifers are often aerobic systems. This is because, under pristine conditions, the input of dissolved oxygen from percolating recharge typically exceeds the input of organic carbon substrates. Thus, microorganisms are more carbon-limited than oxygen-limited. With the advent of organic chemical contamination, however, the situation often rapidly changes. With the sudden influx of potential carbon substrates, carbon limitations disappear and oxygen is rapidly consumed. Where aerobic microbial metabolism might have dominated before contamination, anaerobic metabolism often dominates after contamination. Thus, in addition to acclimating to new carbon substrates, indigenous microorganisms must acclimate to electron acceptors other than oxygen.

One well-documented example of changing electron acceptor availability and its influence on the degradation of organic chemicals is provided by studies of a crude oil spill near Bemidji, Minnesota conducted by the U.S. Geological Survey (Baedecker et al., 1988; Cozzarelli, Eganhouse, and Baedecker, 1988). In August 1979, a pipeline burst and spilled about 100,000 gallons of crude oil onto a glacial-outwash aquifer. This aquifer was under water table conditions and, due to a lack of natural organic matter, contained ground water saturated with dissolved oxygen

(~ 10 mg/L O_2). However, with the introduction of hydrocarbons, the aquifer in the vicinity of the spill rapidly became anaerobic.

Because of the density difference between the crude oil and ground water, the oil formed a lens that floated on the water table. However, soluble components of the oil, such as volatile aromatic hydrocarbons (benzene, toluene, xylene) readily leached into the ground water, forming a plume. Microbial oxidation of these hydrocarbons consumed oxygen near the oil lens creating anaerobic conditions. Downgradient of the oil lens, mixing with ambient ground water gradually reestablished aerobic conditions.

The change in the availability of dissolved oxygen near the oil lens imposed a new set of environmental conditions on ambient microbial populations. Clearly, in the absence of dissolved oxygen, oxygen-based respiration was no longer an advantageous metabolic strategy. However, the glacial outwash sediments were characterized by the presence of ferric oxyhydroxides that occurred as coatings on mineral grain sediments. This Fe(III) represented a pool of potential electron acceptor that was available to microorganisms with the appropriate enzymatic capacity. Measurements of dissolved ferrous iron, the soluble product of microbial Fe(III) reduction, increased over time at the site (Fig. 11.3), indicating the initiation of microbial Fe(III) reduction. This interpretation was confirmed by latter studies (Lovley et al., 1989) of Fe(III) respiration at the site.

It is of historical interest that these studies produced the first documented evidence that a particular anaerobic bacteria (in this case, the Fe(III) reducer GS-15) was capable of completely oxidizing aromatic compounds to carbon dioxide and water (Lovley and Lonergan, 1990). Although such mineralization had been indicated by numerous field and microcosm studies, it had not previously been shown to occur with a pure culture of a strict anaerobic microorganism.

Like oxygen, the availability of Fe(III) oxyhydroxides to support microbial respiration near the oil lens was limited. As Fe(III) became depleted in the sediments, methanogenic bacteria were able to establish themselves. This is indicated by the increase of methane concentrations in ground water near the oil lens (Fig. 11.3).

This case study is important for a number of reasons. First of all, it is one of the few cases in which studies were undertaken soon enough after the spill occurred to document the transition from an oxygen-reducing to an Fe(III)-reducing and methanogenic environment. Similar transitions occur at many spills or leaks of petroleum hydrocarbons. Secondly, it gives some indication of the time scales involved in such transitions. With the initial spill occurring in 1979, evidence of Fe(III)-reducing and methanogenic activity were apparent within a few years. Finally, it shows that even strongly aerobic aquifers harbor the potential for developing active populations of anaerobic bacteria. Studies of pristine aerobic aquifers have shown the presence of a small and relatively inactive population of anaerobic bacteria (Balkwill and Ghiorse, 1985; Jones, Beeman, and Suflita, 1989). It is thought that these microorganisms maintain minimal activity by inhabiting anaerobic microenvironments in the sediments. Clearly, once anaerobic habitats and suitable substrates become more available due to organic chemical contamination, anaerobic microorganisms are capable of rapid growth and of establishing themselves much more widely.

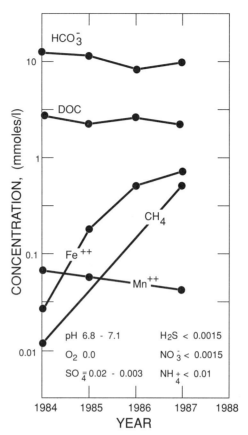

FIGURE 11.3. Increase of dissolved iron and methane over time in ground water near a crude oil spill, indicating the initiation of Fe(III) reduction and methanogenesis in a previously aerobic aquifer. (Modified from Baedecker et al., 1988.)

11.3 FACTORS AFFECTING MICROBIAL ACCLIMATION

The ability of microorganisms to adapt to new conditions imposed by chemical contamination is influenced by a number of factors. Because induction and catabolite repression are important mechanisms in the acclimation response, it can be anticipated that the kinds of organic compounds available and their relative concentrations will be important factors. In addition, because the characteristic lag time associated with microbial acclimation varies widely with the kinds of organic compounds available, the time of exposure to contamination will likely influence the degree of adaptation attained. Finally, the similarity of introduced contaminants to substrates normally utilized by microorganisms will influence microbial acclimation. If they are very similar, then acclimation may entail only the induction of appropriate degradative enzymes. On the other hand, if there is little similarity, than acclimation may involve serendipitous genetic mutations. This, in turn, probably would vastly increase the length of time involved in microbial acclimation. Evaluating the factors involved in microbial acclimation to chemical stress is an important component of designing bioremediation strategies.

11.3.1 Rates of Acclimation

Rates of microbial acclimation to xenobiotic compounds typically are extremely variable. In many cases, it is difficult to determine the precise reasons for the rate at which a particular microbial community adapts to a xenobiotic compound. Nevertheless, it is possible to make some generalizations.

Spain and van Veld (1983) conducted a study of microbial acclimation of aquatic sediment microbial communities to a number of xenobiotic compounds including p-nitrophenol (PNP). In this study, sediment and water samples were taken from a number of sites along the Escambia River near Pensacola, Florida. They prepared three treatments of samples for comparison. First, initial cores were taken from five sites along the river and spiked with 100 nmol/L of [^{14}C]PNP and incubated for eight days. Second, control cores were incubated for the same length of time with no addition. Finally, a set of cores were spiked with 100 nmol of unlabeled PNP and incubated for eight days.

The results of this study are shown in Figure 11.4. Of the five sampling sites, the lag time before significant mineralization of [^{14}C]PNP varied between 40 and 120 hours. After the lag time, PNP was rapidly mineralized (Fig. 11.4a). The control samples, which had been incubated without PNP for eight days prior to [^{14}C]PNP addition, showed virtually identical lag periods at the same sampling sites. This showed that the acclimation response was due to exposure to PNP rather than being an artifact of being handled in the laboratory. Finally, the sediments that had been preexposed to PNP showed no lag before mineralizing PNP (Fig. 11.4b).

The general pattern of acclimation response shown by Spain and van Veld (1983) has been observed by a number of investigators. This pattern includes a wide range of acclimation times to a particular xenobiotic that cannot be predicted a priori. Sediment samples from the different locations on the same river exhibited acclimation times that varied as much as four-fold. This variability is even more pronounced in sediments from ground-water systems (Aelion, Swindoll, and Pfaender, 1987).

For acclimation times on the order of days, such as those shown by Spain and van Veld (1983), enzyme induction is probably the sole acclimation mechanism. If this is the case, it would be expected that acclimation to a particular xenobiotic would persist only as long as that chemical was available for microbial utilization. Once the chemical was no longer available, inducible enzymes would no longer be synthesized and acclimation would disappear. This pattern was in fact observed by Spain and van Veld (1983). Acclimation was defined as the initial rate of 14CO$_2$ production from [^{14}C]PNP. If the initial ^{14}CO$_2$ rate was low, as with unacclimated sediments, the degree of acclimation was considered to be correspondingly low. By this measure, the sediments reached maximal acclimation at about 200 hours (8 days) following exposure, after which there was a steady decrease in acclimation. After 50 days of nonexposure, there was no remaining acclimation.

These results illustrate that microbial acclimation to xenobiotics does not necessarily confer a permanent capacity for degradation. This finding is of more importance to aquatic surface sediments, where pulses of contamination tend to be much more transient than in subsurface environments. Nevertheless, where enzymatic induction is a principal acclimation process, similar behavior will probably be observed.

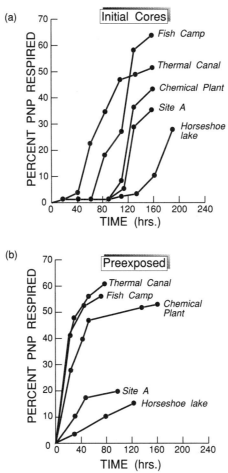

FIGURE 11.4. (a) The presence of a lag time prior to PNP mineralization in pristine sediments and (b) the absence of a lag time in preexposed sediments. (Modified from Spain and van Veld, 1983, with permission courtesy of the American Society for Microbiology.)

11.3.2 Concentration Effects

Concentration of xenobiotic compounds has been shown to have important effects on achievement of acclimation and on the length of the lag time. For example, in a study of sediments from Wintergreen Lake, Michigan, Linkfield, Suflita, and Tiedje (1989) showed that the length of acclimation prior to dehalogenation of 3-chlorobenzoate (3-Cl-BZ) was directly related to the concentrations added to the sediments (Fig. 11.5). At low initial concentrations of 3-Cl-BZ (20 μM), acclimation periods of about 60 days were observed before significant dehalogenation occurred. However, at progressively higher initial concentrations of 3-Cl-BZ acclimation

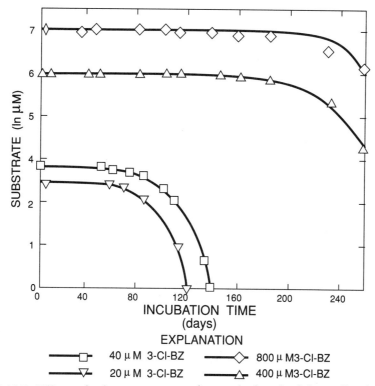

FIGURE 11.5. Effects of substrate concentration on the length of the acclimation period prior to dehalogenation. (Modified from Linkfield, Suflita, and Tiedje, 1989, with permission courtesy of the American Society for Microbiology.)

periods increased, reaching a maximum on 200 days at 800 μM. This pattern apparently reflects the toxicity of chlorinated benzoates. Thus, acclimation times not only reflect the time required for microorganisms to induce appropriate degradative enzyme systems but also may reflect adjustment to unfavorable chemical conditions.

Lower concentrations of xenobiotics are not always favorable to degradation. For example, Linkfield, Suflita, and Tiedje (1989) observed that acclimation to relatively high concentrations (400 μM) of 4-NH2-3-5-diCl-BZ was evident with subsequent mineralization occurring. However, if the initial concentration of 4-NH2-3-5-diCl-BZ were relatively low (20uM), no acclimation response or mineralization was evident. This suggests that a minimum threshold exists below which induction of the necessary enzyme systems does not occur.

The mechanism of this minimum threshold effect was not systematically investigated by Linkfield, Suflita, and Tiedje (1989). However, they suggest that a concentration-dependent receptor mechanism may explain the observed effect. According to this hypothesis, a xenobiotic compound must be present in sufficient concentration before binding to inducer proteins regulating gene expression. Alternatively, this effect may simply reflect competitive utilization (i.e. diauxie) of alternative substrates in the sediment. If the xenobiotic compound is present below a particular threshold, then alternative carbon sources may be more attractive to indigenous microorganisms and preclude induction of degradative enzyme systems.

The minimum threshold effect of acclimation has been observed in sediments cored from ground-water systems. In a study of acclimation of pristine aquifer sediments to xenobiotic compounds, Aelion, Swindoll, and Pfaender (1987) showed that relatively low concentrations (14 ng/g sediment) of p-nitrophenol induced little or no acclimation and mineralization. In contrast, sediments exposed to higher concentrations (529 ng/g sediment) exhibited rapid mineralization of p-nitrophenol following a 40-day acclimation period. It would appear that concentration effects on acclimation of subsurface microbial populations are similar to those observed in surface sediments.

11.3.3 Cross-Acclimation of Xenobiotic Compounds

If induction of enzymatic systems is a principal mechanism in the acclimation of microbial populations to xenobiotic compounds, it might be expected that acclimation to one particular xenobiotic compound would confer acclimation to related compounds. This would be the case if several xenobiotics were degraded via the same enzymatic pathway.

This behavior has, in fact, been reported in the case of reductive dehalogenation of halobenzoates (Horowitz, Suflita, and Tiedje, 1983). In a series of experiments, these investigators showed that acclimation to reductive dehalogenation of one halobenzoate reduced acclimation times to other halobenzoates (Table 11.1). For example, the acclimation time of lake sediments to 3-iodobenzoate was on the order of 2–3 weeks. Once acclimation was achieved, however, these sediments were able to degrade other halobenzoates completely in less than one week. This effect is important to bioremediation technology since it implies that acclimation to one class of xenobiotic may confer enhanced degradation rates to a number of related compounds.

11.3.4 Chemical Structure of Xenobiotics

The chemical properties of organic compounds depend largely on their chemical structures. Because of this, it is reasonable to expect that the acclimation of microbial communities to xenobiotic compounds will be affected by chemical structure.

There is evidence that this is the case for microbial communities in aquatic sediments adapting to reductive dehalogenation of halobenzoates (Linkfield, Suflita, and Tiedje, 1989). In this study, it was shown that acclimation periods depended largely on the kinds of halogens substituting on benzoate and on their position on the molecule. Figure 11.6a shows the chemical structure of 3-chlorobenzoate, 3,5 dichlorobenzoate, and 4-NH2-3-5-diclorobenzoate. As might be expected, the rates of microbial acclimation to these compounds is significantly different. For example, at a substrate concentration of 800 uM ln 800 μM = 6.7 μM in Fig 11.6b, 3,5-diCl-BZ exhibited a consistent acclimation period of 29-35 days (Fig. 11.6b). 4-NH2-3,5-diCl-BZ exhibited a slightly longer acclimation period of about 37 days and 3-Cl-BZ a much longer acclimation period of 125-148 days.

TABLE 11.1 Cross-Acclimation of Anaerobic Sediment Microorganisms to Halobenzoate Decomposition

Substrate Tested for Degradation	Lag time in anacclimated sediment (wk)	Time (wk) for Complete Degradation with Substrate Used to Acclimate Sediment:		
		4-amino-3,5-dichlorobenzoate	3-bromobenzoate	3-iodobenzoate Benzoate
4-amino-3,5-dichlorobenzoate	3–8	2–3	>1[a]	ND[b]
3,5-dichlorobenzoate	2–3	<1	>1[a]	ND
3-iodobenzoate	2–3	2–3	<1	>4[a]
3-chlorobenzoate	32–40	>6[a]	>1[a]	>6[a]
3-bromobenzoate	0.4–4	<1	>1[a]	>2[a]

[a] Lag time before the onset of degradation.
[b] ND, Not determined.
Source: From Horowitz, Suflita, and Tiedje, 1983. Reprinted courtesy of the American Society for Microbiology.

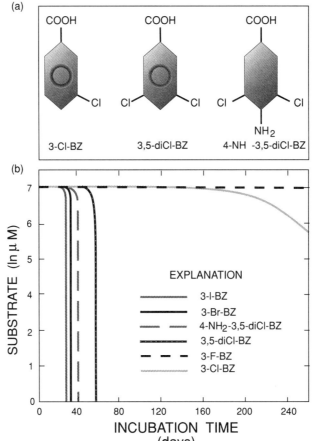

FIGURE 11.6. Relationship of chemical structure to acclimation times. (Modified from Linkfield, Suflita, and Tiedje, 1989, with permission courtesy of the American Society for Microbiology.)

It is interesting to note the differences in acclimation periods imposed by different halogen substitutions on the benzoate molecule. For example, if bromine substituted for chlorine, forming 3-bromobenzoate, the acclimation period would be shorter, on the order of 30 days (Fig. 11.6b). On the other hand, if fluorine substituted into the same position, forming 3-fluorobenzoate, no acclimation at all to the substrate was evident (Fig. 11.6b).

In addition to different halogen substitutions, Linkfield, Suflita, and Tiedje (1989) investigated changes in acclimation period depending on the position of halogen substitution. In the case of bromobenzoates, the effect of ring position was particularly clear. The para isomer exhibited greater acclimation periods than did either the ortho or meta forms. For an initial concentration of 900 μM, the acclimation periods for the ortho, meta, and para isomers were 20, 23, and 39 days respectively. Clearly, differences in chemical structure imposed significant differences in the biochemistry of dehalogenation and corresponding differences in acclimation rates.

11.4 ACCLIMATION TO XENOBIOTICS IN GROUND-WATER SYSTEMS

Much of the research concerning microbial acclimation to xenobiotic chemicals has been focused on aquatic surface sediments or soils. However, there is a small but significant literature concerning microbial acclimation in ground-water systems. Two general approaches have been used in these studies. One approach has been to study sites that exhibit contamination by particular classes of compounds (Wilson et al., 1985). Differences of xenobiotic utilization between contaminated and pristine zones of the same aquifer are then used to document the presence or absence of acclimation effects. Another approach, which is similar to studies of aquatic sediments, is to obtain sediments from pristine subsurface environments, amend them with xenobiotics, and observe the lag period before mineralization commences (Aelion, Swindoll, and Pfaender, 1987; Swindoll, Aelion, and Pfaender, 1988).

11.4.1 Acclimation Response in a Contaminated Aquifer

Wilson et al. (1985) described a case study of microbial acclimation to polyaromatic hydrocarbons (PAH) in a shallow water-table aquifer. The aquifer had been contaminated from creosote wastes over a 23-year period beginning in 1952 and ending in 1975. Creosote is a complex mixture of organic chemicals composed of about 85 % PAH such as naphthalene, anthracene, and phenanthrene. In addition, creoste contains about 15 % phenolic compounds. As the contamination had been in place for a considerable length of time when the study took place, it was reasonable to hypothesize that microorganisms in the plume of contamination would have become acclimated to PAH mineralization.

In order to test this hypothesis, sediments were cored from within the contamination plume and in nearby pristine sediments. Two sites within the contaminated area (sites 5 and 16, Table 11.2) and a site outside the contaminated area (site 14) were examined for differences in mineralizing various PAH compounds.

The results (Table 11.2) showed that sediments from the contaminated site exhibited significantly more mineralization activity than sediments from the pristine site. Even though sediments from one contaminated site (site 5) contained no measurable PAH initially, amending these samples with PAH and oxygen showed rapid consumption relative to autoclaved controls. In contrast, sediments from the pristine site showed little or no consumption of PAH. These results are important because they document the importance of the acclimation process to degradation rates of pollutants in aquifer systems.

While these experiments provided documentation of acclimation in the contaminated sediments, it is apparent that, if in situ degradation rates were as high as those measured in the laboratory, the contamination would have been entirely degraded since operations at the creosote plant closed (approximately ten years). Since contamination was still evident, it was clear that the degradation experiments overestimated actual in situ rates. Wilson et al. (1985) attributed this effect to the addition of oxygen to the experimental vials and suggested that the delivery of oxygen to the contaminated zone was the limiting factor for degradation rates under in situ conditions. The tendency for laboratory incubation experiments to

TABLE 11.2 Biological Activity Against Organic Pollutants Inferred from the Decrease in the Quantity (% Per Week) of Compound Extracted from Subsurface Material from Conroe, Texas, after incubation at 25°C

Sampling site	Naphthalene	2-Methyl-naphthalene	Dibenzofuran	Fluorene	Acenaphthene	1-Methyl-naphthalene
Contaminated site						
5	160[a]	>160[a]	>180[a]	>92[a]	>130	>160[a]
5 (autoclaved)	8.2 ± 6.3[b]	6.5 ± 4.6	4.9 ± 3.6	3.4 ± 4.0	5.0 ± 3.1	5.6 ± 4.6
16	>257[a]	8.0 ± 2.5	No data	No data	32.5 ± 11.8	No data
16 (autoclaved)	7.7 ± 2.7	6.1 ± 2.9	4.9 ± 3.3	11.1 ± 6.9	7.9 ± 3.2	5.9 ± 3.2
Pristine site						
14	3.7 ± 4.0	3.5 ± 2.9	3.9 ± 3.6	6.4 ± 4.7	6.6 ± 2.1	3.6 ± 3.0
14 (autoclaved)	11.3 ± 6.2	11.1 ± 5.7	9.9 ± 6.2	9.4 ± 6.5	9.2 ± 6.9	10.9 ± 5.2

The material was aquired from the sites in June 1983; triplicate samples were analyzed after 0, 1, 2, 4, and 8 weeks of incubation.
[a] Below detection limits (1 ng/g) before first sampling interval.
[b] 95% confidence interval ($s_b t_{0.05}$) on the zero-order rate of decay.

Source: Wilson et al., 1985. Reprinted courtesy of Pergamon Press, LTD.

overestimate actual biodegradation rates has important implications for adapting experimental results to field conditions.

11.4.2 Acclimation Response in Pristine Aquifer Sediments

The study of Wilson et al. (1985) showed that acclimation to xenobiotics occurred in response to chemical contamination of a shallow aquifer system. The next logical question to be addressed was at what rate does such acclimation occur in pristine sediments exposed to various kinds of contamination? These questions were addressed by Aelion, Swindoll, and Pfaender (1987) and Swindoll, Aelion, and Pfaender (1988).

The aquifer sediments studies by these investigators were recovered from the pristine Lula, Oklahoma, site that previously had been described by Wilson et al. (1983) and Balkwill and Ghiorse (1985). As this aquifer is aerobic under ambient conditions, all experiments were conducted under aerobic conditions. A number of compounds were selected for study including chlorobenzene, 1,2,4-trichlorobenzene, m-cresol, phenol, and p-nitrophenol.

Not surprisingly, several different patterns of response were observed (Fig. 11.7). Some compounds, notably m-cresol (Fig. 11.7a), exhibited a linear increase in percent mineralization over time. Other compounds, such as phenol, showed rapid initial mineralization that slowed over time (Fig. 11.7b). The only compound tested that showed a lag time followed by rapid mineralization, the response most often associated with acclimation, was p-nitrophenol (PNP) (Fig. 11.7c). Interestingly, PNP showed a similar classic acclimation response in the studies of Spain and van Veld (1983) using aquatic surface sediments. Other compounds, notably chlorobenzene and 1,2,4-trichlorobenzene showed no mineralization over time.

These results are important because they suggest that the response of microbial communities to a pulse input of xenobiotic compounds will not be uniform. In the case of compounds such as m-cresol and phenol, the indigenous microflora appear to maintain the ability for degradation without induction of specific enzyme systems. In the case of p-nitrophenol, induction of enzyme systems appears necessary before significant degradation can occur. As was shown to be the case in aquatic surface sediments (Spain and van Veld, 1983), this induction response was concentration-dependent. If sediments were treated with low concentrations (14-31 ng/g sediment) of p-nitrophenol, no acclimation response was observed. However, for higher concentrations (452-529 ng/g sediment) the acclimation response was evident. This effect was reproducible between different sample depths.

In real instances of chemical contamination, it is rare that one compound or even one class of compounds is involved; rather, a complex mixture of xenobiotics, as well as nonxenobiotic chemical compounds, typically are delivered simultaneously. The nature of the mixture would be expected to affect acclimation of indigenous microorganisms. In addition, if it were possible to increase rates of the acclimation response by means of chemical addition, this would have important implications for bioremediation technology. These were the topics of investigation reported by Swindoll, Aelion, and Pfaender (1988), which also utilized water table aquifer sediments recovered from the Lula, Oklahoma, site.

Microbial communities in the Lula aquifer sediments exhibited a classic acclima-

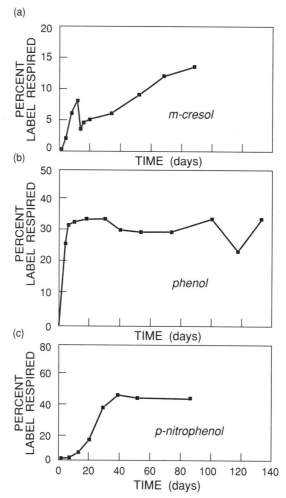

FIGURE 11.7. Different patterns of acclimation response to xenobiotic compounds: (a) linear mineralization of m-cresol, (b) rapid initial mineralization of phenol, and (c) a lag time followed by rapid mineralization of p-nitrophenol. (Modified from Aelion, Swindoll, and Pfaender, 1987, with permission courtesy of the American Society for Microbiology.)

tion response to p-nitrophenol. Could this acclimation response be altered by means of additions of nutrients? In one experiment, sediments amended with p-nitrophenol received either no other amendment, or were amended with nitrogen, or phosphorous, or nitrogen and phosphorous. The results of this experiment showed no increase in the acclimation response for any of the amendments. Clearly, the enzyme induction response for microorganisms in these sediments were not either nitrogen or phosphorous-limited.

Other experiments did show some impact of particular nutrients on the acclimation of microbial communities to p-nitrophenol. For example, addition of vitamins to the sediments did appear to increase acclimation rates whereas addition of glucose did not. The most dramatic increase in acclimation rate was obtained by adding a mixture of nitrogen, phosphorous, mineral salts, and amino acids. In the case of toluene mineralization, nutrient addition actually decreased mineralization.

This may be attributed to preferential utilization (diauxie) of glucose and amino acids by indigenous microorganisms.

11.4.3 Acclimation of Eucaryotic Microorganisms

Eucaryotic microorganisms such as fungi and protozoa are present in low numbers in many ground-water systems (Sinclair and Ghiorse, 1987; Sinclair et al., 1990). Fungi, which are usually aerobic heterotrophs, compete directly with bacteria in many environments. Thus, oligotrophic conditions that limit the population density of bacteria in ground-water systems also limit populations of fungi. The low population density of protozoa in pristine aquifer systems is also related to low populations of bacteria. Because protozoa prey upon bacteria, substrate limitations for bacterial populations translate to substrate limitations for protozoa.

It has been observed that one manifestation of microbial acclimation to polluted ground water is an increase in the number of protozoa (Madsen, Sinclair, and Ghiorse, 1991). This implies that microbial acclimation to xenobiotics is not limited to the bacteria but includes acclimation of higher eucaryotic microorganisms as well.

The pattern of microbial distributions, including bacteria, actinomycetes, fungi, and protozoa, shown by Madsen, Sinclair, and Ghiorse (1991) at a coal tar disposal site documents patterns of eucaryotic acclimation to ground-water contamination. The coal tar leached large amounts of polyaromatic hydrocarbons (PAHs) into the ground water at this site. In response to this chemical stress, counts of viable bacteria were higher than (Fig. 11.8a) were areas outside the zone of contamination. There was, however, no measurable difference in total counts of bacteria (Fig. 11.8b). Interestingly, numbers of actinomycetes were higher in the unsaturated zone (UN) and at the water table (WT) in the contaminated zone but not in the shallow saturated (SS) or deep saturated (SD) zones (Fig. 11.8c). There was no measureable increase in numbers of fungi in the contaminated zone relative to pristine areas (Fig. 11.8d), but numbers of protozoa were significantly higher in the contaminated zone (Fig. 11.8e).

These results suggest that bacterial acclimation and utilization of PAH leads to greater viability, and presumably greater reproduction rates, in the zone of contamination. Fungi, which may be limited by oxygen availability, are not evidently able to increase their population. Protozoa, on the other hand, which feed upon bacterial cells, show a noticeable population increase. This, in turn, suggests that these microorganisms are capable of acclimating to the chemical stress as are the bacteria. The grazing effects of the protozoa are apparently the principal reason that total numbers of bacteria do not increase.

These results are important for two reasons. First of all, they suggest that the ability to acclimate to chemical stress is present in microorganisms other than just the bacteria. The mechanisms by which protozoa acclimate to xenobiotic chemical stress are presently unknown, but are presumably similar to those of the bacteria. Secondly, these results suggest that grazing by protozoa limits bacterial populations in contaminated aquifer systems. This, in turn, may place limits on the ability of bacterial populations to degrade xenobiotics in contaminated environments. The extent or importance of these limits may presently only be guessed at. This is, however, an important topic that warrants future research.

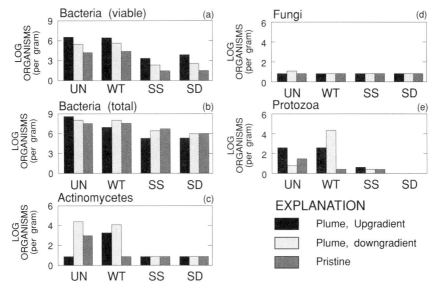

FIGURE 11.8 Comparison of microbial abundance in the unsaturated (UN), water table (WT), shallow saturated (SS), and deep saturated (SD) zones. Each cluster of bars represents numbers of microorganisms in the upgradient (left) and downgradient (center) parts of the contamination plume, as well as in pristine (right) sediments. (Modified from Madsen, Sinclair, and Ghiorse, 1991, with permission courtesy of the American Association for the Advancement of Science.)

11.4.4 Acclimation in Bioremediation Technology

Available evidence clearly shows that microbial populations exposed to xenobiotic chemicals exhibit an acclimation response that, in general, results in increased rates of xenobiotic degradation. This is shown by field studies in which microorganisms in sediments exposed to contamination show greater biodegradation than unexposed sediments (Wilson et al., 1985) and by laboratory studies (Aelion, Swindoll, and Pfaender, 1987; Swindoll, Aelion, and Pfaender, 1988) that show degradation rates increase over time.

The extent to which the acclimation response can be utilized in bioremediation technology is much less clear. Numerous practical considerations intervene. Available evidence shows that significant microbial acclimation occurs in time periods of less than a year. Considering that many instances of ground-water contamination are discovered tens of years after the fact, implementing measures to increase acclimation rates does not seem particularly advantageous. Furthermore, other factors, such as electron-acceptor availability, often are much more limiting to degradation rates than the extent of microbial acclimation (Wilson et al., 1985). Thus, while the acclimation response certainly affects degradation of xenobiotics in ground-water environments, its utilization in bioremediation technology has not been clearly demonstrated.

11.5 ACCLIMATION TO METAL TOXICITY

Metals play important roles in the activity of many enzymatic systems in procaryotic and eucaryotic cells. Obvious examples include the utilization of iron in electron transport mechanisms, the presence of molybdenum in nitrate reductase, or the role of zinc in carbonic anhydrase. However, when one particular metal interferes with enzyme function by competing with or replacing another particular metal, toxicity results. Metal toxicity in this context simply refers to the slowing or complete cessation of cellular growth as a direct result of the presence or absence of a particular metal in the cell's immediate environment.

It is reasonably common for microorganisms, particularly those inhabiting soils or aquatic sediments, to encounter potentially toxic metals. For example, microorganisms inhabiting soils developed on metal ore bodies would be exposed to fairly high metal concentrations. Microorganisms inhabiting aquatic sediments or aquifer systems developed from erosion of ore bodies would suffer similar metal exposure with potentially toxic effects. As a result of such exposure, some microorganisms have developed elaborate biochemical mechanisms for dealing with metal toxicity. Because many of these mechanisms involve induction of enzymatic systems, microbial communities having such biochemical capabilities often exhibit an acclimation response to metal exposure.

The acclimation of microorganisms to metal toxicity and mechanisms of metal resistance have been widely studied (Trevors, Oddie, and Belliveau, 1985). Much of this work has been motivated by the observation that microbial transformations of metals greatly affect metal mobility in the environment. For this reason, microbial acclimation to metal toxicity is relevant to the microbiology of ground-water systems as well.

11.5.1 Metal Detoxification Mechanisms

Microorganisms employ several different biochemical mechanisms in detoxifying metals in their immediate surroundings. These include (1) binding metals to the cell surface in order to prevent transport into the cell, (2) biotransformation of the metal to a less toxic form, and (3) depositing the metal in an insoluble form. Some of these mechanisms, such as metal deposition as sulfides, are gratuitous in that they result from other metabolic functions, such as sulfide production. Other mechanisms, such as biotransformations, are highly specific and are determined at the genetic level. The ability to transform metals to less toxic or more volatile forms often are coded on plasmids or transposons.

Metal Binding The outer layers of bacterial cells are characterized by numerous anionic sites that are available to bind positively charged metal ions of metabolic importance, such as Mg^{2+}, Fe^{3+}, Cu^{2+}, Na^+, and K^+. This binding, which often is the first step in active transport into the cell, results largely from interactions with anionic sites on the cell wall such as phophodiesters, carboxyl groups of peptidoglycan, and hydroxyl groups of carbohydrates. Peptidoglycan is especially efficient at metal binding.

Such heavy metals such as Hg^{2+} and Pb^{2+} are also gratuitously bound by anionic sites on a cell wall. However, the cell wall typically shows less affinity for the heavy metals than for lighter metals such as Mg^{2+}. Once bound to the cell's surface, a potentially toxic metal is immobilized and is effectively prevented from entering the cell where it might interfere with metabolic functions. Thus, metal binding confers some level of metal resistance to all microorganisms.

Biotransformations An important class of metal resistance mechanisms available to microorganisms is that of biotransformation. These biotransformations include oxidation-reduction reactions and organic-inorganic conversions. In both cases, the mobility or volatility of the metal, or both, may be greatly affected. For example, reduction of Hg^{2+} to Hg^{o} greatly increases the volatility of mercury and may contribute to its transport away from the microorganism's immediate environment. Similarly, the attachment of alkyl groups to metals may greatly change its solubility and volatility. For example, some methylated metals are highly volatile and virtually insoluble in aqueous solution. Thus, they may more easily escape from an aquatic environment. There is some evidence that alkylation of metals serves as a primary metal-resistance mechanisms.

Conversely, alkyl substitutions greatly increase the affinity of the metal complex for lipids. This increased lipophilicity is an important factor in metal toxicity in microorganisms and in higher organisms. It has been shown that some microorganisms have the inducible ability to dealkylate metals . In the case of methyl mercury compounds, the first step in this process is demethylation by means of organomercurial lyases followed by the reduction of Hg^{2+} to Hg^{o}. The genetic basis of mercuric reduction is one of the best-documented of all metal-resistance mechanisms and will be discussed in detail in the following section.

Metal Deposition The ability to precipitate metals in an insoluble form is an attractive mechanism for metal resistance. Most metal sulfide compounds are highly insoluble, and deposition of a metal in this form effectively prevents its transport into the cell. In some cases, as in *Desulfovibrio,* sulfide production and metal immobilization is gratuitous and is simply a by-product of the microorganism's normal metabolism. In other cases, as in the case of some *Clostridia,* sulfide production is inducible and is used as a metal resistance mechanism.

The ability of some microorganisms to precipitate metals has applications in biotechnology. Metal-precipitating microorganisms have been used to remove metals from industrial waste solutions. Conversely, metal-solublizing microbial mechanisms have been used to leach and recover metals from ores.

11.5.2 Plasmid-Encoded Metal Resistance Mechanisms

Metal resistance mechanisms present in procaryotic microorganisms are often plasmid-encoded. This is to say that the genetic material coding for the relevant enzyme systems are present on extra-chromosomal DNA. However, as many of these genetic sequences are transposable, they may also be found in chromosomal DNA.

Plasmid-encoded resistance mechanisms have been documented for a wide range of metals including cobalt and zinc (Duxbury and Bicknell, 1983), cadmium (El Solh and Ehrlich, 1982), silver (Haefeli, Franklin, and Hardy, 1984), and mercury (Silver and Misra, 1984). Of these, the mechanism for mercury resistance in *Staphylococcus aureus*, an opportunistic human pathogen, has been particularly well documented. As such, it serves as a convenient model for illustrating the genetic basis for metal resistance in other microorganisms as well.

A number of different plasmids and transposons, each exhibiting slightly different details, have been shown to confer mercury resistance in *Staphylococcus aureus*. These functions are located on plasmid DNA and are referred to collectively as the "mer operon". The basic components of the mer operon and their functions in conferring mercury resistance are shown in Figure 11.9. The regulatory gene mer R is inducible and codes for a protein that, when bound to the promoter-operator (OP) region, initiates synthesis of the mercury resistance enzyme system. This enzyme system is composed of several different functions. The mer T and mer P regions code for a system that traps extracellular Hg^{2+} and transports it through the cell wall and cell membrane to the cytoplasm. The mer A region codes for mercuric reductase, which is present in the cytoplasm. Hg^{2+} delivered to the cytoplasm by the transport mechanism is then reduced to the volatile and less toxic Hg^o form. The mer B region codes for an organomercurial lyase system which serves to split Hg^{2+} from organic complexes so that it can be subsequently reduced by mercuric reductase.

Depending on the components of the mer operon actually present in plasmid DNA, a wide variety of responses to mercury are possible. For example, if the mer A region, which codes for mercuric reductase, is removed from the operon, the resulting cells have the ability to trap Hg^{2+} but not to detoxify it. Such cells are actually hypersensitive to mercury stress. Cells that have broad spectrum resistance to mercury contain both the mer A and mer B genes. Alternatively, cells containing mer A but not mer B exhibit resistance to Hg^{2+}, but less resistance to organomercuric compounds such as methylmercury.

While the mer operon is certainly the most studied of the mercury-resistance mechanisms, others have been documented. For example, *Clostridium cochlearium* has an entirely different plasmid-encoded mercury resistance mechanism. This mechanism does not include mercuric reduction but, rather, the precipitation of mercuric sulfide. Organomercuric compounds are split, presumably by means of a lyase, and sulfide is generated. Both of these functions appear to be plasmid mediated in the case of *C. cochlearium*. This serves to emphasize the wide range

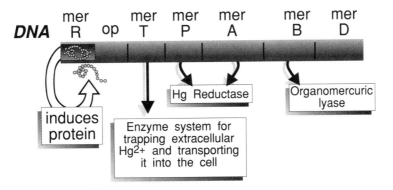

FIGURE 11.9. Basic components of the mer operon.

of resistance mechanisms available to microorganisms and illustrates their flexibility in dealing with metal stress.

11.5.3 Acclimation to Mercury Toxicity

The fact that the mer operon, and other plasmid-mediated metal resistance mechanisms, are inducible enzymatic systems suggests that an acclimation period would be expected in response to mercuric stress. Furthermore, this suggests that microbial populations acclimated Hg^{2+} should be much more efficient at Hg^{2+} reduction than non-acclimated populations.

It has been widely observed that aquatic sediments exposed to mercuric stress exhibit a higher incidence of mercury-resistant bacteria than non-exposed sediments. In one such study, Barkay and Olson (1986) showed that lake sediments exposed to high concentrations of mercury up to 40 $\mu g/g$, showed higher counts of mercury resistant bacteria and a higher incidence of the mer operon than did sediments that had not been exposed. Subsequent research has demonstrated the inducible nature of mercury resistance in aquatic-sediment microbial communities. For example, Barkay (1987) showed that loss of Hg^{2+} from water samples of estuarine showed a distinct lag time. However, samples preexposed to Hg^{2+} showed no such lag time. Furthermore, the half-life of Hg^{2+} in preexposed samples was between two and four-fold shorter than in unexposed samples.

Most of the scientific literature on microbial acclimation to metal stress resides in the aquatic sediment, microbial genetics, or industrial microbiology literature. From that literature, it can be reasonably surmised that subsurface microorganisms behave similarly. However, the scientific literature on subsurface microorganisms and metal stress is virtually nonexistent. Because of the important transformations that can be carried out by means of enzymatic processes, the potential importance of microbial processes and metal mobility in aquifer systems should not be overlooked.

11.6 SUMMARY

In this chapter, we have seen that microorganisms employ a wide variety of mechanisms for acclimating themselves to changing conditions. In many cases, acclimation consists of inducing appropriate enzyme systems or, alternatively, repressing systems that are inappropriate. These changes are effected at the genetic level but do not involve fundamental changes of the microorganism's genome. It is also possible that genetic changes, initiated by point mutations or wholesale deletions of genetic material, may result in advantageous metabolic capabilities. However, these changes are unpredictable and have an uncertain impact on xenobiotic degradation in natural systems.

The hallmark of the acclimation response is the presence of a characteristic lag time in the microbial growth curve that follows exposure to new substrates. These acclimation periods have been observed in microbial communities from numerous environments responding to a broad spectrum of contaminants, including xenobiotic organic compounds and metals. Microbial communities in shallow ground-water systems appear to acclimate to a variety of chemical stresses in time periods as short as several months. The acclimation time for particular compounds can be reduced somewhat by judicious addition of trace elements or vitamins. It remains an open question, however, as to whether that phenomenon can be used practically in bioremediation strategies.

Microorganisms have been shown to possess a number of different mechanisms for acclimating to metal exposure including metal binding to the cell wall, biotransformation of metal species, and metal precipitation reactions. Metal-resistance mechanisms are often plasmid encoded and are often found to be inducible. It is likely that microbial acclimation to metal stress has important impacts on the potential mobility of metals in ground-water systems.

CHAPTER 12

BIODEGRADATION OF PETROLEUM HYDROCARBONS IN SUBSURFACE ENVIRONMENTS

The uncontrolled release of petroleum hydrocarbons to ground-water systems is a serious and very widespread environmental problem in the developed nations of the world. Fuel tanks, either the aboveground or underground variety, are prone to corrosion, which often causes leaks. Human error, such as not opening or closing the right valves at the right times, frequently results in spills. Finally, transporting petroleum products in ships, trucks, and pipelines regularly leads to accidents of chance and human judgment. The inevitable result of handling the huge volumes of these materials consumed by industrialized societies—about 800 billion gallons per year—is the frequent contamination of shallow ground-water systems by petroleum hydrocarbons.

Once petroleum hydrocarbons are delivered to ground-water systems, they are acted on by a combination of physical, chemical, and biologic processes that tend to disperse the contamination. Ground-water flow moves water-soluble components away from the spill site, and the configuration of the water table influences the movement of the nonsoluble components. Chemical processes, such as sorption and volatilization, alternately immobilize and mobilize particular components. Finally, biologic processes immediately begin degrading some components to carbon dioxide. These degradation processes are varied and depend on the chemistry and hydrology of a particular ground-water system. Because they act to transform potential contaminants from relatively toxic forms (cyclic hydrocarbons) to relatively nontoxic forms (carbon dioxide, methane) of carbon, these biological processes are extremely important in the eventual fate of petroleum hydrocarbons in ground-water systems. This chapter describes particular aspects of microbial degradation of petroleum hydrocarbons in ground-water systems.

12.1 COMPOSITION OF CRUDE OIL

The term "crude oil" is strictly one of convenience and refers only to what is actually produced from an oil well. It carries no meaning at all with regard to its actual composition. In fact, crude oils produced from different oil fields differ radically in their organic chemistry. Pennsylvania crude oil, for example, consists almost exclusively (90-95%) of n-alkanes. Crude oils of California and western Mexico, on the other hand consist of less than 50% n-alkanes. The relatively uniform composition of gasoline, kerosene, and jet fuels used worldwide owes much to refinery technology and little to a uniform composition of "crude" oil.

Crude oil consists of n-alkanes-often termed "paraffins" (because of their low aqueous solubility) in the petroleum literature-branched alkanes, cycloalkanes, and aromatic hydrocarbons (Fig. 12.1). The number of carbon atoms in organic molecules of crude oil range in length from 1 carbon (methane) to over 400 carbons. In spite of the enormous number of possible isomers generated by this variety, crude oils actually contain surprisingly few compounds. Other than the n-alkanes and branched alkanes, most of the cycloalkanes consist of alkyl-cyclopentanes and alkyl-cyclohexanes. The aromatic fraction of most crude oils consists of alkylbenzenes and alkyl napthalenes. In addition to hydrocarbons, crude oils may contain small amounts of phenols, alcohols, ketones, and fatty acids. The amount of sulfur-bearing asphaltic components, such as the mercaptans, varies widely between crude oils. Pennsylvania-grade crude oils contain less than 0.1% sulfur whereas some California crude oils contain as much as 5% sulfur. These sulfur-bearing compounds are used in preparing asphaltic materials for surfacing roads.

As one might expect, crude oils are classified on the basis of the relative amounts of paraffins, aromatics, and asphaltic components present in them. Thus, Pennsylvania-grade crude is referred to as "paraffinic" whereas the high-sulfur crude oil of California is referred to as "asphaltic." Various descriptive combinations such as "paraffinic-aromatic" are also commonly used.

12.2. PETROLEUM REFINING AND FUEL BLENDING

Crude oil can be separated into carbon compounds of particular length by means of fractional distillation. By progressively heating the crude oil from room temperature up to 400 °C, carbon compounds of progressively greater boiling points are volatilized. These various fractions can then be condensed for utilization as gasoline, heating oils, or lubricating oils.

Gasoline, for example, consists of hydrocarbons which boil from about 30 °C to 200 °C. Heavy lubricating oils, on the other hand, have very high boiling points of more than 400 °C. The amount of gasoline, or any other component, available by means of fractional distillation is limited and depends on the composition of the crude oil. Many crude oils, for example, contain less than 20 % of the gasoline fraction.

Because gasoline is one of the most desirable products of crude oils, elaborate chemical processes have been developed for increasing the amount of gasoline that can be recovered from crude oil. Hydrocarbons are very stable molecules and

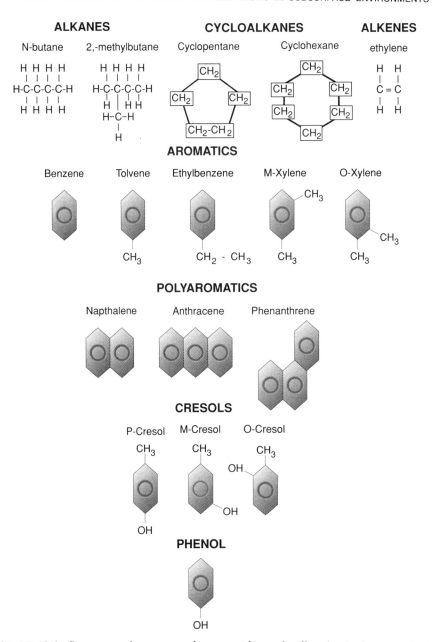

FIGURE 12.1. Some organic compounds present in crude oil and petroleum products.

can be distilled without decomposition. However, if they are heated well beyond their boiling points in the absence of oxygen, the molecules can be caused to disintegrate or "crack" into smaller molecules. Often these hydrocarbon-cracking reactions are assisted by the addition of catalysts which enable the reactions to occur at lower temperatures. These procedures greatly increase the amount of gasoline and other fuels that can be recovered from crude oil.

A consideration that is just as important as the quantity of fuel that can be recovered from a crude oil is the quality of that fuel. It was discovered very early on that some organic compounds make much better components of fuels than other compounds. In order to compare various mixtures of hydrocarbons in terms of their quality as gasoline, an arbitrary "knock-rating" scale has been devised. Gasoline engines burning a poor-quality fuel will exhibit a characteristic "knocking" noise that is absent with good-quality fuels.

In order to construct the knock-rating scale, a pure hydrocarbon with excellent knock properties, isooctane (2,2,4 trimethylpentane), was assigned an *octane number* of 100. Furthermore, a pure hydrocarbon with very poor knock properties, normal heptane, was assigned an octane number of 0. The octane number of a particular gasoline, therefore, refers to its knock properties compared to a mixture of isooctane and n-heptane.

The burning properties of fuels can be improved by increasing the percentage of efficiently burning hydrocarbons that are present. A number of techniques are employed to this end. It is generally observed, for example, that branch-chained hydrocarbons burn more efficiently than straight-chained hydrocarbons. *Isomerization* reactions can be used to accomplish this process. In the presence of aluminum chloride and at a temperature of 100 °C, n-pentane isomerizes to isopentane. This process converts a poor-burning hydrocarbon to a much better one. Another reaction that is used to improve the burning properties of fuels is *aromatization*. When heated over special catalysts, alkanes can be cyclized and dehydrogenated to aromatic hydrocarbons. For example, heptane reacts with a platinum catalyst and heat to form toluene. Toluene, in turn, has much better burning characteristics than heptane.

Gasoline, and other specialized fuels, are a complex mixture of natural and manufactured hydrocarbons derived from crude oil. The composition of these fuels varies with the use for which they were designed. Diesel and jet fuels, for example, can use large percentages of straight-chained hydrocarbons. High-octane gasolines, however, require a higher percentage of branched alkanes and aromatic hydrocarbons. These differences in fuel composition not only affect their properties as fuels but they also affect the manner in which microorganisms can use them as carbon substrates. Thus, it is important to be aware of the different compositions of different fuels when one is addressing ground-water contamination by petroleum hydrocarbons.

12.3 MOVEMENT AND SEPARATION OF PETROLEUM HYDROCARBONS IN GROUND-WATER SYSTEMS

Crude oil and refined hydrocarbon products have much different physical and chemical properties than does water. The mean density of gasoline, for example, is 0.9 gm/cm^3, as opposed to 1.0 for fresh water. In addition, water is a highly polar molecule, whereas most components of refined hydrocarbon products are nonpolar molecules. Both of these differences serve to separate water from petroleum hydrocarbons in ground-water systems. The density difference insures that undissolved hydrocarbons cannot penetrate significantly below the water table, and the different polarity greatly reduces the solubility of hydrocarbons in water.

These physical and chemical properties largely determine the movement and separation of petroleum hydrocarbons in subsurface environments.

12.3.1 Density-Driven Migration of Hydrocarbons

An uncontaminated shallow ground-water system may be viewed as being composed of two immiscible fluids (air and water) that are separated by an interface. The water is present in the saturated zone, the air is present in the overlying unsaturated zone, and the interface between the two fluids is, of course, the water table. When a hydrocarbon is spilled at land surface, the net effect is the introduction of a third immiscible fluid to the system.

Once a spill has occurred at land surface, the movement of the hydrocarbon is determined by the sum of the forces acting on it. Gravity tends to pull it downward, whereas the air in the unsaturated zone and capillary forces tend to buoy it up. In general, if the pressure at the leading edge of the hydrocarbon is greater than air pressure, the hydrocarbon phase will move downward, displacing air as it goes. If the pressure of the hydrocarbon drops below the displacement pressure of the air, as often happens with small spills, movement of the hydrocarbon will cease and the hydrocarbon will be immobilized in the unsaturated zone. If the pressure is sufficient, the hydrocarbon fluid will continue displacing air and moving downward until it encounters the water table.

Once it encounters the water table, the pressure relationships moving the hydrocarbon fluid change. In order to continue moving downward, there must be sufficient pressure to displace water. On the other hand, hydrocarbon that is still above the water table can continue to displace air laterally. The net effect of this is that a lens of hydrocarbon develops on the surface of the water table, and spreads laterally over time (Fig. 12.2). Lateral spreading continues until the hydrocarbon is at or just below the displacement pressure of hydrocarbon into air and water at all points.

The illustration shown in Figure 12.2 represents a lens development in an aquifer that is isotropic, homogeneous, and lacks a hydrologic gradient. Such idealized conditions do not, for all intents and purposes, exist in nature. In real systems, aquifer heterogeneities impose considerable complexity on the distribution of hydrocarbons.

From the point of view of microbial habitats, Figure 12.2 does illustrate the different environments created by a hydrocarbon spill. Perhaps the most stressed environment from a microorganism's point of view is in the hydrocarbon-saturated sediment. Water is not abundant, which slows delivery of dissolved nutrients and electron acceptors needed for microbial growth. There is the additional stress of the toxicity of various hydrocarbon components. Microbial metabolism and growth within the free-product lens of most hydrocarbon spills is limited.

Above the lens, the air tends to be at or near saturation with the more volatile components present in the hydrocarbon. In Figure 12.2, this is referred to as the zone where the air is at "residual saturation" with respect to the hydrocarbon. This is an environment that is much more conducive to microbial growth. The volatile components of hydrocarbons, short-chained alkanes, cyclic alkanes, and aromatic components are readily oxidized by many kinds of bacteria. The presence of air allows oxygen to diffuse into the system with the consequent potential

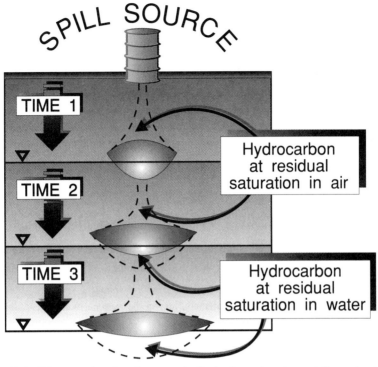

FIGURE 12.2. Diagram showing the spread of a hydrocarbon lens at the water table and the development of hydrocarbon-saturated air and water.

for aerobic metabolism. Because this is occurring in the unsaturated zone, water may be a limiting factor under some conditions. In general, however, the unsaturated zone above a hydrocarbon lens exhibits considerable microbial activity.

The water below the lens is the third general type of habitat created by a hydrocarbon spill. This zone is referred to in Figure 12.2 as being where the hydrocarbon is "at residual saturation in water." In other words, the more soluble components of the hydrocarbon dissolve and reach aqueous saturation near the lens. In fact, actual saturation probably occurs only very near the lens, with a hydrocarbon gradient extending away from it. This also is a relatively favorable habitat for microbial growth. Water is abundant, electron acceptors may be present, and the toxic effects of the hydrocarbons are diminished by the concentration gradient away from the lens.

Each of these three environments is exploited to some degree by microorganisms. Certainly a period of acclimation must follow the influx of hydrocarbon (Chapter 11); however, this period of acclimation appears to be fairly brief. It is likely that patterns of microbial degradation of petroleum hydrocarbons differ in each of these particular habitats.

12.3.2 Solubility and Hydrocarbon Separation in Ground-Water Systems

Many hydrocarbon compounds that are present in fuels have very limited solubility in water. For example, analysis of fuels using gas chromatography shows that

TABLE 12.1 Concentration of Hydrocarbons in Water After Contact with Distillate Products

		Detection Limit	Kerosene	Diesel	Fuel Oil
Benzene	$\mu g/l$	0.5	294	344.0	203.0
Ethylbenzene	$\mu g/l$	0.5	19	139.0	100.0
Toluene	$\mu g/l$	0.5	870	777.0	509.0
Xylenes	$\mu g/l$	0.5	1260	875.0	592.0
M-Tert-But. Ether	$\mu g/l$	0.5	BDL[a]	BDL	BDL
Ethylene-Dibromide	$\mu g/l$	0.005	BDL	BDL	BDL
Tot. Pet. Hydrocarbons	$\mu g/l$	0.2	40	16.8	21.1
Napthalene	$\mu g/l$	0.5	356	6.6	40.0
1-methyl napthalene	$\mu g/l$	0.5	193	66.2	107.0
2-methyl napthalene	$\mu g/l$	0.5	225	108.0	152.0
Total BTEX[b]	$\mu g/l$		2440	2140.0	1400.0
Total Napthalene[c]	$\mu g/l$		774	181.0	299.0

[a] BDL - below detection limit
[b] Summation of benzene, toluene, ethylbenzene, and xylenes
[c] Summation of napthalene, 1-methyl napthalene, and 2-methylnapthalene
Source: Dunlap and Beckman, 1988. Reprinted courtesy of the National Water Well Association.

the relatively water soluble light aromatics-benzene, toluene, ethylbenzene, and xylene (BTEX)-comprise only 2% or 3% of the fuel as a whole. Most of the fuel is composed of C_9 and larger hydrocarbons with fairly limited solubility.

The difference in solubility of the BTEX portion of most fuels compared to the C_9 and larger hydrocarbons was emphasized by Dunlap and Beckmann (1988) using a simple experiment. First, they filled containers with 4 liters of deionized water. Next, they placed 400 mls of various fuels on top of the water, sealed the containers, and let them stand for 22 days. At the end of this time, samples of water were withdrawn from the bottom of the containers and analyzed for BTEX, total petroleum hydrocarbons, and several napthalene isomers.

The results of these analyses serve to illustrate the disproportionate effect that these relatively soluble compounds have on the water (Table 12.1). While aliphatic hydrocarbons (alkanes, etc.) were the major component of each fuel, they were just a tiny fraction of what actually dissolved in the water. Conversely, while the BTEX compounds comprised less than 3% of the fuel, they accounted for most of the dissolved fraction in the water.

This general behavior is widely observed in fuel spills that affect ground-water systems. Most of the effect in terms of dissolved components comes from the BTEX fraction. Conversely, the heavier components do not dissolve readily and become trapped at or near the water table. These differences in relative mobility have important effects on their relative susceptibility to biodegradation.

12.4 MICROBIAL DEGRADATION OF ALIPHATIC HYDROCARBONS

Aliphatic hydrocarbons, the straight-chained alkanes, alkenes, and their branched derivatives are important substrates for a variety of microorganisms. The biochemical mechanisms involved in degradation-those that have been studied-are diverse and depend on the chemical properties of the specific hydrocarbon. Methane oxidation by aerobic bacteria, a special case of alkane oxidation, involves a specialized enzyme system that includes *methane monooxidase*. Longer-chained alkanes and alkenes are degraded by converting these compounds to long-chained fatty acids and subsequent degradation via beta-oxidation. In addition to these well-characterized aerobic processes, there are undoubtedly numerous anaerobic pathways for degradation that have yet to be elucidated.

Beginning with the early work of Zobell (1950), it has been possible to enumerate several generalizations concerning aliphatic hydrocarbon oxidation. These include:

1. Long-chained n-alkanes, those greater than C_{10}, are assimilated into cells and oxidized more readily than short n-alkanes. An important exception to this rule is methane oxidation, which employs a different biochemical pathway for oxidation.
2. Saturated aliphatic hydrocarbons are degraded more readily than unsaturated ones.
3. Branch-chained hydrocarbons are degraded less readily than are straight-chained ones.

The chief reason for the first observation appears to be related to compound toxicity. Aliphatic hydrocarbons in the C_2–C_{10} range are more soluble in water than the longer paraffinic compounds. This solubility seems to confer greater toxicity, probably because the cell has greater difficulty regulating hydrocarbon transport through the cell membrane. The longer aliphatics are too insoluble for such transport, and cellular uptake can be more readily controlled. The mechanisms of aliphatic uptake are complex and are still a matter of controversy and active research. However, it is often observed in nature that microorganisms are found attached to hydrocarbon droplets. The mechanism of cellular attachment to hydrocarbons seems to involve lipopolysaccharides present on the cell wall. In addition, microorganisms acclimated to growth on aliphatics often produce emulsifying compounds that may be involved with regulation of cellular hydrocarbon uptake.

The reasons for the second and third observations are related directly to the biochemistry of aliphatic degradation, which will be considered in some detail in the following sections. In general, however, normal saturated hydrocarbons are more readily converted to long-chained fatty acids for subsequent beta-oxidation than unsaturated or branch-chained hydrocarbons. This, in turn, is the source of considerable recalcitrance to degradation exhibited by some unsaturated and branch-chained aliphatic hydrocarbons.

12.4.1 Methane Oxidation

Methane-oxidizing microorganisms, including bacteria, fungi, and yeasts are exceedingly common in surface waters, aquatic sediments, soils, and subsurface environments. Among the bacteria, methane-oxidizing bacteria have been isolated and described and classified in various genera including *Methylobacter*, *Methylocystis*, and *Methylobacterium*. Because methane-oxidizing autotrophs are so common, considerable effort has been invested in understanding the biochemical mechanisms involved.

Methane oxidation under aerobic conditions proceeds via a sequence of steps that involves methanol, formaldehyde, and formate as intermediate products (Fig. 12.3). The final products of methane oxidation are, of course, carbon dioxide and water. The first step in methane oxidation involves the partial oxidation of methane to methanol with NADH acting as the electron carrier. This reaction is catalyzed by the complex enzyme methane monooxygenase. Methane monooxygenase has a very broad substrate specificity and can partially oxidize numerous carbon compounds including some higher alkanes and alkenes. In addition, methane monooxygenase has been shown to be involved in serendipitous oxidation of chlorinated hydrocarbons, a reaction of considerable importance that will be discussed in Chapter 13.

The next steps in this process are the conversion of methanol to formaldehyde by methanol dehydrogenase and the conversion of formaldehyde to formate by means of formaldehyde dehydrogenase. The last step is the oxidation of formate by means of formate dehydrogenase, this time with NAD^+ acting as the electron carrier. From a biochemical point of view, this and other C_1 oxidations are interesting because of the central position of formaldehyde, a relatively uncommon molecule in the environment. From the viewpoint of hydrocarbon degradation, this pathway is very important in the aerobic oxidation of methane and other C_1 compounds of environmental interest.

The anaerobic oxidation of methane is a subject that has generated much speculation in the literature but little concrete information. For years, marine sediment geochemists have suggested that the oxidation of methane may be coupled to sulfate reduction. Similarly, it has been suggested that methane can be oxidized by Fe(III)-reducing bacteria. However, no microorganism or consortia of microorganisms have been recovered that will perform these processes under laboratory conditions. Considering the strong circumstantial evidence that anaerobic oxidation of methane does occur in natural systems, it is possible that longer-chained aliphatic compounds may also be oxidized anaerobically. This interesting question will doubtlessly continue to receive attention.

12.4.2 Oxidation of n-alkanes

Given the complexity of the biochemical mechanisms involved in oxidizing relatively stable organic compounds, it's not particularly surprising that oxidative

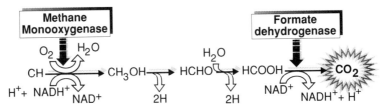

FIGURE 12.3. Aerobic oxidation of methane in which the initial step is catalyzed by methane monooxygenase.

mechanisms for many petroleum hydrocarbons make use of mechanisms for oxidizing more "normal" substrates such as organic acids. This is the case for the aerobic oxidation of n-alkanes. The strategy for oxidation is to convert the n-alkane to a type of organic molecule that can feed into an already established oxidation pathway. In the case of n-alkane oxidation, n-alkanes are converted to long-chain fatty acids that can then be fed to the beta-oxidation pathway.

Beta-oxidation A large number of respirative bacteria are capable of utilizing long-chained fatty acids as substrates for energy production and growth. These include members of the genera *Pseudomonas*, *Acinetobacter*, and *Bacillus*. The primary pathway for long-chain fatty-acid degradation is via beta-oxidation.

The biochemistry of beta-oxidation is well-understood and is summarized in Figure 12.4. The first step in this process is the conversion of the organic acid to a coenzyme A (CoA) ester, a reaction catalyzed by acyl-CoA synthetase and requiring energy in the form of ATP. This enzyme is not specific to particular fatty acids and exhibits activity with acids containing as few as six and as many as 20 carbon atoms. Next, the CoA ester is oxidized in the beta position (hence the term "beta-oxidation"), and the two carbon atoms on the end of the molecule are cleaved to yield acetyl-CoA. Acetyl-CoA can then be fed directly to the tricarboxylic acid (TCA) cycle for subsequent complete oxidation. The remainder of the long-chain molecule, which is now shorter by two carbon atoms, is sequentially oxidized until all of the carbon is converted to acetyl-CoA.

Methyl Group Oxidation Oxidation of a terminal methyl group of a n-alkane results in the formation of a long-chained fatty acid. This fatty acid can then be

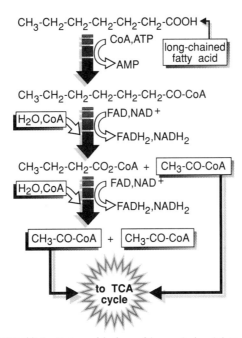

FIGURE 12.4. Beta-oxidation of long-chained fatty acids.

fully degraded via beta-oxidation. Terminal methyl group oxidation is apparently accomplished by means of several mechanisms. One sequence of oxidations and hydroxylations producing either a monocarboxylic or dicarboxylic fatty acid is shown in Figure 12.5a. The sequence begins with the oxidation of the terminal methyl group to an alcohol. Subsequent oxidations then form the carboxylic acid. This carboxylic acid may undergo a hydroxylation reaction with the formation of a dicarboxylic acid. The mono- and dicarboxylic acids are then degraded by beta-oxidation.

n-alkanes may also be fed to beta-oxidation by means of subterminal oxidations (Fig. 12.5b). In this case, a monooxygenase catalyzes the oxidation of a subterminal group in the alkane chain. The chain is then split to form a carboxylic acid and alcohol. The alcohol is further oxidized, and both components are fed to beta-oxidation. There is some evidence that methane monooxygenase may contribute gratuitously to the initial oxidation step. However, it is just as likely that microorganisms acclimated to growth on n-alkanes utilize a different monooxygenase.

12.4.3 Alkene-Oxidation

Alkenes, unsaturated hydrocarbons containing a carbon-to-carbon double bond (C=C), may be degraded via terminal oxidation reactions similar to those found in n-alkanes. In addition to terminal oxidation, there appear to be a large number of possible pathways by which alkenes undergo microbially-mediated oxidation. Typically, a large number of intermediate products are detected when even a single culture of cells are grown on alkenes. Although only a few of these pathways have been studied in detail, it seems reasonable to presume that multiple degradation pathways occur in ground-water systems exposed to alkenes. In addition, few studies on the anaerobic decomposition of alkenes have been undertaken. In spite of this lack of information, it is likely that anaerobic processes are involved in alkene degradation in natural systems.

12.4.4 Branched Aliphatics

Given the central role played by beta-oxidation in the complete oxidation of alkanes and alkenes, it might be expected that the presence of branched chains on hydrocarbon molecules would hinder biodegradation. This is, in fact, widely observed. Most branch-chained aliphatics exhibit much greater resistance to microbial oxidation then is exhibited by non-branched hydrocarbons.

In spite of this recalcitrance, however, microbial mechanisms for the degradation of branched-chain hydrocarbons have been described. One of these, the citronellol degradation pathway, was described by Seubert and Fass (1964). As one might expect, the basis of this pathway is the removal of one of a branch methyl group, production of a carboxylic acid, and subsequent degradation via beta-oxidation. The biochemistry of such removal mechanisms are complex and specific for specific alkyl groups in specific configurations. Thus, there is not a broad spectrum of compounds that can be degraded by any one pathway. Nevertheless, microorganisms are capable, at least to some degree, of circumventing the biochemical difficulties inherent in branch-chained hydrocarbon degradation.

The relative recalcitrance of branched-chain hydrocarbons to biodegradation

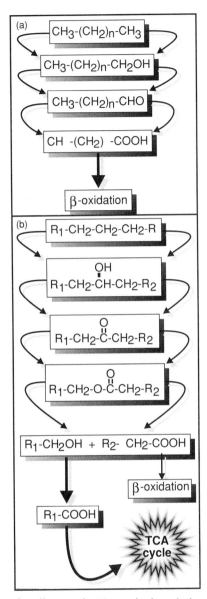

FIGURE 12.5. Oxidation of n-alkanes via (a) terminal methyl oxidation or (b) subterminal methyl oxidation.

has important implications for degrading petroleum hydrocarbons in ground-water systems. As discussed earlier, the burning properties of branched-chain alkanes and alkenes are superior to those of n-alkanes and alkenes. Thus, considerable effort is applied during petroleum refining in order to increase the proportion of branch-chained to unbranched compounds. The net effect of this, of course, is the production of petroleum products that are less easily degraded if introduced to the ground-water environment. Considering the vast number of compounds present, it seems inevitable that some portion of petroleum products—probably many branch-chained aliphatics—will resist full oxidation by microorganisms. Thus, complete oxidation of petroleum hydrocarbon products in ground-water systems may not occur.

12.5 MICROBIAL DEGRADATION OF ALICYCLIC HYDROCARBONS

Alicyclic hydrocarbons such as cyclopentane, cyclohexane, and various substituted cycloalkanes are present in petroleum and petroleum-derived fuels. As such, they are regularly delivered to the environment due to natural hydrocarbon seepage as well as fuel spills and leaks. In such environments, alicyclic compounds are not observed to persist. From this it would seem logical that alicyclic hydrocarbon-oxidizing microorganisms are fairly common in the environment. However, it took several decades of patient work by a number of teams of microbiologists before particular strains of microorganisms were found that could utilize alicyclic compounds as a carbon source in cell growth.

For many years it was assumed that some microorganisms capable of aliphatic hydrocarbon oxidation could also metabolize alicyclic hydrocarbons. To this end, the search for alicyclic-oxidizing hydrocarbons focused on microorganisms that had been previously acclimated to various aliphatic hydrocarbons. This approach met with little success and was eventually abandoned. Finally, Beam and Perry (1974) reported that complete oxidation of cyclohexane could be achieved by consortia of microorganisms in marine sediments. Subsequent to this, de Klerk and van der Linden (1974) showed that while neither a n-alkane nor a cyclohexanol-oxidizing Pseudomonad could oxidize cyclohexane individually, when both species were present cyclohexane was completely degraded to carbon dioxide. Apparently, the observed oxidation of alicyclic compounds in the environment is carried out predominantly by a consortia of microorganisms via commensal (mutually beneficial) relationships.

The study of de Klerk and van der Linden (1974) was important not only because it emphasized the importance of commensal microbial relationships in alicyclic degradation, but also because it gave insight into the mechanisms of degradation. The *Pseudomonas aeruginosa* strain alone did not completely degrade cyclohexane; rather, much of the cyclohexane accumulated as cyclohexanol, a cyclic alcohol compound. Strains of a cyclohexanol-degrading microorganism were known (i.e. *Pseudomnonas* "cyclohexanol"), but by itself this strain could not effect complete degradation of cyclohexane. The next logical step was to combine the two kinds of microorganisms. When this was done, the result was the complete oxidation of cyclohexane to CO_2 with the generation of biomass. In addition to elucidating this particular commensal relationship—*P. aeruginosa* partially oxidized cyclohexane to cyclohexanol, providing the substrate for growth of *Pseudomonas* "cyclohexanol"—it also established the importance of partially oxidized intermediate products in cyclohexane degradation.

12.5.1 Pathways for Cyclohexanol Degradation

Once it became clear that formation of cyclohexanols from alicyclic hydrocarbons was an important step in complete degradation, it became necessary to elucidate pathways of cyclohexanol transformations. A number of research groups contributed to understanding these transformations, but work by Trudgill and coworkers (Trudgill, 1984) on a strain of cyclohexanol-degrading Acinetobacter illustrates the important features of the biochemistry involved.

The first step in the aerobic degradation of cyclohexanol is a dehydrogenation to form an alicyclic ketone. Next, an alicyclic ketone 1,2 monooxygenase catalyzes the insertion of an oxygen into the ring. The ring is then cleaved by means of a hydrolase, and subsequent dehydrogenation reactions result in the formation of a dicarboxylic acid. This acid may then be further degraded via beta-oxidation or other metabolic pathways.

The enzyme systems involved in cyclohexanol degradation are sufficiently nonspecific so that a range of substrates, including cis-cyclohexane-1,2 diol, and trans-cyclohexane-1,2 diol were degraded in a similar fashion.

Application to Environmental Studies The cyclohexane-cyclohexanol example serves to illustrate some of the processes involved in alicyclic degradation by microorganisms. Given the chemical and biological variety of alicyclic molecules, it is certain that degradation proceeds by other pathways as well. Although the anaerobic degradation of alicyclic compounds has apparently received little attention, it is also possible that these compounds are transformed under anaerobic conditions as well.

It is clear, however, that alicyclic compounds are relatively easily degraded by consortia of microorganisms in aquatic sediments. Furthermore, the relationships between microorganisms that effect degradation appear to be both commensal and cooxidative. While degradation of alicyclic compounds in ground-water systems has received little direct attention in the literature, it can reasonably be assumed that similar processes are involved.

12.6 MICROBIAL DEGRADATION OF AROMATIC HYDROCARBONS

Of all the many components of petroleum hydrocarbons, the presence of aromatic compounds is perhaps the most important in terms of mobility in ground-water systems and of environmental impacts.

Even though aromatic compounds comprise only a minor portion (2-8 % by weight) of gasoline, these compounds are relatively soluble in water. The solubility of benzene in aqueous solution is about 1700 mg/L, toluene about 500 mg/L, and the various xylene isomers about 135 mg/L. This compares with the solubility of n-hexane of about 9 mg/L. So, even though normal and branched alkanes are a much larger component than aromatics of most fuels, the relative solubility of the aromatics leads to a large impact on ground-water quality.

In addition to being more readily mobilized into ground water, many aromatic hydrocarbons are now known to be potent carcinogenic (cancer-causing) agents. As early as 1930 it was shown that applying coal tar to mouse skin could induce cancer. Later, this carcinogenic activity was attributed to a particular cyclic hydrocarbon, benzo[a]pyrene. Benzene also has been shown to be carcinogenic and is now considered to be a hazardous substance. The maximum allowable concentration of benzene in drinking water is 0.005 mg/L. In assessing the risk associated with fuel spills, regulatory agencies conider concentrations of benzene to be of primary importance.

Aromatic hydrocarbon compounds are, by definition, characterized by the presence of one or more benzene rings in their structures (Fig. 12.1). The benzene ring itself consists of six carbon atoms linked to form a circle. Unlike cyclohexane, however, the benzene ring contains three C=C double bonds. This forms a very stable chemical structure that poses significant biochemical demands on microorganisms using it as a carbon and energy source. Because of the prevalence of aromatic hydrocarbons in the biosphere, however, many microorganisms have evolved biochemical pathways for degrading them. In this section, some of what is known about the biochemistry of bacterial aromatic hydrocarbon degradation is summarized.

12.6.1 Benzene Degradation

Benzene is not observed to accumulate in the hydrosphere following spills or leaks of petroleum hydrocarbons; rather, benzene appears to readily transformed by microorganisms in a variety of hydrologic settings. Prior to the 1970's, however, little was known about the biochemical mechanisms involved in benzene biodegradation.

Studies by David Gibson and colleagues at the University of Texas resulted in the isolation of a strain of *Pseudomonas putida* that was capable of growth on ethylbenzene and toluene and that also was able to oxidize benzene. The progression of these studies gives an interesting example of the logic involved in understanding biodegradation mechanisms. The first step was to determine some of the intermediate compounds formed during degradation. It was quickly determined that the cells grew on catechol, on cis-benzene dihydrodiol, but not on trans-benzene dihydrodiol. This suggested that catechol and cis-benzene dihydrodiol were intermediate products of benzene oxidation, but that the trans isomer of benzene dihydrodiol was not. Based on this evidence, benzene oxidation appeared to involve sequential oxidations of the benzene ring to catechol (Fig. 12.6).

The presence of a benzene dioxygenase, which performed the initial oxidation of benzene, could be inferred from the observed sequence of reactions. However, this obscures somewhat the biochemical complexity of the oxidation step. Studies by Axcell and Geary (1975) resulted in the isolation of a *Pseudomonas putida* strain that was capable of growth with benzene as the sole carbon and energy source. This, in turn, led to the characterization of the benzene dioxygenase enzyme. This enzyme turned out to be composed of several subunits: a flavoprotein and two iron-sulfur proteins. The proposed organization of this enzyme complex indicates that oxidation is initiated by electron transfer from NADH to the flavoprotein, through one of the iron-sulfur proteins, to the second iron-sulfur protein where the terminal oxidation of benzene apparently takes place. This sequence of reactions also requires, for reasons that are not clear, the presence of dissolved Fe^{2+} in order to proceed.

The oxidation of benzene by *P. putida* has been studied in some detail and gives an example of how complex the biochemical mechanisms that are involved really are. It's tempting sometimes to think of microbial biodegradation processes as being somewhat magical. In reality, when the biochemical processes are subjected

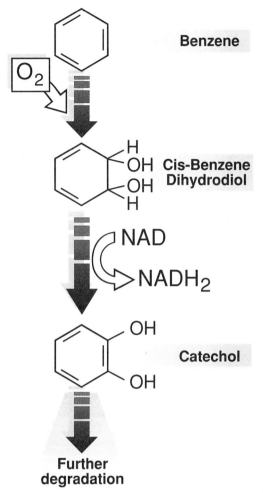

FIGURE 12.6. Aerobic oxidation of benzene to catechol.

to careful study, it's apparent that the presence of the appropriate enzymatic systems is the real key to efficient biodegradation.

Ortho and Meta Cleavage of Catechol Catechol is a central intermediate product in benzene degradation (Fig. 12.6) as well as in the degradation of many other aromatic hydrocarbons. Two pathways-the ortho and meta cleavage pathways-have been identified that effects the breakdown of catechol. *P. putida* was the microorganism in which these biochemical mechanisms were first identified. However, these pathways have been described in a number of genera, including *Acinetobacter, Bacillus, Alcaligenes,* and *Nocardia*.

In the ortho pathway, also termed the "beta-ketoadipate pathway," catechol is cleaved between the carbon atoms that are carrying the hydroxyl groups (Fig. 12.7a), forming cis,cis-muconate. This compound is then further oxidized forming beta-ketoadipate. The meta cleavage pathway differs in that the cleavage is initiated between two carbon atoms, only one of which carries a hydroxyl group (Fig.

12.7b). The enzyme catechol 2,3-dioxygenase catalyzes the insertion of two oxygen atoms. The ring is then cleaved by means of a hydrolase. The final products are organic acids that can be further degraded to carbon dioxide.

Most strains of *Pseudomonas* utilize either the meta or the ortho pathway for degrading catechol. The meta pathway is thought to be somewhat more flexible and involved in the degradation of substituted isomers of catechol. Some strains, interestingly enough, have been shown to induce enzymes for both pathways. In any case, these biochemical mechanisms are equally important in the complete degradation of aromatic hydrocarbons, as are the initial oxidation steps leading to the formation of catechol.

12.6.2 Degradation of Alkyl Benzenes

Alkyl benzenes, such as toluene and xylene, are also important components of gasolines and other petroleum hydrocarbon fuels. The sequence of reactions involved in the oxidation of these compounds by *Pseudomonas* are analogous to

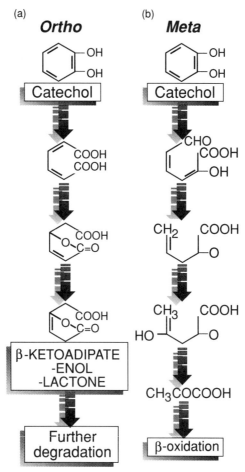

FIGURE 12.7. The (a) ortho and the (b) meta pathways for the aerobic oxidation of catechol.

those involved in the oxidation of benzene. In the case of toluene, the initial oxidation leads to formation of a methylcatechol that then undergoes cleavage via the meta pathway. For xylene, the process is similar except that a dimethylcatechol serves as the intermediate product that is fed to the meta pathway. Ethylbenzene appears to follow a similar pathway for oxidation.

From the biochemistry of alkylbenzene degradation, it might be predicted that the ease of degradation is related to the complexity of the catechol intermediate that is formed. This behavior is, in fact, observed. Larger alkyl substituents have been observed to slow down or even to stop degradation-in some cases with the accumulation of substituted catachols. Thus, as was observed with the degradation of branched alkanes, more complex molecules are degraded less efficiently. This, in turn, is related to the biochemical limitations of the degradative pathways available to microorganisms and leads to recalcitrance exhibited by some compounds.

Aerobic Degradation of Aromatic Hydrocarbons by Subsurface Bacteria Although most microorganisms that are capable of degrading aromatic hydrocarbons have been isolated from soils or aquatic sediments, microorganisms isolated from ground-water systems that carry out these processes have been described. Fredrickson et al. (1991) reported a bacterium, designated strain F199, that was capable of using alkyl benzenes, such as toluene and xylene, as well as other aromatic compounds, as a sole carbon and energy source. This microorganism was recovered from the Middendorf aquifer of South Carolina from a depth of 410 m below land surface.

While the role of various *Pseudomonas* isolates in the catabolism of alkyl benzenes is well-known, strain F199 did not belong to this genera. F199 was gram-positive and exhibited variable cellular morphology that depended upon the particular growth substrate. The ability to degrade aromatic compounds in F199 was inducible with toluene inducing the ability to degrade *14*C-toluene and *14*C-napthalene. Curiously, however, napthalene did not induce the degradation of *14*C-toluene. This in turn, suggested the presence of at least two separate degradative pathways, both of which were capable of degrading napthalene. F199 was shown to harbor two large plasmids which may code for the observed degradative capacity.

There is no particular reason to suppose that aerobic aquifer systems, including such deeply buried aerobic aquifers as the Middendorf aquifer, would not harbor microorganisms with the ability to degrade petroleum hydrocarbons. However, the isolation of F199 is direct evidence that these systems do contain such microorganisms.

12.6.3 Degradation of Polycyclic Aromatic Compounds

In addition to single-ring aromatic hydrocarbons such as the BTEX compounds, petroleum hydrocarbon fuels also contain various polyaromatic hydrocarbons (PAH) compounds such as napthalene and phenanthrene (Fig. 12.1). As with benzene, the presence of PAHs are of environmental concern because they have been shown to be carcinogenic. Microorganisms that degrade these compounds are widely found in soils. The mechanisms of degradation are analogous to those that are characteristic of benzene degradation.

For the case of napthalene, an initial oxidation step results in the formation of a dihydroxynaphthalene. Further oxidation results in cleavage of one of the rings with the production of salicylic acid. Degradation of salicylic acid then proceeds through the familiar catechol intermediate. Phenanthrene degradation appears to follow similar mechanisms. The overall strategy is to decouple the rings and to form intermediate products that can feed into previously established degradation pathways.

As with other aromatic hydrocarbons, there is evidence that an increase in branched substitutions on one or more of the fused rings leads to increased recalcitrance to bacterial degradation. For example, Heitkamp and Cerniglia (1988) studied PAH degradation by using a strain of bacteria isolated from sediments chronically exposed to aromatic hydrocarbons near an oil field. This bacterium, which had been pre-exposed to pyrene, readily degraded napthalene, phenanthrene and pyrene. Curiously, an acclimation period of about one day was required before significant naphthalene degradation was observed. This suggests that the bacterium degrades napthalene via an enzyme system that is separately induced from the pyrene degradation system. The important point, however, is that the more highly substituted (and more carcinogenic) compounds 1-nitropyrene, 6-nitropyrene, and 3-methylcholanthrene were degraded at a much slower rate.

12.6.4 Anaerobic Degradation of Aromatic Hydrocarbons

Much of what is known about microbial degradation of petroleum hydrocarbons centers on aerobic bacteria. Because many aromatic hydrocarbons, the BTEX compounds for instance, do not contain molecular oxygen in their structures, it is easy to assume that oxidation requires aerobic conditions. For this reason, it was once thought that aromatic hydrocarbons were resistant to microbial attack under anaerobic conditions. If this were true, this would have important impacts upon the fate and transport of these compounds in ground water. Many ground-water systems, even shallow water table aquifers, are anaerobic. Furthermore, the introduction of petroleum hydrocarbons to ground-water systems is invariably followed by the development of anaerobic conditions near the free product. Thus, the absence of anaerobic degradation of petroleum hydrocarbons would greatly decrease attenuation of these compounds in ground-water systems.

It is interesting from a historical perspective that one of the first observations of anaerobic degradation of aromatic hydrocarbons was made in a contaminated ground-water system (Reinhard, Goodman, and Barker, 1984). In this study, the selective removal of xylenes from contaminated anaerobic ground water was documented. Furthermore, this removal could not be attributed to hydrodynamic processes such as dispersion or mixing. Apparently, aromatic hydrocarbons could be degraded under anaerobic conditions. Direct experimental evidence of these processes was soon provided. Wilson, Smith, and Rees (1986) showed that ^{14}C-labeled toluene was converted to $^{14}CO_2$ in microcosms of aquifer materials incubated under methanogenic conditions.

As might be expected, however, the microbiology and biochemistry involved with the anaerobic degradation of aromatic hydrocarbons turned out to be much different than that of aerobic degradation. For example while a large number of aerobic microorganisms that degrade aromatics have been isolated in pure culture, very few anaerobic microorganisms have been described that perform this function. This is partly due to the greater difficulty involved with isolating anaerobic bacteria. However, it also reflects the importance of mixed cultures and microbial food chains in the complete degradation of aromatics. Furthermore, as might be expected, the biochemical pathways involved in anaerobic degradation are much different than those of aerobic degradation.

Degradation of Benzoate In much of the experimental work related to the anaerobic degradation of aromatic compounds, benzoate has been used as a model aromatic compound. These studies have served to document the different biochemical pathways involved with anaerobic degradation as compared to aerobic degradation. For example, Williams and Evans (1975) proposed a series of steps in the degradation of benzoate under nitrate-reducing conditions. Furthermore, a number of investigators have proposed pathways of benzoate degradation under methanogenic conditions. The biochemistry of these anaerobic pathways is much different then that of the aerobic pathways discussed earlier. In particular, the central position of catechol in aerobic decomposition is absent in the anaerobic pathways.

Degradation of Toluene and Benzene Under Methanogenic Conditions While benzoate has provided a convenient model compound for laboratory studies of aromatic degradation, degradation of BTEX compounds are of more direct environmental interest. One study of toluene and benzene degradation under methanogenic conditions was provided by Grbic-Galic and Vogel (1987) who used mixed cultures of microorganisms derived from enrichment of sewage sludges. The strategy of this study, which is representative of strategies used to elucidate biochemical pathways, centered on documenting the production of $^{14}CO_2$ from radiolabeled toluene and benzene as well as identifying intermediate compounds produced during degradation.

$^{14}CO_2$ was produced from both ^{14}C-benzene and ^{14}C-labeled toluene, demonstrating that the aromatic ring was being completely degraded (Fig. 12.8). This information by itself, however, gives little clue as to the pathways being utilized to effect degradation. By identifying some of the intermediate compounds produced during degradation, however, it was possible to construct a model of how the degradation of each compound occurs. In this case, numerous intermediate compounds were detected and assigned to particular steps in the degradation process. For example, the appearance of phenol (Fig. 12.8a) suggests that ring oxidation initiates degradation of benzene. Similarly, the appearance of cyclohexene suggests that benzene may also undergo ring reduction. As suggested by studies on benzoate, benzene degradation proceeds through a number of oxidation steps to carboxylic acids. These carboxylic acids are then further oxidized to CO_2 and CH_4.

Toluene, with its single methyl group, apparently may be oxidized in a number of ways. For example, if the molecule initially undergoes ring oxidation (Fig. 12.8b), p-cresol or o-cresol is produced. Depending on which isomer of cresol is generated, the subsequent oxidation steps are different. Alternatively, the methyl group may be oxidized in the initial step with the formation of benzyl alcohol (Fig. 12.8b). In this sequence, the oxidation proceeds though benzoate and to carboxylic acids. Reduction of the toluene ring is also possible, with the production of methylcyclohexane.

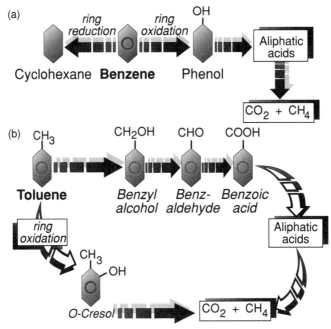

FIGURE 12.8. Degradation pathways of (a) benzene and (b) toluene under methanogenic conditions. (Modified from Grbic-Galic and Vogel, 1987, with permission courtesy of the American Society for Microbiology.)

One question that may reasonably be asked when these proposed sequences of oxidations are examined, is "What is the source of oxygen inserted into the benzene ring to form phenol or into the toluene ring to form cresol?" In the case of aerobic oxidation, insertion of oxygen into ring structures comes exclusively from molecular oxygen. Under anaerobic conditions, however, that seems unlikely. Studies by Vogel and Grbic-Galic (1987) tracing the ^{18}O composition of the intermediate compounds, suggest that the oxygen inserted into the ring structures came from water. Thus, oxygen-bearing carboxylic acids can be produced from carbon compounds lacking any molecular oxygen under anaerobic conditions.

Degradation of Toluene Under Fe(III)-Reducing Conditions The study of aromatic hydrocarbon oxidation by aerobic bacteria was accomplished largely by studies done with pure cultures. The strategy was to use selective enrichment of water, soil, or sediment samples to isolate and purify cultures and then to use these purified cultures to understand the biochemical mechanisms of degradation. The literature on the anaerobic degradation of aromatic hydrocarbons, in contrast, focuses largely on the activity of mixed cultures, often termed "consortia," which effect degradation. To a degree, this approach reflects the greater importance of food chains and commensal relationships in anaerobic degradation mechnanisms. However, it also reflects the greater difficulty of working with cultures under strict anaerobic conditions.

In spite of this, pure cultures of microorganisms that are capable of complete oxidation of aromatic hydrocarbons have been described. The first strictly anaerobic microorganism that was shown to oxidize toluene completely to CO_2 grew with Fe(III) as the sole electron acceptor. From a historical point of view this might not be expected since methanogenic and nitrate-reducing microorganisms were implicated in aromatic hydrocarbon oxidation first. Also, systematic studies of Fe(III) reduction have only recently been undertaken (Lovley and Phillips, 1988). In the course of these studies on Fe(III) reduction, which included consideration of a crude oil spill, it was discovered that toluene, as well as phenol and p-cresol, could be completely oxidized to CO_2 with the reduction of Fe(III) by the dissimilatory Fe(III) reducer strain GS-15 (Lovley and Lonergan, 1990).

The inferred biochemistry of aromatic hydrocarbon degradation followed by GS-15 is shown in Figure 12.9. Toluene is degraded by a series of sequential oxidations in which Fe(III) acts as the sole electron acceptor. As in aromatic oxidation by methanogenic cultures, oxygen is incorporated into the ring structure. This oxygen may come from water, as was the case with methanogenic cultures. Alternatively, the oxygen may be derived from Fe(III) oxyhydroxides. The sequence of toluene degradation proposed for GS-15 (Fig. 12.9) is remarkably similar to that inferred for mixed methanogenic cultures (Fig. 12.8). In both cases, alcohols, aldehydes, and benzoic acid are important intermediate products.

The recognition that Fe(III)-reducing microorganisms can anaerobically oxidize aromatic hydrocarbons is especially important for many ground-water systems. Many shallow sand aquifers-those that are most susceptible to hydrocarbon contamination-contain abundant Fe(III) oxyhydroxides and usually lack significant concentrations of nitrate. Thus, once hydrocarbons are delivered to ground-water systems with the initiation of anaerobic conditions, Fe(III) reduction will usually be the first anaerobic process to actively degrade the hydrocarbon contamination.

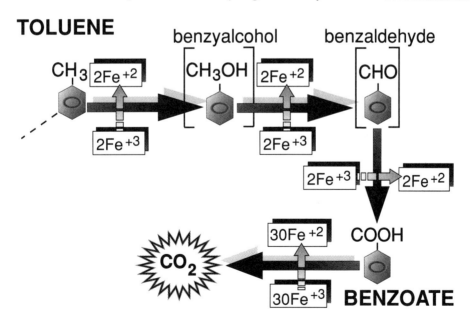

FIGURE 12.9. Degradation of toluene under Fe(III)-reducing conditions. (Modified from Lovley and Lonergan, 1990, with permission courtesy of the American Society for Microbiology.)

Degradation of Alkyl Benzenes Under Denitrifying Conditions Nitrate concentrations in most aquifer systems-either pristine or contaminated-are generally too low for denitrification to be significant relative to other anaerobic microbial processes. Nevertheless, degradation of petroleum hydrocarbons coupled to denitrification is important in bioremediation technology. Because of the limited solubility of oxygen in water, adding oxygen to ground water to increase rates of degradation is difficult. However, because nitrate is relatively soluble, the addition of nitrate is an attractive alternative to the addition of oxygen. Furthermore, the by-product of denitrification, molecular nitrogen gas, is environmentally innocuous. This contrasts with the production of methane, which may create an explosive hazard, or the with production of hydrogen sulfide which creates a noxious odor. Thus, artificially establishing denitrifying conditions may be a useful tool in enhanced biodegradation.

There is evidence that microorganisms present in ground-water systems are capable of degrading benzene, toluene, and xylene (BTX) under denitrifying conditions. Major, Mayfield, and Barker (1988) showed that, when aquifer solids were incubated under denitrifying conditions, metabolism of BTX compounds were significantly enhanced (Table 12.2). Furthermore, the efficiency of BTX removal under denitrifying conditions approached that of aerobic conditions. Although the results of this experiment can, in the strictest sense, only be applied to the aquifer from which the materials were obtained, they are nevertheless important. If it is assumed that most shallow anaerobic aquifers behave in a similar manner, then nitrate addition could significantly enhance biodegradation rates of petroleum hydrocarbons.

Using a separate experimental approach, Kuhn et al. (1988) placed aquifer material in laboratory columns and acclimated the indigenous bacteria to degrade m-xylene with nitrate as the sole electron acceptor. Next, additional alkyl benzenes were introduced and degradation was monitored over a six-day period. It was found that toluene and m-xylene were totally degraded in that period and 3-ethyltoluene was partially degraded. However, little change in the concentrations of benzene, ethylbenzene, and o-xylene were found. Because of the short exposure time, it cannot be concluded that the microorganisms present were unable to metabolize these compounds; rather, they may simply not have had enough time to induce enzymatic systems to degrade these new compounds. In any case, these

TABLE 12.2 Percentage of BTX Remaining After Incubation for 62 Days with Nitrate Under Aerobic or Anaerobic Conditions

	Condition				
Compound	Sterile	Nitrate	Anerobic	Nitrate + Oxygen	Oxygen
Benzene	79	5	66	0	1
Toluene	86	2	65	0	0
O-Xylene	80	15	73	19	15
m-Xylene	80	12	59	8	11

(Initial concentration of compound was 3 mg/l; that is, benzene = 3.84×10^{-5}m; O-, and m-Xylene = 2.83×10^{-5}M.)

Source: Major, Mayfield, and Barker, 1988. Reprinted courtesy of the National Water Well Association.

results are consistent with the interpretation that microorganisms indigenous to aquifer systems are capable of degrading some aromatic hydrocarbons under denitrifying conditions.

Another series of laboratory experiments that evaluated BTEX degradation under denitrifying conditions using aquifer material was reported by Hutchins et al., (1991). These experiments are interesting because they were conducted using uncontaminated aquifer material recovered from a shallow ground-water system, as well as material from the same aquifer that had been previously contaminated with JP-4 jet fuel.

In these experiments, degradation of toluene, ethylbenzene, and xylene was observed in both contaminated and uncontaminated aquifer material. Interestingly, however, degradation of benzene was not observed during 160 days of incubation. Perhaps most important of all, rates of alkylbenzene degradation were consistently greatest in uncontaminated aquifer materials. On one hand, this would seem to be counterintuitive as one might expect contaminated sediments to harbor microorganisms acclimated to alkylbenzene degradation. However, at least two other factors may influence alkylbenzene degradation under these conditions. First of all, the contaminated sediments contained aliphatic hydrocarbons in addition to the added alkylbenzenes. This raises the possibility that the microorganisms preferentially use aliphatic hydrocarbons rather than alkylbenzenes. This possibility is plausible, since nitrate was consumed by contaminated sediments even in the absence of alkylbenzenes. Secondly, the extent of contamination may have been sufficient to cause toxic inhibition of metabolism.

The common thread of studies concerning alkylbenzene degradation under denitrifying conditions is that many compounds, particularly toluene and xylenes, are readily degraded. The behavior of benzene is more uncertain, but its degradation is certainly a possibility. Thus, the addition of nitrate to petroleum hydrocarbon-contaminated systems as an oxidant appears to be justified on the basis of available evidence.

Degradation of Phenols and Cresols Phenols and cresols (Fig. 12.1) are usually only trace components of most petroleum products, being more often associated with municipal landfill leachates (Smolenski and Suflita, 1987), coal tar derivatives (Ehrlich et al., 1982), or cresote wastes (Troutman et al., 1884). Nevertheless, these compounds occur in ground water associated with petroleum hydrocarbon spills as well. In natural systems, it is sometimes difficult to determine if these compounds were present as a primary component of the hydrocarbon product or if they were the intermediate products of microbial degradative processes. Under anaerobic conditions, phenols are an intermediate oxidation product of benzene metabolism and cresols are intermediate products of toluene degradation (Fig. 12.8). As such, the pathways of phenol and cresol degradation may be closely intertwined with other pathways for aromatic hydrocarbon degradation.

Degradation of phenols and cresols in ground-water systems often occurs under anaerobic conditions. The anaerobic degradation of cresols in sediments from a shallow anaerobic aquifer have been extensively studied by J.M. Suflita and colleagues at the University of Oklahoma. Their studies have shown that the position of the hydroxyl group on the cresol isomer strongly influences observed degradation rates. For example, the lag times for the degradation of para-, meta-, and ortho-cresol in anaerobic slurries of aquifer material varied widely (Table 12.3). The para isomer was the most readily degraded compound under both sulfate-reducing and methanogenic conditions. The meta isomer required somewhat more time for degradation to be initiated, and the ortho isomer was largely recalcitrant to degradation. These data suggest that different biochemical pathways are involved in degrading the different cresol isomers.

In a series of experiments, Smolenski and Suflita (1987) investigated pathways of p-cresol degradation under sulfate-reducing conditions. These experiments are interesting in that they illustrate the kinds of logic used to deduce pathways of degradation. First, enrichments of aquifer material were obtained that were acclimated to p-cresol degradation. During degradation of p-cresol, traces of p-hydroxybenzoate were detected. This implied that the sequence of degradation steps proceeded through p-hydroxybenzoate. However, the *lack* of other compounds detected could not be interpreted as evidence that p-hydroxybenzoate was the only intermediate product. It was perfectly possible that other intermediate products were simply below detectable limits.

This problem was addressed using simultaneous adaptation experiments. These techniques involve postulating possible intermediate products, and testing acclimated cultures for their ability to use these compounds without a lag time. For example, when p-cresol, p-hydroxybenzaldehyde, p-hydroxybenzyl alcohol, p-hydroxybenzoate, and benzoate were added to acclimated sulfate-reducing slurries, all the substrates *except* p-hydroxybenzyl alcohol were metabolized without a lag. This suggested that all of these compounds except p-hydroxybenzyl alcohol were intermediate products. However, it was also observed that when slurries were amended with both p-cresol and p-hydroxybenzaldehyde, p-hydroxybenzyl alcohol was observed to accumulate. These results were subject to conflicting interpretations. One experiment seemed to indicate that p-hydroxybenzyl alcohol was not an intermediate product, whereas another experiment indicated that it was. Because the experimental evidence was equivocal, p-hydroxybenzyl alcohol was included in the proposed pathway, however it was inclosed in brackets to emphasize its questionable status. Similarly, benzoate was bracketed, but for another reason. Because benzoate is so readily degraded under anaerobic conditions and because it was not detected as a transient compound in the slurries, its degradation without a lag time was considered equivocal. Logic similar to that

TABLE 12.3 Comparison of Lag Times for the Degradation of Cresol Isomers in Anaerobic Aquifer Slurries

	Lag Time (Days) for:	
Substrate	Sulfate-reducing Incubations	Methanogenic Incubations
p-cresol	<10	46
m-cresol	43	46–90
o-cresol	>100	>90

Source: Smolenski and Suflita, 1987. Reprinted courtesy of the American Society for Microbiology.

followed by Smolenski and Suflita (1987) is behind all proposed degradative pathways.

These experiments are interesting for another reason. The biochemistry of sulfate reduction and methanogenesis are very different (Chapter 4). Therefore it might be expected that degradation rates of aromatic hydrocarbons by these two processes might be very different. In fact, there was a large difference between the acclimation times required under the two conditions. In all cases, the sulfate-reducing slurries require less acclimation time than did the methanogenic slurries. Later work (Ramanand and Suflita, 1991) showed a similar trend during degradation of m-cresol. In this case, m-cresol was actively degraded under sulfate-reducing and nitrate-reducing conditions, but degradation was not evident under methanogenic conditions. Thus, the dominant electron-accepting process in anaerobic systems appears to be important in determining biodegradation rates.

12.7 MICROBIAL DEGRADATION OF PETROLEUM HYDROCARBONS IN GROUND-WATER SYSTEMS

Ground-water contamination by petroleum hydrocarbons was not an issue during much of the twentieth century in the industrialized world. It wasn't so much that people were unaware of petroleum spills as that the effects of such spills on ground-water quality were not immediately apparent. With the slow rates of ground-water flow, it typically took years or even decades for the effects of petroleum contamination to become apparent. When they did become apparent, it was usually because of massive contamination of wells used for drinking water. Because many instances of contamination occurred in areas lacking water wells, it's almost certain that the impacts of most spills were simply never detected.

The effects of microbial processes on petroleum hydrocarbons in ground-water systems, particularly the role of these processes in degrading and attenuating contamination, were even slower to be appreciated. As recently as 1981, papers using quantitative modeling techniques for evaluating hydrocarbon mobility made no mention of microbial processes or the effects of such processes on hydrocarbon attenuation (Yazicigil and Sendlein, 1981).

By 1985, microbiologists had gathered clear evidence that biodegradation of petroleum hydrocarbons occurred in both aerobic and anaerobic aquifer systems. However, the most telling evidence that such processes were important came from numerous field studies showing that petroleum hydrocarbons were selectively removed under a variety of conditions. One study in California, for example (Hadley and Armstrong, 1991), surveyed 7,167 water-supply wells for evidence of chemical contamination. Because gasoline spills had occurred from thousands of leaking underground tanks during the previous 50 years, these researchers anticipated that petroleum hydrocarbon contamination, particularly benzene contamination, would be prevalent. To the surprise of the investigators, very little evidence of benzene contamination of ground water was found. After reviewing the possible processes that could remove benzene from ground water, including sorption and evaporation, they concluded that biodegradation processes were the most likely reason for the puzzling "absence" of benzene.

The realization that microbial processes were important in the fate and transport of petroleum hydrocarbons in ground-water systems led to a number of investigations during the late 1980's. These studies form a basis for evaluating biodegradation as a natural attenuation process and as a possible strategy for remediating hydrocarbon contamination in ground-water systems.

12.7.1 Aerobic Degradation of BTX Compounds

The ability of microorganisms to degrade petroleum hydrocarbons has been widely studied and many of the biochemical pathways involved in degradation have been worked out in some detail (see Section 12.6). However, it is difficult to predict how complex natural systems, particularly inaccessible ground-water systems, will react to hydrocarbon contamination. The "shrinking" of BTEX plumes in shallow ground-water systems had been widely reported during the 1980's, often with little documenting evidence. However, Barker, Patrick, and Major (1987) reported the results of an experiment in which BTX (benzene, toluene, and xylene) compounds were injected into a sandy water-table aquifer and the resulting plume was monitored for the next year and a half. This study provided reliable evidence for the presence of BTX-degrading microbial activity in water-table aquifers.

In the field experiment reported by Barker, Patrick, and Major (1987), about 1800 liters of ground water were spiked with BTX to give average concentrations of 2.4 mg/L benzene, 1.8 mg/L toluene, 1.1 mg/L each of p-xylene, m-xylene, and o-xylene, and 1280 mg/L chloride. The chloride was added so that it would act as a conservative tracer. The spiked water was injected into a sandy water-table aquifer and the resulting plume monitored by means of a dense network of multi-level piezometers. Samples from the monitoring wells were collected each day for three days after injection and each month for 4 months.

The results of this experiment for days 3, 53, and 108 of the experiment are shown in Figure 12.10a. By day 53, the combined effects of advection and dispersion had moved the chloride tracer plume between 1 and 5 meters along the natural gradient prevailing in the aquifer, spreading to about three times its initial size as it moved. The benzene component of the plume behaved similarly to chloride in the first 53 days, spreading as it moved downgradient. The toluene component, however, was actually smaller at day 53 than at the beginning of the experiment. By day 108, the toluene component of the plume had virtually disappeared. Furthermore, by day 108, the benzene component of the plume had shrunk noticeably relative to the chloride tracer. Note, however, that the relative *position* downgradient of the BTX compounds was similar to that of chloride. This observation rules out the possibility that sorption of BTX to aquifer solids was the main attenuating process and implies that biodegradation was the principal process. This conclusion was supported by laboratory scale microcosm experiments.

By vertically integrating concentrations of BTX in the plume, Barker, Patrick, and Major (1987) were able to calculate the mass of each compound present as a function of time (Fig. 12.10b). The mass of chloride, as expected, changed very little. The masses of each BTX compound, however, decreased rapidly. By 108 days, the mass of toluene and the xylene isomers had decreased practically to zero. Although benzene was more resistant to degradation than the other compounds, the mass of remaining benzene was essentially zero by 400 days after injection.

12.7 MICROBIAL DEGRADATION OF PETROLEUM HYDROCARBONS

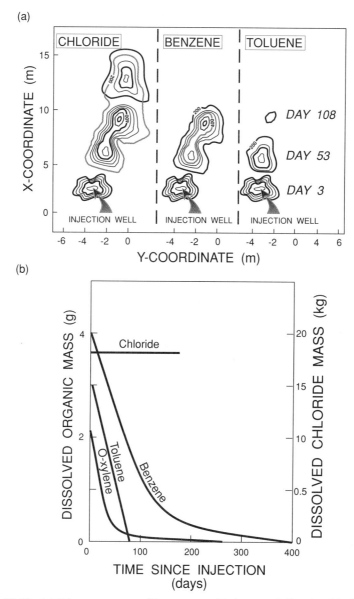

FIGURE 12.10. (a) Disappearance of benzene and toluene relative to chloride in a field scale injection test, and (b) decrease in mass of organic compounds relative to chloride in the injection test. (Modified from Barker, Patrick, and Major, 1987).

This experiment differs from real petroleum hydrocarbon spills in several ways. First of all, the large mass of aliphatic and alicyclic hydrocarbons present in all fuels were not included in the experiment because of their lack of aqueous solubility. Thus, any effects of these compounds on BTX degradation were absent from the experiment. However, as great quantities of aliphatic compounds are not transported downgradient of petroleum spills, the experiment accurately simulates conditions downgradient of petroleum spills. In part because of the lack of aliphatic hydrocarbons in the experiment, aerobic conditions prevailed in the aquifer throughout the experiment. As aerobic bacteria may be more efficient hydrocarbon utilizers than anaerobic bacteria, this experiment probably records degradation rates as high as any that would be observed in ground-water systems.

The chief value of the experiment reported by Barker, Patrick, and Major (1987) is that it verified the numerous observations that BTX compound were actively biodegraded in ground-water systems. However, because of the aerobic conditions and the lack of an immiscible hydrocarbon phase capable of leaching BTX compounds over time, the time required for complete degradation in this experiment was probably less than that required in most petroleum-product spills.

12.7.2 Anaerobic Degradation of Aromatic Hydrocarbons

One of the first consequences of petroleum hydrocarbon contamination of shallow ground-water systems is the establishment of anaerobic conditions near the plume. One well-documented example of this was a crude oil spill that occurred near Bemidji, Minnesota (Section 11.2.4). Prior to the spill, the aquifer was aerobic (DO~10 mg/L). Soon after the spill, however, increased levels of microbial respiration created anaerobic conditions near the oil lens. This, in turn, provided an opportunity for studying the anaerobic degradation of petroleum hydrocarbons in an anaerobic ground-water system under field conditions (Baedecker et al., 1988; Cozzarelli, Eganhouse, and Baedecker, 1988).

Concentrations of aromatic hydrocarbons in the ground water near at the edge of the lens are shown in Figure 12.11a and concentrations 10 meters further downgradient are shown in Figure 12.11b. Concentrations of all of the BTEX compounds were observed to decrease significantly. In fact, all of the alkylbenzene compounds were observed to decrease. Because of the proximity to the oil lens, these concentration decreases could not be attributed to hydrodynamic processes and were interpreted as reflecting anaerobic microbial degradation (Cozzarelli, Eganhouse, and Baedecker, 1988).

Further evidence that microbial processes are involved with degradation are provided by the observed distribution of organic acids in the anaerobic plume. Near the oil lens, fairly high concentrations of organic acids, particularly acetic acid, were observed. These acids have been shown to be intermediate products of anaerobic benzene and toluene degradation in laboratory experiments (Fig. 12.10) and their presence is strong evidence that anaerobic microbial processes are actively degrading the hydrocarbons in this field situation. Downgradient of the oil lens, concentrations of the organic acids decrease because of consumption by a combination of methanogenesis (Baedecker et al., 1988) and Fe(III) reduction (Lovley et al., 1989).

It's interesting to compare the relative degradation rates of each BTEX compound under aerobic conditions (Fig. 12.10) with those observed under anaerobic conditions. Under aerobic conditions, it was observed that the relative rates of degradation were xylene > toluene > benzene (Barker et al., 1987). Furthermore, the relative rates of xylene isomer degradation were observed to be p-xylene > m-xylene > o-xylene. Under anaerobic conditions at the Bemidji site, the relative rates of BTEX degradation were similar to those observed under aerobic conditions. However, the order of xylene isomer degradation rates were reversed,

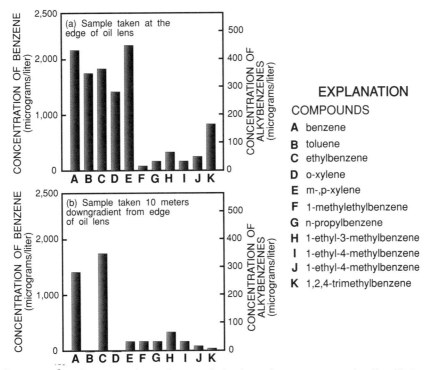

FIGURE 12.11. Concentrations of aromatic hydrocarbons near a crude oil spill, Bemidji, Minnesota (a) at the edge of the oil lens, and (b) 10 meters downgradient of the oil lens. (Modified from Cozzarelli, Eganhouse, and Baedecker, 1988.)

with o-xylene apparently degrading faster than m-xylene or p-xylene (Cozzarelli, Eganhouse, and Baedecker, 1988).

Ethylbenzene was not considered in the study by Barker, Patrick, and Major (1987). However, under the anaerobic conditions of the Bemidji site, ethylbenzene appeared to be relatively resistant to microbial degradation. In the more downgradient part of the plume, where molecular oxygen becomes available, rapid loss of ethylbenzene, as well as benzene, was observed.

12.8 MICROBIAL PROCESSES AND THE REMEDIATION OF PETROLEUM HYDROCARBON CONTAMINATION IN GROUND-WATER SYSTEMS

There is presently no doubt that most ground-water systems contain populations of microorganisms with the metabolic capability for oxidizing petroleum hydrocarbons. This capability has been demonstrated in numerous locations and under diverse hydrologic conditions. Furthermore, a great deal is known about the pathways, either aerobic or anaerobic, by which microorganisms effect hydrocarbon oxidation. From a practical point of view, the question is not whether microbial processes degrade petroleum hydrocarbons—there is unequivocal evidence that they do—but rather, (1) how rapidly can microbial processes completely oxidize particular instances of hydrocarbon contamination, and (2) what technologies can be utilized to increase rates of microbial oxidation?

12.8.1 Field Implementation of Hydrocarbon Bioremediation

Demonstrating biodegradation of petroleum hydrocarbons, such as BTEX compounds, in aquifer sediment microcosms under laboratory conditions is a difficult but manageable task. Sterilized controls can be used to separate abiotic from biotic effects, the uptake of hydrocarbons and electron acceptors can be accurately tracked, and intermediate products of the microbial transformations can be identified (Table 12.3). Demonstrating biodegradation under field conditions, and in particular separating biotic from abiotic effects, is a much more formidable undertaking. The primary reasons for this are (1) the added complexity of dealing with a natural hydrologic system and it's inherent uncertainties, and (2) dealing with the larger scale of the experiment.

The overall strategies of bioremediation procedures for petroleum hydrocarbons are conceptually simple. For example, Suflita (1989) outlined several steps that are common to most bioremediation procedures (Fig. 12.12). First, a contaminated area is identified and its boundaries accurately delineated (Fig. 12.12a). Secondly, ground water is recirculated in such a way as to envelope the contaminated area (Fig. 12.12b). Because the recirculated water is often itself contaminated, it is often treated with airstripping or carbon filtration to avoid reintroducing contaminants to the system. Thirdly, nutrients such as appropriate electron acceptors and trace growth factors are added to the recirculating water creating a zone of high bioactivity. As the zone of bioactivity spreads (Fig. 12.12 d), the hydrocarbons are progressively oxidized and removed. The ideal outcome of the bioremediation system is the complete oxidation of the soluble and residual hydrocarbons.

This overall strategy (Fig. 12.12) illustrates the hydrologic steps that must be followed in order to implement a bioremediation program effectively:

1. An accurate delineation of the contaminated zone is critical to the success of the system. Such delineation may be accomplished by installing monitoring wells, by performing geophysical surveys, or by performing soil gas surveys. In many instances a combination of techniques is needed to delineate the contaminated zone with sufficient confidence.
2. Information on the physical hydrologic system, including water levels, aquifer properties, the presence or absence of confining beds, and hydrologic boundaries is assimilated and used to predict the system's response to pumping stress. This normally is accomplished by constructing and calibrating a digital-flow model of the system and by using this model to predict system response to pumping senarios. Based on the modeling results, the locations of pumping and recharge wells are specified so as to ensure nutrient delivery to all of the contaminated zone. Furthermore, the rates of pumpage and recharge are specified in order to maximize the efficiency of nutrient delivery.
3. After design and construction, the system is started up without nutrient addition in order to verify the hydraulic behavior of the system. Because the results of digital ground-water flow models are approximate, this stage

12.8 PETROLEUM HYDROCARBON CONTAMINATION IN GROUND-WATER SYSTEMS 353

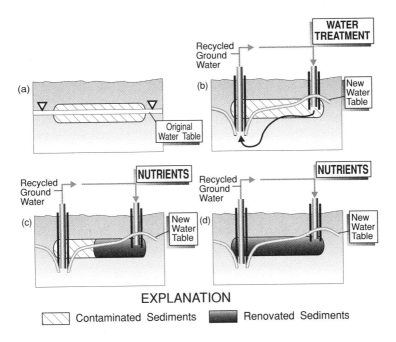

FIGURE 12.12. Steps involved in field-scale bioremediation of shallow aquifer systems.

generally results in fine tuning of the distribution and rates of recirculating water.

4. After nutrient addition commences, concentrations of hydrocarbons in the effluent and, if possible, in the sediment are closely monitored as a function of time. This step is critical because it allows an evaluation of system efficiency.

It's clear from these steps that the most important factor in successfully designing and operating a bioremediation system is the quality of the hydrologic characterization of the system. If the location of contaminant is incompletely known, or if there is large uncertainty regarding the hydrologic properties of the aquifer, adequate design of the remediation system becomes an impossible task. Furthermore, without adequate hydrologic information, it is not possible to monitor the outcome of the remediation system adequately.

Remediation of Jet Fuel Contamination One of the best-documented examples of hydrocarbon contamination bioremediation, and one of the few that has been reported in the peer-reviewed scientific literature, was an instance of JP-4 jet fuel contamination in Michigan (Hutchins et al., 1991). In 1985, several large underground storage tanks at a U.S. Coast Guard facility were found to be leaking. Before the spill was detected, several thousand gallons of fuel leaked into the shallow water-table aquifer.

The aquifer sediments, as one might expect from the location near Lake Michigan, are lacustrine in origin and exhibit the coarsening-upward lithology that is characteristic of such sediments. The section is underlain by fine-grained sediments of deep-water origin. These fine-grained sediments are overlain by sands and gravels deposited in progressively shallower water. The net result is a highly permeable water-table aquifer with a well-defined lower boundary. Water levels in this aquifer vary with recharge events and range from 12 to 18 feet below land surface.

When the jet fuel leaked into the aquifer, it moved downward until it reached the water table and then spread laterally. The seasonal raising and lowering of the water table, however, had the effect of "smearing" contamination over a fairly thick interval. This smearing effect created a difficult hydrologic problem with regard to considering alternative bioremediation strategies. In order to supply nutrients to the entire zone of contamination, relatively large quantities of water would have to be recirculated. For the plot chosen for the bioremediation demonstration, a digital ground-water flow analysis indicated that about 200 gallons per minute of water would be needed to raise the water table to the top of the contaminated zone.

In order to deliver the quantity of ground water needed to raise the water table sufficiently, an infiltration gallery was constructed over the test plot. The infiltration gallery was composed of gravel-packed, perforated PVC pipes buried five feet underground. Ground water was supplied by a line of interdiction wells that had been previously constructed to keep contamination from moving offsite. This water was routed through an activated carbon treatment system prior to injection in order to avoid reintroducing contaminants to the system. A series of multi-screened cluster wells were installed within the test plot to monitor the progress of the remediation system.

As is common in fuel spills, this previously aerobic aquifer was entirely anaerobic in the contaminated zone. Because it was deemed unfeasible to deliver sufficient quantities of oxygen to effect hydrocarbon degradation, nitrate (~ 62 mg/L) was added as an electron acceptor. In addition, quantities of potassium (10 mg/L), phosphate (10 mg/L), and ammonium (20 mg/L) were added to supply other nutrients. Laboratory investigations had previously demonstrated the ability of indigenous microorganisms to degrade petroleum hydrocarbons under denitrifying conditions (Hutchins et al., 1991).

The system was started up in early 1989. First, treated ground water was recirculated for a period of 42 days to establish hydraulic equilibrium and to perform a series of tracer tests for tracking the delivery of recharge water. Dissolved oxygen was present in the recharge water during the initial phase of the experiment. At 42 days, circulation of nitrate and other nutrients commenced.

The progress of the system for removing BTEX compounds from ground water, from the time of the initial nutrient-free addition to about 90 days after the nutrient addition, is shown in Figure 12.13. In Figure 12.13, concentrations of benzene and toluene are shown in the injection water and at different levels within the recharge gallery. Interestingly, benzene was the first compound to be completely removed, and this occurred prior to any nutrient addition (Fig. 12.13a). Toluene was also rapidly removed in the recharge gallery, although it persisted in the injection water throughout the experiment (Fig. 12.13b). Para- and meta-xylenes were removed gradually, with the most noticeable removal occurring after nutrient addition. The ortho-xylene isomer was degraded most slowly.

In addition to following concentration changes with time, it was possible to

12.8 PETROLEUM HYDROCARBON CONTAMINATION IN GROUND-WATER SYSTEMS

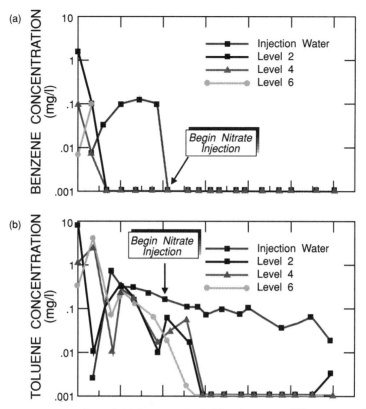

FIGURE 12.13 Degradation of (a) benzene and (b) toluene at different sampling levels during operation of the bioremediation system, Traverse City, Michigan. (Modified from Hutchins et al., 1991, with permission courtesy of the National Water Well Association.)

sum the total mass of oxygen, nitrate, and BTEX compounds removed during recirculation. A large amount of oxygen (55 Kg) was used prior to initiation of nutrient addition indicating that aerobic degradation was a significant process during the experiment. After nutrient injection, about 23 Kg of nitrate were consumed. Much of the benzene present in the system was removed during the initial nutrient-free stage. However, the amount of toluene, ethylbenzene, and xylenes increased, reflecting transport in with injection water. Evidently, most of the benzene was removed by abiotic leaching processes. After nutrient addition commenced, a net loss of toluene, ethylbenzene, and xylenes from the ground water was documented.

In addition to tracing contaminant concentrations in the recirculating ground water, cores of sediment were obtained before and after the treatment. All of the BTEX compounds studied showed significant reductions. Benzene concentrations in the sediment decreased from 0.8 to less than 0.01 mg/Kg and toluene concentrations decreased from 32.6 to 0.03 mg/Kg of sediment. Clearly, there was substantial removal of the contaminant compounds.

The fact that substantial reductions of contaminant levels were documented clearly showed the efficiency of this particular remediation strategy. Less clear is the relative impact of microbial and hydrologic processes on the observed contaminant removals. On one hand, the consumption of dissolved oxygen and nitrate shows that microbial oxidation of hydrocarbons was taking place. On the other hand, the behavior of benzene suggests that simple abiotic leaching following water treatment was the chief removal mechanism. This is one of the most important findings documented by this study. Contaminant removal results from the complex interaction between hydrologic and microbial processes. Separating the relative effects of each is virtually impossible to achieve under field conditions.

The study of Hutchins et al., (1991) was interesting from another perspective. While removal of BTEX compounds was the main focus of the study, it was also documented that about ten times more electron acceptor (oxygen and nitrate) were consumed than could be accounted for by the loss of BTEX compounds. The obvious conclusion is that compounds other than BTEX, such as aliphatic, alicyclic, and polyaromatic compounds were also being degraded. However, as concentrations of these compounds were not studied directly, this conclusion remains equivocal. Based on the observed ability of microorganisms to degrade these other compounds, however, it is the most reasonable interpretation of the data. Thus, the remediation strategy developed by these investigators appears applicable to a wide range of petroleum hydrocarbons in natural environments.

12.9 SUMMARY

Crude oils are composed of a wide variety of organic components and are classified based on the relative amounts of paraffins (alkanes), aromatics, and asphaltic (sulfur-bearing) compounds present. Pennsylvania crude oil consists mainly of paraffins and has less than 0.5% sulfur. In contrast, some of the California crude oils contain up to 5% sulfur. Refined fuels are derived from crude oil by a variety of processes. Fractional distillation of crude oil yields a variety of products including gasoline, the components of which boil between 30°C and 200°C, and lubrication oils which boil at temperatures greater than 400 °C. In addition to distillation, numerous other processes such as high-temperature "cracking" and catalytic processes are employed to increase the yield of gasoline from crude oil. The burning efficiency of gasoline, as measured by the octane rating scale, may be enhanced by catalytically increasing the number of branch-chained hydrocarbons present.

Refined gasolines consist mainly of aliphatic (16-25%), cycloalkane (35-55%), and aromatic (10-20%) hydrocarbons. The aromatic fraction normally consists of 1-3% benzene, toluene, ethylbenzene, and xylene (BTEX). Although they are a minor component of most fuels, BTEX compounds have a disproportionately large environmental impact on ground-water systems because they are the most water soluble of the petroleum hydrocarbons.

The aerobic degradation of aliphatic hydrocarbons often proceeds by oxidizing a terminal methyl group, forming a long-chain fatty acid, followed by beta-oxidation of the organic acid. The presence of branched chains on an aliphatic molecule interferes with beta-oxidation so that these compounds are normally more refractory than straight-chained aliphatic compounds. Degradation of cycloalkanes under aerobic conditions appears to be often mediated by communities of microorganisms, each of which derives some benefit from catalyzing a portion of the process. The first step in cycloalkane oxidation is the oxidation of one of the carbons to form cyclic alcohols such as cyclohexanol. The cyclic ring is then cleaved with the formation of fatty acids that are further degraded via beta-oxidation.

Aromatic hydrocarbons are degraded by a number of aerobic and anaerobic mechanisms. Under aerobic conditions, benzene and its derivatives are partially oxidized to catechols with subsequent ring cleavage and beta-oxidation. Under anaerobic conditions, phenols and cresols are partially oxidized to intermediate products. Most anaerobic biodegradation of aromatic hydrocarbons is effected by communities of microorganisms. However, one strictly anaerobic bacterium, the Fe(III)-reducer GS-15, has been shown to oxidize toluene completely in pure culture under anaerobic conditions.

Much of what is known about the biochemistry of petroleum hydrocarbon oxidation has been learned from aquatic sediment or soil microorganisms. However, subsurface microorganisms appear to follow many of the same degradative pathways. Under aerobic conditions, BTEX compounds have been shown to quickly (1-2 years) degrade. Similar behavior has been observed in anaerobic ground-water systems.

In many cases, rates of microbial degradation are limited mainly by the lack of suitable electron acceptors. Artificially adding electron acceptors and other essential nutrients to ground-water systems is termed "in situ bioremediation" and is a technology that can be effective in restoring hydrocarbon-contaminated shallow aquifer systems. The keys to effective bioremediation are often more hydrologic than biologic and include (1) accurate assessment of the contamination, (2) reliable information on the hydrologic properties of the aquifer, and (3) adequate control of recharging water so that the entire contaminated area can be remediated. In the few documented instances in which these factors have been adequately considered, bioremediation of petroleum hydrocarbon contamination has been highly successful.

CHAPTER 13

MICROBIAL DEGRADATION OF HALOGENATED ORGANIC COMPOUNDS IN GROUND-WATER SYSTEMS

Halogenated organic compounds include some of the most useful and economically important chemicals available to industry and agriculture. Alkyl halides-aliphatic organic compounds that contain chloride, bromide, iodide, or fluoride-are used in a number of applications. Chlorinated alkanes and alkenes, such as trichloroethane and trichloroethylene, are used as dry-cleaning fluids, refrigerants, degreasing agents, solvents, and in the production of caffeine-free coffee. Bromated alkyl compounds, such as ethylene dibromide (EDB), are used as pesticides. Aryl halides-aromatic compounds such as chlorobenzenes, chlorophenols and chlorinated biphenyls-are extensively used as herbicides, insecticides, fungicides, heat transfer media, insulators, and lubricants.

Some of the uses of these compounds, pesticides for example, involve intentional release to the environment. Pesticides used in agriculture may be mobilized from the soil zone and carried to shallow ground-water aquifers by percolating recharge. Many uses for these compounds, solvents or refigerants for example, do not involve intentional release to the environment. Nevertheless, improper disposal techniques (seepage pits) and inadvertent chemical spills have led to extensive release of these compounds to ground-water systems. In either case, the potential for subsurface microorganisms to degrade these compounds influences their mobility in ground-water systems. This chapter considers the various microbially mediated processes that affect the fate and transport of halogenated organic compounds in ground-water systems.

13.1 CHEMISTRY AND USES OF HALOGENATED ORGANIC COMPOUNDS

There are several classes of halogenated organic compounds that are commonly used in industry and agriculture. It is useful to distinquish between these broad classes of compounds because each one has characteristic properties and uses. In addition, each class exhibits characteristic patterns of microbial degradation. In this discussion, it is useful to distinguish between these compounds on the basis of chemical structure, chemical properties, and uses. Thus, the halogenated aliphatic, monoaromatic, and biphenyl compounds are considered separately. In addition, chlorinated insecticides and herbicides are considered separately.

13.1.1 Aliphatic Compounds

The chemical structure of some representative halogenated aliphatic compounds are shown in Figure 13.1.

Halogenated aliphatic compounds, which are often refered to as alkyl halides, are produced from alkanes and alkenes by means of *halogenation reactions*. For example, if methane and chlorine gas are heated together, a small proportion of the chlorine molecules will split into two highly reactive chlorine atoms.

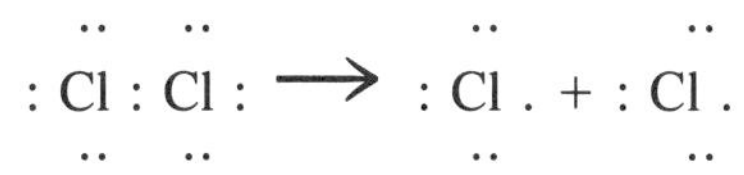

$$:\!\overset{..}{\underset{..}{Cl}}\!:\!\overset{..}{\underset{..}{Cl}}\!: \longrightarrow :\!\overset{..}{\underset{..}{Cl}}\!.\; +\; :\!\overset{..}{\underset{..}{Cl}}\!.$$

Chlorine molecule   Chlorine atoms

These chlorine atoms, which lack the full complement of eight orbital electrons, tend to remove hydrogens from the methane molecules forming a methyl radical.

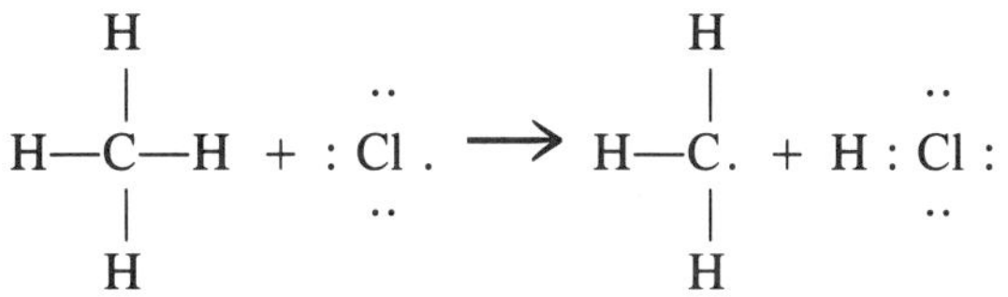

$$\underset{H}{\overset{H}{H-\underset{|}{\overset{|}{C}}-H}} \;+\; :\!\overset{..}{\underset{..}{Cl}}\!. \longrightarrow \underset{H}{\overset{H}{H-\underset{|}{\overset{|}{C}}.}} \;+\; H:\!\overset{..}{\underset{..}{Cl}}\!:$$

Methyl radical + Hydrogen chloride

This methyl radical then reacts with chlorine to form methyl chloride.

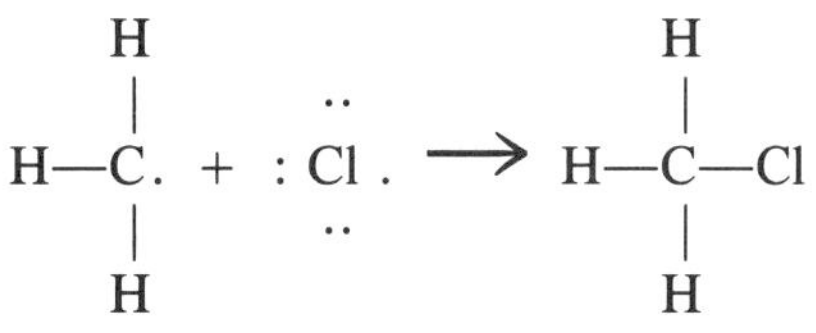

$$\underset{H}{\overset{H}{H-\underset{|}{\overset{|}{C}}.}} \;+\; :\!\overset{..}{\underset{..}{Cl}}\!. \longrightarrow \underset{H}{\overset{H}{H-\underset{|}{\overset{|}{C}}-Cl}}$$

Halogenation reactions are exothermic and, once the reaction sequence begins, tend to continue as a chain reaction. A common source of halogenated methanes in drinking water supplies is the chlorination of methane-containing ground water. Trihalomethanes, which are potentially carcinogenic and which are regulated in water supplies, are produced by sequential halogenation when methane-containing ground water is chlorinated as shown above.

Halogenation reactions can be initiated with alkanes of any length and they do not necessarily terminate after one halogenation reaction has occurred. For example, if ethane reacts with chlorine gas, the first product is chloroethane.

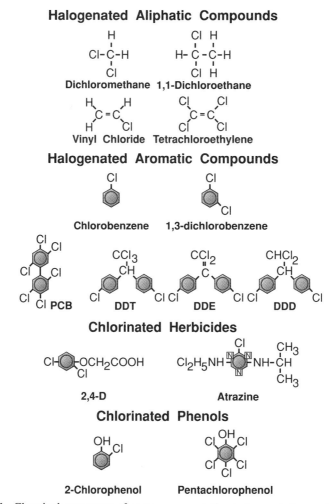

FIGURE 13.1. Chemical structures of some representative halogenated organic compounds.

$$\text{Cl}^- + \text{H}-\underset{\underset{\text{H}}{|}}{\overset{\overset{\text{H}}{|}}{\text{C}}}-\underset{\underset{\text{H}}{|}}{\overset{\overset{\text{H}}{|}}{\text{C}}}-\text{H} \rightarrow \text{H}-\underset{\underset{\text{H}}{|}}{\overset{\overset{\text{H}}{|}}{\text{C}}}-\underset{\underset{\text{H}}{|}}{\overset{\overset{\text{H}}{|}}{\text{C}}}-\text{Cl}$$

With a subsequent chlorination, two alternatives are possible.

$$\text{Cl}^- + \text{H}-\underset{\underset{\text{H}}{|}}{\overset{\overset{\text{H}}{|}}{\text{C}}}-\underset{\underset{\text{H}}{|}}{\overset{\overset{\text{H}}{|}}{\text{C}}}-\text{Cl} \rightarrow \text{H}-\underset{\underset{\text{H}}{|}}{\overset{\overset{\text{H}}{|}}{\text{C}}}-\underset{\underset{\text{H}}{|}}{\overset{\overset{\text{Cl}}{|}}{\text{C}}}-\text{Cl} \quad \text{1-dichloroethane}$$

or

$$\text{Cl}^- + \text{H}-\overset{\overset{\text{H}}{|}}{\underset{\underset{\text{H}}{|}}{\text{C}}}-\overset{\overset{\text{H}}{|}}{\underset{\underset{\text{H}}{|}}{\text{C}}}-\text{Cl} \longrightarrow \text{H}-\overset{\overset{\text{Cl}}{|}}{\underset{\underset{\text{H}}{|}}{\text{C}}}-\overset{\overset{\text{H}}{|}}{\underset{\underset{\text{H}}{|}}{\text{C}}}-\text{Cl} \quad \text{1,2-dichloroethane}$$

Thus, halogenation reactions typically yield a mixture of products that must be subsequently separated.

Alkyl halides have chemical properties that are useful for a wide range of industrial applications: having a high vapor pressure, having a high aqueous solubility, and serving as excellent solvents for non-polar organic compounds. The solubility of selected alkyl halides, which contributes to their mobility in ground-water systems, is given in Table 13.1.

Chloromethane, dichloromethane, and trichloromethane (chloroform) are used in manufacturing silicones, synthetic rubbers, and refrigerants. Carbon tetrachloride is used in the manufacture of fire extinguishers and in dry-cleaning operations. 1,2-dichloroethane is used as an insecticidal fumigant and in the manufacture of paints. Vinyl chloride (1-chloroethene) is widely used in the manufacture of polyvinyl polymers. Ethylene dibromide (EDB) is used as a solvent, in the manufacture of waterproofing agents, and as a soil fumigant for controlling nematodes in agriculture.

Because of the many uses to which these compounds are put, large amounts are manufactured yearly in the industrialized nations. The yearly production of some alkyl halides in the United States is given in Table 13.2. The amounts of these chemicals that are produced are staggering. The production of just dichloroethane, for example, exceeded 4.5 million metric tons in 1981.

With such large quantities of alkyl halides being produced, it is inevitable that some will be released into the environment. When released to soils or to surface water, the primary fate of most alkyl halides is volatilization. This is due to the relatively high vapor pressure of these compounds. In addition, the relatively low octanol-water partition coefficients (Table 13.1) of these compounds results in their having relatively little affinity for sorption onto organic matter and minerals in soils. Experimental studies have shown that evaporative losses of alkyl halides from soils and surface water generally exceeds 80% in less than ten days.

Once these compounds have been delivered to ground-water systems, however, rates of volatilization are greatly diminished. In subsurface environments, the principal attenuating mechanisms for alkyl halides are sorption, biodegradation, and volatilization to the unsaturated zone.

13.1.2 Monocyclic Aromatic Compounds

The chemical structure of some common halogenated monocyclic aromatic compounds are shown in Figure 13.1.

Halogenated monocyclic aromatic compounds, which are often termed "aryl halides," may be produced from benzene by means of halogenation reactions. Although these halogenation reactions are analogous to those of the aliphatic

TABLE 13.1 Octanol-water Partition Coefficients (K_{ow}) and Solubilities of Selected Halogenated Hydrocarbons

Compound	Log K_{ow}	Solubility (mg/L)
Dichloromethane	—	20,000
Chloroform	1.97	7,950
Carbon tetrachloride	2.64	800
1,2-Dichloroethane	—	8,450
1,1,1-Trichloroethane	—	1,360
1,1,2,2-Tetrachloroethane	—	3,230
1,2-Dibromoethane	—	3,520
Vinyl chloride	—	2,700
1,1,2,2-Tetrachloroethylene	2.60	400
Bromoform	—	3,190
1,2-Dichloropropane	—	3,570
Chlorobenzene	2.84	448
1,2-Dichlorobenzene	3.38	148
1,4-Dichlorobenzene	3.38	79
1,2,4-Trichlorobenzene	4.02	19
Fluorobenzene	2.27	1,540
Bromobenzene	2.99	446
Iodobenzene	3.25	340

Source: From Smith, Witkowski, and Fusillo, 1987.

compounds, the details of reaction are different. Aryl halides, such as chlorobenzene, may be produced from benzene and chlorine gas in the presence of a ferric chloride catalyst. The ferric chloride serves to produce a chloronium ion

$$: \ddot{\underset{..}{Cl}} : \ddot{\underset{..}{Cl}} : \; + \; FeCl_3 \longrightarrow \; : \ddot{\underset{..}{Cl}} \; + \; FeCl_4^-$$

that is highly electrophilic and attacks the electrons of the carbon-carbon double bond present in benzene. The net result is the exchange of the chloronium ion for a hydrogen ion in the benzene structure, with the consequent formation of chlorobenzene. Similar reactions are involved in producing a variety of substituted benzene structures.

Halogenated benzenes are used for a variety of industrial and agricultural applications. Chlorobenzene is used as an insecticide, as well as in the manufacture of dyes and more complex aryl halides. Dichlorobenzene (DCB) is used as a fumigant and insecticide (DCB is the active ingredient in mothballs), as well as in industrial odor control and in metal polishing. DCB is also used as an active ingredient in toilet bowl cleaners-a use that delivers large amounts of DCB to sewage treatment plants and septic systems. Brominated benzene structures, such as bromobenzene, are used as solvents for waxes and fats and as additives to motor oil and fuels.

Aryl halides are manufactured in large quantities in the United States (Table 13.2) for these and other uses. Although the quantities of aryl halides produced are large, they are substantially less than the amounts of alkyl halides produced.

TABLE 13.2 U.S. Production (metric tons × 10⁴) of Selected Halogenated Aliphatic and Monocyclic Aromatic Hydrocarbons

Compound	1961–65	1966–70	1971–75	1976	1977	1978	1979	1980	1981	1982
Chloromethane	5.5	14.9	20.8	17.1	21.6	20.6	21.0	16.4	18.4	16.0
Dichloromethane	6.9	14.5	19.0	24.4	21.7	25.9	28.7	25.5	26.8	23.8
Chloroform	3.6	9.1	11.6	13.2	13.7	15.8	16.1	16.0	18.3	13.5
Carbon tetrachloride	21.8	36.5	45.9	38.9	36.7	33.4	32.4	32.1	32.9	nd
Chloroethane	26.6	23.1	28.1	30.4	27.8	24.5	26.4	17.9	14.7	13.1
1,2-dichloroethane	82.1	234.8	379.2	264.6	498.7	498.9	534.9	503.7	452.3	nd
1,1,1-trichloroethane	nd	14.0	21.9	28.6	28.8	29.2	32.5	31.4	27.8	25.0
Chlorodifluoromethane	nd	nd	nd	7.7	8.1	9.3	9.6	10.3	11.4	nd
Trichlorofluoromethane	5.6	10.4	13.6	11.6	9.7	8.8	7.6	10.3	7.4	nd
Dichlorodifluoromethane	9.6	15.1	19.9	17.8	16.2	14.8	13.3	13.4	14.7	nd
Vinyl chloride	63.8	142.1	223.1	257.5	271.5	315.8	289.7	293.2	311.7	nd
Trichloroethylene	13.9	24.5	18.8	14.3	13.5	13.6	14.5	12.1	11.7	8.1
Tetrachloroethylene	14.2	27.0	32.3	16.3	27.8	32.9	35.1	34.7	31.3	nd
Bromoform	0.6	0.9	1.2	nd	1.6	1.6	nd	nd	nd	nd
Ethylene dibromide	nd	13.8	14.0	9.1	11.1	10.4	13.0	8.8	7.6	nd
Monochlorobenzene	25.0	25.0	17.2	14.9	14.8	13.4	14.7	12.8	2.9	nd
1,2-dichlorobenzene	1.9	2.7	2.7	2.2	2.1	1.9	2.6	2.2	2.3	nd
1,4-dichlorobenzene	3.2	2.9	2.9	1.7	3.0	1.9	3.8	3.4	3.3	nd
Alkyl benzenes	28.3	29.4	23.6	nd	23.6	23.8	28.4	40.6	24.2	nd
2,4 + 2,6-dinitrotoluene	nd	12.6	18.6	18.0	nd	29.8	31.2	nd	22.8	nd

Source: From Smith, Witkowski, and Fusillo, 1987.

Nevertheless, large quantities of these compounds are released to the environment by means of accidental spills or pesticide applications.

The aryl halides are less soluble than the alkyl halides and are less volatile. However, the environmental fate of these two classes of compounds are similar. The aryl halides, when spilled at land surface or in surface-water bodies, tend to volatilize to the atmosphere. However, because they are moderately soluble, once they have been delivered to ground-water systems, they tend to be fairly mobile. The relatively low octanol-water partition coefficients of these compounds indicate that they are not strongly sorbed by mineral particles or organic material. Once they have reached ground-water systems, biodegradation processes are principal attenuation mechanisms for these compounds.

13.1.3 Polychlorinated Biphenyls

The chemical structure of one common polychlorinated biphenyl (PCB), 2,2′,4,4′,5,5′ PCB, is shown in Figure 13.1.

PCBs are produced by chlorination, a chemical reaction that requires catalysis, of the biphenyl structure. PCBs are chemically and thermally quite stable. In addition, they exhibit excellent heat transfer and dielectric properties. Because of these properties, PCBs have been extensively utilized as dielectric fluids in electrical capacitors and transformers, in heat transfer systems, and as lubricants in turbine engines and vacuum pumps.

PCBs were first used in industrial applications during the 1930's. Production of PCBs peaked during the 1960's and early 1970's with about 35,000 metric tons being produced per year in the United States alone. Because of their environmental persistence and toxicity, PCBs were first regulated in 1976, and in 1979 the manufacture and sale of PCBs was discontinued in the United States. In spite of this action, a large number of operating transformers still contain PCBs. Furthermore, because of their persistence in the environment, PCBs remain a serious environmental problem in some areas.

PCBs exhibit significantly different characteristics of aqueous solubility and mobility than do halogenated monoaromatic compounds. Table 13.3 shows the solubility and octanol-water partition coefficients of selected PCBs. The observed octanol-water partition coefficients are much higher and the solubilities are much lower than monoaromatic compounds. Because of these properties, PCBs are strongly adsorbed by mineral and organic materials in soils and are seldom found in significant aqueous concentrations in ground-water systems. Nevertheless, their presence in sediments associated with shallow water-table aquifers is a matter of environmental concern in some hydrologic systems. As such, the microbial degradation of PCBs in these environments is an important issue.

13.1.4 Organochlorine Insecticides

The structure of dichlorodiphenyltrichloroethane (DDT), the first organochlorine insecticide ever developed, is shown in Figure 13.1. Since the introduction of DDT in 1939, numerous other chlorinated insecticides have been developed and used to varying degrees. Chlordane was developed in 1945 followed by heptachlor, aldrin, dieldrin, and toxaphene in 1948. Many of these chlorinated compounds do

TABLE 13.3 Octanol-water Partition Coefficients (K_{ow}) and Solubilities of selected PCB's and Organochlorine Insecticides

Compound	Log K_{ow}	Solubility (mg/L)
Biphenyl	4.09	7.5
2,4'-PCB	5.10	—
4,4'-PCB	5.58	0.062
2,5,2',5'-PCB	5.81	0.016
2,4,5,2'5'-PCB	6.11	0.010
2,4,5,2'4',5'-PCB	6.72	0.00095
p,p'-DDT	6.36	0.0031
o,p'-DDT	—	0.026
p,p'-DDD	5.99	0.020
o,p'-DDD	6.08	—
p,p'-DDE	5.69	0.040
o,p'-DDE	5.78	—
Lindane	3.70	7.87
Aldrin	—	0.017
Dieldrin	—	0.195
Chlordane	5.48	0.056
Toxaphene	—	0.74
Endrin	—	0.26
Heptachlor	—	0.056
α-Endosilfan	—	0.53

Source: From Smith, Witkowski, and Fusillo, 1987.

not contain the alicyclic ring, but are treated together in this discussion because of their common usage.

Organochlorine insecticides have been widely distributed in the environment by agricultural spraying. However, because of their persistence in the environment, toxic effects, and tendency to bioaccumulate in higher organisms, their use has been greatly decreased over the last 15 years. During the 1960's, the annual production of these compounds in the United States exceeded 70,000 tons. However, by the mid-1990's, U.S. production was virtually nonexistent.

The solubility and octanol-water partition coefficients of organochlorine insecticides are similar to those of PCBs (Table 13.3). Lindane, with an aqueous solubility of 7.87 mg/L is the most soluble of these compounds, but most exhibit solubilities of under 0.01 mg/L. This low solubility, as well as the high affinity of these compounds for minerals and organic matter in soils, greatly limits their delivery to ground-water systems. However, because these compounds were extensively applied to agricultural areas, some transport to ground-water systems has occurred. Because of this, the biodegradation characteristics of these compounds in ground-water systems are important in some systems.

13.1.5 Chlorinated Herbicides

Herbicides are extensively used in the United States to control weed growth in agriculture and to control aquatic vegetation in navigable surface-water bodies.

The structures of atrazine and 2,4-dichlorophenoxyacetic acid (2,4-D), the most widely used herbicides in the United States, are shown in Figure 13.1.

Chlorinated herbicides are characterized by relatively high aqueous solubilities (Table 13.4). Because of this, they do not tend to sorb strongly to mineral surfaces or soil organic matter. Because these chemicals are applied extensively to agricultural areas, they are often found in underlying water-table aquifers. As such, the ability of subsurface microorganisms to degrade these substances is a topic of considerable interest.

Most herbicides are rapidly degraded in surface-water systems. Investigations have shown that the half-lives of 2,4-D and atrazine in surface-water systems are on the order of 3 to 12 days. 2,4-D has been reported to degrade rapidly under the aerobic conditions characteristic of most surface-water environments. However, the degradation of 2,4-D appears to be slowed under anaerobic conditions. In contrast, atrazine appears to degrade under either aerobic or anaerobic conditions.

13.1.6 Chlorinated Phenols

The phenols are a group of organic compounds characterized by a benzene ring with one or more hydroxyl groups attached. Chlorinated phenols have one or more chlorines attached (Fig. 13.1). Monochlorophenols are seldom used directly in industrial applications but are important intermediates in the production of di-, tri-, and pentachlorophenols. 2,4-dichlorophenol is used as an antimildew agent in the textile industry and in the manufacture of wood preservatives. In addition, 2,4-dichlorophenol is used to manufacture the herbicide 2,4-D and 2,4,5 trichlorophenol is used to manufacture the herbicide 2,4,5-T. The total production of di-, tri-, tetra-, and pentachlorophenol is about 25,000 metric tons.

TABLE 13.4 Aqueous solubilities of some common herbicides

Compound	Solubility (mg/L)
2,4-D	890
2,4,5-T	278
Atrazine	70
Simazine	5
Monuron	230
Diuron	42
Linuron	75
Monolinuron	735
Propanil (Stam F-34)	50–225
Picloram	430
Benefin	0.1
Profluralin	0.1
Trifluralin	4
Silvex	140
Diquat	700,000
Paraquat	miscible

Source: Smith, Witkowski, and Fusillo, 1987.

Phenolic compounds are important components of creosote, which is used extensively as a wood preservative and particularly as a preservative of telephone poles. Creosote is a volatile, oily substance that is obtained by distillation of wood tar. Prior to about 1950, creosote was the principal material used as a preservative for telephone poles and related products. After 1950, chemical engineering technologies for producing pentachlorophenols (PCPs) became more efficient, and PCPs began to supplant raw creosote as a wood-preserving material.

Chlorophenols have very high solubilities and moderately low octanol-water partition coefficients (Table 13.5). As such, sorption to mineral grain and organic matter in aquatic sediments and ground-water systems is generally a minor factor in chlorophenol mobility. On the other hand, chlorophenols are susceptible to microbial degradation under certain conditions. Thus, microbial degradation processes are probably the most important process affecting chlorophenol mobility in subsurface environments.

13.2 MICROBIAL DEGRADATION OF HALOGENATED ORGANIC COMPOUNDS

Because environmental contamination by halogenated organic compounds is so common, considerable efforts have been made to understand the microbial processes that affect their fate and transport in aquatic systems. In general, the kinds of operative microbial degradation processes depend upon the chemical characteristics of the individual compounds and upon the aerobic or anaerobic nature of the environment. It is useful, therefore, to consider the microbial degradation of each class of compound separately.

13.2.1 Aliphatic Compounds

Halogenated aliphatic compounds are subject to biodegradation under both aerobic and anaerobic conditions. However, the microbial processes and the biochemistry involved are very different. As such, it is convenient to consider the aerobic and anaerobic degradation of these compounds separately.

TABLE 13.5 Solubilities and Octanol-Water Partition Coefficients (K_{ow}) of Selected Phenolic Compounds

Compound	Solubility (mg/L)	Log K_{ow}
Phenol	93,000	1.46
2-Chlorophenol	28,500	2.17
2,4-Dichlorophenol	4,500	2.75
2,4,6-Trichlorophenol	800	3.38
Pentachlorophenol	14	5.01
2-Nitrophenol	2,100	1.76
4-Nitrophenol	16,000	1.91
2,4-Dinitrophenol	5,600	1.53
2,4-Dimethylphenol	17,000	2.50
p-Chloro-m-cresol	3,860	2.95
4,6-Dinitro-o-cresol	—	2.85

Aerobic Degradation The wide usage of halogenated aliphatic compounds such as trichloroethylene (TCE) in dry-cleaning operations and in metal-degreasing processes has led to widespread contamination of shallow water-table aquifers. During the 1960's and early 1970's, simply dumping these used chemicals into unlined trenches was considered to be an acceptable disposal practice. As a result of such disposal practices, countless instances of TCE and other halogenated aliphatic compound contamination have been documented. In California, for example, a water-quality survey of 7,167 water supply wells revealed that 812, or about 11%, contained measurable concentrations of organic contaminants. By far the most common contaminants found were trichloro- and tetrachloroethylene.

The widespread occurrence of halogenated aliphatic compounds in shallow ground-water systems reflects not only multiple contamination sources but also their relative resistance to microbial attack under aerobic conditions. Because halogenation of alkanes and alkenes is accompanied by the loss of electrons (oxidation), many halogenated aliphatic compounds are relatively oxidized. Because they have few electrons available to donate, they contain little useful energy for microorganisms to exploit via oxidation. There are presently no known microorganisms that are capable of growth using halogenated aliphatic compounds as a sole electron or carbon source. The inability of aerobic microorganisms to utilize compounds such as TCE as an energy source is an important reason for their observed persistence in aerobic aquifer systems.

Even though aerobic microorganisms cannot obtain energy for growth by oxidizing halogenated aliphatic compounds, it does not follow that aerobic transformations cannot occur. Wilson and Wilson (1985) presented evidence that TCE was degraded to carbon dioxide by soil microorganisms under aerobic conditions if methane was also present. Subsequent studies revealed that TCE was being degraded gratuitously via the activity of the methane monooxygenase system of methanotrophic bacteria.

Methane monooxygenase (MMO) is an enzyme system that catalyzes the incorporation of molecular oxygen into methane to produce methanol. Two types of MMO have been identified. One is an insoluble membrane-bound form, and the other is a soluble form. While performing the same function, these forms of MMO are thought to differ in their chemical structures as well as in their modes of catalysis. At present, only the soluble form of MMO has been purified and characterized (Colby and Dalton, 1978). At first it was thought that only the membrane-bound MMO was involved in TCE degradation. Subsequent work, however, has demonstrated that TCE is degraded by the soluble MMO as well (Tsien et al., 1989).

The degradation of TCE and other chlorinated alkenes by MMO is not a straightforward process. Methane oxidation is initiated by the MMO-mediated incorporation of oxygen. This requires reducing power in the form of NADH (Fig. 12.3). In methane oxidation, NADH is regenerated in the final step by the oxidation of formate. This NADH is then recycled to produce more methanol, and the oxidation is continuous. There is evidence, however, that the cometabolic oxidation of TCE does not result in the regeneration of NADH. The process, therefore, is self-limiting and does not continue indefinitely. Methane can be provided as an electron donor, but methane is a competitive inhibitor of TCE oxidation (Semprini et al., 1991). Furthermore, there is the possibility that some of the oxidized intermediate compounds may be toxic to microorganisms (Henry and Grbic-Galic, 1991). Thus, there are numerous possible limitations to TCE cometabolism by methanotrophic bacteria.

The potential limitations of TCE degradation by methanotrophic bacteria can be relieved under laboratory conditions by a number of manipulations. One strategy is to deliver limited amounts of methane to the degrading microbial populations. For example, Lanzarone and McCarty (1991) showed that methane concentrations of 1.5 mg/L in the influent of a methanotrophic column was more effective in removing TCE than a column fed 4.5 mg/L methane. Another strategy is to build up intracellular NADH levels and MMO activity by feeding the microorganisms on pulses of methane. In between methane pulses, TCE is gratuitously degraded, resulting in its removal. This process has been tested as a strategy for bioremediating TCE-contaminated aquifers under field conditions (Semprini et al., 1990).

Methane-oxidizing bacteria, and thus MMO activity, are very common in soils and aquifer systems. However, because methane availability in aerobic water-table aquifers and soils is so limited, the activity of MMO is typically too low to effect TCE degradation in most aerobic aquifers. For this reason, and because other heterotrophic oxidation of TCE does not occur under most conditions, TCE is ubiquitously observed to accumulate in shallow aerobic water-table aquifers.

Anaerobic Degradation Although the degradation of alkyl halides under aerobic conditions is limited to specialized enzyme systems such as MMO, degradation under anaerobic conditions is a much more general process. Early experience with spills of alkyl halides into the environment indicated that compounds such as tetrachloroethylene and trichloroethylene were progressively dechlorinated under anaerobic conditions. The microbial processes involved in these reductive dehalogenation reactions were subsequently elucidated by studies at Stanford University during the early 1980's (Bouwer, Rittmann, and McCarty, 1981; Bouwer and McCarty, 1983). These studies utilized batch experiments and continuous-flow column experiments to trace the transformation of alkyl halides under methanogenic conditions.

Since those early studies, a number of investigations have examined reductive dehalogenation of alkyl halides under anaerobic conditions and a consistent picture of the microbial processes involved has emerged. In one such study, Vogel and McCarty (1985) studied the behavior of trichloroethylene (TCE) and tetrachloroethylene (PCE) in a 180-cm methanogenic column. Methanogenic conditions were maintained by feeding the column a number of organic compounds, including acetone, isopropanol, phosphate, and ammonia. Concentrations of TCE and PCE decreased rapidly along the direction of flow in the column indicating that these compounds were degraded fairly efficiently (Fig. 13.2a). However, vinyl chloride was shown to accumulate in the column. This behavior, which has been observed in many studies, indicated that dehalogenation occurred in a sequential fashion (Fig. 13.2b). Furthermore, this evidence indicated that the final dehalogenation step was the most rate limiting; hence the observed accumulation of vinyl chloride.

The tendency of reductive dehalogenation to transform TCE and PCE to vinyl chloride causes considerable environmental problems. While TCE and PCE are suspected of being carcinogenic, there is no doubt about the carcinogenic properties of vinyl chloride. Considerable effort has been applied to determining how

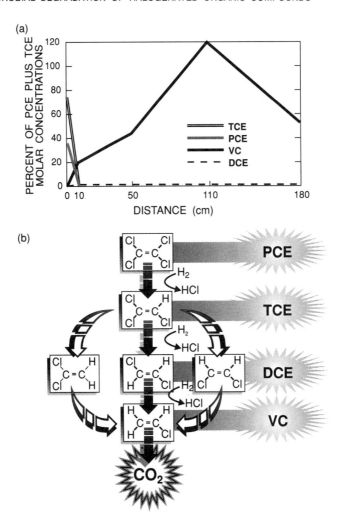

FIGURE 13.2. (a) Degradation of trichloroethylene (TCE) and tetrachloroethylene (PCE), with the accumulation of vinyl chloride (VC) under methanogenic conditions, and (b) the sequential reductive dehalogenation of PCE. (Modified from Vogel and McCarty, 1985, with permission courtesy of the American Society for Microbiology.)

this problem can be overcome. For example, Freedman and Gossett (1989) showed that acclimated methanogenic cultures could completely dehalogenate PCE and TCE, provided sufficient electron donor, such as methanol or gaseous hydrogen, were supplied. These results suggest that ex situ (i.e., above ground) bioreactors could be developed to dechlorinate alkyl halides completely. However, given the electron donor-poor condition of most ground-water systems, these results also suggest that complete reductive dehalogenation in subsurface environments is difficult to achieve.

In contrast to TCE, complete reductive dehalogenation of alkyl halide pesticides such as ethylene dibromide (EDB), 1,2-dibromo-3-chloropropane (DBCP), and 2,3-dibromobutane have been reported (Castro and Belser, 1968). In one experiment using anaerobic soils, about 97% of radiolabeled EDB was transformed to ethylene via reductive dehalogenation in eight weeks of incubation. DBCP proved more recalcitrant to reductive dehalogenation, but significant conversion of DBCP to n-proponol was observed.

13.2.2 Monocyclic Aromatic Compounds

Halogenated monocyclic aromatic compounds, like halogenated aliphatic compounds, are subject to biodegradation under both aerobic and anaerobic conditions. Under aerobic conditions, however, there is little similarity between the biochemistry of alkyl halide and aryl halide degradation. As discussed in Section 13.2.1, aerobic heterotrophs cannot grow on alkyl halides. In contrast, a number of microorganisms have been described that can grow on aryl halides as a sole carbon and energy source.

Under anaerobic conditions, there is more similarity between microbial processes degrading alkyl and aryl halides. Both classes of compounds are subject to reductive dehalogenation which can lead to complete mineralization. However, the biochemical pathways involved differ due to the different properties of the compounds.

Aerobic Degradation A number of microorganisms have been described that are capable of utilizing aryl halides as primary substrates. For example, Reineke and Knackmuss (1984) described a bacterium capable of utilizing chlorobenzene as a sole carbon and energy source. Studies with this chlorobenzene-utilizing bacterium led to the elucidation of the operative degradation pathway.

The general pathway of chlorobenzene degradation as indicated by the studies of Reineke and his colleagues is shown in Figure 13.3. The general strategy is the formation of chlorocatechol followed by *ortho* cleavage of the ring. The substituted muconic acid is then dechlorinated and the non-chlorinated intermediate products metabolized to 3-oxoadipate which then enters the cell's tricarboxylic acid cycle. Note that this general pathway is virtually identical to the pathway of aerobic benzene degradation. The only substantial difference is the necessary elimination of the chloride from the organic compound at some point in the pathway.

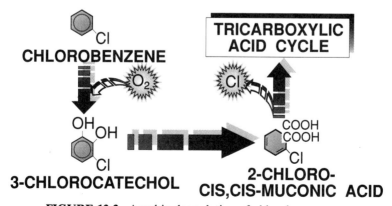

FIGURE 13.3. Aerobic degradation of chlorobenzene.

More recent work on two *Pseudomonas* strains (Sander et al., 1991) has demonstrated that similar mechanisms are involved in the degradation of trichloro- (TCB) and tetrachlorobenzene (TeCB). In the case of 1,2,4-TCB, degradation is initiated by dioxygenation of the aromatic nuclei forming 3,4,6-trichlorocatechol. Subsequent *ortho* cleavage results in formation of the corresponding 2,3,5-trichloromuconate, which channels into the tricarboxylic cycle, as was the case with chlorobenzene (Fig. 13.3).

The similarity of chlorobenzene degradation to that of benzene has not been lost on microbial ecologists. Because most aryl halides have only been produced in the last few decades, it seems unlikely that degradation pathways existed prior to widespread availability of aryl halides to microbial populations via environmental contamination. To some investigators, this suggests that the enzyme systems involved in aryl halide degradation evolved from preexisting systems. This, in turn, implies an extremely rapid rate of microbial evolution to xenobiotic compounds.

To other investigators, however, such rapid evolution seems unlikely. They point to the fact that there are numerous naturally occurring chlorinated aromatic compounds, 2,4-dichlorophenol for instance, that could form the basis for the observed degradation pathways. The important issue in this unresolved controversy deals with understanding the rate at which microorganisms adapt genetically to new potential substrates. If, as some would argue, microorganisms have evolved the ability to grow on aryl halides during the last 20 years, this implies that new degradative pathways may evolve fairly quickly. On the other hand, if microorganisms already have the enzymatic capability to deal with aryl halides, then the implication is that degradative pathways may not evolve as quickly as some would like to think. The fact that aerobic microorganisms have not evolved mechanisms for growing on alkyl halides, despite long exposure in contaminated environments, suggests that one cannot count on microorganisms evolving degradative mechanisms to any xenobiotic compound in just a few years.

Anaerobic Degradation Aryl halides are subject to reductive dehalogenation mediated by microorganisms in the same manner as alkyl halides. These processes have been closely studied in anaerobic lake sediments using halogenated benzoates as model aryl compounds (Horowitz, Suflita, and Tiedje, 1983; Suflita, Robinson, and Tiedje, 1983). These studies show that anaerobic microorganisms were capable of first dehalogenating benzoates and subsequently converting the dehalogenated benzoate to methane.

The dehalogenation and utilization of chlorobenzoates observed in methanogenic lake sediments follows a characteristic pattern. After addition of 2-bromobenzoate there is a lag time (see Chapter 10) before initiation of dehalogenation. Once dehalogenation is initiated, benzoate begins to accumulate and is subsequently converted to methane and carbon dioxide. Dehalogenation was not observed in heat- or formaldehyde-killed controls suggesting that the dehalogenation process was enzymatic. Furthermore, dehalogenation did not occur if the sediments were incubated above 39 °C, further indicating that dehalogenation was microbially mediated.

The fact that a well-defined lag time is required prior to the onset of dehalogenation is significant. This implies that the biochemical mechanisms involved in dehalogenation are inducible. This, in turn, suggests that the genetic capability for dehalogenation is present in the microbial population but that it remains unexpressed in the absence of the appropriate substrate. The fact that the dehalogenated substrate is subsequently mineralized suggests that these mechanisms enable the microorganisms to transform these potential substrates to a form that is more readily assimilated.

Because reductive dehalogenation of aryl halides is microbially-mediated, several unexpected features of these processes have been noted (Suflita, Robinson, and Tiedje, 1983). For example, when anaerobic sediments acclimated to 3-chlorobenzoate were exposed to 3,5-dichlorobenzoate, it was expected that both compounds would be degraded simultaneously. Instead, it was observed that the 3,5-dichlorobenzoate was preferentially metabolized with the accumulation of 3-chlorobenzoate. Apparently these two substrates were utilized competitively, with the presence of 3,5-dichlorobenzoate inhibiting the utilization of 3-chlorobenzoate. Based on these results, similar unexpected patterns of aryl halide reductive dehalogenation in aquatic or subsurface systems may be reasonably anticipated.

13.2.3 Polychlorinated Biphenyls

Polychlorinated biphenyls (PCBs) were used extensively as transformer fluids between about 1935 and 1978 in the United States. The Monsanto Company, the sole U.S. manufacturer, marketed these products under the trade name Aroclor. Several varieties of Aroclor were produced that differed primarily in the weight percent of chlorine per molecule. In Monsanto's nomenclature, Aroclor was denoted by a four-digit number in which a "12" referred to a biphenyl and the second two digits referred to the weight percent of chlorine in the molecule. Thus, Aroclor 1254 was a product consisting largely of biphenyls that were 54% chlorine, which translates to about 5 chlorine atoms per molecule. In practice, however, Aroclors were mixtures of many different PCBs containing anywhere from one to ten chlorines per molecule.

The resistance of PCBs to microbial degradation is one property that made these compounds so useful in industrial applications. Nevertheless, they are subject to degradation under certain conditions. PCBs with relatively few chlorines per molecule are subject to oxidation reactions under aerobic conditions. For example, Clark, Chian, and Griffin (1979) described degradation of PCBs by mixed microbial cultures taken from soils and aquatic sediments.

Furukawa et al. (1978) described an *Acinetobacter* species capable of degrading 2,4,4'-trichlorobiphenyl. The pathway of degradation implied by these studies involves the familiar pattern of hydroxylation followed by *meta* cleavage of the alicyclic ring. These studies have documented the relative recalcitrance of heavily chlorinated PCBs to aerobic microbial oxidation. Presumably this reflects interference in the hydroxylation and ring cleavage steps.

Under anaerobic conditions, the predominant microbially mediated degradation process is reductive dehalogenation. In anaerobic sediments of the Hudson River, which were extensively contaminated by Aroclor 1242, a marked increase in the proportion of biphenyls exhibiting 1, 2, and 3 chlorines relative to the parent product was noted (Brown et al., 1984). These authors suggest that the observed reductive dehalogenation is analogous to the reductive processes documented for halogenated benzoic acids.

The picture that emerges from consideration of microbial attack on PCBs under aerobic and anaerobic conditions is that highly chlorinated molecules are resistant to aerobic oxidation but are subject to reductive dehalogenation. Conversely, lightly chlorinated PCBs are resistant to reductive dehalogenation, but may be aerobically oxidized. This pattern has not been lost on researchers investigating possible bioremediation strategies. Specifically, it has been suggested that a sequence of anaerobic degradation followed by aerobic degradation can result in complete mineralization of PCBs. Unfortunately, the sequence normally followed in natural aquatic sediments, initial shallow burial and aerobic degradation followed by deeper anaerobic degradation, is just the reverse of the pattern that would be most effective. However, this shows that microbial processes have the potential to completely degrade these compounds under the proper conditions.

13.2.4 Organochlorine Insecticides

Organochlorine insecticides such as DDT, lindane, and chlordane were extensively used in agriculture during the 1940's, 1950's, and 1960's. However, because of their environmental persistence and because of their tendency to accumulate in lipid reservoirs of aquatic organisms, their use has been largely discontinued in the United States. They have been replaced by such insecticides as the pyrethroids, permethrin, and fenvalerate with require much lower application rates. Nevertheless, because of DDT's environmental persistence it remains an issue in some areas.

Although DDT is very persistent in the environment, degradation does occur. The first step that must occur in order for degradation to proceed is the reductive dechlorination of DDT to DDD (dichlorodiphenyldichloroethane). This process occurs only under highly reducing conditions and does not appear to require microbial mediation. However, this reaction may be facilitated by reduced iron porphyrins present in mixed populations of anaerobic bacteria. In addition, DDT may be reduced to DDE (dichlorodiphenyldichlorethylene). This appears to be a dead-end pathway as further degradation of DDE is not generally observed.

Like DDT, other organochlorine insecticides are rather recalcitrant to degradation. The cyclodiene insecticides aldrin, chlordane, endrine, and heptachlor are generally considered to be persistent in most environments. Heptachlor may be oxidized to heptachlor epoxide by soil microorganisms and endrin is metabolized to form ketones and aldehydes that retain the attached chlorines. Lindane, the most soluble of the chlorinated insecticides, is also one of the most degradable, forming chlorinated benzenes with the mediation of fermentative soil bacteria such as *Bacillus* and *Clostridium*.

13.2.5 Chlorinated Herbicides

While PCBs and organochlorine insecticides are relatively toxic and resistant to microbial degradation, their mobility in ground-water systems is limited by their tendency to partition onto sediments and organic matter. Chlorinated herbicides such as atrazine and 2,4-D are much more soluble and tend to be transported more easily in the unsaturated and saturated zones of ground-water systems. Because of this relative mobility, relative rates of microbial degradation have a significant impact on their fate and transport.

Atrazine belongs to the general class of herbicides known as the s-triazines, a name that refers to the central ring structure that contains three nitrogen groups. Attached to the ring structure of atrazine is a chlorine atom, an isopropylamino group, and an ethylamino group. Atrazine is degraded by a number of biotic and abiotic mechanisms that involve removal of one or more of the functional groups as well as ring cleavage.

The most important mechanisms in atrazine degradation are hydroxylation (H) and dealkylation (D) reactions (Fig. 13.4). The dechlorination and hydroxylation

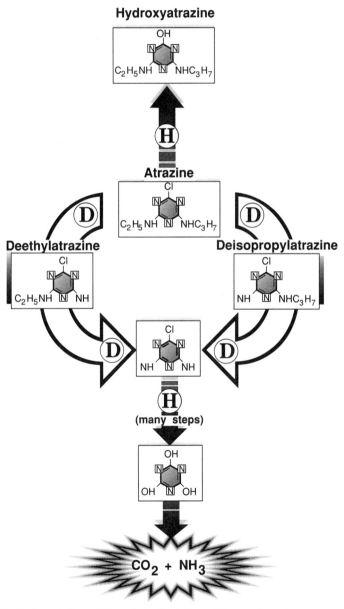

FIGURE 13.4 Hydroxylation (H) and dealkylation (D) reactions in the degradation of atrazine.

of atrazine to produce hydroxyatrazine occurs both biotically and abiotically. When occurring abiotically, this process has been shown to be pH-dependent with maximal rates occurring at a pH of about 8.0. The formation of hydroxyatrazine results in detoxification of atrazine. Dealkylation of both the isopropyl and ethyl groups have been reported. Behki and Khan (1986) showed that particular strains of *Pseudomonas* were capable of utilizing atrazine as a sole carbon source and that atrazine was metabolized via dealkylation reactions. This study also suggested that formation of deisopropylatrazine was favored over deethylatrazine and that particular species were capable of dechlorinating either of these metabolites when another carbon source was provided. It was also observed that dechlorination of atrazine tended to proceed after dealkylation.

13.2.6 Chlorinated Phenols

Chlorinated phenols released into surface and subsurface sediments from wood-treatment facilities or as degradation products of chlorinated herbicides are subject to both aerobic and anaerobic biodegradation. As with other chlorinated organic compounds, however, the mechanisms and pathways involved differ substantially under aerobic and anaerobic conditions. As such, they will be considered separately.

Aerobic Degradation Under aerobic conditions, two distinct mechanisms for the metabolism of chlorophenols have been extensively studied. The first of these mechanisms is the methylation of one or more hydroxyl groups associated with the phenolic compound. Neilson, Allard, and Hynning (1983) showed that a species of *Arthrobacter* carries out this process. The chloroanisoles that are produced by this process are volatile and tend to escape to the atmosphere at land surface. As chlorophenols are typically present in the wood shavings used for litter in poultry-raising operations, the production of volatile chloroanisoles may result in contamination of chicken eggs or meat. However, such volatilization is probably a minor process in most subsurface environments.

The second mechanism for the metabolism of chlorophenols that has been extensively studied involves the formation of chlorocatechols from chlorophenols. A number of *Pseudomonas, Alcaligenes,* and *Nocardia* species have been described that carry out this and similar biochemical transformations. For example, Knackmuss and Hellwig (1978) showed that a *Pseudomonas* species could grow using 4-chlorophenol as a sole carbon and energy source with the complete degradation of the 4-chlorophenol. The initial step in this metabolic pathway was the conversion of 4-chlorophenol to 4-chlorocatechol with subsequent cleavage of the alicylic ring via either the *ortho* or the *meta* pathways.

Whether these or other biochemical mechanisms are involved in the aerobic degradation of chlorophenols in natural systems is difficult to say. However, the aerobic degradation of chlorinated phenols has been observed in soils and this process has been suggested as a means for the bioreclamation of chlorophenol-contaminated soils.

Anaerobic Degradation The anaerobic degradation of chlorinated phenols has been extensively studied because of the delivery of these compounds to sewage treatment plants and their subsequent degradation by activated-sludge microorganisms. Mikesell and Boyd (1986) showed that chlorinated phenols were completely degraded by a combination of reductive dehalogenation and anaerobic metabolism.

The reductive dehalogenation of chlorophenols was studied by Bryant, Hale, and Rogers (1991) who showed that the order of chlorine removal from pentachlorophenol during reductive dehalogenation was predictable. In the case of one adapted sediment community, the chlorine in the *para* position tended to be removed preferentially to that in the *ortho* position and that the chlorine in the *meta* position tended to be removed last. These or similar reductive processes are probably the principal mechanisms involved in chlorophenol degradation in anaerobic aquifer systems.

13.3 DEGRADATION OF HALOGENATED ORGANIC COMPOUNDS IN GROUND-WATER SYSTEMS

Microbiologic studies such as those outlined in the previous section have demonstrated the potential for microorganisms to transform a number of halogenated organic compounds. More importantly, however, these studies form the basis for understanding the biochemical mechanisms involved in these transformations. The importance of such understanding is difficult to overestimate. On one hand, it provides a framework for evaluating how microbial processes may affect contaminant fate and transport in a variety of environments, including ground-water systems. On the other hand, such understanding forms the basis for developing practical in situ and ex situ bioremediation strategies.

Although laboratory studies are a necessary first step in understanding the microbial degradation of chlorinated hydrocarbons under particular conditions, field studies are also necessary. Laboratory studies can never perfectly reproduce the variety of conditions present in even the simplest ground-water system. Thus, determining the extent and rates of biodegradation processes for a particular system is best accomplished using ground-water and/or sediment chemistry data. Ideally, laboratory studies and field studies can be used in conjunction to form a better understanding of microbial processes in contaminated ground-water systems.

13.3.1 Aliphatic Compounds

Chlorinated aliphatic compounds such as trichloroethylene (TCE), trichloroethane (TCA), tetrachloroethylene (PCE), and their degradation products are among the most commonly observed contaminants found in shallow ground-water systems. This reflects, in part, the shear volume of these chemicals used in industry and the laxity of disposal practices earlier in this century. It also reflects the relative recalcitrance of these compounds to microbial degradation under conditions commonly found in shallow ground-water systems.

There have been literally thousands of reported instances of alkyl halide contamination of ground water in the United States alone. It is very seldom that these are "clean" spills, in the sense that alkyl halides are the only compounds involved; rather, it is most commonly observed that other contaminants, such as petroleum hydrocarbons or metals, accompany the alkyl halides. The net effect is that it is often difficult to use the results of laboratory degradation experiments as analogs for evaluating in situ degradation patterns and rates. Nevertheless, it is possible to make some generalizations and the following two case studies serve as examples.

TCE Contamination at a Metal-Plating Plant One well-documented case of TCE contamination occurred at Picatinny Arsenal in New Jersey (Imbrigiotta et al., 1988), and this instance of contamination is in many ways typical of what is observed at numerous other spills.

From 1960 to 1981, Building 24 at the Picatinny Arsenal housed a metal-plating operation. Prior to applying the metal-plating treatment, unfinished metal was subjected to degreasing, with TCE and TCA being the principal agents used. This being a metal-plating operation, however, large quantities of trace metals, including cyanide, cadmium, copper, lead, nickel, and zinc, were mobilized in the treatment process. The wastewater stream that came from the operation included large quantities of metal-bearing solutions as well as TCE and TCA. The wastewater was discharged into a dry well where it was allowed to percolate into the local ground-water system.

The hydrology of this site is unique and consists of stratified and unstratified glacial sediments that fill a valley incised into metamorphic bedrock. Recharge is provided to the glacial aquifer primarily at the valley margins, and ground-water discharges from the system at Green Pond Brook. This is primarily an aerobic aquifer with uncontaminated ground-water containing 8.0 mg/L dissolved oxygen. As would be expected, the contamination has oriented itself around the predominant directions of ground-water flow, so that significant amounts of contaminants are transported and ultimately discharged to Green Pond Brook. A zone of particularly high TCE contamination is present at the base of the alluvial aquifer. This configuration reflects the fact that TCE is denser than water and suggests that downward free-phase flow of TCE has occurred in the past.

Evaluating how microbial processes may impact the fate and transport of TCE in this system is not straight-forward. On one hand, given the fact that TCE is not readily degradable via aerobic metabolism, one might predict that TCE will behave conservatively in this system. However, petroleum hydrocarbons such as BTEX compounds are present in the contamination plume as well, apparently having been removed from metal surfaces during the degreasing operations or delivered from separate petroleum spills. An additional unknown in evaluating microbial processes at this site is the effects of the potentially toxic trace metals. Concentrations ranged up to 65 μg/L for cadmium, 48 μg/L for chromium, 40 μg/L for cyanide, 70 μg/L for copper, and 190 μg/L for lead.

Given these considerable uncertainties, evaluating ground-water chemistry data is the only real alternative for deducing the nature of ongoing microbial processes at this site. Although the aquifer is naturally aerobic, the contamination plume is anaerobic indicating the presence of significant aerobic metabolism. It would not be expected that aerobic metabolism of TCE would be the principal oxygen-depleting process, and this has been confirmed by laboratory incubations of cored sediments. However, the presence of BTEX compounds, which are readily degraded by aerobic bacteria, apparently explains the presence of anaerobic conditions (Wilson, 1988).

The presence of BTEX compounds with the consequent production of anaerobic conditions, has led to substantial reductive dehalogenation of TCE. Concentrations of *cis*-dichloroethylene (*cis*-DCE) and vinyl chloride (VC) range up to 200 and 30 μg/L respectively. Thus, the mixed nature of the waste has contributed substantially to the operative microbial processes. It is difficult to evaluate the potentially toxic effects of the trace metals in this systems. However, it can be concluded that these effects are not sufficient to preclude active reductive dehalogenation. As the products of reductive dehalogenation (DCE and VC) are more susceptible to aerobic oxidation than TCE, it is possible that this particular combination of microbial processes contributes substantially to the degradative fate of TCE in this system.

Degradation Products of TCA in Landfill Leachate Another example of how microbiologic studies can provide a framework for interpreting microbial degradation patterns of TCA in a ground-water system was given by Broede and Passero (1988). The system under study was a leaking landfill in Kalamazoo, Michigan. Between 1980 and 1986, an extensive ground-water monitoring program revealed a plume of organic chemical contamination emanating from the landfill. Some of the compounds present included phenols, BTEX, as well as low concentrations ($\sim$ 10 μg/L) of 1,1,1-trichloroethane (TCA). Surprisingly, concentrations of 1,1-dichloroethane (1,1-DCA) and 1,2-dichloroethane (1,2-DCA) were fairly substantial and ranged between 200 and 1,500 μg/L.

As is typical of this kind of contamination event, the observed behavior of the contamination was erratic and puzzling. For example, the concentrations of contaminants varied widely over time at any given location. In one well near the source of contamination, concentrations of 1,1-DCA and 1,2-DCA varied by a factor of 5 over several years. Just as puzzling, however, was the fact that TCA, and not DCA, had been the compound that was disposed of in the landfill. TCA was present in the plume of contamination but always in very low concentrations. Furthermore, chloroethane (CLA) was observed near the landfill but not downgradient of the landfill.

Having this water-chemistry data in hand, but lacking a detailed evaluation of microbial processes at the site presented Broede and Passero (1988) with a very common dilemma. Reductive dehalogenation of TCA to DCA had been demonstrated in published laboratory studies, but such studies had not been performed at the Kalamazoo landfill. If on one hand they made the obvious interpretation that microbial reductive dehalogenation produced 1,1-DCA and 1,2-DCA from TCA, then they were in the position of invoking a microbial process that had not been demonstrated at that site. Conversely, if they did not invoke this microbiologic interpretation, then they were left without an explanation for their water chemistry data.

Broede and Passero (1988) chose to interpret the water-chemistry data based on the model of microbial reductive dehalogenation. Specifically, they proposed that 1,1,1-TCA was transformed in the sequence.

$$1,1,1\text{-TCA} \xrightarrow{\text{Fast}} \begin{matrix} 1,1\text{-DCA} \\ 1,2\text{-DCA} \end{matrix} \xrightarrow{\text{Slow}} \text{CLA} \xrightarrow{\text{Fast}} CO_2 + H_2O$$

The interpretation that 1,1,1-TCA was transformed quickly to DCA was based entirely upon the observation that 1,1,1-TCA was largely absent in the ground water. Similarly, the interpretation that DCA was transformed slowly was based on the observed high concentrations of these compounds. Finally, the absence of CLA in downgradient areas was taken as evidence that CLA was rapidly dechlorinated.

The details of this proposed sequence are subject to criticism on the grounds that evidence for these processes independent of the water chemistry data are not provided. Broede and Passsero justified this approach based on the fact that it was the only rational alternative. Given the fact that reductive dehalogenation is so ubiquitously observed in anaerobic sediments, they argued, such evidence was not necessary. These data, however, raise questions that can only be answered by means of detailed laboratory studies. For example, the interpretation that the DCA-CLA transformation is slower than the CLA-CO_2 transformation is at variance with the observation that reductive dehalogenations tend to slow with decreasing chlorine content. It is perfectly appropriate for water chemistry data to suggest such hypotheses. However, such hypotheses would need independent verification before they could be reliably accepted.

13.3.2 In Situ Bioremediation of Chlorinated Aliphatic Compounds

Chlorinated aliphatic compounds such as TCA and TCE present special technical challenges to in situ methods of bioremediation. In the case of petroleum hydrocarbons, degradation is typically limited by the availability of appropriate electron acceptors or trace nutrients. However, under the aerobic conditions commonly encountered in shallow water-table aquifers, the degradation of alkyl halides are limited by the availability of particular degradative enzyme systems. Inducing indigenous microorganisms to produce the enzyme systems that are capable of actively degrading alkyl halides, and using this as the basis for in situ bioremediation has received considerable attention.

The most promising bioremediation strategy for alkyl halides, and the one that has attracted the most attention in recent years, is the stimulation of indigenous populations of methanotrophic bacteria with consequent induction of their methane monooxygenase enzyme systems. As MMOs are capable of gratuitously oxidizing alkyl halides, such stimulation could result in the large-scale degradation of these compounds in ground-water systems. The microbiologic basis for this has been well documented (Wilson and Wilson, 1985; Fogel, Taddeo, and Fogel, 1986); however, scaling up this process so that it can be applied to ground-water systems creates special challenges.

A field-scale evaluation of in situ bioremediation of chlorinated ethenes, which tested the feasibility of MMO induction to degrade alkyl halides, has been conducted at the Moffett Naval Air Station in California (Roberts et al., 1990; Semprini et al, 1990; Semprini et al., 1991). The hydrologic system selected for study was a shallow (4-6 meter), highly permeable (K ~ 200 ft/d) sand-gravel aquifer that was confined (S approximately 0.00013) by overlying and underlying silts and clays. Hydrologic control of the system was maintained by a series of injection-extraction wells, and the system was monitored by a series of sampling wells.

In a series of tests that were conducted over a three year period, the feasibility of stimulating indigenous methanotrophic bacteria by injecting oxygen and methane into the aquifer, and the efficiency of TCE, c-DCE, and t-DCE degradation were evaluated. In order to stimulate indigenous populations of methanotrophic bacteria, alternating pulses of first oxygen (~20 mg/L) followed by methane (~5 mg/L) were introduced at a rate of 1 L/min into the injection well. Water was withdrawn continuously from the extraction well at a rate of 10 L/min, and concentrations monitored at the intervening sampling wells.

The results of the first biostimulation experiment performed at this site are shown in Figure 13.5a. Initially, concentrations of both methane and dissolved oxygen increased in an intermediate sampling well (S2) in response to the pulsed additions. After a lag time of about 200 hours, concentrations of both DO and methane began to decrease. This indicated the growth of methanotrophic bacteria and consequent methane oxidation. The decrease of methane concentrations to near zero after 400 hours of injection at sampling well S2 indicated that methane oxidation was most active near the injection well. This created concern that the injection well might become biofouled, so the pulse cycle was increased to 12 hours. This, in turn, increased DO and methane concentrations at well S2 at about 500 hours (Fig. 13.5a).

Once an active population of methanotrophs was established, the next phase of the experiment was to evaluate the efficiency of alkyl halide removal. This was accomplished by injecting TCE, t-DCE, c-DCE, and VC into the aquifer superimposed on the alternating pulses of DO and methane. Some results of this experiment are shown in Figure 13.5b. Concentrations of TCE decreased the least (data not shown), with measured concentrations relative to injected concentrations (C/C_o) near 0.9 at the sampling point. c-DCE was also transformed relatively slowly. However, t-DCE and VC exhibited relatively rapid rates of transformation. In addition to demonstrating t-DCE removal, it was also shown that the epoxide intermediate of t-DCE oxidation, trans-dichloroethylene oxide, appeared during these experiments.

The results of these experiments demonstrate that some degree of alkyl halide degradation could be effected by stimulating indigenous populations of methanotrophic bacteria. In addition, these experiments demonstrate that the more heavily halogenated compounds (TCE) are degraded less efficiently than lightly halogenated compounds (VC). This is in general agreement with the results of laboratory experiments. The maximum TCE degradation observed was between 20-30% over a flowpath distance of 2 meters and a residence time of one to two days. The maximum degradation of VC, however, was 90-95% over the same flowpath segment.

Given the results of this important field experiment, it becomes possible to evaluate the efficiency of methanotrophic stimulation as an in situ bioremediation strategy. Careful consideration of these results suggests several limitations of this technology. First of all, it is apparent that primary, highly halogenated compounds such as tetrachlorinated and trichlorinated ethenes and probably ethanes, are the least susceptible to remediation. If even under the most closely controlled

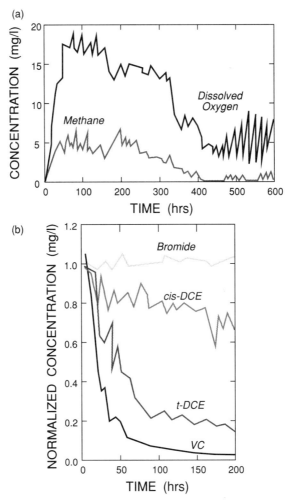

FIGURE 13.5. (a) Concentrations of dissolved oxygen (DO) and methane, and (b) concentrations of bromide, cis-dichloroethylene (C-DCE), trans-dichloroethylene (t-DCE), and vinyl chloride (VC) in a sampling well during in situ bioremediation. (Modified from Semprini et al., 1991, with permission courtesy of the National Water Well Association.)

conditions removal efficiencies are on the order of 20 to 30 %, it becomes questionable that such treatment could effect practical amounts of contaminant removal under less favorable conditions.

The second consideration is economic practicality. It is clear that the efficiency of treatment will be based on extensive hydrologic site characterization, construction of elaborate networks of injection-withdrawal wells, and construction of technologically sophisticated injection equipment. Furthermore, as not all contaminants are likely to be removed at withdrawal wells, some provisions for either effluent treatment and/or reinjection of the waste water must be made. Given the level of technology involved in alkyl halide in situ bioremediation strategies, there may be no clear economic advantage of in situ treatment over conventional pump-and-treat or vacuum-extraction technologies. In any case, the economic pluses and minuses of this possible technology must be considered when selecting appropriate treatment methodologies.

Savannah River Site Demonstration Project   The Savannah River Site in South Carolina is owned by the Department of Energy (DOE) and has been involved in nuclear weapons manufacture since it was built in the early 1940's. For many years, large amounts of degreasers and solvents, predominantly TCE and TCA, were disposed of by percolation into the ground from unlined settling ponds. These waste disposal practices, which have since been discontinued, have extensively contaminated the Tertiary and Cretaceous aquifers that underlie the site.

Because TCE contamination is so prevalent at many DOE sites due to past waste disposal practices, DOE has invested considerable resources into investigating innovative technologies for remediating this problem. One of these technologies currently being tested is bioremediation by methanotrophic bacteria.

The bioremedation system, which is being tested at the SRS by Terry C. Hazen and his associates, differs from the Moffett Naval Air Station experiments in several respects. The scale of the hydrologic system is much larger, for example. The Tertiary aquifers that have been contaminated are about 140 feet below land surface, and the zone of contamination is several hundred feet in diameter. The larger scale of this system created several technical problems. The most serious of these was to design a system of wells that could efficiently deliver oxygen and methane to the zone of contamination. Rather then drilling a large number of standard vertical wells to deliver nutrients and remove contaminated water, two long horizontal wells were constructed instead. The deeper of the two wells was used to inject air and methane into the contaminated aquifer, and the shallow one was used to extract air and contaminants from the unsaturated zone overlying the aquifer. A network of vertical monitoring wells was installed in order to assess the effectiveness of the system.

The bioremediation system was started up in the spring of 1992 with the injection of air into the saturated zone surrounding the deeper well, and the withdrawal of air from the shallower well in the unsaturated zone. Two months later, methane was added to the stream of injected air. It was quickly realized that a major challenge was to assess the relative effectiveness of biologic and nonbiologic processes in remediating the contamination. Because of the volatility of TCE, simply injecting and retrieving air will effectively remove contamination. Thus, simply documenting decreasing concentrations of TCE in the aquifer is not sufficient to show effective bioremediation.

This difficulty of assessing the relative importance of biologic from nonbiologic effects in contamination removal is common to all in situ bioremediation systems. Hazen's group addressed this problem by instituting a comprehensive microbiologic and hydrologic sampling program. Some early results of this sampling program (Terry C. Hazen, written communication, 1992) show that concentrations of TCE and numbers of methanotrophic bacteria were fairly constant during the air injection campaign. However, with the addition of methane, numbers of methanotropic bacteria increased and concentrations of TCE declined noticeably.

This data suggests that methane addition contributed to a buildup of methanotrophic bacteria with a consequent increase in TCE biodegradation rates. However, the evidence is indirect and is as yet inconclusive. This difficulty of clearly establishing the effectiveness of biologic relative to nonbiologic processes is characteristic of all bioremediation systems. As such, methods for evaluating the effectiveness of bioremediation systems is and will continue to be an active research topic.

13.3.3 Degradation Patterns of Alkyl Halide Insecticides

Particular alkyl halides, such as 1,2-dibromo-3-chloropropane (DBCP) and ethylene dibromide (EDB), are widely utilized as fumigants for controlling nematodes in agricultural areas. In addition, compounds such as 1,2,3-trichloropropane (TCP) (a compound more widely used as paint remover) are present as contaminants in soil fumigant preparations. Since these compounds are applied directly to the soil, it is possible for them to be leached by vertically percolating recharge and delivered to ground-water systems. Because of the relatively high solubility of EDB and DBCP relative to other chlorinated insecticides such as DDT, these compounds are more commonly found in ground water.

EDB Contamination in Hawaii One particularly well-documented case of alkyl halide contamination of ground water occurred on the island of Oahu in Hawaii (Oki and Giambelluca, 1987). DBCP and EDB were used extensively for nematode control in pineapple agriculture. In addition, TCP was present as a contaminant of the fumigant DD (dichloropropane-dichloroproene) and applied as well. Concentrations of EDB appear to have increased between 1983 and 1985 whereas concentrations of TCP appear to have decreased. Since these compounds are considered to be readily degradable by aerobic microorganisms, their presence might not be expected. However, a combination of geologic, hydrologic, and horticultural factors appears to speed delivery of these chemicals to the ground-water system and to depress microbial degradation once the chemicals have entered the ground water.

Pineapples must be fairly large and have a high water content (i.e., be juicy) for them to fetch high prices in agricultural markets. In order to produce such high-quality pineapples, pests such as nematodes must be effectively suppressed prior to planting and the fields must be liberally irrigated. To control nematodes, the soils are fumigated once every three to five years. Because of the high rates of irrigation, however, these fairly soluble compounds are readily leached to the underlying aquifer. This aquifer consists of unweathered basalts that are highly permeable. Because of the high temperature of lava that formed the basalts (1200 °C), the rocks contain virtually no organic carbon and microbiologic activity is probably correspondingly low. Thus, even though EDB is readily degradable, degradation rates are less than ground-water flow rates. The net effect is the rapid transport of these chemicals in the subsurface and their appearance in ground-water supplies.

Pesticide Contamination, Long Island Although DDT is relatively insoluble, the large amounts of this compound applied to agricultural areas has resulted in it's being found in ground-water supplies. On Long Island, New York, for example,

DDT, DDD, and DDE have been reported in ground water sampled in Suffolk County. Reported concentrations, however, were very low and in the parts per billion range. The most significant aspect of these data is that they demonstrate that production and accumulation of the DDD and DDE, which would be expected based on laboratory studies of DDT degradation, are actually observed in field studies.

13.3.4 Degradation Patterns of Chlorobenzenes

Chlorobenzenes are distributed to ground-water systems by a number of different avenues and are readily degraded under aerobic conditions (via hydroxylation to chlorocatechol, with subsequent *ortho* cleavage) and anaerobically via reductive dehalogenation.

The behavior of chlorobenzenes in aerobic aquifers seems to be consistent with what would be expected based on laboratory studies. An excellent example of this was given by Schwarzenbach et al. (1983) who studied the infiltration of contaminated surface water into a shallow ground-water system in Switzerland.

A series of observation wells were placed so that concentration changes of trichloroethylene (TCE) and 1,4-dichlorobenzene (DCB) could be monitored as ground-water flowed downgradient. Concentrations of TCE changed very little along the flowpath of this aerobic aquifer, which is consistent with laboratory studies showing that nonmethanotrophic aerobic metabolism does not appreciably transform TCE. Concentrations of DCB, however, showed a consistent and significant trend of decrease along the flowpath. Since DCB is chemically stable and not strongly sorbed on aquifer material, these patterns were attributed to aerobic microbial degradation. It is likely that degradation of chlorobenzenes is widespread in aerobic aquifers, as illustrated by this example.

The behavior of chlorobenzenes in anaerobic aquifers is more problematic. While reductive dehalogenation has been closely studied using halobenzoates as model compounds, much less is known about the behavior of chlorobenzenes. Based on the benzoate model, however, reductive dehalogenation would be expected to be the predominant degradative process. However, this behavior has not been observed in field studies. In one study of a sewage effluent-contaminated anaerobic aquifer, DCB was shown to persist for a minimum of 20 years (Barber, 1988). However, because denitrification was the predominant terminal electron-accepting process, it is difficult to generalize these results for other terminal electron-accepting processes such as Fe(III) reduction or methanogenesis. It may simply be that microbially mediated reductive dehalogenation of DCB is less efficient under denitrifying conditions. However, this question has yet to be systematically studied in the laboratory.

13.3.5 Degradation of Chlorinated Herbicides

Atrazine is one of the most commonly detected contaminants of ground water in agricultural areas. For example, in a study of 2,263 wells in Nebraska, 13.5% were found to contain atrazine concentrations ranging from 10 to 0.01 $\mu g/L$ (Spalding, Burbach, and Exner, 1989). The occurrence of atrazine in ground water reflects both high rates of usage (about 20 million lbs/yr in Nebraska alone) and it's relatively high solubility (32 mg/L).

In spite of the tremendous usage of atrazine as a preemergent herbicide for corn production, it is significant that average concentrations in intensely irrigated areas do not seem to be increasing over time. This implies that atrazine transport to these ground-water systems is approximately at steady-state with rates of degradation. Atrazine degradation in aquifer systems can be considered as the sum of both biologic and abiologic processes such as irreversible sorption. However, based on its relatively low octanol-water partition coefficient, it seems unlikely that sorption is the sole process leading to removal from aqueous solution. This raises the possibility that microbial degradation in ground-water systems may be significant.

The microbial degradation of atrazine in soils has been well-documented (Behki and Khan, 1986). Curiously, however, the potential for atrazine degradation in aquifer sediments is less clear. The first reported study of atrazine degradation by subsurface sediments (Sinclair and Lee, 1990) showed little or no atrazine degradation in aquifer sediments from Oklahoma. Similarly, Konopka and Turco (1991) were unable to detect atrazine degradation in slurries of aquifer sediment from Indiana that were incubated for as long as 128 days.

The first laboratory evidence for microbial degradation of atrazine in aquifer sediments was provided by McMahon et al. (1992). In a series of experiments, it was shown that ^{14}C ethyl-labeled atrazine was slowly degraded to $^{14}CO_2$ under aerobic conditions, whereas ^{14}C ring-labeled atrazine was not significantly degraded. The aquifer sediments used in this study were cored from a coarse-grained alluvial aquifer underlying an active cornfield in Ohio. Despite the long-term heavy usage of atrazine on the cornfield, monitoring well data never reported measurable concentrations of atrazine. However, measurable concentrations of deethylatrazine were found. Thus, based on both laboratory and field evidence, it appears that atrazine in this ground-water system is subject to partial microbial deethylation. Complete degradation of the atrazine molecule, however, appears to be very sluggish.

Taken together, the available evidence from field studies (Spalding et al., 1989) and laboratory studies (Sinclair, 1990; McMahon et al., 1992) suggests that microbial processes in ground-water systems are partially capable of degrading atrazine at a finite but slow rate. Because of the apparently slow rates, steady-state concentrations of atrazine may be fairly high (1-10 $\mu g/L$) in aquifers receiving high rates of atrazine-laden recharge. However, for most aquifer systems, atrazine contamination of ground water is unlikely to persist due to microbially mediated deethylation reactions (Fig. 13.4).

13.3.6 Degradation of Chlorophenolic Compounds

The microbial degradation of phenolic compounds in ground-water systems was one of the first aspects of subsurface microbiology to be systematically investigated. Ehrlich et al. (1982) reported on the methanogenic degradation of phenols associated with coal tar wastes in a shallow aquifer system and Godsy, Goerlitz, and Ehrlich (1983) investigated the methanogenic degradation of these compounds in laboratory digesters. Similarly, the degradation of chlorinated phenols associated with creosote wastes in a shallow sand and gravel aquifer was described by Troutman et al. (1984).

An interesting aspect of these early studies, particularly that of Troutman et al. (1984), is that consideration of ground-water chemistry and hydrology clearly indicated the importance of microbial processes in degrading the phenolic and chlorophenolic compounds prior to laboratory evidence to that effect. The site studied by these investigators was the American Creosote Works (ACW) site in Pensacola, Florida. At this site, telephone poles were treated with wood preservatives. Prior to 1950, creosote was used exclusively for this purpose. After 1950, pentachlorophenol (PCP) was increasingly utilized and reached about 25,000 gallons per month by the time the plant was closed in 1981.

After being used to treat logs, the creosote and PCP were collected in a recirculation impoundment. An overflow impoundment served to provide additional storage when needed. Leakage from these impoundments resulted in extensive contamination of ground water at this site (Fig. 13.6). Near the impoundments, concentrations of total phenols exceeded 20 mg/L, and concentrations of PCP were in the 1 mg/L range.

Troutman et al. (1984) closely examined the ground-water hydrology of the ACW plant and its vicinity. They determined that the mean horizontal groundwater flow velocities ranged from a maximum of 700 ft/yr in the shallow parts of the system to 200 ft/yr in the deeper parts. Using the minimum flow velocities and assuming that waste disposal had continued for 80 years, then a plume of phenolic compounds up to 16,000 feet long could be expected downgradient of the impoundments. Instead, however, extensive ground-water monitoring documented that phenolic compounds were undetectable in ground water more than 1,000 feet downgradient (Fig. 13.6). This considerable discrepancy could not be explained by either hydrologic or geochemical mechanisms. It was concluded, therefore, that microbial degradation of phenolic compounds was the principal attenuation mechanism.

In support of this hypothesis, Troutman et al. (1984) showed that ground water from downgradient of the impoundments contained elevated counts of methanogenic bacteria. In addition, the ground water contained relatively high concentrations of methane, suggesting that methanogenesis was the predominant terminal electron accepting process at this site. Finally, it was shown that phenols were degraded in laboratory digesters maintained under methanogenic conditions. Interestingly, however, PCP was not degraded in the digesters, and it was suggested that PCP inhibited phenol degradation when present in concentrations greater than 0.25 mg/L.

More recent work by Mueller et al. (1991) reached similar conclusions regarding the fate of PCP at the ACW site. These investigators incubated samples of ground water in shake flasks and monitored the degradation of a number of organic compounds over time. The results of these studies were largely consistent with the results of Troutman et al. (1984). Whereas phenols, polyaromatic hydrocarbons (PAH; groups 1,2, and 3 PAHs contain 2,3, and 4 fused benzene rings respectively), and N-, S-, and O-bearing heterocyclic compounds were actively degraded, there was no measurable degradation of PCP. The degradation of PCP under anaerobic conditions in ground-water systems has, therefore, yet to be demonstrated in

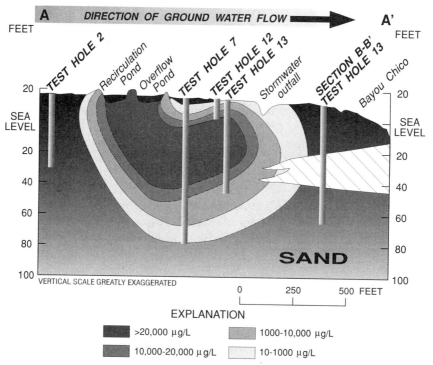

FIGURE 13.6. Phenol contamination at the American Cresote Works Plant, Pensacola, Florida. (Modified from Troutman et al., 1984.)

the laboratory. Circumstantial evidence, such as PCP not extending beyond the phenolic plume, suggests that degradation does occur.

13.4 SUMMARY

Chlorinated organic compounds are widely used for industrial and agricultural purposes. Alkyl halides such as trichloroethylene (TCE) are used as solvents and cleaning agents. Aryl halides such as dichlorobenzene are used for household and industrial deodorizing. Other chlorinated organic compounds are used as insecticides, herbicides, antifungal fumigants, dielectric agents, lubricants, and a host of other applications. It is fairly safe to assert that modern industrial society could not function without the use of these chemicals. As a result of their wide usage, however, these compounds have contaminated a large number of groundwater systems and represent a significant health hazard in some cases.

Microorganisms are capable of transforming chlorinated organic compounds under a wide range of environmental conditions. However, the mechanisms, pathways, and rates of degradation vary widely and depend largely on the structure of the chemical. Alkyl halides such as TCE are not significantly degraded by aerobic heterotrophic bacteria. Consequently, alkyl halides tend to persist in aerobic ground-water systems. These compounds are, however, degraded by the methane monooxygenase enzyme present in methanotrophic bacteria. Methanotrophs cannot obtain energy for growth by TCE oxidation, and methane competitively inhibits this reaction. Nevertheless, stimulation of methanotrophic activity is presently being evaluated as a potential in situ and ex situ bioremediation strategy. Under anaerobic conditions, reductive dehalogenation reactions are the most important alkyl halide-degrading processes. In electron donor-poor ground-water systems, however, reductive dehalogenations tend to be incomplete, resulting in the accumulation of vinyl chloride.

Reductive dehalogenation reactions may also be important microbial processes involved in degrading aryl halides. Unlike alkyl halides, however, aryl halides are capable of supporting the growth of heterotrophic aerobic bacteria. Consequently, aryl halides tend to be rapidly degraded in aerobic ground-water systems. While reductive dehalogenation reactions have been shown in laboratory studies to degrade halogenated benzoates, the importance of reductive dehalogenation of aryl halides in anaerobic ground-water systems is not clear. Studies have reported that dichlorobenzenes persist for long periods of time in some anaerobic ground-water systems.

Chlorinated pesticides vary widely in their susceptibility to microbial attack in ground-water systems. Some insecticides such as DDT are recalcitrant and degrade only partially. As a result of this recalcitrance, DDT has been virtually banned from agricultural practice in the United States. Herbicides such as atrazine are widely used in agricultural practice and are biodegradable in the soil zone. However, degradation rates of atrazine in nutrient-poor aquifer systems is relatively slow and appear to be limited to deethylation reactions. Thus, if rates of atrazine delivery to shallow ground-water systems exceeds rates of degradation, atrazine can accumulate in solution.

Contamination of ground-water systems by phenolic compounds is fairly common and often accompanies wood-treatment operations. Creosote, and later pentachlorophenol (PCP) were widely used as wood preservatives, and wastes from these operations often reach shallow ground-water systems. Phenolic compounds are rapidly degraded in ground-water systems under either aerobic or anaerobic conditions. The most common chlorinated phenol, pentachlorophenol, however, appears to be much more recalcitrant. Nevertheless, field evidence suggests that degradation does occur in some systems.

Available microbiologic and biochemical evidence is sufficient to explain many patterns of chlorinated hydrocarbon degradation observed in ground-water systems. However, because field conditions typically involve multiple contaminants, the presence of inhibiting chemicals or competing substrates, and other factors that make microbial processes difficult to evaluate, it is clear that much more field-oriented research is needed before a consistent picture of degradation in ground-water systems can emerge.

REFERENCES

Aelion, C. M., C. M Swindoll, and F. K. Pfaender. 1987. Adaptation to and biodegradation of xenobiotic compounds by microbial communties from a pristine aquifer. *Applied and Environmental Microbiology* 53(9):2212–2217.

Alexander, M. 1977. *Soil Microbiology*. New York: John Wiley and Sons.

Al'tovskii, M. E., E. L. Bykova, Z. I. Kuznetsova, and V. M Schvets. 1962. Organic substances and microflora of groundwaters and their significance in the processes of oil and gas origin. Moscow: Gostoptekhizdat.

Axcell, B. C., and P. J. Geary. 1975. Purification and some properties of a soluble benzene-oxidizing system from *Psedomonas*. *Biochemical Journal* 146:173–183.

Back, W., and I. Barnes. 1965. Relation of electrochemical potentials and iron content to ground-water flow patterns. U.S. Geological Survey Professional Paper 498-C.

Back, W. 1966. Hydrochemical facies and ground-water flow patterns in northern part of Atlantic Coastal Plain. U.S. Geological Survey Professional Paper 498-A.

Back, W. 1986. Role of aquitards in hydrogeochemical systems: a synopsis. *Applied Geochemistry* 1:427–437.

Baedecker, M. J., and W. Back. 1979. Hydrogeological processes and chemical reactions at a landfill. *Ground Water* 17(5):429–437.

Baedecker, M. J., F. J. Franks, D. F. Goerlitz, and J. A. Hopple. 1986. Geochemistry of a shallow aquifer contaminated with creosote products. In Ragone, S. E., ed., *U.S. Geological Survey Program on Toxic Waste- Ground-Water Contamination*. Proceedings of the Second Technical Meeting, Cape Cod, Massachusetts, October 21- 25, 1985. U.S. Geological Survey Open-File Report 86–481, pp. A-17–20.

Baedecker, M. J., D. I. Siegel, P. C. Bennett, and I.M. Cozzarelli. 1988. The fate and effects of crude oil in a shallow aquifer: I. The distribution of chemical species and geochemical facies, In Mallard, G. E., and S. E. Ragone eds., *U.S. Geological Survey Toxic Substances Hydrology Program*. Proceedings of the Technical Meeting, Phoenix, Arizona, September 26-30, 1988. U.S. Geological Survey Water-Resources Investigations Report 88-4220, pp. 13–20.

Balkwill, D. L. 1989. Numbers, diversity, and morphological characteristics of aerobic, chemoheterotrophic bacteria in deep subsurface sediments from a site in South Carolina. *Geomicrobiology Journal* 7:33–52.

Balkwill, D. L., and W. C. Ghiorse. 1985. Characterization of subsurface bacteria associated with two shallow aquifers in Oklahoma. *Applied and Environmental Microbiology* 50:580–588.

Balkwill, D. L., F. R. Leach, J. T. Wilson, J. F. McNabb, and D. C. White. 1988. Equivalence of microbial biomass measures based on membrane lipid and cell wall components, adenosine triphosphate, and direct counts in subsurface aquifer sediments. *Microbial Ecology* 16:73–84.

Balkwill, D. L., J. K. Fredrickson, and J. M. Thomas. 1989. Vertical and horizontal variations in the physiological diversity of the aerobic chemoheterotrophic bacterial microflora in deep southeast coastal plain sediments. *Applied and Environmental Microbiology* 55(5):1058–1065.

Barber, L. B., II. 1988. Dichlorobenzene in ground water: Evidence for long-term persistence. *Ground Water* 26(6):696–702.

Barkay, T., and B. H. Olson. 1986. Phenotypic and genotypic adaptation of aerobic heterotrophic sediment bacterial communities to mercury stress. *Applied and Environmental Microbiology* 52:403–406.

Barkay, T. 1987. Adaptation of aquatic microbial communities to Hg^{2+} stress. *Applied and Environmental Microbiology* 53(12):2725–2732.

Barker, J. F., G. C. Patrick, and D. Major. 1987. Natural attenuation of aromatic hydrocarbons in a shallow sand aquifer. *Ground Water Monitoring Review*, Winter, pp. 64–71.

Bastin, E. S. 1926. The presence of sulphate-reducing bacteria in oil-field waters. *Science* 63:21–24.

Beam, H. W., and J. J. Perry. 1974. Microbial degradation and assimilation on n-alkyl-substituted cycloparaffins. *Journal of Bacteriology* 118:394–399.

Beeman, R. E., and J. M. Suflita. 1987. Microbial ecology of a shallow unconfined ground water aquifer polluted by municipal landfill leachate. *Microbial Ecology* 14:39–54.

Beeman, R. E., and J. M. Suflita. 1988. Evaluation of deep subsurface sampling procedures using serendipitous microbial contaminants as tracer organisms. *Geomicrobiology Journal* 7:223–233.

Behki, R. M., and S. U. Khan. 1986. Degradation of atrazine by *Pseudomonas*: N-Dealkylation and dehalogenation of atrazine and its metabolites. *Journal of Agricultural Food Chemistry* 34:746–749.

Belyaev, S. S., and M. V. Ivanov. 1983. Bacterial Methanogenesis in underground waters. In Hallberg, R., ed. *Environmental Biogeochemistry and Ecology Bulletin*. (Stockholm) 35:273–280.

Betz, J. L., and P. H. Clarke. 1972. Selective evolution of phenylacetamide-utilizing strains of Pseudomonas aeruginosa. *Journal of General Microbiology* 73:161–174.

Bouwer, E. J., B. E. Rittmann, and P. L. McCarty. 1981. Anaerobic degradation of halogenated 1- and 2-carbon organic compounds. *Environmental Science and Technology* 15:596–599.

Bouwer, E. J., and P. L. McCarty. 1983. Transformations of 1- and 2-carbon halogenated aliphatic organic compounds under methanogenic conditions. *Applied and Environmental Microbiology* 45(4):1286–1294.

Broede, L.D., and R.N. Passero. 1988. Spatial and temporal variations of synthetic organics in the contaminant plume from the KL landfill, Kalamazoo County, Michigan. Proceedings of the Ground Water Chemistry Conference, Denver, Colorado. The National Water Well Association, p. 579–594.

Brown, J. F., R. E. Wagner, D. L. Bedard, M. J. Brennan, J. C. Carnahan, R. J. May, and T. J. Tofflemire. 1984. PCB transformations in upper Hudson sediments. *Northeastern Environmental Science* 3:167–179.

Bryant, F. O., D. D. Hale, and J. E. Rogers. 1991. Regiospecific dechlorination of pentachlorophenol by dichlorophenol-adapted microorganisms in freshwater, anaerobic sediment slurries. *Applied and Environmental Microbiology* 57(8):2293–2301.

Bulger, P. R., A. E. Kehew, and R. A. Nelson. 1989. Dissimilatory nitrate reduction in a waste-water contaminated aquifer. *Ground Water* 27(5):664–671.

Burton, J. F., M. J. Day, and A.T. Bull. 1982. Distribution of bacterial plasmids in clean and polluted sites in a South Wales river. *Applied and Environmental Microbiology* 44:1026–1029.

Bush, P. W., and R. H. Johnston. 1986. Floridan regional aquifer-system study. In R.J. Sun, Ed., Regional Aquifer-System Analysis Program. U.S. Geological Survey Circular 1002, p. 17–29.

Castro C. E., and N. O. Belser. 1968. Biodehalogenation: Reductive dehalogenation of the biocides ethylene dibromide 1,2-dibromo-3-choropropane, and 2,3- dibromobutane in soil. *Environmental Science and Technology* 2:779–783.

Cedarstrom D. J. 1946. Genesis of groundwaters in the coastal plain of Virginia. *Economic Geology* 41(3):218–245.

Champ, D. R., J. Gulens, and R. E. Jackson. 1979. Oxidation-reduction sequences in ground-water flow systems. *Canadian Journal of Earth Sciences* 16(1):12–23.

Chapelle, F. H. 1983. Groundwater geochemistry and calcite cementation of the Aquia aquifer in Southern Maryland. *Water Resources Research* 19(2):545–558.

Chapelle, F. H., and T. M. Kean. 1985. Hydrogeology, digital solute-transport simulation, and geochemistry of the Lower Cretaceous aquifer system near Baltimore, Maryland. *Maryland Geological Survey Report of Investigations* No. 43, 120 pp.

Chapelle, F. H., and L. L. Knobel. 1985. Stable carbon isotopes of DIC in the Aquia aquifer, Maryland: Evidence for an isotopically heavy source of CO_2. *Ground Water* 23(5):592–599.

Chapelle, F. H., J. L. Zelibor, D. J. Grimes, and L. L. Knobel. 1987. Bacteria in deep coastal plain sediments of Maryland: A possible source of CO_2 to ground water. *Water Resources Research* 23(8):1625–1632.

Chapelle, F. H., J. T. Morris, P. B. McMahon, and J. L. Zelibor, Jr. 1988. Bacterial metabolism and the del-13C composition of ground water, Floridan aquifer, South Carolina. *Geology* 16:117–121.

Chapelle, F. H., and D. R. Lovley. 1990. Rates of bacterial metabolism in deep coastal-plain aquifers. *Applied and Environmental Microbiology* 56:1865–1874.

Chapelle, F. H., and P. B. McMahon. 1991. Geochemistry of dissolved inorganic carbon in a coastal plain aquifer: 1. Sulfate from confining beds as an oxidant in microbial CO_2 production. *Journal of Hydrology,* 127:85–108.

Chapelle, F. H., and D. R. Lovley. 1992. Competitive exclusion of sulfate-reduction by Fe(III)-reducing bacteria: A mechanism for producing discrete zones of high-iron ground water. *Ground Water* 30(1):29–36.

Clark, R. R., E. S. K. Chian, and R. A. Griffen. 1979. Degradation of polychlorinated biphenyls by mixed microbial cultures. *Applied and Environmental Microbiology* 37:680

Clarke, P. A. 1984. The evolution of degradative pathways. In D.T. Gibson, ed., Microbial Degradation of Organic Compounds, New York: Marcel Dekker. p. 11–27.

Colby, J., and H. Dalton. 1978. Resolution of the methane monooxygenase of *Methylococcus capsulatus* (Bath) into three components: Purification and properties of component C, a flavoprotein. *Biochemistry Journal* 171:461–468.

Colwell, F. S. 1989. Microbiological comparison of surface soil and unsaturated subsurface soil from a semiarid high desert. *Applied and Environmental Microbiology* 55(9):2420–2423.

Cozzarelli, I. M., R. P. Eganhouse, and M. J. Baedecker. 1988.The fate and effects of crude oil in a shallow aquifer: II.Evidence of anaerobic degradation of monoaromatic hydrocarbons. In Mallard, G. E., and S. E. Ragone eds., *U.S. Geological Survey Toxic Substances Hydrology Program.* Proceedings of the Technical Meeting, Phoenix, Arizona, September 26-30, 1988. U.S. Geological Survey Water-Resources Investigations Report 88-4220, p. 21–33.

Dalhoff, A., and H. J. Rehm. 1976. Studies on regulation of tetradecane oxidation in *Pseudomonas aeruginosa:* II. The effect of glucose on tetradecane oxidation. *European Journal of Applied Microbiology* 3:203–211.

Davis, J.B. 1967. *Petroleum Microbiology.* Amsterdam, The Netherlands: Elsevier.

De Klerk, H., and A. C. van der Linden. 1974. Antonie van Leeuwenhoek 40:7

Dockins, W. S., G. J. Olson, G. A. McFeters, and S. C. Turbak. 1980. Dissimilatory bacterial sulfate reduction in Montana groundwaters. *Geomicrobiology Journal* 2:83–97.

Dunlap, L. E., and D.D. Beckmann. 1988. Soluble hydrocarbons analysis from kerosene/diesel. In Proceedings of the Conference on Petroleum Hydrocarbons and Organic Chemicals in Ground Water: Prevention, Detection and Restoration. National Water Well Association, Dublin, Ohio.

Dunlap, W. J., and J. F. McNabb. 1973. U.S. Environmental Protection Agency Report No. EPA-660/2-73-014. R.S. Kerr Environmental Research Laboratory, Ada, Oklahoma.

Dunlap, W. J., J. F. McNabb, M. R. Scalf, and R. L. Cosby. 1977. Sampling for organic chemicals and microorganisms in the subsurface. U.S. Environmental Protection Agency Report No. EPA-600/2-77-176.

Duxbury, T., and B. Bicknell. 1983. Metal-tolerant bacterial populations from natural and metal-polluted soils. *Soil Biology and Biochemistry* 15:243–250.

Edmunds, W. M. 1973. Trace element variations across an oxidation-reduction barrier in a limestone aquifer. Proceedings of the Symposium on Hydrogeochemistry and Biochemistry, Tokyo, Vol. 1. New York: Clarke, pp. 500–526.

Ehrlich, G. G., R. A. Schroder, and P. Martin, 1984. Microbial populations and nutrient concentrations in a jet fuel-contaminated shallow aquifer a Tustin, California. U.S. Geological Survey Open-File Report 86-324, Chapter G.

Ehrlich, G. G., D.F. Goerlitz, E.M. Godsy, and M.F. Hult. 1982. Degradation of phenolic contaminants in ground water by anaerobic bacteria, St. Louis Park, Minnesota. *Ground Water* 20(6):703–710.

Ekzertsev, V. A. 1951. Microscopic investigations of bacterial flora in the oil-bearing facies of second Baku. *Mikrobiologiya.* Russia. 20:324–329.

Ekzertsev, V.A. and S.I. Kuznetsov. 1954. Investigatons of the microflora of the oil fields of second Baku. *Mikrobiologiya.* Russia. 93:3–14.

El Solh, N., and S. D. Ehrlich. 1982. A small cadmium resistance plasmid isolated from *Staphylococcus aureus. Plasmid* 7:77–84.

Federle, T. W., D. C. Dobbins, J. R. Thornton-Manning, and D. D. Jones. 1986. Microbial biomass, activity, and community structure in subsurface soils. *Ground Water* 24(3):36574.

Flipse, W. J., Jr., and F. T. Bonner. 1985. Nitrogen-isotope ratios of nitrate in ground water under fertilized fields, Long Island, New York. *Ground Water* 23(1):59–67.

Fogel, M. M., A. R. Taddeo, and S. Fogel. 1986. Biodegradation of chlorinated ethenes by a methane-utilizing mixed culture. *Applied and Environmental Microbiology* 51:720–724.

Foster, M. D. 1950. The origin of high sodium bicarbonate waters in the Atlantic and Gulf Coastal Plains. *Geochimica et Cosmochimica Acta* 1(1):33–48.

Fredrickson, J. K., R. J. Hicks, W. W. Li, and F. J. Brockman. 1988. Plasmid incidence from deep subsurface sediments. *Applied and Environmental Microbiology* 54:2916–2923.

Fredrickson, J. K., T. R. Garland, R. J. Hicks, J. M. Thomas, S. W. Li, and S. M. McFadden. 1989. Lithotrophic and heterotrophic bacteria in deep subsurface sediments and their relation to sediment properties. *Geomicrobiology Journal* 7:53–66.

Fredrickson, J. F., D. L. Balkwill, J. M. Zachara, S. W. Li, F. J. Brockman, and M. A. Simmons. 1991. Physiological diversity and distributions of heterotrophic bacteria in deep Cretaceous sediments of the Atlantic Coastal Plain. *Applied and Environmental Microbiology* 57(2):402–411.

Fredrickson, J. K., F. J. Brockman, D. J. Workman, S. W. Li, and T. O. Stevens. 1991. Isolation and characterization of a subsurface bacterium capable of growth on toluene, naphthalene, and other aromatic compounds. *Applied Environmental Microbiology* 57(3):796–803.

Freedman, D. L., and J. M. Gossett. 1989. Biological reductive dechlorination of tetrachloroethylene and trichloroethylene to ethylene under methanogenic conditions. *Applied and Environmental Microbiology* 55(9):2144–2151.

Freeze, R. A., and P. A. Witherspoon. 1967. Theoretical analysis of regional groundwater flow: 2. Effect of Water-table configuration and subsurface permeability variation. *Water Resources Research* 3(2):623–634.

Freeze, R. A., and J. A. Cherry. 1979. *Groundwater*. Englewood Cliffs, NJ. Prentice-Hall.

Froelich, P. N., G. P. Klinkhammer, M. L. Bender, N. A. Luedtke, G. R. Heath, D. Cullen, P. Dauphin, D. Hammond, B. Hartman, and V. Maynard. 1979. Early oxidation of organic matter in pleagic sediments of the eastern equatorial Atlantic: suboxic diagenesis. *Geochimica et Cosmochimica Acta* 44:1075–1090.

Furukawa, K., K. Tonomura, and A. Kamibayashi. 1978. Effect of chlorine substitution on the biodegradability of polychlorinated biphenyls. *Applied and Environmental Microbiology* 35:223–227.

Garland, T., E. Murphy, and K. McFadden. 1991. Evaluation of core segment pore water contamination from tracers, well water samplings, and sediment extractions, p.7177. In Fliermans, C. B., and Hazen, T. C. eds., Proceedings of the First International Symposium on Microbiology of the Deep Subsurface, January 15-19, 1990, Orlando, Florida. WSRC Information Services, Aiken, South Carolina.

Garrels, R. M., and F. T. MacKenzie. 1967. Origin of the chemical compositions of some springs and lakes. In *Equilibrium Concepts in Natural Water Chemistry: Advances in Chemistry Series 67*. Washington, D.C., American Chemical Society, pp. 222–242.

Gehron, M. J., and D. C. White. 1983. Sensitive measurements of phospholipid glycerol in environmental samples. *Journal of Microbiological Methods* 1:23–32.

Gerhart, J. M. 1985. Ground-water recharge and its effects on nitrate concentration beneath a manured field site in Pennsylvania. *Ground Water* 24(4):483–489.

Ghiorse, W. C., and D. L. Balkwill. 1983. Enumeration and characterization of bacteria indigenous to subsurface environments. *Developments in Industrial Microbiology* 24:213–224.

Ghiorse, W.C., and J. L. Wilson. 1988. Microbial ecology of the terrestrial subsurface. *Advances in Applied Microbiology* 33:107–172.

Gillan, R. T. 1983. Analysis of complex fatty acid methyl ester mixtures on non-polar capillary GC columns. *Journal of Chromatography Science* 21:293–297.

Godsy, E. M., and G. G. Ehrlich. 1978. Reconnaissance for microbial activity in the

Magothy aquifer, Bay Park, New York, four years after artificial recharge. *U.S.Geological Survey Journal of Research* 6(6):829–836.

Godsy, E. M. 1980. Isolation of *Methanobacterium bryantii* from a deep aquifer by using a novel broth-antibiotic disk method. *Applied and Environmental Microbiology* 39:1074–1075.

Godsy, E. M., D. F. Goerlitz, and G. G. Ehrlich. 1983. Methanogenesis of phenolic compounds by a bacterial consortium from a contaminated aquifer in St. Louis Park, Minnesota. *Bulletin of Contaminant Toxicolology* 30:261–268.

Gormly, J. R., and R. F. Spalding. 1979. Sources and concentrations of nitrate-nitrogen in ground water of the Central Platte region, Nebraska. *Ground Water* 17(3):291–301.

Gray, E. M., and M. Morgan-Jones. 1980. A comparative study of nitrate levels at three adjacent ground-water sources in a chalk catchment area west of London. *Ground Water* 18(2):159–167.

Grbic-Galic, D., and T. M. Vogel. 1987. Transformation of toluene and benzene by mixed methanogenic cultures. *Applied and Environmental Microbiology* 53(2):254–260.

Gurevich, M. S. 1962. The role of microorganisms in producing the chemical composition of ground water. In Kuznetsov, S.I. ed., *Geologic Activity of Microorganisms*. Transactions (Trudy) of the Institute of Microbiology No. IX, New York, Consultants Bureau, pp. 65–75.

Hadley, P. W., and R. Armstrong. 1991. "Where's the benzene?"-Examining California ground-water quality surveys. *Ground Water* 29(1):35–40.

Haefeli, C., C. Franklin, and K. Hardy. 1984. Plasmid- determined silver resistance in *Pseudomonas stutzeri* isolated from a silver mine. *Journal of Bacteriology* 158:389–392.

Harvey, R. W., R. L. Smith, and L. George. 1984. Effect of organic contamination upon microbial distributions and heterotrophich uptake in a Cape Cod, Mass., aquifer. *Applied and Environmental Microbiology* 48:1197–1202.

Harvey, R. W., and L. H. George. 1987. Growth determinations for unattached bacteria in a contaminated aquifer. *Applied and Environmental Microbiology* 53(12):2992–2996.

Harvie, C. E., Moller, and J. H. Weare. 1984. The prediction of mineral solubilities in natural waters: The Na-K-Mg-Ca-H-Cl-SO4-OH-HCO3-CO3-CO2-H2O system to high ionic strengths at 25°C. *Geochimica et Cosmochimica Acta* 48:723–752.

Hazen, T. C., L. Jimenez, G. L. de Victoria, and C. B. Fliermans. 1991. Comparison of bacteria from deep subsurface sediment and adjacent groundwater. *Microbial Ecology* 22:293–304.

Heitkamp, M. A., and C. E. Cerniglia. 1988. Mineralization of polycyclic aromatic hydrocarbons by a bacterium isolated from sediment below an oil field. *Applied and Environmental Microbiology* 54(6):1612–1614.

Helgeson, H. C. 1968. Evaluation of irreversible reactions in geochemical processes involving minerals and aqueous solutions, I: Thermodynamic relations. *Geochimica et Cosmochimica Acta* 32:853–877.

Helgeson, H. C. R. M. Garrels, and F. T. Mackenzie. 1969. Evaluation of irreversible reactions in geochemical processes involving minerals and aqueous solutions, II: Applications. *Geochimica et Cosmochimica Acta* 33:455–481.

Helgeson, H. C. T. H. Brouwn, A. Nigrini, and T. A. Jones. 1970. Calculation of mass transfer in geochemical processes involving aqueous solutions. *Geochimica et Cosmochimica Acta* 34:569–592.

Hem, J. D. 1985. Study and interpretation of the chemical characteristics of natural water. U.S. Geological Survey Water-Supply Paper 1473.

Henry, S. M., and D. Grbic-Galic. 1991. Influence of endogenous and exogenous electron donors and trichloroethylene oxidation toxicity on trichloroethylene oxidation by methanotrophic cultures from a groundwater aquifer. *Applied and Environmental Microbiology* 57(1):236–244.

Hicks, R. J., and J. K. Fredrickson. 1989. Aerobic metabolic potential of microbial population indigenous to deep subsurface environments. *Geomicrobiology Journal* 7:67–78.

Hirsch, P., and E. Rades-Rohkohl. 1983. Microbial diversity in a groundwater aquifer in northern Germany. *Developments in Industrial Microbiology* 24:183–200.

Hirsch, P., and E. Rades-Rohkohl. 1990. Microbial colonization of aquifer sediment exposed in a groundwater well in northern Germany. *Applied and Environmental Microbiology* 56(10):2963–2966.

Horowitz, A., J. M. Suflita, and J. M. Tiedje. 1983. Reductive dehalogenation of halobenzoates by anaerobic lake sediment microorganisms. *Applied and Environmental Microbiology* 45:1459–1465.

Hult, M. F. 1988. Mobilization, transport, and fate of hydrocarbon vapors in the unsaturated zone. In Mallard G. E., and S. E. Ragone eds., *U.S. Geological Survey Toxic Substances Hydrology Program*. Proceedings of the technical Meeting, Phonenix, Arizona, September 26-30, 1988: U.S. Geological Survey Water-Resources Investigations Report 88-4220, p. 53.

Hutchins, S. R., W. C. Downs, J. T. Wilson, G. B. Smith, D. A. Kovacs, D. D. Fine, R. H. Douglass, and D. J. Hendrix. 1991. Effect of nitrate addition on biorestoration of fuel-contaminated aquifer: Field demonstration. *Ground Water* 29(4):571–580.

Imbrigiotta, T. E., M. Martin, B. P. Sargent, and L. M. Voronin. 1988. Preliminary results of a study of the chemistry of ground water at the Building 24 research site, Picatinny Arsenal, New Jersey. In Mallard, G. E. and S. E. Ragone, eds, *U.S. Geological Survey Toxic Substances Hydrology Program*. Proceedings of the Technical Meeting, Phoenix, Arizona, September 26-30, 1988. U.S. Geological Survey Water-Resources Investigations Report 88-422, pp. 351–360.

Jeenes, D. J., W. Reineke, H. J. Knackmuss, and P. A. Williams. 1982. TOL plasimid pWW0 in constructed halobenzoate-degrading Pseudomonas strains: Enzyme regulation and DNA structure. *Journal of Bacteriology* 150:180–187.

Jimenez, L. 1990. Molecular analysis of deep-subsurface bacteria. *Applied and Environmental Microbiology* 56:2108–2113.

Jones, R. E., R. E. Beeman, and J. M. Suflita. 1989. Anaerobic metabolic processes in deep terrestrial subsurface. *Geomicrobiology Journal* 7:117–130.

Jorgensen, B. B. 1990. A thiosulfate shunt in the sulfur cycle of Marine Sediments. *Science* 249:152–154.

Katz, B. G., J. B. Lindner, and S. E. Ragone. 1980. A comparison of nitrogen in shallow ground water from sewered and unsewered areas, Nassau County, New York, from 1952 through 1976. *Ground Water* 18(6): 607–616.

Kelly, F. X., K. J. Dapsis, and D. A. Lauffenburger. 1988. Effect of bacterial chemotaxis on dynamics of microbial competition. *Microbial Ecology* 16:115–131.

Keswick, B. H., D. Wang, and C. P. Gerba. 1982. The use of microorganisms as groundwater tracers: A review. *Ground Water* 20(2):142–149.

Kharaka, Y. K., and I. Barnes. 1973. SOLMNEQ: Solution-mineral equilibrium computations. NTIS Technical Report PB214-899, Springfield, Virginia.

Kinner, N. E., A. L. Bunn, L. D. Meeker, and R. W. Harvey. 1990. Enumeration and variability in the distribution of protozoa in an organically contaminated aquifer. (Abstract) EOS, Transactions, *American Geophysical Union* 71(43):1319.

Knackmuss, H.-J., and M. Hellwig. 1978. Utilization and cooxidation of chlorinated phenols by *Pseudomonas* sp. B13. *Archives of Microbiology* 177:1–7.

Kolbel-Boelke, J., E. Anders, and A. Nehrkorn. 1988. Microbial communities in the saturated groundwater environment, II: Diversity of bacterial communities in a Pleistocene sand aquifer and their in vitro activities. *Microbial Ecology* 16:31–48.

Konopka, A., and R. Turco. 1991. Biodegradation of organic compounds in vadose zone and aquifer sediments. *Applied and Environmental Microbiology* 57(8):2260–2268.

Kreitler, C. W., and D. C. Jones. 1975. Natural soil nitrate: The cause of the nitrate contamination of ground water in Runnels County, Texas. *Ground Water* 13(1):53–61.

Kuhn, E. P., J. Zeyer, P. Eicher, and R. P. Schwarzenbach. 1988. Anaerobic degradation of alkylated benzenes in denitrifying laboratory aquifer columns. *Applied and Environmental Microbiology* 54(2):490–496.

Kuznetsov, S. I. 1950. Investigation of the possibility of contemporaneous formation of methane in gas-petroleum formations in the Saratov and Buguruslan regions. *Mikrobiologiya* 19:193–202.

Langmuir, D. 1969. Geochemistry of iron in a coastal plain aquifer of the Camden, New Jersey area. U.S. Geological Survey Professional Paper 650-C, pp.224–235.

Lanzarone, N. A., and P. L. McCarty. 1991. Column studies on methanotrophic degradation of trichloroethene and 1,2-dichloroethane. *Ground Water* 28(6):910–919.

Larrick, S. R., J. R. Clark, D. S. Cherry, and J. Cairns, Jr. 1981. Structural and functional changes of aquatic heterotrophic bacteria to thermal heavy and fly ash effluents. *Water Research* 15:875–880.

Lauffenburger, D. A., and B. Calcagno. 1983. Competition between microbial populations in a non-mixed environment: Effect of random botility. *Biotechnology and Bioengineering* 25:2103.

Leach, L. E., F. P. Beck, J. T. Wilson, and D. H. Kampbell. 1989. Aseptic subsurface sampling tecniqques for hollow-stem auger drilling. In Proceedings, Second National Outdoor Action Conference on Aquifer Restoration, Ground Water Monitoring and Geophysical Methods. Vol. 1, pp. 31–51.

Leahy, P. P., and M. Martin. 1986. Northern Atlantic Coastal Plain Regional aquifer-system study: Simulation of ground-water, Flow. In Sun, R. J. ed., *Regional Aquifer-System Analysis Program of the U.S. Geological Survey, 1978-84*. U.S. Geological Survey Circular 1002, pp.169–175.

Leenheer, J. A., R. L. Malcolm, and W. R. White. 1976. Physical, chemical, and biological aspects of subsurface organic waste injection near Wilmington, North Carolina. U.S. Geological Survey Professional Paper 987.

Lindberg, R. D., and D. D. Runnells. 1984. Ground-water redox reactions: An analysis of equilibrium state applied to Eh measurements and geochemical modeling. *Science* 225:925–927.

Linkfield, T. G., J. M. Suflita, and J. M. Tiedje. 1989. Characterization of the acclimation period before anaerobic dehalogenation of halobenzoates. *Applied and Environmental Microbiology* 55(11):2773–2778.

Lovley, D. R., and M. J. Klug. 1983. Sulfate reducers can outcompete methanogens at freshwater sulfate concentrations. *Applied and Environmental Microbiology* 45:187–192.

Lovley, D. R., and M. J. Klug. 1986. Model for the distribution of methane production and sulfate reduction in freshwater sediments. *Geochimica et Cosmochimica Acta* 50:11–18.

Lovley, D. R., and S. Goodwin. 1988. Hydrogen concentrations as an indicator of the predominant terminal electron-accepting reactions in aquatic sediments. *Geochimica et Cosmochimica Acta* 52:2993–3003.

Lovley, D. R., and E. J. P. Phillips. 1988. Novel mode of microbial energy metabolism: organic carbon oxidation coupled to dissimilatory reduction of iron or manganese. *Applied and Environmental Microbiology* 54(6):1472–1480.

Lovley, D. R., M. J. Baedecker, D. J. Lonergan, I. M. Cozzarelli, E. J. P. Phillips, and D. I. Siegel. 1989. Oxidation of aromatic contaminants coupled to microbial iron reduction. *Nature* (London) 339:297–299.

Lovley, D. R., Chapelle, F. H., and E. J. P. Phillips. 1990. Recovery of Fe(III)-reducing bacteria from deeply buried sediments of the Atlantic Coastal Plain. *Geology* 18:954–957.

Lovley, D. R., and D. J. Lonergan. 1990. Anaerobic oxidation of toluene, phenol, and p-cresol by the dissimilatory iron-reducing organism, GS-15. *Applied and Environmental Microbiology* 56(6):1858–1864.

Lovley, D. R., E. J. P. Phillips, and D. J. Lonergan. 1991. Enzymatic versus nonenzymatic mechanisms for Fe(III) reduction in aquatic sediments, *Environmental Science and Technology* 26(6):1062–1067.

Madsen, E. L., J. L. Sinclair, and W. C. Ghiorse, 1991. In situ biodegradation: Microbiological patterns in a contaminated aquifer. *Science* 252:830–833.

Major, D. W. C. I. Mayfield, and J. F. Barker. 1988. Biotransformation of benzene by denitrification in aquifer sand. *Ground Water* 26(1):8–14.

Matthess, G. 1982. *The Properties of Groundwater*. New York: John Wiley.

McMahon, P. B. 1990. Role of bacterial CO_2 production in the formation of high-bicarbonate ground water in the Black Creek and Middendorf aquifers of South Carolina. Unpublished Ph.D. Dissertation, the University of South Carolina, Columbia, South Carolina.

McMahon, P. B., D. F. Williams, and J. T. Morris. 1990. Production and isotopic composition of bacterial CO_2 in deep coastal-plain sediments of South Carolina. *Ground Water* 28:693–702.

McMahon, P. B., and F. H. Chapelle. 1991. Microbial production of organic acids in aquitard sediments and its role in aquifer geochemistry. *Nature* 349:233–235.

McMahon, P. B., F. H. Chapelle, and M.L. Jagucki. 1992. Atrazine biodegradation potential of alluvial aquifer sediments under aerobic conditions. *Environmental Science and Technology*

McNabb, J. F., and W. J. Dunlap. 1975. Subsurface biological activity in relation to groundwater pollution. *Ground Water* 13:33–44.

Mikesell, M.D., and S.A. Boyd. 1986. Complete reductive dechlorination and mineralization of petachlorophenol by anaerobic microorganisms. *Applied and Environmental Microbiology* 52:861–865.

Mueller, J. G., D. P. Middaugh, S. E. Lantz, and P. T. Chapman. 1991. Biodegradation of creosote and pentachlorophenol in contaminated groundwater: Chemical and biological assessment. *Applied and Environmental Microbiology* 57(5):1277–1285.

Neilson, A. H., A.-S Allard, and P. K. Hynning. 1983. Bacterial methylation of chlorinated phenols and guaiacols: formation of veratroles from guaiacols and high-molecular-weight chlorinated lignin. *Applied and Environmental Microbiology* 45:774–783.

Nelson, J. D., and R. R. Colwell. 1975. The ecology of mercury-resistant bacteria in Chesapeake Bay. *Microbial Ecology* 1:191–218.

Nordstrom et al. 1979. A comparison of computerized chemical models for equilibrium calculations in aqueous systems. In Jenne, E.A., ed. *Chemical Modeling in Aqueous Systems*. American Chemical Society Symposium Series 93:857–892.

Norvatov, A. M, and O. V. Popov. 1961. Laws of the formation of minimum stream flow. *Bulletin of the International Association of Science and Hydrology* 6(1):20–27.

Ogunseitan, O. A., E. T. Tedford, D. Pacia, K. M. Sirotkin, and G. S. Saylor. 1987. Distribution of plasmids in groundwater bacteria. *Journal of Industrial Microbiology* 1:311–317.

Oki, D. S., and T. W. Giambelluca. 1987. DBCP, EDB, and TCP contamination of ground water in Hawaii. *Ground Water* 25(6):693–702.

Olson, G. J., W. S. Dockins, G. A. McFeters, and W. P. Iverson. 1981. Sulfate-reducing and methanogenic bacteria from deep aquifers in Montana. *Geomicrobiology Journal* 12:327–340.

Parkhurst, D. L., D. C. Thorstenson, and L. N. Plummer. 1980. PHREEQE-A computer program for geochemical calculations. U.S. Geological Survey Water Resources Investigations Report 80-96.

Parkhurst, D. L., L. N. Plummer, and D. C. Thorstenson. 1982. BALANCE-A computer program for calculation of chemical mass balance. U.S. Geological Survey Water Resources Investigations Report 82-14.

Pearson, F. J., Jr., and D. E. White. 1967. Carbon 14 ages and flow rates of water in Carrizo Sand, Atascosa County, Texas. *Water Resources Research,* 3:251–261.

Pearson, F. J., Jr. and B. B. Hanshaw. 1970. Sources of dissolved carbonate species in ground water and their effects on carbon-14 dating. In *Isotope Hydrology 1970,* International Atomic Energy Agency, Vienna, pp. 271–286.

Pedersen, K., and S. Ekendahl. 1990. Distribution and activity of bacteria in deep granitic groundwaters of southeastern Sweden. *Microbial Ecology* 20:37–52.

Phelps, T. J., E. G. Raione, D. C. White, and C. B. Fliermans. 1989a. Microbial activities in deep subsurface environments. *Geomicrobiology Journal* 7(1-2):79–92.

Phelps, T. J., C. B. Fliermans, T. R. Garland, S. M. Pfiffner, and D. C. White. 1989b. Methods for recovery of deep terrestrial subsurface sediments for microbiological studies. *Journal of Microbiological Methods* 9:267–279.

Pitzer, K. S. 1973. Thermodynamics of Electrolytes-1: Theoretical basis and general equations. *Journal of Physical Chemistry* 77(2):268–277.

Plummer, L. N., B. F. Jones, and A. H. Truesdell. 1976. WATEQF: a computer program for calculating chemical equilibria of natural waters. U.S. Geological Survey Water Resources Investigations Report 76-13.

Plummer, L. N. 1977. Defining reactions and mass transfer in part of the Floridan aquifer. *Water Resources Research* 13:801–812.

Plummer, L. N., and W. Back. 1980. The mass balance approach: Application to interpreting the chemical evolution of hydrologic systems. *American Journal of Science* 280:130–142.

Plummer, L. N., D. L. Parkhurst, and D. C. Thorstenson. 1983. Development of reaction models for ground-water systems. *Geochimica et Cosmochimica Acta* 47:665–686.

Plummer, L. N. 1984. Geochemical modeling: A comparison of forward and inverse methods. In Hitchon, B. and E.I. Walick, eds., First *Canadian/American Conference on Hydrogeology, Practical Applications of Ground Water Geochemistry,* pp. 149-177, National Water Well Association, Dublin, Ohio.

Plummer, L. N., J. F. Busby, R. W. Lee, and B. B. Hanshaw. 1990. Geochemical modeling of the Madison aquifer in parts of Montana, Wyoming, and South Dakota. *Water Resources Research* 26(9):1981–2014.

Pucci, A. A., Jr., J. P. Owens. 1989. Geochemical variations in a core of hydrogeologic units near Freehold, New Jersey. *Ground Water* 27:802–812.

Ramanand, K., and J. M. Suflita. 1991. Anaerobic degradation of m-cresol in anoxic aquifer slurries: Carboxylation reactions in a sulfate-reducing bacterial enrichment. *Applied and Environmental Microbiology* 57(6): 1689–1695.

Reineke, W., and H.-J. Knackmuss. 1978. Chemical structure and biodegradability of halogenated aromatic compounds. Substituent effects on the 1,2-dioxygenation of benzoic acid. *Biochimica Biophysics Acta* 542:412–423.

Reinhard, M.,N. L. Goodman, and J. F. Barker. 1984. Occurrence and distribution of organic chemicals in two landfill leachate plumes. *Environmental Science and Technology* 18:953–961.

Rightmire, C. T., and B. B. Hanshaw. 1973. Relationship between the carbon isotope composition of soil CO_2 and dissolved carbonate species in groundwater. *Water Resources Research*, 9(4):958–967.

Rightmire, C. T., F. J. Pearson, Jr., W. Back, R. O. Rye, and B. B. Hanshaw. 1974. Distribution of sulfur isotopes of sulfates in ground waters from the principle artesian aquifer of Florida and the Edwards aquifer of Texas, USA. In *Isotope Techniques in Groundwater Hydrology*, 2. Vol. II. International Atomic Energy Agency, Vienna, 191–207.

Roberts, P. V., G. D. Hopkins, D. M. MacKay, and L. Semprini. 1991. A field evaluation of in-situ biodegradation of chlorinated ethenes: Part 1, methodology and field site characterization. *Ground Water* 28(4):591–604.

Robertson, F. N. 1979. Evaluation of nitrate in the ground water in the Delaware Coastal Plain. *Ground Water* 17(4): 328–337.

Rogers, G. S. 1917. Chemical relations of the oil-field waters in San Joaquin Valley, California. *U.S. Geological Survey Bulletin* 653:93–99.

Rosnes, J. T., T. Torsvik, and T. Lien. 1991. Spore-forming thermophilic sulfate-reducing bacteria isolated from North Sea oil field waters. *Applied and Environmental Microbiology* 57(8):2302–2307.

Rozanova, E. P., and A. I. Khudyakova. 1974. A new nonspore-forming thermophilic sulfate-reducing organism, *Desulfovibrio thermophius*. *Mikrobiologiya* 43:908–912.

Rozanova, E. P,. and T. N. Nazina. 1979. Occurrence of thermophilic sulfate-reducing bacteria in oil-bearing strata of Apsheron and Western Siberia. *Mikrobiologiya* 48: 807–911.

Sander, P., R. M. Wittich, P. Fortnagel, H. Wilkes, and W. Francke. 1991. Degradation of 1,2,4-Trichloro- and 1,2,4,5-Tetrachlorobenzene by *Pseudomonas* strains. *Applied and Environmental Microbiology* 57(5):1430–1440.

Schnurer, J., and T. Rosswall. 1982. Fluorescein diacetate hydrolysis as a measure of total activity in soil and liter. *Applied and Environmental Microbiology* 43:1256–1261.

Schwarzenbach, R. P., W. Giger, E. Hoehn, and J. D. Schneider. 1983. Behavior of organic compounds during infiltration of river water to ground water: field studies. *Environmental Science and Technology* 17:472–479.

Semprini, L., P. V. Roberts, G. D. Hopkins, and P. L. McCarty. 1990. A field evaluation of in-situ biodegradation of chlorinated ethenes: Part 2, results of biostimulation and biotransformation experiments. *Ground Water* 28(5):715–727.

Semprini, L., G. D. Hopkins, P. V. Roberts, D. Grbic-Galic, and P. L. McCarty. 1991. A field evaluation of in-situ biodegradation of chlorinated ethenes: Part 3, studies of competitive inhibition. *Ground Water* 29(2):239–250.

Senum, G. I., and R. N. Dietz. 1991. Perfluorocarbon tracer tagging of drilling muds for the assessment of sample contamination, pp. 7–145. In Fliermans, C.B., and T.C. Hazen, eds., Proceedings of the First International Symposium on Microbiology of the Deep Subsurface. Orlando, Florida. January 15–19, 1990,. WSRC Information Services, Aiken, South Carolina.

Seubert, W., and E. Fass. 1964. Untersuchungen uber den bakteriellen Abbau von Isoprenoiden. V. Der Mechanismus des Isoprendoid-Abbaues. *Biochemistry Zeitung* 341:35–44.

Silver, S., and T. K. Misra, 1984. Bacterial transformations of and resistances to heavy metals, In Omenn, G.S. and A. Hollaender, eds., *Genetic Control of Environmental Pollutants*. New York, Plenum.

Sinclair, J. L., and W. C. Ghiorse. 1987. Distribution of protozoa in subsurface sediments of a pristine groundwater study site in Oklahoma. *Applied and Environmental Microbiology* 53(5):1157–1163.

Sinclair, J. L., and W. C. Ghiorse. 1989. Distribution of aerobic bacteria, protozoa, algae, and fungi in deep subsurface sediments. *Geomicrobiology Journal* 7:15–32.

Sinclair, J. L., S. J. Randtke, J. E. Denne, L. R. Hathaway, and W. C. Ghiorse. 1990. Survey of microbial populations in buried-valley aquifer sediments from northeastern Kansas. *Ground Water* 28(3):369–377.

Sinclair, J. L., and T. R. Lee. 1990. Biodegradation of atrazine in surface and subsurface environments. Abstract, *American Geophysical Union*, EOS 71(43):1329.

Smith, R. L., and J. H. Duff. 1988. Denitrification in a sand and gravel aquifer. *Applied and Environmental Microbiology* 54(5):1071–1078.

Smolenski, W. J., and J. M. Suflita. 1987. Biodegradation of cresol isomers in anoxic aquifers. *Applied and Environmental Microbiology* 53:710–716.

Spain, J. C., and P. A. Van Veld. 1983. Adaptation of natural microbial communities to degradation of xenobiotic compounds: Effects of concentration, exposure time, inoculum, and chemical structure. *Applied and Environmental Microbiology* 45(2):428–435.

Spalding, R. F., M. E. Burbach, and M. E. Exner. 1989. Pesticides in Nebraska's ground water. *Ground Water Monitoring Review* (Fall) pp. 126–133.

Sparling, G. P. 1981. Microcalorimetry and other methods to assess biomass and activity in soil. *Soil Biology and Biochemistry* 13:93–98.

Speiran, G. K. 1987. Relation of aqueous geochemistry to sedimentary depositional environments. In Vecchioli, J. and A.I. Johnson, eds. *American Water Resources Association Monograph Series* No. 9:79–96.

Steel, J. H. 1838. An analysis of the mineral waters of Saratoga and Ballston. New York, G.M. Davis.

Stetzenbach, L. D., L. M. Delley, and N. A. Sinclair. 1986. Isolation, identification, and growth of well-water bacteria. *Ground Water* 24(1):6–10.

Stevenson, R. J., and E. F. Stoermer. 1982. Abundance patterns of diatoms on *Cladophora* in Lake Huron with respect to a point source of wastewater treatment plant effluent. *Journal of Great Lakes Research* 8:184–195.

Suflita, J. M. 1989. Microbiological principles influencing the biorestoration of aquifers. In *Transport and Fate of Contaminants in the Subsurface*, EPA 625/4-89/019, Robert S. Kerr Environmental Research Laboratory, pp. 85–99.

Suflita, J. M., J. A. Robinson, and J. M. Tiedje. 1983. Kinetics of microbial dehalogenation of haloaromatic substrates in methanogenic environments. *Applied and Environmental Microbiology* 45:1466–1473.

Swindoll, C. M., C. M. Aelion, and F. K. Pfaender. 1988. Influence of inorganic and organic nutrients on aerobic biodegradation and on the adaptation response of subsurface microbial communities. *Applied and Environmental Microbiology* 54(1):212–217.

Thorn, P. M. and R. M. Ventullo. 1988. Measurement of bacterial growth rates in subsurface sediments using the incorporation of tritiated thymidine into DNA. *Microbial Ecology* 16:3–16.

Thorstenson, D. C. 1984. The concept of electron activity and its relation to redox potentials in aqueous geochemical systems. U.S. Geological Survey Open File Report 84-072.

Thorstenson, D. C., D. W. Fisher, and M. G. Croft. 1979. The geochemistry of the Fox Hills-Basal Hell Creek aquifer in southwestern North Dakota and northwestern South Dakota. *Water Resources Research* 15:1479–1498.

Thorstenson, D. C., E. P. Weeks, H. Haas, and D. W. Fisher. 1983. Distribution of gaseous $^{12}CO_2$, $^{13}CO_2$, and $^{14}CO_2$ in the sub-soil unsaturated zone of the Western U.S. Great Plains. *Radiocarbon* 25(2):315–346.

Thurman, E.M. 1985. *Organic Geochemistry of Natural Waters*. Dordrecht, The Netherlands. Martinus Nijhoff/DR W. Junk Publishers.

Toth, J. 1963. A theoretical analysis of groundwater flow in small drainage basins. *Journal of Geophysical Research* 68(16)4795–4812.

Trapp, H., Jr., L. L. Knobel, H. Meisler, and P. P. Leahy. 1984. Test well DO-CE 88 at Cambridge, Dorchester County, Maryland. U.S. Geological Survey Water Supply Paper 2229.

Trevors, J. T., K. M. Oddie, and B. H. Belliveau. 1985. Metal resistance in bacteria. *FEMS Microbiology Reviews* 32:39–54.

Troutman, D. E., E. M. Godsy, D. F. Goerlitz, and G. G. Ehrlich. 1984. Phenolic contamination in the sand-and-gravel aquifer from a surface impoundment of wood treatment wastes, Pensacola, Florida. U.S. Geological Survey Water-Resources Investigations Report 84-4230. U.S. Geological Survey, Denver, Colorado.

Trudgill, P. W. 1984. Microbial degradation of the alicyclic ring, In Gibson, D.T., ed., *Microbial Degradation of Organic Compounds*, New York, Marcel Dekker.

Truesdell, A. H., and B. F. Jones. 1974. WATEQ, a computer program for calculating chemical equilibria of natural waters. *Journal of Research of the U.S. Geological Survey*, 2, pp. 233–274.

Tsien, H. C., G. A. Brusseau, R. S. Hanson, and L. P. Wachett. 1989. Biodegradation of trichloroethylene by *Methylosinus trichosporium* OB3b. *Applied and Environmental Microbiology* 55(12):3155–3161.

van Bavel, C. H. M. 1951. A soil aeration theory based on diffusion. *Soil Science* 72:33–46.

van Eyk, J., and T. J. Bartels. 1968. Paraffin oxidation in *Pseudomonas aeruginosa*. 1. Induction of paraffin oxidation. *Journal of Bacteriology* 96:706–712.

Veldkamp, H., H. van Gemerden, W. Harder, and H. J. Laanbroek. 1984. Competition among bacteria: an overview. In Klug, M. J., and C. A. Reddy, eds, *Current Perspectives in Microbial Ecology*. Washington, D.C: American Society for Microbiology.

Vogel, T. M., and P. L. McCarty. 1985. Biotransformation of tetrachloroethylene to thrichloroethylene, dichloroethylene, vinyl chloride, and carbon dioxide under methanogenic conditions. *Applied and Environmental Microbiology* 49:1080–1083.

Waksman, S. A. 1932. *Principles of Soil Microbiology*. Baltimore: Williams & Wilkins.

Weeks, J. B. 1986. High Plains regional aquifer-system study. In Sun, R.J., ed. *Regional Aquifer-Systems Analysis Program of the U.S. Geological Survey Summary of Projects, 1978–84*, pp. 30–50.

Wellings, F. M., A. L. Lewis, C. W. Mountain, and L. V. Pierce. 1975. Demonstration of virus in groundwater after effluent discharge onto soil. *Applied and Environmental Microbiology* 29:751–757.

White, A. F., M. L. Peterson, and R. D. Solbau. 1990. Measurement and interpretation of low levels of dissolved oxygen in ground water. *Ground Water* 28(4):584–590.

White, D.C. 1983. Analysis of microorganisms in terms of quantity and activity in natural environments. In Slater, J.H., R. Whittenbury, and J. W. T. Wimpenny, eds., *Microbes in Their Natural Environment,* pp. 37–66. New York: Cambridge University Press.

White, D. C., Fredrickson, J. F., Gehron, M. H., Smith, G. A., and R. F. Martz. 1983. The groundwater aquifer microbiota: Biomass, community structure, and nutritional status. *Developments in Industrial Microbiology* 24:189–199.

Williams, R. J., and W. C. Evans. 1975. The metabolism of benzoate by *Moraxella* species through anaerobic nitrate respiration. *Biochemistry Journal* 148:1–10.

Wilson, B. H. 1988. Biotransformation of chlorinated hydrocarbons and alkylbenzenes in soil at Picatinny Arsenal, New Jersey. In Mallard, G.E., and S.E. Ragone, eds. *U.S. Geological Survey Toxic Substances Hydrology Program.* Proceedings of the Technical Meeting, Phoenix, Arizona, September 26–30, 1988. U.S. Geolgical Survey Water-Resourses Investigations Report 88-4220, p. 389–396.

Wilson, B. H., G. B. Smith, and J. F. Rees. 1986. Biotransformantions of selected alkylbenzenes and halogenated aliphatic hydrocarbons in methanogenic aquifer material: a microcosm study. *Environmental Science and Technology* 20:997–1002.

Wilson, J. T., J. F. McNabb, D. L. Balkwill, and W. C. Ghiorse. 1983. Enumeration and characterization of bacteria indigenous to a shallow water-table aquifer. *Ground Water* 21:134–142.

Wilson, J. T., J. F. McNabb, J. W. Cochran, T. H. Wang, M. B. Tomson, and P. B. Bedient. 1985. Influence of microbial adaptation on the fate of organic pollutants in groundwater. *Environmental Toxicology and Chemistry* 4:721–726.

Wilson, J. T., and B. H. Wilson. 1985. Biotransformation of trichloroethylene in soil. *Applied and Environmental Microbiology* 49(1):242–243.

Winograd, I., and F. Robertson. 1982. Deep oxygenated ground water:anomaly or common occurrence? *Science* 216:1227–1229.

Woese, C. R. 1981. Archaebacteria. *American Journal of Science,* 244:98–122.

Wolery, T.J. 1983. EQ3NR, A computer program for geochemical aqueous speciation-solubility calculations: Users guide and documentation. UCRL-53414. Lawrence Livermore National Laboratory, Livermore, California.

Wolters, N., and W. Schwartz. 1956. Untersuchungen uber Vorkommen und Verhalten von Mikroorganismen in reimum Grundwasser. *Archive von Hydrobiologie* 51:500–541.

Wood, W. W., and M. J. Petraitis. 1984. Origin and distribution of carbon dioxide in the unsaturated zone of the Southern High Plains of Texas. *Water Resources Research* 20(9):1193–1208.

Wunsch, D. R. 1988. High barium concentrations in ground water in eastern Kentucky. Proceedings of the Ground Water Geochemistry Conference, February 16-18, 1988 Denver, Colorado. Published by the National Water Well Association, Dublin, Ohio.

Yazicigil, H., and L. V. A. Sendlein. 1981. Management of ground water contaminated by aromatic hydrocarbons in the aquifer supplying Ames, Iowa. *Ground Water* 19 (6):648–665.

Zinger, A. S. 1966. Microflora of underground waters in the lower Volga region with reference to its use for oil prospecting. *Microbiologiya* 35:305–311.

Zobell, C. E. 1947. Microbial transformation of molecular hydrogen in marine sediments, with particular reference to petroleum. *Bulletin of the American Association of Petroleum Geologists* 31:1709–1751.

Zobell, C. E. 1950. Assimilation of hydrocarbons by microorganisms. *Advances in Enzymology* 10:443–486.

Zobell, C. E. 1952. Bacterial life at the bottom of the Philippine Trench. *Science* 63:507–508.

Zobell, C. E. 1958. Ecology of sulfate-reducing bacteria. *Producers Monthly* 22:12–29.

INDEX

Abiotic processes, 19, 21
Abundance and distribution (of bacteria), 174–207
  classification of subsurface environments, 174–178
  gram-negative bacteria, 40
  intermediate flow systems, 195–203
    bacterial distribution in, 198–202
    described, 195–197
    microbial habitat, 197–198
    microbial processes in confining beds, 202–203
  local flow systems, 189–195
    bacterial distribution, 191–195
    described, 189
    as microbial habitat, 190–191
  regional flow systems, 203–207
    bacterial distributions in, 205–207
    described, 203
    petroleum reservoirs, early observations, 204–205
  unsaturated zone, 179–189
    bacterial distribution in, 186–189
    biomass measurements in soil microbiology, 182–186
    described, 179–180
    microbial habitat, 180–182
Acclimation, 296–321
  factors affecting, 304–310
    concentration effects, 306–308
    rates of acclimation, 305–306
    xenobiotic chemical structures, 308, 310
    xenobiotic compounds, cross-acclimation, 308, 309
  mechanisms of, 298–304
    catabolite repression, 300–301
    electron acceptors, 302–304
    genetic mutation, 301–302
    induction, 299–300
  metal toxicity, 317–320
    mechanisms of, 317–318
    mercury toxicity, 320
    plasmid-encoded metal resistance mechanisms, 318–320
  microbial response to environmental changes, 297–298
  overview of, 296
  xenobiotics in ground-water systems, 311–316
    bioremediation technology, 316
    contaminated aquifer, 311–313
    eucaryotic microorganisms, 315
    pristine aquifer sediments, 313–315
Acetate fermentations, bacterial metabolism and ground water, 81–85
Acid mine drainage, sulfide oxidizers, 98–99
Acinetobacter:
  benzene degradation, 337
  biodegradation, 373
  described, 42
  oxidation, 331
Acquired immune deficiency disorder (AIDS), 35, 127
Acridine orange direct count (AODC):
  bacterial growth enumeration techniques, 68–69
  biomass measurement in soil microbiology, unsaturated zone, 183

Activity measurements:
  biomass measurement in soil microbiology, unsaturated zone, 184–185
  in microcosms, microbial ecology of ground water methods, 146–148
Addition mutation, described, 115
Adenosine monophosphate (AMP):
  catabolite repression, 301
  gene expression and regulation, 114
Adenosine triphosphate (ATP):
  autotrophic carbon dioxide fixation, chemolithotrophy, 100–102
  biosynthesis, 92
  culture media design, 60
  gene expression and regulation, 114
  lipids, biosynthesis, 96
  metabolism
    energy storage, 73–74
    energy utilization, 77–78
  oxygen reduction-aerobic metabolism, bacterial metabolism and ground water, 91
Adenosine triphosphate (ATP) assay, biomass measurement in soil microbiology, unsaturated zone, 184, 186
Adenovirus, 35
Aelion, C. M., 305, 308, 311, 313, 316
Aerobacter aerogenes, described, 43
Aerobes, bacteria classification, 39
Aerobic bacteria:
  gram-negative cocci, described, 42
  gram-negative rods, described, 40–42
  oxygen reduction-aerobic metabolism, bacterial metabolism and ground water, 91
Aerobic degradation (halogenated organic compound degradation):
  aliphatic compounds, 368–369
  chlorinated phenols, 376
  monocyclic aromatic compounds, 371–372
Agriculture, 251–253. *See also* Fertilizers
Agrobacterium, described, 41
Air drilling and coring, microbiologic sampling of unsaturated zone, 210–211
Alcaligenes:
  benzene oxidation, 337
  biodegradation, 376
Alexander, M., 180
Algae, eucaryotes, 29–31
Alicyclic hydrocarbons, biodegradation of, 334–335
Aliphatic compounds:
  chemistry and uses of, 359–361
  halogenated organic compound degradation, 367–371
    in ground-water systems, 377–380
    remediation in ground-water systems in situ, 380–384

Aliphatic hydrocarbon biodegradation, 328–334
  alkene-oxidation, 332
  branched aliphatic, 332–334
  generally, 328–329
  methane oxidation, 329–330
  n-alkanes, 330–332
Aliphatic petroleum hydrocarbons, pollution control, genetic engineering applications, 122–123
Alkene-oxidation, aliphatic hydrocarbon biodegradation, 332
Alkyl benzenes:
  anaerobic degradation processes, aromatic hydrocarbon biodegradation, 344–345
  aromatic hydrocarbon biodegradation, 338–339
Alkyl halide insecticides, halogenated organic compound degradation, in ground-water systems, 384–385
Allard, A.-S., 376
Allosteric effector sites, metabolism, enzyme regulation, 80–81
Allosteric inhibitor, metabolism, enzyme regulation, 81
Al'tovskii, M. E., 205
Ames test, chemical mutagen, 115
Amino acids, biosynthesis, 93–94
Ammonia oxidizing (nitrifying) bacteria, chemolithotrophy, 100
Anaerobes, bacteria classification, 39
Anaerobic bacteria:
  carbon cycle and, 238
  geochemistry and, 20
  gram-negative rods, described, 43–44
  iron reduction and, 88–89
Anaerobic conditions, microbial ecology of ground water, 133–134
Anaerobic degradation:
  aromatic hydrocarbon biodegradation, 340–347
  halogenated organic compound degradation
    aliphatic compounds, 369–371
    chlorinated phenols, 377
    monocyclic aromatic compounds, 372–373
Anders, E., 136, 137, 138, 139, 140, 146, 156
Animal excrement, nitrogen cycle, ground-water systems, 247–248
Antagonism, population mechanics, 167–168
Anthrax, 15–16, 17
Anthropogenic compounds, unsaturated zone, 179
Antibiotic disc method, bacterial abundance and distribution, intermediate flow systems, 198
Antibiotics, procaryotes, 26
Antiquity:
  geology and, 3
  microbiology and, 13
Aquatic sediments, iron cycle, 254–255

Aqueous model, geochemical modeling methodology, speciation calculations, 272–274
Aquifers:
  bacterial reproduction and, 48–49
  geochemical modeling and, 21–22
  microbial ecology, 22–24
  sulfur cycle and, 260–263
Archaebacteria, described, 31–33
Aristotle, 14
Armstrong, R., 347
Aromatic hydrocarbon(s), pollution control, genetic engineering applications, 123
Aromatic hydrocarbon biodegradation, 335–347
  alkyl benzenes, 338–339
  anaerobic degradation processes, 340–347
  benzene, 336–338
  generally, 335–336
  polycyclic aromatic compounds, 339–340
Arthrobacter:
  biodegradation, 377
  described, 45
Aseptic procedures:
  development of, 22–23
  split-spoon and push-tube sampling with, local flow systems, 214–215
Atrazine, chlorinated herbicides, degradation, 375–376
Autotrophic carbon dioxide fixation, chemolithotrophy, 100–102
Axell, B. C., 336
Azotobacter:
  described, 40, 41
  nitrogen cycle and, 245

Bacillus:
  described, 45–46
  oxidation, 331, 337
Back, W., 21, 56, 132, 198, 250, 257, 264
Bacon, Francis, 11
Bacteria, 35–46
  classification of, 36–40
  eucaryotic microorganisms, xenobiotics in ground-water systems (acclimation), 315
  gram-negative bacteria, 40–44
    aerobic gram-negative cocci, 42
    aerobic gram-negative rods, 40–42
    anaerobic gram-negative rods, 43–44
    distribution of, 40
    facultatively anaerobic gram-negative rods, 42–43
  gram-positive bacteria, 44–46
    coryneform bacteria, 45
    gram-positive cocci, 45
    spore-forming rods, 45–46

  ground water chemistry and, 20–21
  host-parasite interaction with viruses, 168
  hydrocarbons and, 19–20
  microbial ecology of ground water, 135
  overview of, 35–36
  procaryotes, 26–29
Bacterial abundance and distribution. *See* Abundance and distribution
Bacterial genetics. *See* Genetics
Bacterial growth, 47–70
  culture techniques, 57–66. *See also* Culture methods
    growth media design, 58–63
    isolation techniques, 63–66
  ecological modeling, microbial ecology of ground water methods, 151–155
  enumeration techniques, 66–69. *See also* Abundance and distribution (of bacteria)
    direct counting, 68–69
    viable counting, 67–68
  environmental conditions, 51–57
    molecular oxygen, 55
    osmotic pressure, 56–57
    pH, 55–56
    temperature, 52–54
    water, 54–55
  population kinetics, 49–51
  reproduction, 47–49
Bacterial metabolism. *See* Metabolism
Bacteriophagic virus, 33, 35
Bacteroides, described, 44
Baedecker, M. J., 56, 250, 302, 350, 351
BALANCE program (computer software), 280
Balkwill, D. L., 22, 23, 31, 40, 43, 45, 65, 125, 140, 141, 143, 145, 146, 154, 156, 157, 169, 187, 192, 303, 313
Barber, L. B., II, 385
Barium:
  drilling fluid contamination of cored sediments, evaluation of, 227
  osmotic pressure, environmental conditions (bacterial growth), 57
Barkay, T., 320
Barker, J. F., 340, 348, 350
Barnes, I., 257, 276
Bartels, T. J., 301
Bastin, E. S., 19–20, 44, 204, 205, 208
Beam, H. W., 334
Beckmann, D. D., 328
Beeman, R. E., 44, 99, 150, 167, 194, 230, 303
Behki, R. M., 376, 386
Beijerinck, Martinus, 22, 33
Belliveau, B. H., 317
Belser, N. O., 370
Belyaev, S. S., 205

Benzene:
  alkyl benzenes, aromatic hydrocarbon biodegradation, 338–339
  anaerobic degradation processes, 341–342
  aromatic hydrocarbon biodegradation, 336–338
  microbial degradation, ground-water systems, 348
  remediation of ground-water systems, 353–356
Benzoate, anaerobic degradation processes, 341
Bergey, David, 17
Berkley, M. J., 5
Berzelius, Jons Jakob, 4, 14
Beta-ketoadipate pathway, benzene oxidation, 337–338
Beta-oxidation, n-alkanes oxidation, aliphatic hydrocarbon biodegradation, 331
Betz, J. L., 301–302
Bible, geology and, 3, 8
Bicknell, B., 318
Biejerinck, Martinus, 6
Bifidium pathway, lactate and acetate fermentations, bacterial metabolism and ground water, 81–82
Binary fission:
  bacterial reproduction, 47–49, 116
  DNA structure and organization, 107
Biochemical cycling, 233–263
  carbon cycle, 238–244
    described, 238–239
    ground-water systems, 241–244
    integrated carbon, oxygen, and hydrogen cycle, 240–241
  concepts and definitions in, 233–234
  iron cycle, 253–258
    aquatic sediments, 254–255
    described, 253–254
    ground-water systems, 255–258
  nitrogen cycle, 244–253
    described, 244–247
    ground-water systems, 247–253
  oxygen cycle, 234–238
    described, 234–235
    ground-water systems, 234–235
  sulfur cycle, 258–263
    described, 258–259
    ground-water systems, 259–263
Biochemical marker techniques, microbial ecology of ground water methods, 142–146
Biochemistry, microbiology and, 18
Biodegradation. *See* Bioremediation; Halogenated organic compound degradation; Petroleum hydrocarbon biodegradation
Biologic tracers, drilling fluid contamination of cored sediments, evaluation of, 230
Biomass:
  biochemical marker techniques, microbial ecology of ground water methods, 143, 145
  measurements of, in soil microbiology, unsaturated zone, 182–186
Bioremediation. *See also* Halogenated organic compound degradation; Petroleum hydrocarbon biodegradation
  acclimation and, 304
  aliphatic compounds, halogenated organic compound degradation, remediation in ground-water systems in situ, 380–384
  procaryotes, 26–27
  xenobiotics in ground-water systems (acclimation), 316
Biosynthesis, 92–96
  amino acids, 93–94
  carbohydrates, 94–96
  lipids, 96
Biotic processes, 21
Biphasic growth curve, catabolite repression, 301
Black plague, quarantine and, 13
Black smokers, sulfide oxidizers, 98–99
Body temperature, infection, mesophilic microorganisms, 53
Bog iron ores, iron cycle and, 254
Bonner, F. T., 251
Bouwer, E. J., 369
Boyd, S. A., 377
Boyer, Harold, 7
Branched aliphatic, aliphatic hydrocarbon biodegradation, 332–334
Broede, L. D., 379, 380
Bromide, drilling fluid contamination of cored sediments, evaluation of, 229
Brongniart, Alexandre, 9–10
Bryant, F. O., 377
Buchanan, Robert, 17
Buffon, Georges de, 4, 8, 12
Bulger, P. R., 250
Bull, A. T., 126
Burbach, M. E., 385
Burton, J. F., 126

Calvin cycle:
  autotrophic carbon dioxide fixation, chemolithotrophy, 100–102
  discovery of, 18
Capillary fringe (deep unsaturated zone):
  bacterial distribution in, 187, 189
  unsaturated zone, 180
Capsule, procaryotes, 28
Carbohydrates, biosynthesis, 94–96
Carbon:
  culture media design, 58–60
  microbial ecology of ground water, 133
Carbon cycle, 238–244
  described, 238–239
  ground-water systems, 241–244

integrated carbon, oxygen, and hydrogen cycle, 240–241
Carbon dioxide:
  activity measurements in microcosms, microbial ecology of ground water methods, 146–148
  autotrophic carbon dioxide fixation, chemolithotrophy, 100–102
  bacterial abundance and distribution, intermediate flow systems, 202
  carbon cycle and, 238–244
  geochemical modeling, bacterial abundance and distribution, intermediate flow systems, 198–199, 201
  production of
    biomass measurement in soil microbiology, unsaturated zone, 184–185
    microbial processes in aquifer systems, geochemical modeling, 283–285
Cardinal temperature, environmental conditions (bacterial growth), 52
Castro, C. E., 370
Catabolite(s), gene expression and regulation, 113–114
Catabolite repression, acclimation mechanisms, 300–301
Catalese, anaerobic gram-negative rods, 43–44
Catastrophism, geology and, 8, 9
Caulobacter, described, 42
Cedarstrom, D. J., 20, 198–199, 260, 261, 283
Cell coatings, procaryotes, 28–29
Cell membrane, procaryotes, 27–28
Cell wall:
  carbohydrates, biosynthesis, 94
  eucaryotes, 29–30
Cerniglia, C. E., 340
Chakrabarty, Ananda, 122–123
Champ, D. R., 151
Chapelle, F. H., 21, 23, 40, 41, 43, 44, 45, 55, 66, 148, 170, 184, 197, 198, 199, 200, 202, 216, 227, 228, 230, 236, 255, 260, 261, 282, 283
Chemical additive tracers, drilling fluid contamination of cored sediments, evaluation of, 228–229
Chemical mutagen, described, 115
Chemistry. *See also* Biochemical cycling; Biochemistry
  ground water chemistry, microbiology and, 18–24
  microbiology and, 18
Chemolithotrophs, bacteria classification, 39
Chemolithotrophy, 96–102
  ammonia oxidizing (nitrifying) bacteria, 100
  autotrophic carbon dioxide fixation, 100–102
  hydrogen oxidizers, 97
  iron oxidizers, 99–100
  sulfide oxidizers, 98–99

Chemosmosis, metabolism, energy utilization, 77–78
Chemostat, ecological modeling, microbial ecology of ground water methods, 151–152
Cherry, J. A., 132
Chian, E. S. K., 373
Chlordane, halogenated organic compound degradation, 375
Chlorinated herbicides:
  chemistry and uses of, 365–366
  degradation described, 375–376
  degradation in ground-water systems, 385–386
Chlorinated phenols:
  chemistry and uses of, 366–367
  halogenated organic compound degradation, 376–377
Chlorobenzenes, halogenated organic compound degradation, in ground-water systems, 385
Chlorophenolic compounds, halogenated organic compound degradation, in ground-water systems, 386–388
Chloroplasts, eucaryotes, 29
Cholera, 13, 16
Chromabacterium, described, 43
Chromosomal DNA, structure and organization of, 107. *See also* DNA
Citric acid cycle, oxygen reduction-aerobic metabolism, bacterial metabolism and ground water, 91
Clark, R. R., 373
Clarke, P. A., 301–302
Clay content, bacterial abundance and distribution, intermediate flow systems, 199–200
Cleavage, of catechol, aromatic hydrocarbon biodegradation, 337–338
Clostridium:
  described, 45–46
  distribution of, 180
  food chain, 133
  metal toxicity acclimation, 318, 319
  nitrogen cycle and, 245
Coatings, cell coatings, procaryotes, 28–29
Coccus cell, bacteria classification, 37, 38
Codons. *See also* Genetics
  mutation and, 115
  RNA and, 109–112
Coenzymes, metabolism, enzyme regulation, 80
Cohen, Stanley, 7
Colby, J., 368
Colwell, F. S., 187, 210–211
Commensalism, population mechanics, 162–163
Common cold, 35
Competition, population mechanics, 164–167
Competitive inhibition, metabolism, enzyme regulation, 80

Computer, 265
  material balance calculations, 280
  speciation calculations, 275–276
Concentration effects, acclimation, factor affecting, 306–308
Conjugation, genetic recombination, natural genetic exchanges, 117–118
Contamination:
  acclimation to, 296–321. *See also* Acclimation
  aquifers, xenobiotics in ground-water systems (acclimation), 311–313
  aseptic technique, split-spoon and push-tube sampling with, local flow systems, 214–215
  bacterial abundance and distribution, intermediate flow systems, 198
  bacterial reproduction and, 49
  of cored sediments by drilling fluid, 224–230
    core fracture contamination, 225–226
    core seepage contamination, 225
    down-hole contamination, 224–225
    evaluation of, 226–230
  ground-water systems
    local, abundance and distribution (of bacteria), 191–192
    source recognition, nitrogen cycle, 251
  waste disposal
    liquid wastes, geochemical modeling and, 21
    subsurface contamination by, 22
Continental drift hypothesis, geology and, 6, 11–12
Core fracture contamination, microbiologic sampling, drilling fluid contamination of cored sediments, 225–226
Core seepage contamination, microbiologic sampling, drilling fluid contamination of cored sediments, 225
Coring procedures:
  aseptic, development of, 22–23
  contamination, of cored sediments by drilling fluid, 224–230
  mud rotary coring, microbiologic sampling, intermediate and regional flow systems, 220–224
Correns, K. E., 105
Coryneform bacteria, described, 45
Cozzarelli, I. M., 302, 350, 351
Cresols, anaerobic degradation processes, aromatic hydrocarbon biodegradation, 345–347
Crick, Francis, 7, 18, 105
Croft, M. G., 198, 206, 207, 286, 289
Crude oil, composition of, petroleum hydrocarbon biodegradation, 323
Culture methods:
  abundance and distribution (of bacteria), ground water flow systems (regional), 205–206

bacterial growth, 57–66
  growth media design, 58–63
  isolation techniques, 63–66
  diversity, niches and, in aquifer systems, microbial ecology of ground water, 156
  microbial ecology of ground water, 136–141, 146
Cuvier, Georges, 5, 9–10, 12
Cyanobacteria, nitrogen cycle and, 245
Cyclic adenosine monophosphate (AMP):
  catabolite repression, 301
  gene expression and regulation, 114
Cyclohexanol degradation, alicyclic hydrocarbon biodegradation, 334–335
Cytoplasmic enzymes, metabolism, enzyme regulation, 80
Cytoplasmic membrane, procaryotes, 27–28

Dalhoff, A., 301
Dalton, H., 368
Dana, James, 5, 10–11
Dapsis, K. J., 153, 154
Darwin, Charles, 5
Davis, J. B., 44, 191, 204–205, 206
Day, M. J., 126
DDT, 375, 384–385
Deep unsaturated zone. *See* Capillary fringe (deep unsaturated zone)
De Klerk, H., 334
Deletion mutation, described, 115
Delley, L. M., 41, 42, 157, 158, 231
Density, of drilling fluids, 217
Density-driven migration, ground-water systems, movement and separation of petroleum hydrocarbons in, 326–327
Deoxyribonucleic acid (DNA):
  bacteria classification, 36, 37, 39
  bacterial reproduction, 47–49
  gene expression and regulation, 113–114
  genetic engineering and, 118–121, 124–128
  mutation and, 114–116
  procaryotes, 26
  spore-forming rods, 45–46
  structure and organization of, 7, 18, 105, 106–108
  viruses, 33–35
Depth, subsurface environment classification, 174
Desulfovibrio:
  described, 44
  metal toxicity acclimation, 318
  osmotic pressure and, 57
  sulfate reduction, 87
De Vries, Hugo, 105
Diabetes mellitus, insulin production, genetic engineering applications, 122

INDEX   411

Diamond v. Chakrabarty, 123
Dicyclohexylcarbodiimide (DCCD), activity measurements in microcosms, microbial ecology of ground water methods, 147
Dilution solution, ground water flow systems (local), abundance and distribution (of bacteria), 191–192
Direct counting, bacterial growth enumeration techniques, 68–69
Direct observation, microbial ecology of ground water methods, 141–142
Direct viable count (DVC) method, bacterial growth enumeration techniques, 68–69
Distribution (of bacteria). *See* Abundance and distribution (of bacteria)
Diversity, niches and, in aquifer systems, microbial ecology of ground water, 156–161
DNA. *See* Deoxyribonucleic acid (DNA)
Dockins, W. S., 21, 44, 205, 206, 289
Down-hole contamination, 215, 224–225
Drake, E. L., 5, 204
Drilling, mud rotary drilling, microbiologic sampling, intermediate and regional flow systems, 215–216. *See also* Microbiologic sampling
Drilling fluid:
  contamination of cored sediments by, 224–230
  core fracture contamination, 225–226
  core seepage contamination, 225
  down-hole contamination, 224–225
  evaluation of, 226–230
  microbiologic sampling, intermediate and regional flow systems, 216–220
  mud rotary coring and, microbiologic sampling, intermediate and regional flow systems, 224
Dry-land farming practices, nitrogen cycle, ground-water systems, 251–253
Duff, J. H., 249
Dunlap, L. E., 328
Dunlap, W. J., 22, 132, 192, 212, 214
Duxbury, T., 318

Earthquake, 3
Ecological modeling, microbial ecology of ground water methods, 151–155. *See also* Geochemical modeling
Ecology, microbial ecology, ground water and, 22–24, 130–171. *See also* Microbial ecology of ground water
Eganhouse, R. P., 302, 350, 351
Ehrlich, G. G., 21, 22, 23, 345, 386
Ehrlich, P., 6
Ehrlich, S. D., 318
Ekendahl, S., 157, 158, 206, 228, 231
Ekzertsev, V. A., 205

Electron acceptor:
  acclimation mechanisms, 302–304
  bacteria classification, 39
  culture media design, 61
  electron transport systems, metabolism, energy release, 75
Electron balance, material balance calculations, 278–279
Electron carriers, electron transport systems, 76
Electron microscope, 33
Electron transport system:
  energy release, 74–77
  energy utilization, 77–78
El Solh, N., 318
Embden–Meyerhof–Parnas (EMP) pathway, adenosine triphosphate (ATP) metabolism, energy storage, 74, 75
Endoplasmic reticulum, eucaryotes, 29
Endospores, spore-forming rods, 45–46
Energy:
  metabolism and thermodynamics, 71–73
  microbial ecology of ground water, 133
  release of, electron transport systems, 74–77
  storage of, adenosine triphosphate (ATP) metabolism, 73–74
  utilization of, chemosmosis, 77–78
Enterobacteriaceae, described, 42–43
Enterovirus, 35
Environmental conditions (for bacterial growth), 51–57
  molecular oxygen, 55
  osmotic pressure, 56–57
  pH, 55–56
  temperature, 52–54
  water, 54–55
Environmental studies, alicyclic hydrocarbon biodegradation applications, 335
Enzymes:
  acclimation mechanisms, induction, 299–300
  adenosine triphosphate (ATP), metabolism, energy storage, 73–74
  archaebacteria, 31
  DNA structure and organization of, 107
  measurement, biomass measurement in soil microbiology, unsaturated zone, 185
  metabolism and, 78–81
  procaryotes, 26
Epifluorescent microscope:
  bacterial growth enumeration techniques, 68–69
  biomass measurement in soil microbiology, unsaturated zone, 183
Escherichia coli:
  classification of, 36
  described, 42–43, 117–118
  enzymes, 299
  genetics, 113, 127

induction and, 299
metabolism of, 76, 79
reproduction of, 48, 117
Ester linkages, lipids, biosynthesis, 96
Ethics, genetic engineering, release controversy, 127–128
Ethylbenzene, remediation of ground-water systems, 353–356
Eucaryotes:
describhed, 29–31
xenobiotics in ground-water systems (acclimation), 315
Evolution concept, 5
Excrement (animal), nitrogen cycle, ground-water systems, 247–248
Exner, M. E., 385
Exopolymers, biochemical marker techniques, microbial ecology of ground water methods, 143
Experimental method, microbiology and, 14–15
Exponential bacterial growth, population kinetics, 49–51
Extreme environment, acclimation to, 296–321. *See also* Acclimation

Facultative anaerobes:
bacteria classification, 39
gram-negative rods
described, 42–43
growth requirements, 55
Fass, E., 332
Fatty acids, lipids, biosynthesis, 96
Fecal contamination, hepatitis and, 35
Federle, T. W., 183, 185
Fermentation:
bacteria classification, 39
bacterial abundance and distribution, intermediate flow systems, 202
microbial ecology of ground water, 133–134
microbiology and, 4, 5, 13, 14, 15
synergism, 163
Ferredoxins, hydrogen and acetate in fermentation, bacterial metabolism and ground water, 82–85. *See also* Iron
Fertilizers. *See also* Agriculture
ground water flow systems (local), abundance and distribution (of bacteria), 191
nitrogen cycle, ground-water systems, 249–253
Fever, infection, mesophilic microorganisms, 53
Fisher, D. W., 198, 206, 207, 286, 289
Flagella, bacteria classification, 38
Flavobacterium, described, 40, 41
Fleming, Alexander, 6
Flipse, W. J., Jr., 251
Fluid loss control, of drilling fluids, 220
Fluorescein, drilling fluid contamination of cored sediments, evaluation of, 228

Fluorescein isothiocyanate (FITC), biomass measurement in soil microbiology, unsaturated zone, 183
Fogel, M. M., 380
Fogel, S., 380
Food chains, microbial ecology of ground water, 133
Fossil(s):
antiquity and, 3, 8
geology and, 4, 5, 9–10
Fossil fuels, carbon cycle and, 239
Foster, M. D., 20, 260, 283
Fracastoro, Girolamo, 4, 13, 16
Fracture, core fracture contamination, 225–226
Franklin, C., 318
Frederickson, J. K., 41, 97, 99, 100, 125, 126, 127, 150, 156, 157, 167, 169, 192, 194, 199, 200, 216
Freedman, D. L., 370
Freeze, R. A., 132, 176, 189
Freezing, environmental conditions for bacterial growth, 52
Froelich, P. N., 164, 264
Fuel spills. *See* Petroleum spills
Fungi:
eucaryotes, 29–31
xenobiotics in ground-water systems (acclimation), 315
Furukawa, K., 373
Fusobacterium, described, 44

Gallionella:
described, 41–42
iron cycle and, 99, 253–254
Garrels, R. M., 21, 264, 265
Gases:
bacterial abundance and distribution, intermediate flow systems, 198–199
geochemical modeling, 264
ground water flow systems (local), abundance and distribution (of bacteria), 191
unsaturated zone, microbial habitat, 181–182
Gasoline, refining and fuel blending of petroleum, petroleum hydrocarbon biodegradation, 323–325
Gasoline spills. *See* Petroleum spills
Gastrointestinal illness, 35
Geary, P. J., 336
Gehron, M. J., 184
Gel strength, of drilling fluids, 219–220
Genetic engineering, 118–128
applications, 121–128
insulin production, 122
pollution control, 122–124
subsurface microbiology, 124–128
release controversy, 127–128
techniques of, 118–121

Genetic mutation, acclimation mechanisms, 301–302
Genetics, 7, 18, 105–129
  DNA structure and organization, 106–108
  gene expression and regulation, 112–114
    induction, 113
    repression, 113–114
  genetic engineering, 118–128
    applications, 121–128
    techniques of, 118–121
  historical overview, 105
  mutation, 114–116
    agents of, 115–116
    described, 114–115
    transposable material, 115
  natural genetic exchanges, 116–118
  RNA structure and organization, 108–112
    classes of RNA, 108
    transcription, 108–109
    translation-making proteins, 109–112
Geochemical modeling, 264–294
  advantages of, 264
  bacterial abundance and distribution
    intermediate flow systems, 198–199
    regional flow systems, 206–207
  considerations in, 265–271
    generally, 265–266
    hydrologic factors, 266–267
    kinetic factors, 270–271
    mineralogic factors, 267
    thermodynamic factors, 267–270
  historical perspective on, 264–265
  methodology, 271–283
    logic of, 281–283
    material balance calculations, 277–280
    mineral equilibrium calculations, 276–277
    speciation calculations, 271–276
  microbial ecology of ground water methods, 149–151
  microbial processes in aquifer systems, 283–294
    carbon dioxide production, 283–285
    sulfate reduction, 289–294
    sulfate reduction and methanogenesis, 286–289
  quantitative methods, 21
Geochemical processes, metabolic control of, 102–103
Geologic plate tectonics, geology and, 12
Geology:
  microbiology and, 19–22
  milestones in development of, 4–7
  science of, 3–12
George, L. H., 49, 142, 154
Geosyncline, geology and, 10–11
Gerba, C. P., 35, 43
Gerhart, J. M., 248

Germ theory of disease, 4, 6, 13–14, 16
Ghiorse, W. C., 22, 23, 31, 40, 43, 45, 65, 131, 132, 140, 141, 142, 145, 146, 154, 156, 168, 170, 187, 191, 192, 193, 199, 303, 313, 315
Gibson, D., 336
Gillan, R. T., 183
Global carbon cycle. See Carbon cycle
Gluconeogenesis, carbohydrates, biosynthesis, 94
Glucose:
  adenosine triphosphate (ATP) metabolism, energy storage, 74
  carbohydrates, biosynthesis, 94–95
  culture media design, 58–59
  metabolism, enzyme regulation, 78–79
  metabolism and thermodynamics, 72
Glycocalyx, procaryotes, 28–29
Glycolysis:
  adenosine triphosphate (ATP) metabolism, energy storage, 74, 75
  pyruvate and, 81–85
Godsy, E. M., 21, 22, 198, 386
Goerlitz, D. F., 386
Golgi bodies, eucaryotes, 29
Goodman, N. L., 340
Goodwin, S., 23, 149, 150, 151
Gormley, J. R., 251
Gossett, J. M., 370
Gram-negative bacteria, 40–44
  aerobic gram-negative cocci, 42
  aerobic gram-negative rods, 40–42
  anaerobic gram-negative rods, 43–44
  cell wall, procaryotes, 28
  distribution of, 40
  facultatively anaerobic gram-negative rods, 42–43
Gram-positive bacteria, 44–46
  cell wall, procaryotes, 28
  cocci, described, 45
  coryneform bacteria, 45
  gram-positive cocci, 45
  spore-forming rods, 45–46
Gram stain:
  bacteria classification, 36, 37, 38
  techniques, 17
Gray, E. M., 250
Grbic-Galic, D., 342, 369
Greer, Frank, 19
Griffen, R. A., 373
Ground water:
  bacterial metabolism, 81–91
    ferredoxins and hydrogen and acetate in fermentation, 82–85
    iron reduction, 88–89
    lactate and acetate fermentations, 81–82
    methanogenic pathways, 85–87
    nitrate reduction, 89–91

oxygen reduction-aerobic metabolism, 91
sulfate reduction, 87–88
chemistry of, microbiology and, 18–24
contamination of, acclimation to, 296–321. *See also* Acclimation
microbial ecology of ground water, 22–24, 130–171. *See also* Microbial ecology of ground water
microbiology and, 2–3, 18–24
microorganisms in, 25–46
archaebacteria, 31–33
bacteria in ground-water systems, 35–46. *See also* Bacteria
eucaryotes, 29–31
overview of, 25–26
procaryotes, 26–29
viruses, 33–35
sampling of, microbiologic sampling, 230–231. *See also* Microbiologic sampling
Ground-water systems (generally):
biochemical cycling in, 233–263. *See also* Biochemical cycling
halogenated organic compound degradation in, 358–389. *See also* Halogenated organic compound degradation
iron cycle, 255–258
nitrogen cycle and, 247–253
petroleum hydrocarbon biodegradation, 347–351
aerobic degradation of BTX compounds, 348–350
anaerobic degradation of aromatic hydrocarbons, 350–351
generally, 347–348
remediation processes, 351–356
subsurface environment classification, 175–176
sulfur cycle, 259–263
Ground-water systems (intermediate):
abundance and distribution of bacteria, 195–203
bacterial distribution in, 198–202
described, 195–197
microbial habitat, 197–198
microbial processes in confining beds, 202–203
biochemical cycling, carbon cycle, 244
microbiologic sampling, 215–224
drilling fluids, 216–220
mud rotary coring, 220–224
mud rotary drilling, 215–216
Ground-water systems (local):
abundance and distribution of bacteria, 189–195
bacterial distribution, 191–195
described, 189
as microbial habitat, 190–191
biochemical cycling, carbon cycle, 242–244
microbiologic sampling, 211–215
aseptic technique with split-spoon and push-tube sampling, 214–215

push-tube (Shelby Tube) methods, 212–214
split-spoon sampling, 211–212
Ground-water systems (regional):
abundance and distribution of bacteria, 203–207
bacterial distributions in, 205–207
described, 203
petroleum reservoirs, early observations, 204–205
biochemical cycling, carbon cycle, 244
microbiologic sampling, 215–224
drilling fluids, 216–220
mud rotary coring, 220–224
mud rotary drilling, 215–216
Growth factors, culture media design, 61
Growth media design. *See* Bacterial growth, culture techniques; Culture methods
Gulens, J., 151
Gurevich, M. I., 20–21
Gurevich, M. S., 198

Hadley, P. W., 347
Haefeli, C., 318
Hale, D. D., 377
Hall, James, 5, 10
Halogenated organic compound degradation, 358–389. *See also* Petroleum hydrocarbon biodegradation
chemistry and uses of halogenated organic compounds, 358–367
aliphatic compounds, 359–361
chlorinated herbicides, 365–366
chlorinated phenols, 366–367
monocyclic aromatic compounds, 361–364
organochlorine insecticides, 364–365
polychlorinated biphenyls, 364
degradation processes, 367–377
aliphatic compounds, 367–371
chlorinated herbicides, 375–376
chlorinated phenols, 376–377
monocyclic aromatic compounds, 371–373
organochlorine insecticides, 375
polychlorinated biphenyls, 373–375
in ground-water systems, 377–388
aliphatic compounds, 377–380
aliphatic compounds remediation, in situ, 380–384
alkyl halide insecticides, 384–385
chlorinated herbicides, 385–386
chlorobenzenes, 385
chlorophenolic compounds, 386–388
overview of, 358
Halophiles, archaebacteria, 32
Hand auger, microbiologic sampling of unsaturated zone, 209–210
Hanshaw, B. B., 21, 179
Hardy, K., 318

Harvey, R. W., 49, 142, 154
Harvie, C. E., 276
Hazen, T. C., 154
Heat output, biomass measurement in soil microbiology, unsaturated zone, 185
Heitkamp, M. A., 340
Helgeson, H. C., 264
Hellwig, M., 376
Helmholtz, Herman von, 10
Hem, John, 21
Henry, S. M., 369
Hepatitis, 35
Herbicides. *See also* Insecticides; Pesticides
    chlorinated herbicides
        chemistry and uses of, 365–366
        degradation described, 375–376
        degradation in ground-water systems, 385–386
    unsaturated zone, 179
Heredity, 105. *See also* Genetics
Herpes simplex virus types I and II, 35
Heterotrophy, 39, 96
Hicks, R. J., 150, 199
Higgins, P., 204
Hirsch, P., 23, 41, 42, 46, 156, 157, 170, 231
Holley, Robert, 18, 110
Holmes, Arthur, 6, 11
Hook, Robert, 13–14
Horowitz, A., 308, 372
Human immunodeficiency virus (HIV), 35
Hutchins, S. R., 353, 354, 356
Hutton, James, 4, 9, 12
Huygens, Christian, 14
Hydraulic pressure, unsaturated zone, 179, 209
Hydrocarbons:
    biodegradation of. *See also* Petroleum hydrocarbon biodegradation
        alicyclic, 334–335
        aliphatic, 328–334
    ground water and, 19–21
    hydrocarbon-degrading enzymes, acclimation mechanisms, induction, 299–300
Hydrogen, ferredoxins and acetate in fermentation, bacterial metabolism and ground water, 82–85
Hydrogen carriers, electron transport systems, metabolism, energy release, 76
Hydrogen cycle, integrated carbon, oxygen, and hydrogen cycle, 240–241
Hydrogen oxidizers, chemolithotrophy, 97
Hydrogen peroxide, anaerobic gram-negative rods, 43–44
Hydrolases, metabolism, enzyme regulation, 79
Hydrologic considerations, geochemical modeling, 266–267
Hydrologic setting, subsurface environment classification, 175
Hynning, P. K., 376

Immunization, discovery of, 16–17
Induction:
    acclimation mechanisms, 299–300
    gene expression and regulation, 113
Infection:
    immunization from, 16–17
    mesophilic microorganisms, temperature, 53
    viruses, 33–35
Inoculation, 17
Inorganic nutrients, culture media design, 60–61
Insecticides. *See also* Herbicides; Pesticides
    alkyl halide insecticides, halogenated organic compound degradation, in ground-water systems, 384–385
    organochlorine insecticides
        chemistry and uses of, 364–365
        halogenated organic compound degradation, 375
Insulin production, genetic engineering applications, 122
Intermediate flow systems. *See* Ground-water systems (intermediate)
Intermediate subzone (of unsaturated zone), described, 180
Invariant system, geochemical modeling, thermodynamic considerations, 269
Ions, acclimation, metal toxicity mechanisms, 317–318
Iron:
    acclimation to available electron receptors, 303
    bacterial abundance and distribution, 202–203
    carbon cycle and, 241
    chlorobenzenes, halogenated organic compound degradation, in ground-water systems, 385
    culture media design, 60
    methane oxidation, 330
    microbial degradation, ground-water systems, 350
    microbial ecology of ground water, 133–134
    toluene degradation, anaerobic degradation processes, 342–343
Iron cycle, 253–258
    aquatic sediments, 254–255
    described, 253–254
    ground-water systems, 255–258
Iron oxidizers, chemolithotrophy, 99–100
Iron-reducing bacteria:
    anaerobic gram-negative rods, 44
    bacterial metabolism and ground water, 88–89
    competition, microbial ecology, 165–167
Isolation techniques, bacterial growth culture techniques, 63–66
Isostasy, geology and, 10–11
Isotope balance, material balance calculations, geochemical modeling methodology, 279–280
Isotopic dating, geology and, 6, 10

Ivanov, M. V., 205
Iwanowski, O., 33

Jackson, R. E., 151
Jacob, Fran»ois, 7, 18
Jeenes, D. J., 302
Jenner, Edward, 17
Jet fuel contamination, remediation of ground-water systems, 353–356
Jimenez, L., 42, 125
Jones, B. F., 21, 264, 265, 276
Jones, D. C., 252
Jones, R. E., 44, 150, 194, 303
Jorgensen, B. B., 258, 261

Katz, B. G., 248
Kean, T. M., 236, 260
Kehew, A. E., 250
Kelly, F. X., 153, 154
Kelvin, Lord (William Thomson), 10
Keswick, B. H., 35, 43
Khan, S. U., 376, 386
Kharaka, Y. K., 276
Khorana, H. G., 18, 110
Khudyakova, A. I., 54
Kinetic factors, geochemical modeling, 270–271
Kinner, N. E., 168
Klug, M. J., 164
Kluyver, Albert, 17–18
Knackmuss, H.-J., 371, 376
Knobel, L. L., 198
Koch, Robert, 6, 15–16, 18
Kolbel-Boelke, J., 136, 137, 138, 139, 140, 142, 146, 156
Konopka, A., 386
Krebs, Hans, 7
Krebs cycle, oxygen reduction-aerobic metabolism, bacterial metabolism and ground water, 91
Kreitler, C. W., 252
Kuznetsov, S. I., 205

Lactate fermentations, bacterial metabolism and ground water, 81–82
Lactobacillus arabinosus, synergism, 163
Lactose, acclimation mechanisms, induction, 299–300
Langmuir, D., 257
Larochelle, A., 12
Larrick, S. R., 160
Latour, Charles Cagniard de, 5
Lauffenburger, D. A., 153, 154
Leach, L. E., 214
Leahy, P. P., 195, 197
Lee, T. R., 386
Leenheer, J. A., 21, 198
Leeuwenhoek, Anton van, 2, 3, 4, 14, 17
Leprosy, quarantine and, 13

Leptothrix, iron cycle and, 253–254
Liebig, Justus von, 14
Lien, T., 46
Ligases:
  DNA structure and organization of, 107
  metabolism, enzyme regulation, 79
Lignin, microbial ecology of ground water, 133–134
Lindane, halogenated organic compound degradation, 375
Linder, J. B., 248
Linkfield, T. G., 307, 308, 310
Lipids, biosynthesis, 96
Lipopolysaccharides:
  biochemical marker techniques, 143
  carbohydrates, biosynthesis, 94
Liquid wastes, geochemical modeling and, 21
Lister, Joseph, 6
Lithotrophs, bacteria classification, 39
Local flow systems. *See* Ground-water systems (local)
Lonergan, D. J., 253, 303
Lorochelle, A., 7
Lovley, D. R., 23, 44, 55, 88, 89, 149, 150, 151, 164, 170, 198, 200, 202, 227, 253, 255, 303, 350
Lubricity, of drilling fluids, 220
Lyases, metabolism, enzyme regulation, 79
Lyell, Charles, 9
Lysosomes, eucaryotes, 29

MacKenzie, F. T., 21, 264, 265
Madsen, E. L., 31, 297
Magnesium, culture media design, 60
Magnetic field, geology and, 7, 12
Major, D., 348, 350
Malcolm, R. L., 21
Manures, nitrogen cycle, ground-water systems, 247–248
Marker techniques, biochemical, microbial ecology of ground water methods, 142–146
Martin, M., 197
Mass balance, material balance calculations, geochemical modeling methodology, 277–278
Material balance calculations, geochemical modeling methodology, 277–280
Mathews, D. H., 7, 12
Matthess, G., 132
Maxorella, described, 42
Mayer, A. E., 33
McCarty, P. L., 369
McMahon, P. B., 23, 46, 184, 197, 202, 228, 230, 261, 282, 283, 386
McNabb, J. F., 22, 132, 192
Meister, Harold, 260–261
Membrane-bound enzymes, metabolism, enzyme regulation, 80

Membrane filtration method, bacterial growth
    enumeration techniques, 67
Mendel, Gregor, 105
Mercury toxicity, metal toxicity acclimation, 320
Mesophilic microorganisms, temperature,
    environmental conditions (bacterial growth), 53
Messenger RNA (mRNA), function of, 108,
    110–112
Metabolism, 71–104
  acclimation mechanisms
    catabolite repression, 300–301
    genetic mutations, 301–302
    induction, 299–300
  bacteria classification, 39
  biosynthesis, 92–96
    amino acids, 93–94
    carbohydrates, 94–96
    lipids, 96
  chemolithotrophy, 96–102
    ammonia oxidizing (nitrifying) bacteria, 100
    autotrophic carbon dioxide fixation, 100–102
    hydrogen oxidizers, 97
    iron oxidizers, 99–100
    sulfide oxidizers, 98–99
  control of geochemical processes, 102–103
  energy release, electron transport systems,
    74–77
  energy-releasing pathways of geochemical
    importance, 81–91
    ferredoxins and hydrogen and acetate in
      fermentation, 82–85
    iron reduction, 88–89
    lactate and acetate fermentations, 81–82
    methanogenic pathways, 85–87
    nitrate reduction, 89–91
    oxygen reduction-aerobic metabolism, 91
    sulfate reduction, 87–88
  energy storage, adenosine triphosphate (ATP)
    synthesis, 73–74
  energy utilization, chemosmosis, 77–78
  enzymes in, 78–81
  geochemical modeling, bacterial abundance
    and distribution, intermediate flow
    systems, 198–199
  subsurface environments, 23
  thermodynamics and, 71–73
Meta cleavage, of catechol, aromatic
    hydrocarbon biodegradation, 337–338
Metal toxicity (acclimation), 317–320
  mechanisms of, 317–318
  mercury toxicity, 320
  plasmid-encoded metal resistance mechanisms,
    318–320
Methane mono oxidase aliphatic hydrocarbon
    biodegradation, 329, 330
Methane monooxygenase (MMO), aerobic
    degradation, halogenated organic compound
    degradation, aliphatic compounds, 368–369

Methane oxidation, aliphatic hydrocarbon
    biodegradation, 329–330
Methanogenesis:
  conditions for, toluene and benzene anaerobic
    degradation processes and, 341–342
  methanogenic bacteria
    archaebacteria, 32
    carbon cycle and, 238–239
  microbial processes in aquifer systems,
    geochemical modeling, 286–289
Methanogenic pathways, bacterial metabolism
    and ground water, 85–87
Methyl group oxidation, N-alkanes oxidation,
    aliphatic hydrocarbon biodegradation,
    331–332
Methylobacter, methane oxidation, 329
Methylobacterium, methane oxidation, 329
Methylocystis, methane oxidation, 329
Michaelis-Menton equation:
  ecological modeling, microbial ecology of
    ground water methods, 152
  metabolism, enzyme regulation, 79
Microbial acclimation, 296–321. *See also*
    Acclimation
Microbial ecology of ground water, 22–24,
    130–171
  diversity and niches in aquifer systems,
    156–161
    measurements, 156–158
    sources of diversity, 158–160
    stress and diversity, 160–161
  methods in, 135–155
    activity measurements in microcosms,
      146–148
    biochemical marker techniques, 142–
      146
    culture methods, 136–141
    direct observation, 141–142
    ecological modeling, 151–155
    geochemical methods, 149–151
  overview of ecology, 130–132
  population mechanics, 161–168
    antagonism, parasitism, and predation,
      167–168
    commensalism, 162–163
    competition, 164–167
    neutralism, 162
    synergism and mutualism, 163–164
  r and K strategies, 168–170
    aquifer environment, 169–170
    described, 168–169
  scope of, 132–135
Microbial processes in aquifer systems,
    geochemical modeling, 283–294
  carbon dioxide production, 283–285
  sulfate reduction, 289–294
  sulfate reduction and methanogenesis, 286–
    289

Microbiologic sampling, 208–232
  drilling fluid contamination of cored sediments, 224–230
    core fracture contamination, 225–226
    core seepage contamination, 225
    down-hole contamination, 224–225
    evaluation of, 226–230
    ground-water sampling for microorganisms, 230–231
  intermediate and regional flow systems, 215–224
    drilling fluids, 216–220
    mud rotary coring, 220–224
    mud rotary drilling, 215–216
  local flow systems, 211–215
    aseptic technique with split-spoon and push-tube sampling, 214–215
    push-tube (Shelby Tube) methods, 212–214
    split-spoon sampling, 211–212
  problems in, 208–209
  unsaturated zone, 209–211
    air drilling and coring, 210–211
    hand auger, 209–210
Microbiology:
  genetic engineering applications, 124–128
  ground water and, 2–3, 18–24
    geoscience, 19–22
    microbial ecology, 22–24, 130–171. See also Microbial ecology of ground water
  milestones in development of, 4–7
  science of, 13–18
  subsurface, genetic engineering applications, 124–128
Micrococcus:
  bacteria classification, 38
  described, 45
  reproduction of, 48
Microcosms, activity measurements in microcosms, microbial ecology of ground water methods, 146–148
Microorganisms, 25–46
  archaebacteria, 31–33
  bacteria in ground-water systems, 35–46
    classification of, 36–40
    gram-negative bacteria, 40–44
    gram-positive bacteria, 44–46
    overview of, 35–36
  eucaryotes, 29–31
  microbial ecology of ground water, 22–24, 130–171. See also Microbial ecology of ground water
  nitrogen cycle and, 245
  overview of, 25–26
  procaryotes, 26–29
  viruses, 33–35
Microscope:
  bacterial growth enumeration techniques epifluorescent microscope, 68–69
  local flow systems, 193–194
  biomass measurement in soil microbiology, unsaturated zone, 182–183
  development of, 2–3
  electron microscope, 33
  microbial ecology of ground water methods and, 135, 141
Mikesell, M. D., 377
Mineral equilibrium calculations, geochemical modeling methodology, 276–277
Mineralogic considerations, geochemical modeling, 267
Misra, T. K., 319
Mitchell, Peter, 7
Mitochondria, eucaryotes, 29
Modeling, geochemical modeling, bacterial abundance and distribution, intermediate flow systems, 198–199. See also Geochemical modeling
Moisture content, unsaturated zone, microbial habitat, 181–182
Molecular oxygen, environmental conditions (bacterial growth), 55
Moller, first name, 276
Monocyclic aromatic compounds:
  chemistry and uses of, 361–364
  halogenated organic compound degradation, 371–373
Monod, J., 7, 18, 300
Morgan-Jones, M., 250
Morley, L., 7, 12
Morris, J. T., 184
Most-probable-number (MPN) technique, bacterial growth enumeration techniques, 67–68
Motility:
  bacteria classification, 38
  ecological modeling, microbial ecology of ground water methods, 154
Mountains, geology and, 10–11
Mud rotary drilling and coring, microbiologic sampling, intermediate and regional flow systems, 215–216, 220–224
Mueller, J. G., 387
Municipal wastes, nitrogen cycle, ground-water systems, 250–251
Murchison, Roderick, 10
Mutation, 114–116
  agents of, 115–116
  described, 114–115
  transposable material, 115
Mutualism, population mechanics, 163–164
Mycrocyclus, described, 41

n-alkanes, aliphatic hydrocarbon biodegradation, 330–332
Nazina, T. N., 54
Needham, John, 14

Nehrkorn, A., 136, 137, 138, 139, 140, 146, 156
Neilson, A. H., 376
Neisseria, described, 42
Nelson, R. A., 250
Neutralism, population mechanics, 162
Niches, diversity and, in aquifer systems, microbial ecology of ground water, 156–161
Nicotine adenine dinucleotide (NAD), metabolism, energy storage, 73–74
Nirenberg, Marshal, 18, 110
Nitrate reduction, bacterial metabolism and ground water, 89–91
Nitrifying bacteria. *See* Ammonia oxidizing (nitrifying) bacteria
Nitrobacter, ammonia oxidizing (nitrifying) bacteria, 100
Nitrogen cycle, 244–253
  described, 244–247
  ground-water systems, 247–253
Nitrogen fertilizers, nitrogen cycle, ground-water systems, 249–253
Nitrogen fixation:
  aerobic gram-negative rods, 41
  nitrogen cycle and, 245–246
Nitrogen sources, culture media design, 60
Nitrosomanas, ammonia oxidizing (nitrifying) bacteria, 100
Nocardia:
  benzene oxidation, 337
  biodegradation, 376
Noncompetitive inhibition, metabolism, enzyme regulation, 80
Nordstrom, first name, 276
Norvatov, A. M., 175
Nuclear magnetic resonance imaging, bacterial abundance and distribution, intermediate flow systems, 197
Nutrition, bacteria classification, 39

Obligatae anaerobes:
  bacteria classification, 39
  growth requirements, 55
Obligate aerobes:
  bacteria classification, 39
  growth requirements, 55
Oceans, geology and, 8–9
Oddie, K. M., 317
Ogunseitan, O. A., 126
Oil spills. *See* Petroleum spills
Olson, G. J., 22, 54, 205, 206, 231
Optics, microbiology and, 2–3, 13–14
Organochlorine insecticides:
  chemistry and uses of, 364–365
  halogenated organic compound degradation, 375
Ortho cleavage, of catechol, aromatic hydrocarbon biodegradation, 337–338

Osmotic pressure, environmental conditions (bacterial growth), 56–57
Owens, J. P., 261
Oxidases, metabolism, enzyme regulation, 79
Oxidation:
  alkene-oxidation, aliphatic hydrocarbon biodegradation, 332
  methane oxidation, aliphatic hydrocarbon biodegradation, 329–330
  n-alkanes oxidation, aliphatic hydrocarbon biodegradation, 330–332
  xenobiotic, catabolite repression, 301
Oxidation-reduction reactions, geochemical modeling methodology, speciation calculations, 274–275
Oxygen, molecular oxygen, environmental conditions (bacterial growth), 55
Oxygen cycle:
  biochemical cycling, 234–238
  integrated carbon, oxygen, and hydrogen cycle, 240–241
Oxygen reduction-aerobic metabolism, bacterial metabolism and ground water, 91

Parasitism, population mechanics, 167–168
Parkhurst, D. L., 198, 264, 265, 280, 286
Particulate tracers, drilling fluid contamination of cored sediments, evaluation of, 229–230
Passero, R. N., 379, 380
Pasteur, Louis, 5, 14–15, 16, 17, 18
Pasteurization process, 5, 15
Patrick, G. C., 348, 350
Pearson, F. J., Jr., 21
Pedersen, K., 157, 158, 206, 228, 231
Penicillin, 6
Peptidoglycan, carbohydrates, biosynthesis, 94
Perfluorocarbons, drilling fluid contamination of cored sediments, evaluation of, 229
Periplasmic space, procaryotes, 28
Perry, J. J., 334
Pesticides. *See also* Herbicides; Insecticides
  ground water flow systems (local), abundance and distribution (of bacteria), 191
  unsaturated zone, 179
Peterson, M. L., 237
Petroleum hydrocarbon biodegradation, 322–357. *See also* Halogenated organic compound degradation
  alicyclic hydrocarbons, 334–335
  aliphatic hydrocarbons, 328–334
    alkene-oxidation, 332
    branched aliphatic, 332–334
    generally, 328–329
    methane oxidation, 329–330
    n-alkanes, 330–332
  aromatic hydrocarbons, 335–347
    alkyl benzenes, 338–339
    anaerobic degradation processes, 340–347

benzene, 336–338
  generally, 335–336
  polycyclic aromatic compounds, 339–340
 crude oil composition, 323
 in ground-water systems, 347–351
  aerobic degradation of BTX compounds, 348–350
  anaerobic degradation of aromatic hydrocarbons, 350–351
  generally, 347–348
  movement and separation of petroleum hydrocarbons in, 323–325
   density-driven migration, 326–327
   solubility, 327–328
  remediation of, 350–356
 petroleum refining and fuel blending, 323–325
 threat of uncontrolled release, 322
Petroleum reservoirs, abundance and distribution (of bacteria), regional flow systems, 204–205
Petroleum spills:
 acclimation to available electron receptors, 302–303
 aromatic hydrocarbon biodegradation, 335–336
 ground water flow systems (local), abundance and distribution (of bacteria), 191, 192
 microbial degradation, ground-water systems, 347, 350
 threat of uncontrolled release, 322
Pfaender, F. K., 305, 308, 311, 313, 316
pH:
 acclimation mechanisms, 298
 environmental conditions (bacterial growth), 55–56
Phage vectors, genetic engineering and, 121
Phelps, T. J., 21, 23, 55, 194, 202, 216, 228
Phenols:
 anaerobic degradation processes, aromatic hydrocarbon biodegradation, 345–347
 chlorinated phenols
  chemistry and uses of, 366–367
  halogenated organic compound degradation, 376–377
Phillips, E. J. P., 44, 89, 253, 255
Phospholipids, biochemical marker techniques, microbial ecology of ground water methods, 143, 145
Phospholipid technique, biomass measurement in soil microbiology, unsaturated zone, 183–184
Phosphorus nutrients, culture media design, 60–61
Photolithotrophs, bacteria classification, 39
Photosynthesis, 96
 biochemical cycling, oxygen cycle, 234–238
 carbon cycle and, 238–239, 240
 sulfur cycle and, 259
Pitzer, K. S., 276
Plague, quarantine and, 13

Plasmids:
 genetic engineering and, 119–120, 125–127
 plasmid-encoded metal resistance mechanisms, metal toxicity acclimation, 318–320
 procaryotes, 26
Plasmid vectors, genetic engineering and, 121
Plate count, bacterial growth enumeration techniques, 67
Plate tectonics, geology and, 12
Plenciz, Anton von, 16
Plummer, L. N., 21, 198, 201, 262, 264, 265, 267, 268, 276, 277, 280, 281, 283, 286, 289
Polarization, geology and, 12
Polio, 35
Pollution. See also Herbicides; Insecticides; Pesticides; Petroleum spills; Sewage wastes; Waste disposal
 carbon cycle and, 239
 control of, genetic engineering applications, 122–124
 oil spills, threat of uncontrolled release, 322
Polychlorinated biphenyls:
 chemistry and uses of, 364
 halogenated organic compound degradation, 373–375
Polycyclic aromatic compounds, aromatic hydrocarbon biodegradation, 339–340
Polymerases, metabolism, enzyme regulation, 79
Polymeric hydroxyalkanoic (PHA) acids, biochemical marker techniques, microbial ecology of ground water methods, 143, 146
Polysaccharides, carbohydrates, biosynthesis, 94
Popov, O. V., 175
Population kinetics, bacterial growth, 49–51
Population mechanics, 161–168
 antagonism, parasitism, and predation, 167–168
 commensalism, 162–163
 competition, 164–167
 neutralism, 162
 synergism and mutualism, 163–164
Potassium, culture media design, 60
Pouchet, F.-A., 14
Predation, population mechanics, 167–168
Pristine aquifer sediments, xenobiotics in ground-water systems (acclimation), 313–315
Procaryotes. See also Bacteria
 described, 26–29
 plasmid-encoded metal resistance mechanisms, metal toxicity acclimation, 318–320
 reproduction, 116
Protein(s):
 amino acids, biosynthesis, 93–94
 eucaryotes, 29–30
 RNA and, 110–112
Protein coat, viruses, 33–35

Protozoa:
  biomass measurement in soil microbiology, unsaturated zone, 182
  eucaryotes
    described generally, 29–31
    xenobiotics in ground-water systems (acclimation), 315
  local flow systems, 195
  microbial ecology of ground water, 135
  parasitism, population mechanics, 168
Pseudomonas:
  biodegradation, 336, 338, 372, 376
  described, 40–41
  enzymes, 299–300
  genetic engineering, 122, 123, 125, 302
  oxidation, 301, 331, 334
Pseudomonas fluoresecens, microbial ecology, 157
Psychrophilic microorganisms, temperature, environmental conditions (bacterial growth), 52
Pucci, A. A., Jr., 261
Push-tube (Shelby Tube) sampling, local flow systems, 212–214
Putrefaction, sulfur cycle and. *See also* Fermentation
Putrefaction sulfur cycle and, 258
Pyruvate:
  carbohydrates, biosynthesis, 94
  glycolysis and, 81–85
Pyruvate kinase, carbohydrates, biosynthesis, 95

Quantitative methods, geochemical modeling, 21
Quarantine, infectious disease and, 13

Rabies vaccine, 6, 17
Rades-Rohkohl, E., 23, 41, 42, 46, 156, 157, 170, 231
Radiation, genetic mutation and, 115–116
Radioactive dating, geology and, 6, 10
Ragone, S. E., 248
Ramanand, K., 347
Recombination, genetics, natural genetic exchanges, 117–118. *See also* Genetic engineering
Reductases, metabolism, enzyme regulation, 79
Rees, J. F., 340
Refining, of petroleum, petroleum hydrocarbon biodegradation, 323–325
Regional ground-water systems. *See* Ground-water systems (regional)
Rehm, H. J., 301
Reineke, W., 371
Reinhard, M., 340
Release controversy, genetic engineering, 127–128
Religion, geology and, 3, 8
Remediation, of ground-water systems, petroleum hydrocarbon biodegradation, 350–356
Repression, gene expression and regulation, 113–114
Reproduction. *See* Bacterial growth: reproduction
Respiration:
  bacteria classification, 39
  biomass measurement in soil microbiology, unsaturated zone, 184–185
  carbon cycle and, 238–239, 240
  electron transport systems, metabolism, energy release, 75
Rhinovirus, 35
Rhizobium:
  described, 40, 41
  nitrogen cycle and, 245, 246
Rhodamine dye, drilling fluid contamination of cored sediments, evaluation of, 228
Ribitol, biochemical marker techniques, microbial ecology of ground water methods, 145
Ribonucleic acid (RNA):
  archaebacteria, 31–32
  bacteria classification, 40
  gene expression and regulation, 113–114
  genetic information storage, 106
  procaryotes, 27
  structure and organization, 108–112
    classes of RNA, 108
    transcription, 108–109
    translation-making proteins, 109–112
  viruses, 33–35
Ribosomal RNA (rRNA), function of, 108, 109–112
Ribosomes:
  archaebacteria, 31
  eucaryotes, 30
  procaryotes, 27
Rightmire, C. T., 179, 262
Rittmann, B. E., 369
RNA. *See* Ribonucleic acid (RNA)
Roberts, P. V., 380
Robertson, F. N., 201, 237, 247–248
Robinson, J. A., 372, 373
Rock, age of, 10
Rogers, G. S., 204
Rogers, J. E., 377
Rogers, Sherburne, 19
Rosnes, J. T., 46
Rosswall, T., 185
Rozanova, E. P., 54
Rutherford, Ernest, 6, 10

Salmonella typhimurium, Ames test, 115
Sampling. *See* Microbiologic sampling

Sander, P., 372
Saturated zone, subsurface environment classification, 175
Saturation index, geochemical modeling:
  mineral equilibrium calculations, 276
  thermodynamic considerations, 268
Saturation kinetics, metabolism, enzyme regulation, 79
Scale conflict, microbial ecology of ground water methods and, 135–136
Scanning electron microscope, biomass measurement in soil microbiology, unsaturated zone, 183
Schnurer, J., 185
Schulze, Franz, 14
Schwann, Theodor, 14
Schwartz, W., 156
Schwarzenbach, R. P., 385
Scientific method, microbiology and, 14–15
Sedgwick, Adam, 10
Sedimentary rock, age of, 10
Sedimentation:
  fossils and, 9–10
  stratigraphy and, 8
Seepage, core seepage contamination, microbiologic sampling, drilling fluid contamination of cored sediments, 225
Selective growth, culture media design, 61–63
Semprini, L., 368, 380
Sendlein, L. V. A., 347
Septic effluents, ground water flow systems (local), abundance and distribution (of bacteria), 191
Serial dilution, isolation techniques, bacterial growth culture techniques, 64
Seubert, W., 332
Sewage wastes. *See also* Pollution; Waste disposal
  ground water flow systems (local), abundance and distribution (of bacteria), 191
  nitrogen cycle, ground-water systems, 248–249
Shelby Tube. *See* Push-tube (Shelby Tube) sampling
Silver, S., 319
Sinclair, J. L., 31, 66, 157, 158, 168, 193, 195, 199, 315, 386
Sinclair, N. A., 41, 42, 231
Slime layer, procaryotes, 28
Smallpox, 17
Smith, G. B., 340
Smith, R. L., 154, 249
Smith, William, 5, 9, 10, 12
Smolenski, W. J., 345, 346, 347
Snider, Antonio, 11
Soddy, Frederick, 6, 10
Sodium, culture media design, 60

Soil fumigation method, biomass measurement in soil microbiology, unsaturated zone, 183
Soil microbiology, microbial ecology, ground water and, 22–24
Soil subzone:
  intermediate flow systems and, 197
  unsaturated zone, 179–180
Solbau, R. D., 237
Solubility, ground-water systems, movement and separation of petroleum hydrocarbons in, 327–328
Sorby, Henry Clifton, 5
Spain, J. C., 305, 313
Spalding, R. F., 250, 251, 385
Spallanzani, Lazzaro, 4, 14
Sparling, G. P., 185, 186
Speciation calculations, geochemical modeling methodology, 271–276
Spirillum cell, bacteria classification, 37, 38
Split-spoon sampling, local flow systems, 211–212
Spontaneous generation, 14, 15
Spore-forming rods, described, 45–46
Spring waters, integrated carbon, oxygen, and hydrogen cycle, 241–242
Stanley, Wendell, 7, 33
Staphylococcus:
  described, 45
  metal toxicity acclimation, 319
Steel, J. H., 241–242
Steno, Nicholas, 4, 8, 12
Stetzenbach, L. D., 41, 42, 157, 158, 231
Stevenson, R. J., 161
Stoermer, E. F., 161
Stratigraphy, geology and, 8
Streak-late method, isolation techniques, bacterial growth culture techniques, 63–64
Streptococcus:
  bacteria classification, 38
  described, 45
Streptococcus faecalis, synergism, 163
Streptococcus mutans, 29
Stress, diversity and, microbial ecology of ground water, diversity and niches in aquifer systems, 160–161
Substitution mutation, described, 115
Subsurface environments, classification of, abundance and distribution (of bacteria), 174–178
Subsurface microbiology. *See* Microbiology
Suchtelen, Hesselink van, 185
Suflita, J. M., 44, 99, 150, 167, 194, 230, 303, 307, 308, 310, 345, 346, 347, 352, 372, 373
Sulfate, carbon cycle and, 241
Sulfate-reducing bacteria:
  anaerobic gram-negative rods, 44
  bacterial metabolism and ground water, 87–88

competition, microbial ecology, 164–165
gram-positive bacteria, spore-forming rods, 46
mesophilic microorganisms, 53–54
microbial processes in aquifer systems, geochemical modeling, 286–289, 289–294
osmotic pressure, environmental conditions (bacterial growth), 57
Sulfide oxidizers, chemolithotrophy, 98–99
Sulfide-oxidizing bacteria, pH requirements, 55, 56
Sulfur cycle, 258–263
  described, 258–259
  ground-water systems, 259–263
Supreme Court (U.S.), *Diamond v. Chakrabarty*, 123
Swindoll, C. M., 305, 308, 311, 313, 316
Synergism, population mechanics, 163–164
Syphilis, 6

Taddeo, A. R., 380
Taxonomy, bacteria classification, 36–40
Teichos acids, biochemical marker techniques, microbial ecology of ground water methods, 143
Temperature, environmental conditions (bacterial growth), 52–54, 255
Tetrads, bacteria classification, 38
Thermoacidophiles, archaebacteria, 32
Thermodynamics:
  geochemical modeling, 267–270
  metabolism and, 71–73, 78
Thermophilic microorganisms, temperature, environmental conditions (bacterial growth), 53
Thiobacillus:
  iron cycle and, 253–254
  iron oxidizers, 99
  sulfide oxidizers, 98
Thiosulfate, sulfur cycle and, 258–259
Thomas, J. M., 125, 156, 157, 169
Thomson, William (Lord Kelvin), 10
Thorn, P. M., 49
Thorstenson, D. C., 198, 206, 207, 264, 265, 280, 286, 289
Thurman, E. M., 190
Tiedje, J. M., 307, 308, 310, 372, 373
Tobacco mosaic disease, 33
Toluene:
  anaerobic degradation processes, 341–343
  microbial degradation, ground-water systems, 348
  remediation of ground-water systems, 353–356
Tool contamination, aseptic technique, split-spoon and push-tube sampling with, local flow systems, 214

Torsvik, T., 46
Toth, J., 175–176, 177, 189
Toxic chemicals, procaryotes, bioremediation, 26–27. *See also entries under specific toxins and pollutants*
Trace elements, culture media design, 60
Tracers, drilling fluid contamination of cored sediments, evaluation of, 227–230
Transcription, RNA structure and organization, 108–109
Transduction, genetic recombination, natural genetic exchanges, 118
Transferases, metabolism, enzyme regulation, 79
Transfer RNA (tRNA), function of, 108, 109–112
Transformation, genetic recombination, natural genetic exchanges, 117
Translation-making proteins, RNA structure and organization, 109–112
Transmission electron microscopy, microbial ecology of ground water methods, 141
Transposable material, mutation, 115
Trapp, H., Jr., 260
Trevors, J. T., 317
Tricarboxylic acid (TCA) cycle, 18, 91
Trophic pyramid, microbial ecology of ground water, 134
Troutman, D. E., 345, 387
Trudgill, P. W., 335
Truesdell, A. H., 21, 264, 265, 276
Tschermak-Seysenegg, Erich, 105
Tsien, H. C., 368
Turco, R., 386

Ultraviolet light:
  biomass measurement in soil microbiology, unsaturated zone, 183
  genetic mutation and, 115–116
Uniformitarianism, geology and, 9
United States Environmental Protection Agency (EPA), 22, 192
United States Geological Survey, 19, 20, 199, 260, 302
United States Supreme Court, *Diamond v. Chakrabarty*, 123
Unsaturated zone, 179–189
  bacterial distribution in, 186–189
    capillary fringe (deep unsaturated zone), 187, 189
    intermediate subzone, 186–187
    soil subzone, 186
  biomass measurements in soil microbiology, 182–186
  described, 179–180
  microbial habitat, 180–182

microbiologic sampling, 209–211
  air drilling and coring, 210–211
  hand auger, 209–210
subsurface environment classification, 175

Vaccination, 17
van der Linden, A. C., 334
van Eyk, J., 301
van Veld, P. A., 305, 313
Vectors, genetic engineering and, 120–121
Veldkamp, H., 152
Ventullo, R. M., 49
Viable counting, bacterial growth enumeration techniques, 66–68
Vibrio cell, bacteria classification, 37, 38
Vinci, Leonardo da, 4, 8
Vine, F. J., 7, 12
Viruses:
  described, 33–35
  host-parasite interaction with bacteria, 168
Viscosity, of drilling fluids, 218–219
Vitamins, culture media design, 61
Vogel, T. M., 342, 369
Volcanoes, geology and, 8–9

Waksman, S. A., 186
Wang, D., 35, 43
Waste disposal. *See also* Pollution; Sewage waste
  ground water flow systems (local), abundance and distribution (of bacteria), 191
  liquid wastes, geochemical modeling and, 21
  subsurface contamination by, 22
Water, environmental conditions (bacterial growth), 54–55
Watson, James, 7, 18, 105
Weare, J. H., 276
Weeks, J. B., 197
Wegener, Alfred, 6, 11
Wellings, F. M., 35
Wells:
  Gallionella, 42, 99
  iron cycle and, 256–257
  nitrogen cycle, ground-water systems, 247–248
Werner, A. G., 4, 8–9
White, A. F., 237

White, D. C., 22, 23, 31, 143, 145, 146, 156, 183, 184
White, D. E., 21
White, W. R., 21
Wigley, first name, 280
Williams, D. F., 184
Wilson, B. H., 311, 313, 316, 340, 368, 379, 380
Wilson, J. L., 131, 132, 148, 170, 191
Wilson, J. T., 22, 23, 142, 154, 192, 212, 214, 215, 311, 313, 316, 368, 380
Wine production, microbiology and, 15
Winograd, I., 201, 237
Winogradsky, Sergei, 6, 17, 22, 180, 182
Witherspoon, P. A., 176, 189
Woese, Carl R., 31–32
Wohler, Friedrich, 14
Wolery, T. J., 276
Wolter, N., 156
Wunsch, D. R., 57

Xenobiotic compounds, acclimation, factor affecting, cross-acclimation, 308, 309
Xenobiotic oxidation, catabolite repression, 301
Xenobiotics in ground-water systems (acclimation), 311–316
  bioremediation technology, 316
  contaminated aquifer, 311–313
  eucaryotic microorganisms, 315
  pristine aquifer sediments, 313–315
Xylene:
  anaerobic degradation processes, aromatic hydrocarbon biodegradation, 344–345
  microbial degradation, ground-water systems, 348
  remediation of ground-water systems, 353–356

Yazicigil, H., 347
Yeast, 15
Yield point, of drilling fluids, 219
Yoch, D., 41

Zinger, A. S., 53–54
Zobell, C. E., 20, 53, 329
Zonation, subsurface environment classification, 176–177